Metalcutting:

Today's Techniques for Engineers and Shop Personnel

By the editors of American Machinist

Metalcutting:

Today's Techniques for Engineers and Shop Personnel

By the editors of American Machinist

American Machinist
McGraw-Hill Publications Company
New York

LIBRARY OF CONGRESS CATALOGING IN PUBLICATION DATA
Main entry under title:

Metalcutting: today's techniques for engineers and shop personnel.

Includes index.
1. Metal-cutting. 2. Machine-shop practice.
I. American machinist.
TJ1185.M4728 671.5'3 79-14955
ISBN 0-07-001545-7

Second Printing, 1980

Metalcutting

1234567890 HDHD 7865432109

The editors for this book were William M. Stocker, Jr. and Tyler Hicks.

Table of Contents

Table of Contents (continued)

Foreword

It was just over 40 years ago that the editors of American Machinist first applied the term "Special Report" to a long, basic article designed to start at the very beginning of a particular subject and bring it down to the very latest practice. The term has since been used in the magazine more than 700 times and has spread to countless other magazines.

Most of the reports have been written by members of the staff; a few have been contributed by people in industry—all have held to a standard of careful research, depth reporting, and clear presentation that neither confuses the newcomer nor talks down to the initiated.

In recent years, the editors have worked on the production of a group of reports that dealt with the basic metalcutting processes. With the publication of Special Report No. 709, Fundamentals of Drilling, the coverage of the most significant basic processes was completed. They have been combined in this book, along with reports on some of the new metalcutting processes, such as *electrical discharge* and *laser cutting,* to provide a convenient, cohesive collection for study and reference.

Why such a collection of reports just on metalcutting? Because metalcutting is at the heart of most manufacturing operations in all those industries that produce machinery, appliances, instruments, and the other durable products produced by the metalworking industries. *Metalcutting* machine tools represent about 75% of the machine tools used in production plants in these industries. The only products that are not made by machine tools (such as clothing, food, and television programs) are produced with the aid of machinery that was produced on machine tools.

According to a study by the editors of American Machinist, there are currently about 2.5-million metalcutting machine tools in the U.S. The only significant processes not covered in this book are the highly specialized gear-cutting machines and the specialized machines that combine several of the different types of operations that are discussed here. Also omitted are the basic planing and shaping processes, once widely used but now being replaced by milling.

Most of the reports on individual processes deal extensively both with the kinds of equipment that are needed and the special cutting tools that are employed. An exception is the report on boring, and the two reports on turning. These concentrate on the many variations in machine design for handling workpieces and tools. Two reports concerned with cutter geometry and insert-type tools deal with the tools that are today most often used in both boring and turning operations.

Variations in methods of holding work for the different processes are handled in a separate report, while two reports are concerned with the cutting fluids that may be used and the problem, recognized today as never before, of cleaning these fluids so they can be reused.

Twelve of the reports are the work of six editors on the American Machinist staff: Richard T. Berg, John J. Dwyer, Robert L. Hatschek, Edward A. Huntress, Joseph Jablonowski, and George H. Schaffer. The other three, which were contributed from industry, are the work of F. J. McGee, P. Albrecht, H. N. McCalla, Richard P. Cottrell, and James J. Joseph.

Together, these people have provided reports that are remarkable for their compactness and clarity. In this combination they cover the basic metalworking processes in a way that will be useful to both the student and the practicing engineer in industry.

Anderson Ashburn, editor

New York
March, 1979

Fundamentals of drilling

Holemaking is the most common of all machining operations, and drilling is done on a wide range of machine tools in addition to 'drilling' machines

By R. L. Hatschek, senior editor

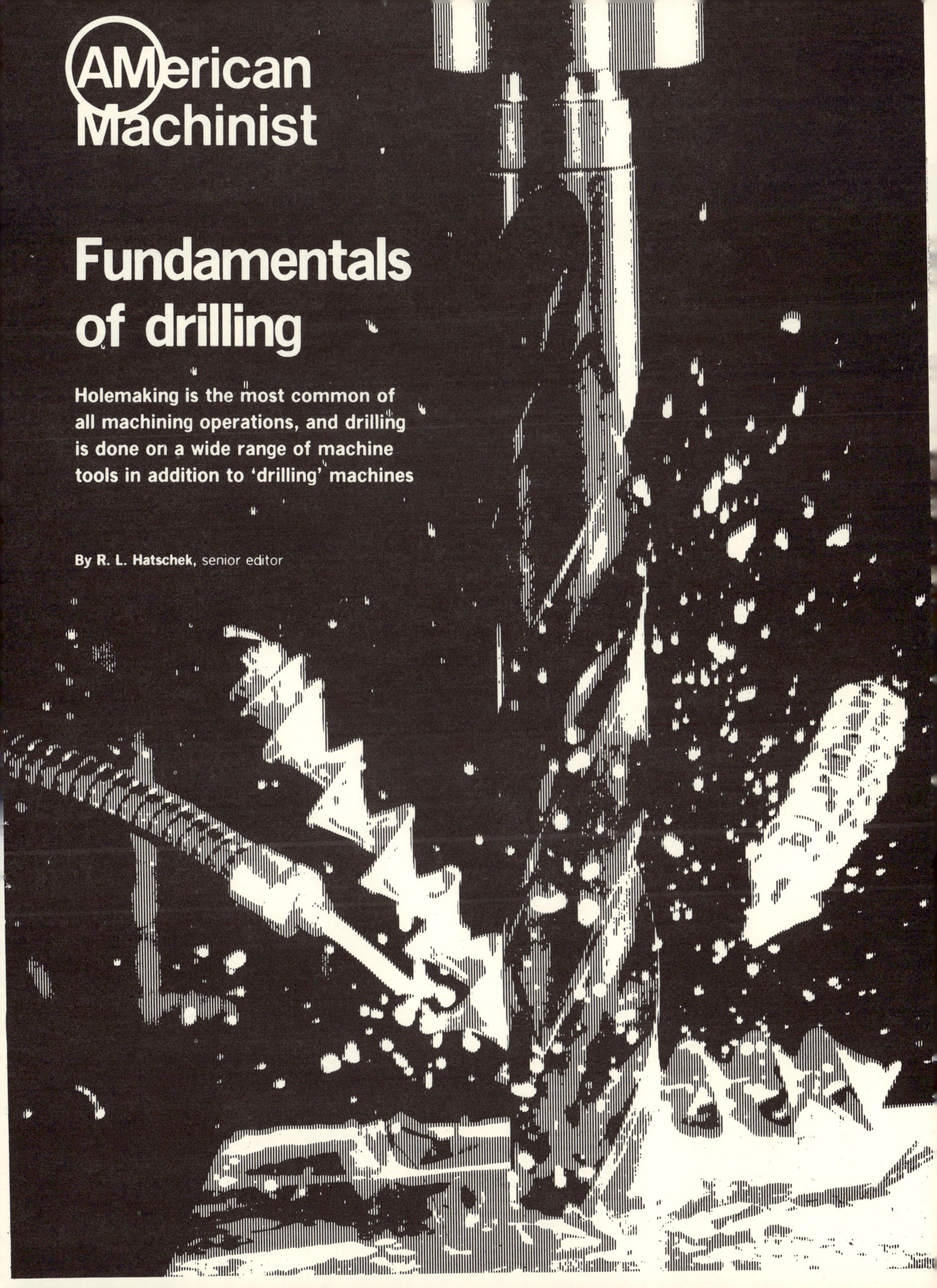

HOLE PRODUCTION is unquestionably the most common of all machining operations, and drilling—itself more a family of processes than a single method—is the most widely used technique for producing holes.

Basically categorized by the types of tools they employ, the individual drilling processes possess overlapping capabilities, yet each is unsurpassed in certain areas of application. Choice of the specific process for any particular application starts with the requirements of the hole—diameter, depth, various tolerances—and then must include consideration of production volume and the specific types of drilling equipment available.

For the purposes of this article, we will define drilling basically as the production of holes in solid material by conversion of the material within that desired cavity into conventional chips by the relative rotation of a cutting tool and the workpiece. Either the tool or the work—and, in some cases, both—may rotate.

There are other methods for making initial holes, including punching of relatively thin stock, coring of a casting, electrical-discharge machining and other electrochemical machining processes, flame cutting, and even zapping the workpiece with a laser or electron beam. These processes, however, are beyond the scope of this article.

Drilling, in many cases, is but the first machining operation in the total production of a hole. Subsequent operations are often required to give that hole its required functional characteristics, such as improved precision of size, location, or cylindricity; finer finish; or special internal configurations exemplified by tapers, threads, keyways, recesses, etc. Typical of these secondary operations are reaming, boring, honing or lapping, counterboring or countersinking, tapping, and broaching. Such processes are also beyond the scope of this article, although some specialized multidiameter drilling tools will be touched upon.

Drilling is a family of processes

As noted initially, the term *drilling* covers a variety of machining processes, each of which provides special attributes or advantages in the production of certain varieties of holes. One difficulty that all drilling tools share, however, is that they must perform under extremely severe machining conditions. They are buried in the work in an environment that obstructs both coolant flow to the cutting site and the chip flow away from it and that tends to retain heat in both the workpiece and the cutting tool. Except for trepanning, the depth of cut is essentially fixed by the radius of the hole being drilled, and (with the same exception) cutting speed varies from an inefficient zero at the center of the hole to whatever maximum has been selected at the periphery of the tool. And the minimal rigidity of what is generally a long and slender tool demands, in most cases, that it obtain lateral support from the hole it is producing.

What are the basic types of drilling tools? By far the most common are helically fluted twist drills. Covering a size range in which twist drills are most common is the half-round drill, one of the simplest of all cutting tools and one that offers some advantages that many of today's manufacturing engineers seem to have forgotten. Both of those two types are used for producing holes of only several thousandths of an inch in diameter, but, as dimensions shrink further (to 0.001 in. and less), the flat "pivot drill" comes into its own. And, as hole sizes grow to 1 in. and more, another type of flat drill becomes prominent—the inserted-blade spade drill.

The most recently added tool in the hole-making arsenal is the indexable-insert drill, now offered by at least half a dozen tool companies. And, finally, an entire subcategory of drilling is termed "deep-hole drilling" by some and "precision drilling" by others. Within this group are gundrilling, the BTA system (Boring & Trepanning Assn), ejector drilling, and trepanning—all of which, though similar in some respects, also involve important differences.

Drilling processes compared

	Twist drill	Half-round	Pivot (micro)	Spade (inserted blade)	Indexable-insert drill	Gundrill	BTA system	Ejector drill	Trepanning
Diameters, in.									
Typical range	0.020-2	0.0059-¼	0.001-0.020	1-6	⅝-3	0.078-1	7/16-8	¾-2½	1¾-10
Min	0.0059	0.003	<0.0001	⅝ spec, 1 std	⅝	0.039	3/16	¾	1¾
Max	3½ std 6 spec	1	⅛	18	3	2½	12	7	>24
Depth/diameter ratio									
Min practical	no min	no min	no min	no min	<1	1	1	1	10
Common max*	5-10	>10 (horiz)	3-10	>40 (horiz)	2-3	100	100	50	100
Ultimate*	>50	>50	20	10 (vert) >100 (horiz)	—	200	>100	>50	>100

* Maximum depth/diameter ratios in this table are estimates of what can be achieved with special attention and under ideal conditions. Equality of tolerances should not be assumed for the different processes

Drilling machines

DRILLING MACHINES" rank only as the third most numerous type of machine tool, totaling only about 324,000 units in US metalworking plants with 20 or more employees, according to the *12th American Machinist Inventory of Metalworking Equipment* (AM—Dec'78,p133). How, then, does one justify the statement that drilling is the most commonly performed of all machining operations? It is simply that drilling operations can also be done on virtually any type of machine tool capable of providing a relative rotation of workpiece and cutting tool (with the major exception of grinding machines).

This includes turning machines (the most numerous of all machine tools), boring machines, many of the most common types of milling machines, multifunction machines, and most trans-

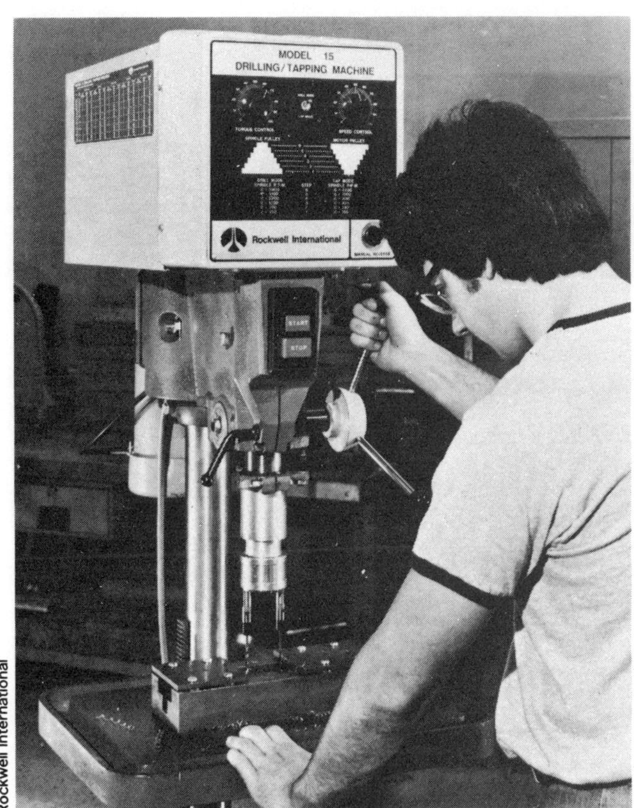

Rockwell International

Simple attachment converts sensitive drillpress for drilling or tapping two holes simultaneously. Center distance is adjustable

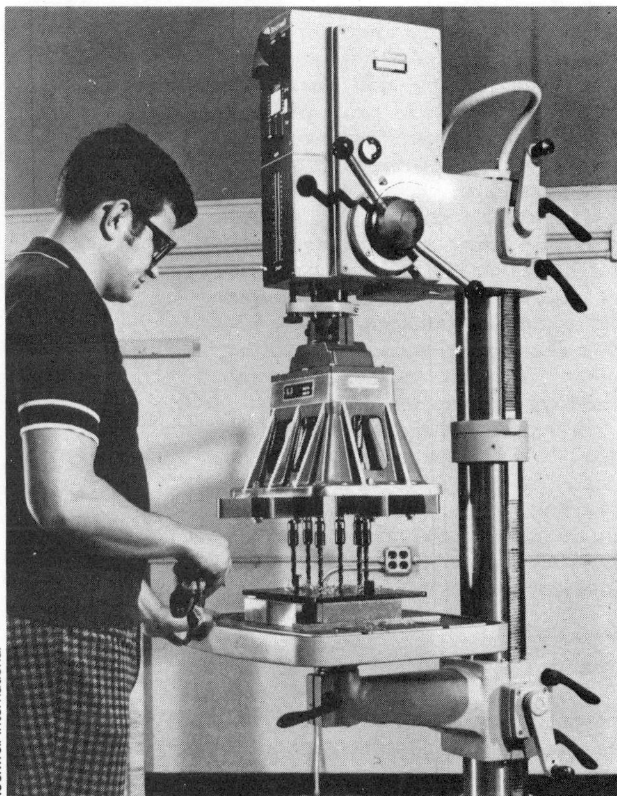

Rockwell International

Geared-head vertical drilling machine is adapted with universal-joint adjustable-center cluster head for six drills

fer-type machines. Thus, of the more than 1.7-million metalcutting machine tools in US plants employing 20 or more workers, considerably more than half—more than a million machines—can be and probably are applied to drilling operations at one time or another.

Again considering only those machines designated as "drilling" machines, the AM Inventory shows that 87% of all metalworking plants have them—the widest distribution of all machine types.

The most common of all drilling machines is also, not surprisingly, the simplest and the least expensive. This is the ubiquitous "sensitive" drillpress, typically providing a vertical base-mounted column, a table that is adjustable in height and tilt angle, and a powered spindle that can be fed down into the work by means of a hand-operated lever. Accounting for 36.5% of all drilling machines, the sensitive drillpress comes in two basic versions, one floor or pedestal mounted and one for bench mounting. The spindle is commonly belt-driven, with either stepped pulleys or variable-pitch pulleys for speed selection, although geared-head versions are also available. The manual feed, giving the operator "feel" (hence the name "sensitive"), is generally provided by a rack-and-pinion arrangement linking the feed

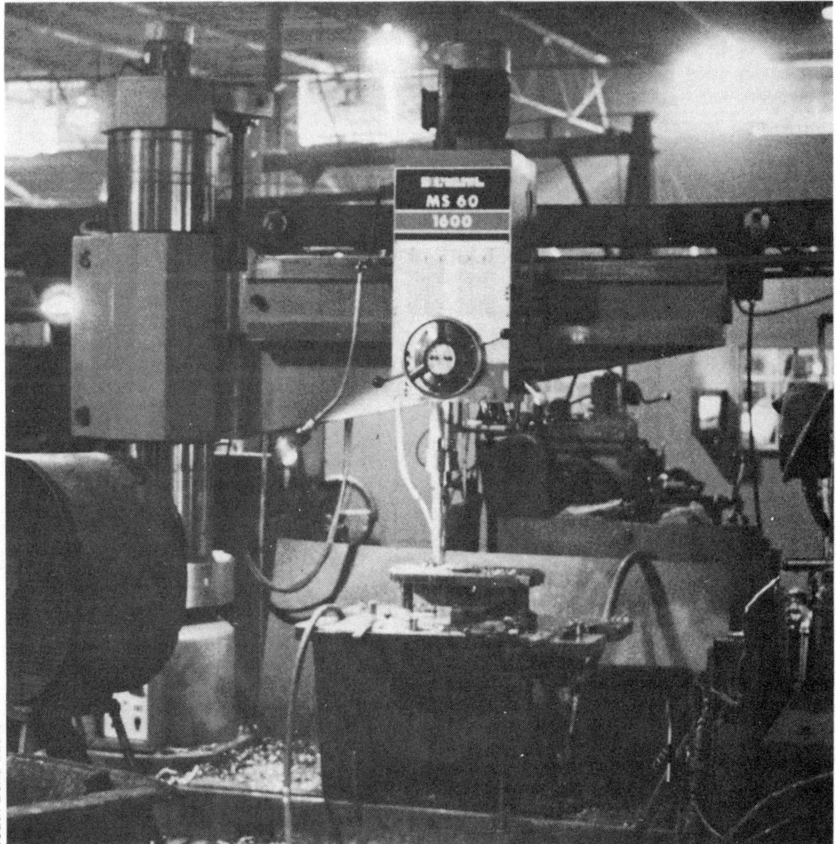

South Bend Lathe

Radial drills are available in a wide range of sizes, typically measured by giving both column diameter and arm length. Model pictured is by South Bend Lathe

lever with a quill in which the spindle rotates. A simple adjustable depth-stop limits total travel in most cases. Toolholding is normally by means of a key-type drill chuck, typically of ½-in. capacity, though many larger machines provide Morse-taper spindles.

The sensitive drillpress is a general-purpose machine most frequently used for drilling one hole at a time in relatively small workpieces with smaller-diameter twist drills—virtually never over 1 in. It is also capable, of course, of applying other tools—most commonly countersinks, counterbores, and reamers.

In production work, the sensitive drillpress is often used with a drill jig or bushed fixture to guide the tool for accurate hole location. Multispindle attachments suitable for use on sensitive drillpresses can also be used effectively in production situations.

Add power feed

As drilling operations become heavier, the upright single-spindle drill with power feed comes into the picture, and, at almost 30% of all drilling machines, this is the second most populous type. Many of the larger machines in this category incorporate box columns for increased rigidity, and worktables may incorporate milling-machine-type knee supports and even leadscrews for table positioning. Power feed not only permits the use of larger drills than an operator can easily feed manually but also gives improved accuracy on diameter, better finish, and longer tool life and prevents such problems as work-hardening of some materials when an underfed drill merely rubs the work. Power feed, of course, is the only way to ensure that a specific selected feedrate is actually achieved.

When workpieces grow to proportions that are impractical to move about on a drilling-machine table, yet still require a number of holes to be drilled, the radial drill is often the most convenient type of general-purpose drilling machine to use. A radial drill incorporates a round vertical column of up to 30 in. dia or larger about which a horizontal arm pivots and elevates. On a large model, the arm may be 12 ft long; this carries the drilling head, which can be positioned as necessary along the arm. Some radial drills, or "radials" as they're often called, also provide a tilt adjustment for the drilling head so that angled holes can be drilled. Spindles are generally of high horsepower for drilling large holes, and power feed is incorporated. Nearly 12% of all drilling machines are radials, according to the AM Inventory.

Ranking next in order of popularity of drilling machines, at just under 10%, are

Burgmaster

Eight-spindle Burgmaster Model 25CH turret drill tooled with multispindle attachments for drilling, countersinking/counterboring, and tapping

Zagar

A 200-spindle special machine for drilling, reaming, and countersinking up to 64 holes at a time in atomic-energy components fixtured on indexing table

Natco

Massive Natco vertical multispindle hydraulic-feed drilling machine incorporates shuttling fixtures for loading/unloading one position during drilling at other

Natco

Inverted multispindle drilling machine built by Natco uses universal-joint drives for drills pointing upward to drill as table is fed downward hydraulically. Chips fall out

upright gang drills, essentially a number of individual drilling machines mounted along single bench or common table or base. These may be either sensitive or power-fed, and the individual units need not be identical. Indeed, a gang-drill setup could be built simply by setting up a number of bench-type machines side by side. The essence of the gang drill is individual action of the spindles in a multispindle setup, with each spindle carrying a different size or type of cutting tool for sequential operations.

This is essentially a production setup for limited-volume use, saving the time of changing tools in a single spindle for successive operations and yet requiring only very modest investment. Depending on the specifics of the situation, one operator might handle one, two, or several more spindles, or there might be individual operators at each with the work passed along from one to the next.

Holes come in bunches

When production volume warrants, however, the multispindle cluster-type drilling machine becomes much easier to justify. Here, of course, major time savings are generated by performing a number of machining operations simultaneously. Machines of this type account for nearly 7% of all drilling machines in US metalworking plants.

Various arrangements are used. Tool-holding spindles may be driven by intermediate shafts and universal joints, by eccentric crank-type arrangements, or by gears (in which case, left- and right-hand drills may be mounted in alternating spindles to eliminate idler gears and permit closer spacing of individual spindles). In many cases, the positions of individual spindles can be adjusted within the overall cluster area almost at random. In other cases, slip spindle plates are made with a fixed spindle pattern. When relatively small numbers of tools are involved—say, up to a dozen or so—standard multispindle attachments are available with which single-spindle machines may be adapted for multispindle use. This, of course, multiplies the requirements for spindle horsepower and feed force and must be evaluated carefully in terms of machine rigidity as well.

Traveling drill-guide bushing plates are often used in multispindle operations. In this arrangement, the bushing plate is attached to the drill head rather than to the part-holding fixture and is spring loaded so that the drills can continue to advance after initial contact has been made with the workpiece or fixture. Guide pins are usually used to ensure proper registration of the bushing plate and work-holding fixture.

Users of such multispindle machines, which, in some cases, may involve hundreds of drills doing their jobs simultaneously, sometimes stagger tool lengths so that total drilling loads are imposed gradually and are relaxed gradually upon drill breakthrough.

Next in order of popularity is the upright turret drill. Excluding numerically controlled turret machines, which are classified as multifunction machining centers, this category accounts for 2% of the machines classed strictly as drilling machines. This type of machine, which again may be either sensitive or power fed, is more or less analogous to a single-spindle gang drill. Typically providing six or eight spindles (anywhere from four- to ten-spindle versions have been built) radiating from an indexing turret, these machines are usually tooled for successive sequential operations, such as drill, ream, and countersink.

Adding numerical control

The turret drill is one of the most effective types to which numerical control of workpiece positioning can be added—essentially for converting it into a turret-type machining center—but it is not the only type for which NC is useful. Virtually all of the basic single-spindle types have had NC added in certain situations. And even multispindle cluster-type drills with NC are not uncommon in certain applications. One specific example is a type that has become almost a standard, if specialized, machine: the circuit-board drill. In fact, some circuit-board drills offer ganged clusters of drilling spindles that are used for holemaking in stacks of workpieces. Other examples include beam drills and tube-sheet drills.

Many installations of NC machining centers are used almost entirely for drilling operations, and virtually all are used for some drilling operations. However, the versatility of a modern NC or CNC system, which typically provides feedrate control and contouring control in the X-Y plane as well as control of the spindle's Z-axis. Thus, the "machining center," whether it requires manual toolchanging, has a turret, or is equipped with an automatic toolchanger, is truly a multifunction machine.

For really large production volumes, especially of complex components requiring many holes on several faces of each part, even more specialized machines are frequently justified. These may be single-station machines with several multispindle heads advancing into the workpiece from different sides or multistation machines in which the

National Jet

Manual microdrilling is done under 20X magnification with 20:1 mechanical advantage in counterbalanced system

Albion

Specials are often used in large-volume manufacturing industries. This multispindle, multihead Drillmation machine is for drilling tractor differential cases

Eldorado Tool

A 5/16-in. hole 2¾ in. deep is gundrilled at 8-ipm feed in Eldorado Mega 50 machine. Diameter is held to 0.0005 in. in mild steel part—a gundrill driver

part is transferred in either linear or rotary fashion for single or multiple operations at each position.

Like NC machining centers, these transfer machines are multifunction units and are thus not classified in the strictly drilling category. It is interesting to note, however, that these two classes are the fastest-growing machine-tool types in American industry; the 12th AM Inventory indicated a 37% increase in the number of NC machining centers over the 11th AM Inventory (1973) and a 26% increase in special way-type and transfer machines.

A still newer type of machine tool, which has only begun to emerge within the past few years, is also often used for drilling operations. This is the headchanger machine, in which individual heads—generally multispindle heads for holemaking applications—are changed automatically for sequential operations on a workpiece at a single station. Such machines are a logical cross between NC machining centers and transfer machines for machining in moderately long runs.

Deep-hole drilling

All of the above types of machines are tooled primarily with twist drills and are used mainly in drilling relatively shallow holes. As holes get deeper—5, 10, 20 times diameter—special forms of twist drills are applied, such as heavy-webbed crankshaft drills, oil-hole drills, or coolant-fed spade drills in larger diameters. But, as holes get really deep—and especially when tolerances on size, finish, concentricity, and straightness become important—more-specialized drilling processes are frequently used. These are covered later in this article in sections devoted to gundrilling, the BTA system, ejector drilling, and trepanning.

Machines using these self-guided drilling tools for deep-hole operations are generally horizontal. However, the use of these processes for production of relatively short, precision holes or for operations on relatively small workpieces also permits vertical spindle orientation. One recent exhibition of machine tools displayed a gundrilling machine that had its spindle angled upward at 30° or so to assist chip removal.

Probably the most common type of deep-hole drilling machine provides a rotating drill that is advanced into the work, which is fixtured on a table at one end of the machine. Where the workpiece itself is a long, slender part—such as a gun barrel, a machine-tool spindle, a piece of oil-field equipment, etc—the workpiece may be rotated and a nonrotating drill may be fed into it. When the ultimate precision in hole alignment is required, both the work and the tool may be rotated in opposite directions.

Some common attributes are required for deep-hole drilling machines. High-pressure coolant systems are mandatory for the elimination of chips, and monitoring of the pressure is a useful way to quickly spot problems. Machines require positive feed systems to ensure production of consistent chips. Successful deep-hole drilling depends on careful attention to all variables.

[Continued]

DeHoff

Both tool and work are rotated in special DeHoff machine for drilling 5/8-in. and 7/8-in. holes through turbine studs up to 96 in. long. Heller BTA tooling is used

VDF Wohlenberg

VDF deep-hole boring machine has swing of 63 in. over bed, maximum hole depth of over 65 ft, and can rotate either the tool or the workpiece or both. Work-spindle power is 220 hp; tool-spindle, 150 hp. Trepans to 25 in.

Twist drills

If holemaking is the most common of all machining operations, twist drills are the most common of all metalcutting tools. They are made by the millions of a wide range of materials from carbon tool steel to solid carbide. They are cataloged in a broad variety of styles to accommodate the machining idiosyncrasies of virtually any material, either metallic or nonmetallic, that can be converted into chips by a cutting tool. They come in standard diameters from 0.0059 in. to 3½ in., which are sized according to four commonly used systems: fractional inch, number (or wire gage), letter, and metric. And they are employed in almost any type of machine tool that can provide relative rotation of a tool and a workpiece (grinders are a major exception) as well as both pneumatic and electric portable power tools (we'll omit crank-type hand-drills, pin vises, and the carpenter's brace).

Based on a statistical sampling of more than 50-million standard twist drills sold commercially (excluding sales to the government), National Twist Drill & Tool Div (Rochester, Mich) has come up with some interesting data on the most-used sizes: The median 90% (5% are larger, 5% are smaller) fall between 0.050 in. and 0.400 in. dia, neither of which is a standard size. The bottom size of the range falls between a 3/64-in. (0.0469 in.) and a No. 55 (0.052 in.) while the top size is between an X (0.397 in.) and a Y (0.404 in.).

Most-used drill is a small one

The most common size is a No. 30 (0.1285 in., a nice size for a ⅛-in. rivet), with the ⅛-in. drill size very close behind. Including these two, there are only 20 individual sizes that each exceed 1% of total drill sales. And only 1% of all twist drills sold exceed ¾ in. diameter.

Although it is manufactured to fairly close tolerances, the twist drill is not a precision tool; it is principally a roughing tool designed to make a hole economically. If hole quality must be better than that which a twist drill can produce, subsequent operations, such as reaming, are the normal scheme.

What the conventional twist drill does is a severe machining operation under cutting conditions that are considerably less than ideal. The tool is typically long and slender, its helical flutes constitute a column eccentricity that further reduces rigidity under axial thrust loading, chip elimination is difficult, heating problems are aggravated, cutting speeds vary from zero at the center to a maximum at the periphery, and depth of cut is fixed at half the diameter.

Complexity in the cut

The cutting end of a twist drill presents two distinctly different features, a pair of cutting lips and the so-called chisel edge. Both are involved in removing metal, but their actions are quite different. The lips are essentially conventional cutting edges with positive axial rake created by the flute helix angle, negative radial rake created by the web-producing offset, a lead angle (typically) of 31°, and an end-clearance (lip relief) that generally varies according to drill size and workpiece material over a range of 6° to 26°. Chip-formation at these lips is closely analogous to the chip-formation by an ordinary lathe tool or a milling cutter.

The chisel edge, the short line obliquely across the web that joins the cutting lips, presents a different picture. As described by Carl Oxford, vice president of National Twist Drill & Tool, "The mechanism of metal removal under the web is extremely complex. At radii near the bottom of the flutes, the chisel edge and the drill clearance surfaces can be thought of as forming a cutting tool with a high negative rake, but, as the center of the drill is approached, the action becomes more akin to that of a blunt, wedge-shaped indentor." Severe deformation takes place in this relatively inefficient region, and these deformation products are wiped or extruded into the drill flute, where they intermingle with the chips produced by the main cutting edges.

As a result, about half the force required to push a ½-in. drill of standard proportions and geometry into metal is contributed by the chisel edge, and the length of this chisel edge, of course, is determined by web thickness at the point (the web is normally tapered, thickening toward the shank). In drills of smaller diameter the web thickness is proportionately larger so that maximum stiffness is retained. Extra-length drills, such as crankshaft drills, and other heavy-duty drills also are designed with heavy webs for increased rigidity.

With conventional point geometry, all such heavy-duty drills would pay the penalty of additional thrust force required to drive the chisel edge into the work. Hence, a number of variations on standard point geometry have been devised under the generic heading of "web thinning." Blend thinning and notch thinning are two such grinds, the former retaining conventional point geometry to the largest extent, the latter a more easily produced point that reduces rake somewhat in the ground area.

Points to end 'walking'

A more radical design is the so-called split point, or crankshaft point, in which the clearance face of each cutting edge is given a sharp (55° is typical) secondary relief to the center of the chisel edge, thus creating a secondary cutting lip on the opposite cutting edge. The angle between these lip segments acts like a chip-splitter, and the narrower chips produced may be somewhat easier to eject. More significant than this effect, however, is that the split point reduces web-generated thrust to the minimum. Also, by rotating the secondary lip toward the primary one, the chisel edge can be modified so that a high point exists at the center; this will eliminate skating or walking of the point when starting the drill without a bushing. Differing slightly in detail, the "four-facet" point is similar.

"Spiral" points, ground on special drill-pointing machines which generate a chisel edge with an S-shaped curve rather than a straight line, also provide a high point at the center that eliminates wandering upon starting without a bush-

Twist drill

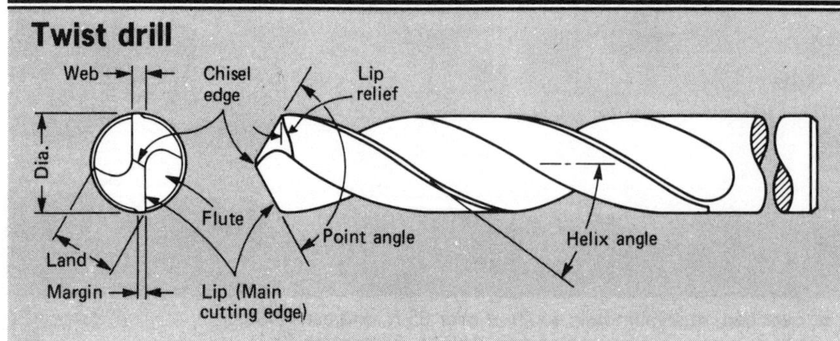

ing. In this case, the drill point is quite conventional in other respects.

In all cases of web thinning, as of drill pointing itself, it is of the utmost importance that the point be as perfectly symmetrical as possible. The geometry can be produced by a skilled operator working manually, but all experts agree that special drill-point grinding equipment or fixturing ensures the best possible results. Lack of symmetry in the length or angles of a twist drill's cutting lips will cause a number of problems, such as poor starting, oversize holes, and accelerated tool wear.

Another variation on standard drill-point geometry is a change of drill-point angle. Twist drills are generally ground to a point angle of 118°, but blunter angles, often 135°, are recommended for better results in harder materials (and deeper holes). The effect of the larger angle, in addition to an increase in absolute rake angle to more nearly that of the helix, is to generate a thicker, narrower chip for a given feedrate, which is also of benefit in drilling materials that have stronger tendencies to work-harden. A negative effect is the concentration of abrasive wear at the outer corner of the drill point.

More acute point angles (60° to 90°) produce an opposite set of effects—thinner chips and reduced abrasive wear at the corners—and are recommended for soft nonferrous metals, plastics, and soft

Web-thinning

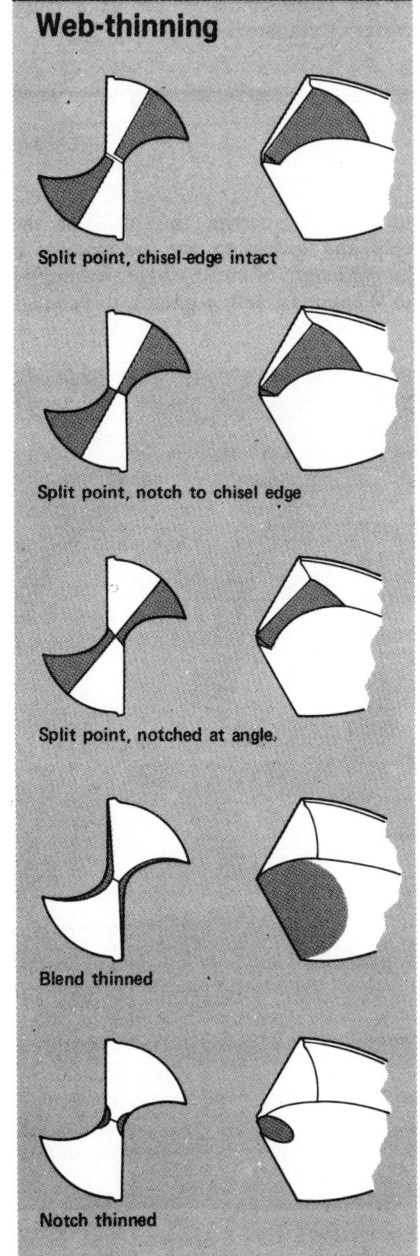

Split point, chisel-edge intact

Split point, notch to chisel edge

Split point, notched at angle.

Blend thinned

Notch thinned

Feeds for hss twist drills

Diameter, in.	Feed, ipr
Under 1/8	0.001-0.002
1/8-1/4	0.002-0.004
1/4-1/2	0.004-0.007
1/2-1	0.007-0.015
1 and over	0.015-0.025

Source: Cleveland Twist Drill

Speeds for hss twist drills

Material	Speed, sfm
Aluminum alloys	200-300
Brass, bronze	150-300
High-tensile bronze	70-150
Zinc-base diecastings	300-400
High-temp alloys, solution-treated & aged	7-20
Cast iron, soft	75-125
medium hard	50-100
hard chilled	10-20
malleable	80-90
Magnesium alloys	250-400
Monel or high-Ni steel	30-50
Bakelite and similar	100-300
Steel, 0.2%-0.3% C	80-100
0.4%-0.5% C	70-80
tool, 1.2% C	50-60
forgings	40-50
alloy, 300-400 Bhn	20-30
High-tensile steel,	
heat treated to 35-40 Rc	30-40
heat treated to 40-45 Rc	25-35
heat treated to 45-50 Rc	15-25
heat treated to 50-55 Rc	7-15
Maraging steel, heat treated	7-20
annealed	40-55
Stainless steel,	
free machining	30-100
Cr-Ni, non hardenable	20-60
Straight-Cr, martensitic	10-30
Titanium, commercially pure	50-60
6Al-4V, annealed	25-35
6Al-4V, solution-trt & aged	15-20
Wood	300-400

Source: Cleveland Twist Drill

Special drilling tools

The subject of drilling is primarily concerned with straight, cylindrical holes of constant diameter in solid material. Not all holes, however, are so simple, and even limited production quantities will often justify the use of special drilling tools analogous to form-tools used in turning and milling.

Probably the most common of these is the combined drill and 60° countersink used to produce the specially shaped holes in the ends of workpieces that are to be turned or ground between centers. Indeed, this "special" tool, the center-drill, is so common that it's listed in a variety of sizes as a standard item in virtually all drill catalogs.

Another special is the step drill, essentially a conventional twist drill that has had a smaller diameter (or diameters) created on it by grinding down a portion of its basic diameter and, usually, thinning the web at the point. Each step can be square, angled, or even contoured as required. At least generally, the smallest diameter of a step drill should not be less than about 35% of the largest diameter. Because they can often be made from standard tools, the use of step drills is fairly common.

Resharpening can be a problem with such step drills, especially when the length of the smaller diameter is relatively short. As each step or shoulder is reground, it is virtually impossible to avoid nicking the margin of the smaller diameter. This does not preclude use of the tool, but repeated regrinding is limited. This problem is prevented by the use of a subland drill, which can be described as two coaxial two-flute drills. In other words, each diameter has its own separate land or margin. In general, the smallest diameter of a subland drill should not be less than about half that of the largest, and about 1/8 in. is the smallest practical size for the small diameter on this type of tool.

Specially formed or stepped tools are also made in most other types of drills, including half-round drills, pivot-type microdrills, both inserted-blade and one-piece spade drills, and the various types of deep-hole drills.

Also, not all "drills" are designed for producing holes in solid material. "Core drills" are available as standard tools in virtually all of the configurations just described. More akin to reamers—and, indeed, the gundrill type is most often referred to as a gun-reamer—these tools are designed for opening up already existing holes, such as cored holes in castings.

cast iron. This leads to the concept of the double point angle: a larger included angle at the center, with a smaller secondary angle for the outer portion to reduce abrasive wear at the corner. And from that concept, it's a fairly short step to the continuously varying point angle, more familiarly called a radius grind, in which the lips and margins of the drill are blended by a smooth curve.

The radius point, for which special grinding machines are available, reduces corner wear—especially in abrasive workpiece materials—and also provides a reamer-like cutting action near the periphery so that a radius-pointed drill tends to produce holes closer in size to the tool diameter than conventionally-pointed twist drills.

Helix angle can also be varied, but not after a twist drill has been made. In addition to the normal helix angle of 25° to 30°, drills with both higher angles (high helix or fast spiral) and lower ones (low helix or slow spiral) are available as standard items. The former, which also provides improved chip-extracting action, is typically used in materials such as aluminum that cut better with sharper rake angles, the latter in shorter-chipping materials such as screw-machine stock. Straight-flute drills (0° helix) are sometimes used for drilling brass.

The one minor exception to the statement that helix angle can't be varied after drill manufacture is that the rake angle of the cutting lips can be reduced by a grind very similar to a web-thinning operation. This is often done to reduce the tendency of a helically fluted tool to pull itself into workpieces of certain materials or shape (especially thin sheet stock) by a screw-like action.

In the same vein, for some applications it is beneficial to grind a chip-curling groove in the rake-face of the cutting lip, and chip-splitting grooves (similar to those of an inserted-blade spade drill) are sometimes ground in either the clearance face (end face) or rake face of larger twist drills.

Flute shapes vary, too

Drill manufacturers also offer a rather wide variety of flute configurations, especially in deep-hole or heavy-duty drills. Their objectives are twofold, though often opposing: first, an increase in tool rigidity; and second, an improved ability to extract chips without any need (or reduced need) for "woodpeckering," the repeated withdrawal of the tool to pull chips out of the hole. One type of flute shape incorporates a discontinuity in the cross-section to act as a chip-curler or chipbreaker.

Special points speed drill penetration, multiply tool life

"Drilling is a very important part of our operation," says Robert H. Woggon, a manufacturing engineer at Trane Co (LaCrosse, Wis), adding, "Properly ground drills can reduce or eliminate downtime, so we're always looking for new ways to improve."

One particular improvement, he says, has built some impressive gains in drilling efficiency: where a conventional drill point may produce 1000 holes before regrinding is necessary, he is now specifying a radiused drill point (Racon) for many applications and is getting as many as 14,000-16,000 holes before resharpening is required.

Woggon's responsibilities lie in the company's compressor plant—Trane is a major manufacturer of air-conditioning systems for commercial, residential, institutional, industrial, and process applications—where he is involved in purchase of new equipment, development of machining processes, and research in cutting-tool technology. Compressor production begins with raw castings, mostly cast iron, and machining involves such operations as boring, drilling, grinding, and deburring. Assembly and testing are also performed in the La Crosse plant.

Drilling operations involve about 50 drillpresses, most of which are multispindle jobs with up to 16 drills used at a time. Drill diameters range from ⅛ in. to ¾ in., with the majority being 7/16 in. and ½ in. About 20% of the drills reground in the plant are ground to the Racon point configuration on a Winslow Model HC grinder built by the Bickford Div of Giddings & Lewis. This point has a curved lip that gradually narrows the included point angle from the chisel edge to the outer corner, thus thinning the chip and spreading out the load on a longer length of cutting edge. According to Woggon (at left in photo), this config-

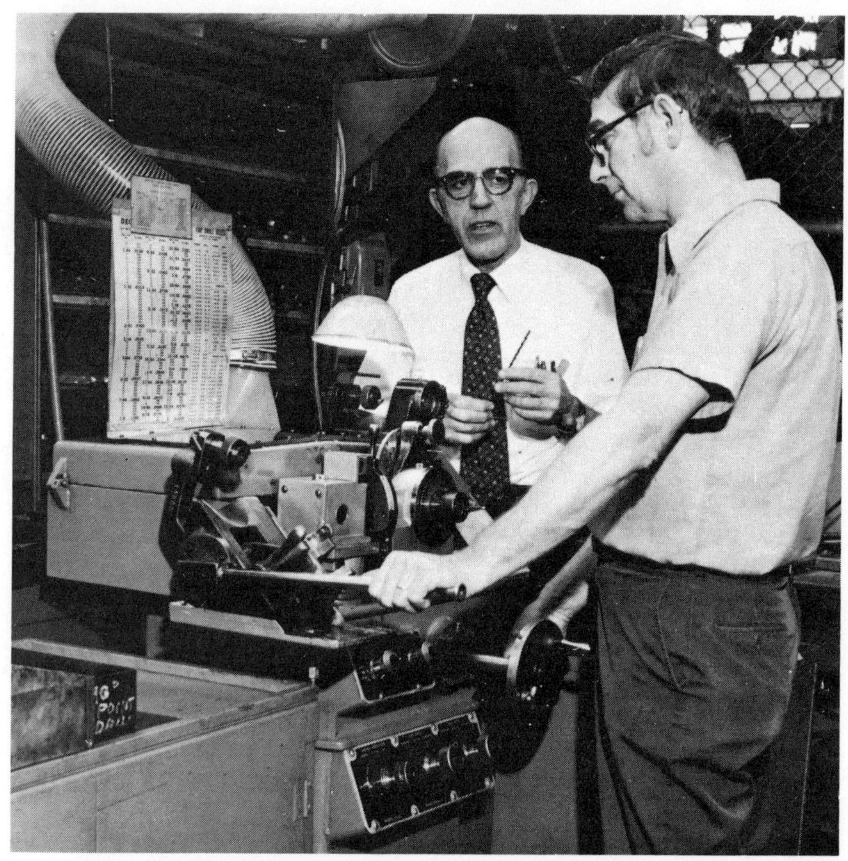

Another available drill variant is the double-margin configuration, in which a secondary margin of full diameter is left at the trailing edge of the land. This provides the extra guidance in bushings or the hole itself of four peripheral contact points approximately equally spaced around the drill, but it has special merit when used with "stubbing" collets that grip the drill on its fluted portion rather than on its shank. This technique gains the advantage of very short tool overhang for high rigidity while using standard length (usually jobbers' length) drills for shallow-hole operations without guide bushings—a frequently encountered situation with numerically controlled machining centers. To avoid any problems that might be caused by the second margin, it can be removed for a short distance above the point (by offhand grinding; it's not a critical area), and at least one drill manufacturer offers such tools with the trailing margin removed for about three diameters.

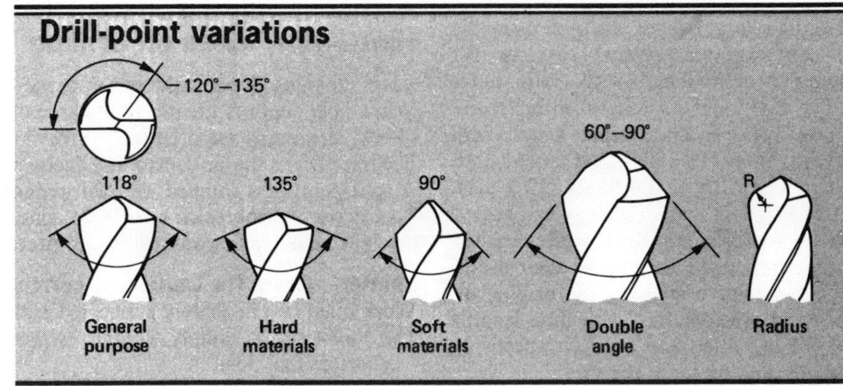

How deep can you drill with a twist drill? One expert sums it up succinctly: "As deep as the flutes, if you use the proper technique." But the longer the flutes, the less rigid is the drill. Therefore twist-drill manufacturers offer a wide range of standard flute lengths to suit most applications, and they'll also supply special lengths when required.

uration generates less heat in drilling and dissipates it more quickly, thus contributing significantly to the working life of the Racon-pointed tool. By comparison, he says, the sharp corner of a conventional drill point tends to break down more rapidly.

In addition, Woggon finds, the special point (upper drill in photo at right) is more free-cutting. "With the Racon drill," he says, "both feed and speed are increased. We can do 20% more cutting in a given time, and we get an even bigger saving from the extended tool life. This ranges from 4-5 times longer than the conventional drill up to as much as 14-16 times longer."

Woggon attributes still another advantage to the curved drill point: "We have seen industry figures saying that a conventional ½-in. drill working in cast iron requires more than 500 lb of thrust force. We use many multiple-spindle drillpresses; with eight drills in a fixture, for example, we may have about two tons of feed pressure being exerted through the drills. When we do through-hole work, the release of this pressure on breakthrough causes a strong reaction on the machine. And, when the tool is ¾-in. dia, the thrust force rises to more than 900 lb on a single, conventionally pointed drill. "We don't have actual figures on the thrust force required for our Racon drills, but, with these special points in through-hole work, the fixture doesn't flex as much, and there is less wear and tear on the machine tool."

Woggon feels that the overall life of any drill depends on how closely its operation is scrutinized and, once its efficiency starts to deteriorate, how quickly the dull tool is returned for grinding. A dull drill doesn't cut metal; it scrapes. And it requires more feed pressure. Thus, Woggon makes the drillpress operator responsible for deciding when regrinding is needed. This determination is made by the sound of the tool, by a visual check, and by noting if hole size is being lost.

In addition to the Racon point, the Winslow grinder can also be used to produce conventional points, helical points, and split points on drills ranging in diameter from 1/16 in. to 1½ in.

Taking straight-shank drills of 3/16-in. dia as an example, typical catalogs list screw-machine length with 1 1/8-in. flutes (6 x dia), jobbers length with 2 5/16-in. flutes (12.3 x dia), taper length with 3 3/8-in. flutes (18 x dia), and taper length long flute with 4 1/8-in. flutes (22 x dia). Also listed in some catalogs are heavy-duty, extra-length drills—often called crankshaft drills because of their use in the automotive industry for drilling oil holes in crankshafts—with flute lengths of 7 1/2 in., 9 in., and 12 in., respectively 40, 48, and 64 times the 3/16-in. dia we're looking at. Flute lengths in other diameters range similarly.

Perhaps one of the most impressive demonstrations of deep-hole drilling is performed in the manufacture of other drills—oil-hole twist drills, produced in larger sizes by drilling two longitudinal holes through a high-speed-steel blank, then twisting it to the desired helix and performing the many other operations necessary to make a drill. Smaller sizes are made by twisting specially extruded stock that incorporates the oil holes—and these are available as small as 1/8-in. diameter.

Advantages of oil-hole drills

More important than how they're made is why. Cutting fluid is delivered in quantity at the cutting surface where it can most effectively perform its cooling and lubrication functions, and the high pressure (100-1000 psi) adds a significant kick in chip ejection. As a result, according to Howard Whalley of the George Whalley Co (a Cleveland supplier of coolant-fed cutting tools), drill penetration rates can often be boosted to as much as triple those of conventional drills with equal tool life.

This is of increasing significance, Whalley points out, because the tripled penetration rates can help justify the high investment required for such machines as NC machining centers and the extended tool life can greatly reduce the economic penalties of toolchanging downtime in transfer machines. In addition, he says, coolant-fed drills produce better holes, "The holes are closer to size they have a better internal finish, and—especially important in the aerospace industry—the remaining material has a higher metallurgical integrity that reduces cracking problems."

Like many drilling experts, Whalley believes conventional drills are too often oversped and underfed. In switching to coolant-fed drilling, he therefore recommends for starting conditions a boost of about 10% in rpm and 15% to 20% in feed per revolution, with further increases as justified by test results. The caution is justified by the high cost of the

Hints, tips, ideas on drilling

Chips clogging flutes is the major cause of twist-drill failure.
When hole depth is greater than three diameters, consider it a deep hole.
"Peck" drilling is required to remove chips when drilling deep holes.
Feedrate is the major controlling factor in breaking up chips.
A split-point or a thinned web will reduce thrust on the drill.
Tool design (chipbreaker) can help control the breaking of chips.
Coolant helps wash away chips and prevents tool and work from overheating.

Chatter—locate the cause, and correct it

Work must be adequately supported and held.
The tool must be amply rigid to withstand heavy feeds without twisting; use the shortest tool possible.
Spindle must have capacity to supply steady power to a multidrill unit.
Mounting of a multispindle head must be adequate to support the unit under the working load.
The entire machine must resist deflection under the working load.
And the fixture should be rigid enough to withstand the work forces.

Drill jigs

Drill bushings should be spaced far enough from the work to allow chips to escape without entering the bushing.
When drilling into an angled surface, the bushing should be very close, and, when the drill has penetrated to at least one half the drill diameter, the bushing should be retracted to provide chip clearance.
Design of the drill jig must not obstruct coolant flow to where it's needed.
Bushing length should be 1.75 to 2.5 times drill diameter.

Deep-hole drilling

Practically all holes 0.050 in. or smaller are in the deep-hole category.
Webs of small drills are proportionately thicker than larger drills.
Lip relief angle should be greater on small drills.
Avoid the tendency to overspeed and underfeed drills—they must produce chips, not powder.
Peck drill just enough to prevent packing chips in the flutes.
Decrease speeds and feeds as holes get deeper:

Depth/dia ratio	Speed reduction	Feed reduction
3	10%	10%
4	20%	
5	30%	20%
6-8	35-40%	

Coolants

Prime purpose is to prevent heat buildup. Flooding also helps wash away chips.
Water is the best coolant known (water-based fluids are preferred).
Coolants must contain rust inhibitor. Avoid coolants that become rancid.
A coolant with good lubricity can help improve hole quality.

Drilling high-tensile steel

Rigidity of tool and work are of utmost importance. Drills should be short.
The drill's point angle should be 130°-150°.
Use split point on materials up to Rc 45 hardness.
Thinned web with short chisel edge is better for hardnesses over Rc 45.
Use low speeds and feeds for hard materials.

Stainless steels

The 200- and 300-Series stainlesses are more difficult to drill than 400-Series (for example, 15-40 sfm for 301 to 316 and 30-75 sfm for 410).
The 400-Series is considered free-machining.
A 135°-140° split-type point is best for stainless steel.

Reconditioning of drills

Remove all of the worn area. Check margin diameter for extent of wear.
Regrind point to proper angle and clearance for the job.
Thin the web or split the point. (Remember, web thickness increases as point is ground back from original length.)
In splitting point, it is important to remove all of the chisel edge.
Do not grind gash beyond the center of the chisel edge.
The gash angle must be at least 55° from the drill end (35° from drill axis).

Source: Lou Yane, Master Mechanic, Zagar Inc, Cleveland, Ohio

Why twist drills fail

Indication	Causes
Outer corners break down	Cutting speed too high, hard spots in work material, no cutting compound at drill point.
Cutting edges chip	Feed too high, lip clearance too great.
Checks or cracks in cutting edges	Overheated or too quickly cooled while sharpening or drilling.
Margin chips	Oversize jig bushing.
Drill breaks	Point improperly ground; feed too heavy; spring or backlash in machine, fixture, or work; drill dull; flutes clogged with chips.
Tang breaks	Imperfect fit between taper shank and socket caused by dirt or chips or by burred or badly worn sockets.
Drill splits up center	Lip clearance too small, feed too heavy.
Drill will not enter work	Dull drill, lip clearance too small, web too thick.
Hole rough	Point improperly ground or dull, no cutting compound at drill point, feed too high. Fixture not rigid.
Hole oversize	Drill point improperly ground.
Chip shape changes while drilling	Drill becoming dull, cutting edges have chipped.
Large chip coming from one flute, small one from other	Point improperly ground and one lip doing all the cutting

Source: National Twist Drill & Tool

tools, which are priced about three times as high as conventional drills in sizes above ½ in. and considerably more than that in the smaller sizes.

One method of feeding coolant through these drills provides a high-velocity pulsating flow that is said to be especially effective in clearing chips without the need for woodpeckering. The Jet Pulser system, available in two sizes from Balcrank Products Div of Wheelabrator-Frye (Mishawaka, Ind), hooks up to a shop compressed-air line and multiplies that pressure by 7 or 10 times (depending on model) and delivers up to 3 gpm or 10 gpm of fluid in three different modes: at constant low pressure, mist, or pulsating high pressure—either through the spindle or externally. Some major manufacturers of NC machining centers are incorporating these coolant systems in their products, and they are also applicable to a wide range of either standard or special machines, both old and new.

Half-round drills

Seen rather infrequently—except in screw-machine shops—is the half-round drill, a hole-making tool of relatively simple geometry that possesses a number of favorable characteristics in many drilling operations. Few handbooks, however, give it more than a passing mention.

Its name is highly descriptive; the half-round drill is essentially a length of round rod with approximately half its diameter ground away at the business end and a conical point that is offset or skewed to provide radial relief for the single cutting lip. The flat, typically about the same length as the flutes of a twist drill of the same size, provides both space for chip egress and a 0°-rake cutting face. Slight back-taper is incorporated behind the tip.

Sizes up to an inch

Listed as a standard catalog item in fractional, wire-gage (number), letter, and metric sizes in a range from at least No. 97 (0.0059 in.) to ½ in., half-round drills are also available in diameters from 0.003 in. or smaller up to at least 1 in. from some suppliers.

Although the half-round drill is somewhat similar to a single-lipped miniature spade drill (or pivot drill), its action and effectiveness are really more analogous to those of a gundrill. Indeed, the half-round drill is historically the predecessor of the gundrill, having been used in early times for producing gun or rifle bores, and gundrill-like point configurations can be and sometimes are ground on half-round drills.

Correctly ground half-round drills have no chisel edge, but rather a conical point on dead center. Thus, they will start accurately in position without bushings and they have no tendency to skate—even in very small diameters that are inherently flexible. Given an assymetrical gundrill point, half-round drills must be either very short and rigid, or they must be started with the aid of bushings or a shallow starting hole predrilled to the same size as the half-round drill's diameter.

Drilling, reaming in one pass

Because of the gundrill-like support provided in the hole being drilled, half-round drills tend to run true to center, and they also exceed the capability of two-lipped drills on diameter tolerances and internal finish. Indeed, proponents

Half-round drill

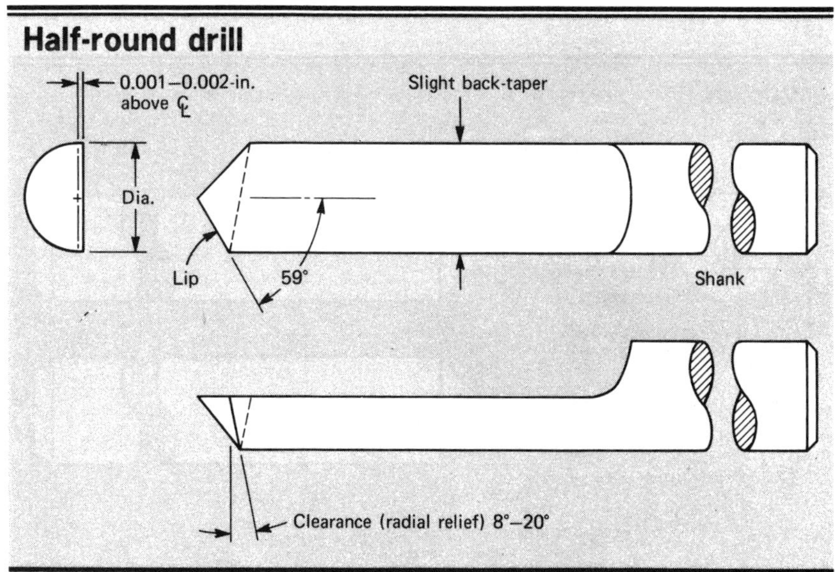

of this type of tool often describe it as the equivalent of drilling and reaming in a single pass.

What's lacking, of course, is the gundrill's high-pressure application of coolant through the tool to the cut to flush out chips or the chip-pulling action of a twist-drill's helical flutes. The only thing pushing out chips is new chips being made. Thus, while it has historically been used for rifle drilling, the depth/diameter ratio of half-round drills is generally held below a maximum of about 7:1 to 10:1 in horizontal applications (such as screw machines) or 2:1 or 3:1 in vertical applications. Woodpeckering helps, but there's still the possibility of leaving some chips that will be in the wrong position in the hole when the drill is advanced again.

Also because of chip-ejection limitations, half-round drills are best applied—especially in relatively deep holes—to workpiece materials that tend to form short or broken chips rather than stringy ones. It must be noted, however, that the 0° rake angle of the cutting lip tends to produce broken chips in many materials that would ordinarily produce long spirals when drilled with helically fluted tools and their inherently high positive rake angles. When half-round drills are given a gundrill point, they tend to produce smaller chips that flow more easily out of the drilled hole.

Half-round drills are frequently used dry. When coolant is desired, its application depends somewhat on the diameter being drilled: in larger holes, a small-diameter nozzle is desirable so that a high-velocity stream of coolant can be effectively directed into the hole. In smaller hole diameters, where this approach may not be practical, a simple, low-pressure flood—essentially to cool the workpiece—is probably best.

Recommended speeds for half-round drills made of high-speed steel are the same as for hss twist drills in the same workpiece material. When the half-round drill is made of carbide or is carbide-tipped, the recommendation is about two thirds the speed of a carbide gundrill.

Because it has only a single cutting lip, instant analysis suggests that a hss half-round drill should be fed at half the ipr of an equivalent two-lipped hss twist drill. This often leads to too conservative feeds because many of the factors that can compromise twist-drill feeds, such as burring and breakthrough problems or the tendency of a helically fluted tool to "screw" itself into the work, either are not present or are minimized with half-round drills. For these reasons, and especially if the half-round drill is short and rigid, it is often possible to feed it more heavily than a twist drill.

Specials are simple

Half-round drills can be produced fairly easily in such special configurations as multidiameter step drills, and these are offered by some suppliers. In addition, the point of a half-round drill can be reground to reverse a right-hand-cutting tool to left-hand cutting.

Half-round drills do not impose any special requirements on the machines in which they are used. They can be effectively applied in ordinary drilling machines, in standard lathes as well as automatics, in NC machining centers (horizontal or vertical), in special transfer machines, and in single- or multispindle setups.

Microdrilling

As THE TERM SUGGESTS, microdrilling is the name for the very special world of miniature-hole machining. It is a world with its own very special problems, involving dimensions at which many workpiece materials no longer exhibit the uniformity and homogeneity normally expected in producing larger-size chips. Grain borders, inclusions, alloy or carbide segregates, and microscopic voids can loom large at working diameters expressed in numbers that generally represent the tolerances in precision drilling operations of more normal proportions.

To put a border around it, microdrilling is generally considered to start at about 0.020-in. ID and progress downwards from there. At the bottom end of the scale, diameters less than 0.0001 in. have been successfully drilled, although some people propose the term *submicrodrilling* for such sizes.

Helically fluted twist drills are cataloged down to No. 97 (0.0059 in.), and half-round drills are available down to about 0.003 in. dia. Thus both of these types of tools can be and are used in the microdrilling range. However, for the smallest diameters—down to 0.0001 in. and less—it is more common to use microdrills of the "pivot-drill" type, a name that acknowledges early application in the manufacture of clocks and watches. These tools are also available in diameters as large as about 1/8 in.

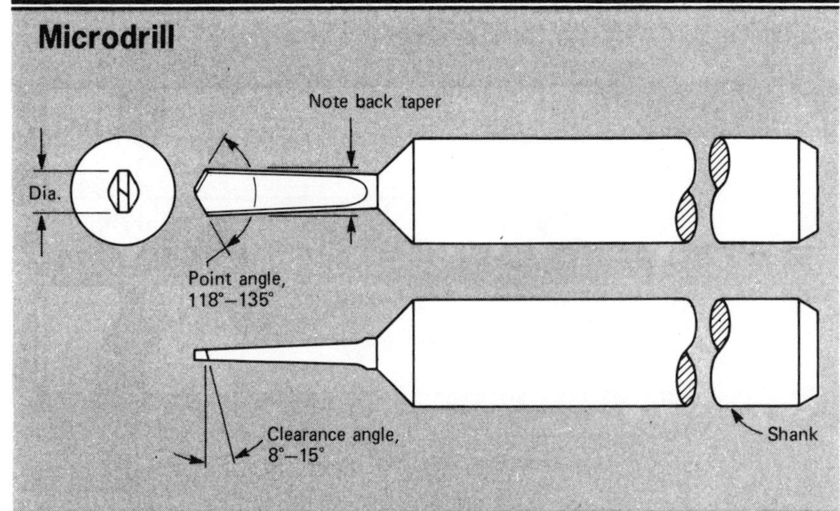

Pivot drills are a form of flat drill or spade drill. They are two-lipped (two-fluted, if you prefer) end-cutting tools of relatively simple geometry that lies somewhere between that of a half-round drill with the second side also flatted (symmetrically) and that of a twist drill without the helical twist (see illustration). Web thickness tapers toward the point, and a rather generous back-taper is incorporated. For softer workpiece materials, point angles are typically 118°, and lip clearance is 15°. For steels and general use in harder metals, 135° points and 8° clearance are recommended. The chisel edge is similar to that of a twist drill. Pivot drills are made of tungsten-alloy tool steel in standard sizes from 0.0001 in. to 0.125 in. and of sintered tungsten carbide from 0.001 in. to 0.125 in.

One of the major problems of using any tool of extremely small size is the physical fact that rigidity is inversely proportional to the fourth power of diameter. Although this is rigorously true only for a cylindrical shape, it is approximately true for such fluted tools as twist drills, half-round drills, and pivot drills. Microdrills of any type, therefore, should be as short as possible; deflection varies as the cube of length.

Jobbers-length twist drills in the range from No. 81 to No. 97 have flute lengths of approximately eight times diameter, and catalog half-round drills in this size range (0.0130 to 0.0059 in.) have flute lengths of 8-10 times diameter. For pivot-type microdrills, National Jet Co (Cumberland, Md) recommends maximum depth-to-diameter ratios of about 20:1 for diameters above 0.004 in., about 14:1 for 0.004-in. diameters, about 7:1 for 0.002 in., and not over 3:1 for diameters under 0.001 in.

Tight tolerances loom large

The second major consideration in microdrilling is that even extremely close tolerances loom large in proportion to the dimensions involved; 0.0001-in. tolerance, after all, is 10% of the diameter of an 0.001-in. drill. Najet offers microdrills in three blade-diameter tolerance categories: +0.0000/−0.0001 in., ±0.0001 in., and ±0.0002 in.

Because workpiece materials vary in their machinability and because drills do not cut to precisely their own diameter—although it's claimed that pivot drills cut closer to diameter than other types because they burnish as they cut—it is recommended that test holes be drilled with various-size tools when hole-diameter tolerances are extremely precise.

But accuracy on blade diameter is only a starting point; a "perfect" blade with a symmetrical point would demonstrate little merit unless it was concentric with its shank and then rotated accurately about the machine's spindle axis. Najet sidesteps the latter problem in its drilling machines by the simple expedient of eliminating the spindle. Instead, the oversize drill shank itself, or a mandrel in which the drill is mounted, is rotated in a pair of diamond-surfaced V-blocks and is both powered and held in place by a resilient drive belt over a small pulley clamped by a collet to the shank or mandrel between the two V-blocks. To ensure concentricity of the working end of the drill with its shank or mandrel, the drills are manufactured in a similar setup.

Specialized microdrilling machines are hand-fed, sensitive units, although *ultrasensitive* might be a more descriptive term. Feed pressure is applied through a compound lever arrangement acting on the upper end of the drill shank or mandrel. This provides a mechanical advantage of 20:1 to eliminate the effects of normal human hand tremor, and the lever system is fully counterbalanced by an adjustable weight so that the operator can truly feel the action of the microdrill biting into the workpiece. A 20X stereomicroscope for observation of the operation "adjusts" the operator's view to the mechanically magnified feed distance required at the input end of the lever system.

Automated microdrilling machines for production work generally incorporate positive feed systems, such as cams, a cam-driven sealed hydraulic system incorporating master and slave bellows, or electronically controlled stepping motor drives with ballscrews.

Don't over-rev microdrills

Consulting standard references for cutting speeds and feeds is not recommended for microdrilling! A modest cutting speed of 50 sfm, for example, would require nearly 64,000 rpm for an 0.003-in. drill; this is simply not practical. Optimum speeds vary with different work materials, of course, but Najet's recommendations for tool-steel microdrills of 0.005 to 0.006 in. dia are to stay below 2800 rpm—and slower for smaller diameters. Higher speeds can be used effectively with carbide drills.

Feedrates also vary with different work materials—as well as with hole-quality requirements—but would typically be in the range of 0.000,05 ipr on the average. What's important in feeding a microdrill is that the tool should cut at all times and should not be allowed to dwell in the bottom of the hole. Frequent withdrawals of the microdrill are used to clear chips, and it is important that both the feed engagements and the retractions be controlled for a predetermined amount of material removal in each. A final recommendation is the use of a light, lard-based cutting oil, preferably containing sulfur.

Spade drills

ANOTHER TYPE of "flat" drill comes into prominence for producing larger hole diameters: the inserted-blade spade drill, in which blade and shank are two separate pieces for convenience and economy.

Spade drills of this type have been made in diameters as small as ⅝ in., though most product lines start at a minimum diameter of 1 in. From there they range upward to 5-in., 6-in., or even 10-in. diameters in catalog listings, and they have been produced as specials in diameters approaching 18 in.

Spade drill

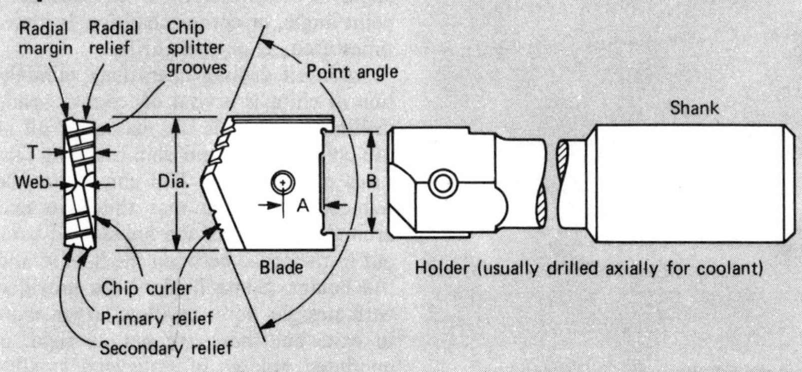

Thus, at the bottom end of their size range, spade drills do compete with the larger end of the twist-drill range. Proponents of inserted-blade spade drills often tout lower costs in this overlap area—typically from 1 in. to 3 or 3½ in. diameter. Waukesha Cutting Tools (Waukesha, Wis), for example, recently offered a comparison covering eight sizes from 2 9/16 in. to 3 in. in 1/16 in. increments. The eight twist drills in the Waukesha example, costing individually from $226 to $340, totaled $2265; in comparison, the company listed one E-size holder at $140 and eight blades at $25.50 each, for a total of only $344. [Continued]

Other advocates of the spade drill, such as Frank Butrick (Butrick Mfg Co, Akron, Ohio), prefer to emphasize its operational advantages, such as greater inherent rigidity, heavier feedrates, and faster production rates possible, in addition to the larger diameters of which it is capable.

The two-piece tool consists basically of a holder that has a shank at one end and is slotted at the other to accept a mating notch on the rear end of the blade. An exception to this is the Doubl-Dex spade drill offered by Madison Industries (Providence, RI), which features a double-ended blade that can be indexed 180° for a new cutting point and therefore requires a special holder.

Holders are available in eight standard sizes (A through H) and several larger sizes (see table), each of which accepts a range of blade diameters. Various lengths are available, as are straight-shank, Morse-taper, and milling-machine-taper holders. Most holders, especially in longer lengths, incorporate oil holes for delivery of cutting fluids through the tool, which is recommended in almost all applications.

Geometry of the blade itself bears many resemblances to the working end of a standard twist drill and to that of the miniature pivot drill—in concept if not in proportions. Because of the large diameters at which they work and the heavy feeds often used, inserted-blade spade drills incorporate a number of special features designed to reduce thrust and horsepower requirements and to facilitate chip removal.

Rake face of each cutting lip is generally ground in a manner that provides a positive rake angle of 5°-12° along most or all of the length of the lip and simultaneously thins the "web." Although a spade drill has no web in the same sense as a twist drill, the analogy is realistic; the web-thinning grind on a spade-drill blade shortens the chisel edge with beneficial effects the same as with a twist drill. The contour of this grind, analogous to a chipbreaker groove on a turning tool, also can assist in the vital function of chip control.

Butrick Mfg

Spade drills are available in a wide range of diameters; these two, by Butrick Mfg, are 10 in. and 1⁷⁄₆₄ in. and do not represent either maximum or minimum

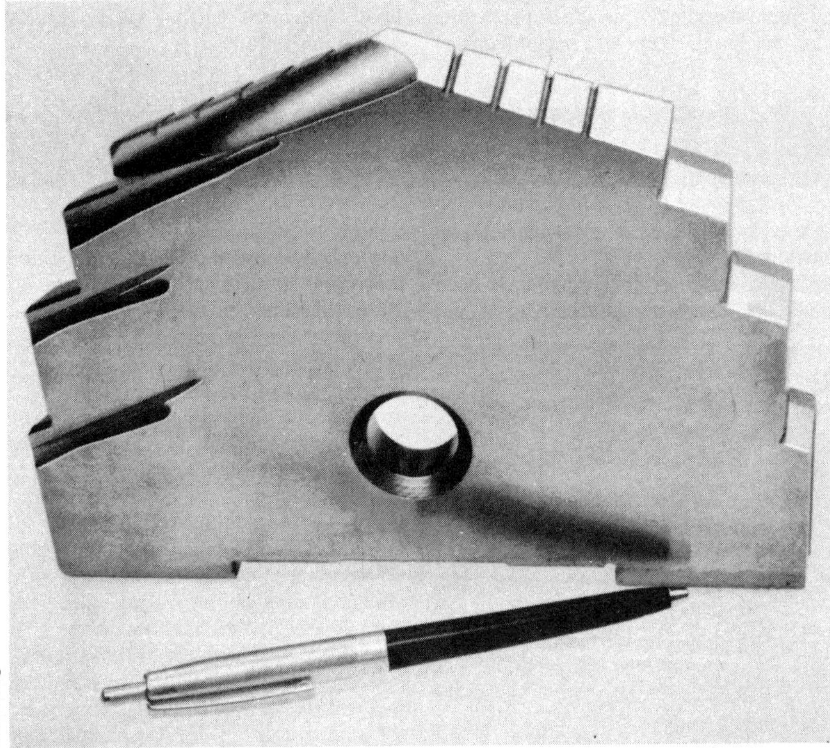

Butrick Mfg

Special multistepped spade-drill has maximum diameter of about 6¾ in. Tool is for approximate roughing of ball socket in steel forging

Notched for chip-splitting

The clearance face, typically ground with a relief angle of 6°-8°, is generally notched to provide a chip-splitting action, the notches being staggered on the blade's opposite cutting edges. Secondary relief is generally not used except in cases requiring modification of the point (four-facet point or crankshaft point) to reduce walking tendencies for better centering on the work or to reduce the required thrust forces to within the machine tool's capability.

Spade-drill point angles are much the same as those used on twist drills—118°-135° being most common—and effects are much the same: smaller angles produce wider, thinner chips and reduce blade wear at the outer corners; larger angles generate thicker, narrower chips, may reduce work-hardening, and tend to cause chips to flow in a more outward radial direction. The 135° point angle is most common. A secondary point angle, or corner chamfer, is sometimes used, as on twist drills.

As in all drilling operations, elimination of chips is a vital concern in spade drilling, and this is the reason for all of the chip-splitting and chip-breaking features of the point. The chips must be reduced in size so that they can exit around the head of the holder and back out in the space between the hole ID and the holder. Some holders are provided with straight flutes to allow larger chips to exit, but these are not as rigid as nonfluted holders of somewhat smaller diameter. The problem is more acute in smaller diameters (A and B Series) because holder diameters are relatively larger, leaving less space for chip flow, and experts emphasize the need for extreme care in spade-drilling deep holes of small diameter.

Except in the shallowest holes, generous flow of coolant through the toolholder is mandatory to flush out chips, and, for very deep holes, it is common practice to supplement this with compressed air at 40-60 psi introduced into the coolant line.

Despite these cautions, spade drilling is an effective technique for producing deep holes. There are cases in which depth-to-diameter ratios exceeding 120:1 have been produced with spade drills, even though the method does not fall into the general category of deep-hole-drilling techniques. Because the chips don't have to fight gravity, horizontal application of spade drills imposes fewer restrictions on the process; in vertical drilling, experts recommend that depth-to-diameter ratios never exceed 10:1 in a single pass.

Use of drill bushings is often recommended in spade-drill applications in order to enhance the accuracy of starting location. When bushings are used, because of the relatively short length of spade-drill blades, they should be placed relatively close to the workpiece, allowing a maximum chip-clearance space of no more than one-quarter to one-third of the bushing's ID. The rear corners of blades used with bushings should be radiused or chamfered to prevent problems in retraction through the bushing.

An alternate technique that is sometimes more practical is to start the hole with a spade drill in a short holder, drilling about one diameter deep before switching to the final drill in its longer holder. Use of centerdrills or small predrilled lead holes should be avoided unless the sharp edge of the hole is chamfered to match the point angle of the spade drill.

Take care at breakthrough

Problems can also arise at the other end of the work as a spade drill breaks through. As the point breaks through, cutting fluid escapes through this newer path of least resistance instead of washing radially outward to the outer corners of the blade. With less coolant at the outer corners, they can overheat, causing metallurgical damage that will result in corner breakdown on the next operation. A supplementary flood coolant line with a solenoid valve triggered to open just before breakthrough is recommended to solve this problem.

Another breakthrough problem, often aggravating the one just mentioned, is

Spade-drilling condition adjustments

Hole depth x dia	Reduce speed	Reduce feed	Coolant application: Horizontal operation — Flood	Through holder	Air assist	Vertical operation — Flood	Through holder	Air assist
3	—	—	OK	Yes	—	OK	Yes	+ Yes
4	—	—	OK	Yes	+ Yes	No	Yes	+ Yes
5	5%	5%	No	Yes	+ Need	No	Yes	+ Need
6	5%	5%	No	Yes	+ Need	No	Yes	+ Must
8	10%	5%	No	Yes	+ Need	No	Yes	+ Vital
10	12%	10%	No	Yes	+ Need	No	Yes	+ Vital
12	15%	10%	No	Yes	+ Must		Too deep	
15	17%	10%	No	Yes	+ Must			
20	20%	20%	No	Yes	+ Vital			
30+	30%	25%	No	Yes	+ Vital			

Source: Universal Engineering Div, Houdaille (Frankenmuth, Mich)

Sleuthing spade-drill problems

Clues	Suspected culprits
The chips	
"C" or "9" shape	Correct
Yellow-brown color	Correct
Clocksprings	Feed too light
Powdery	Feed too light
Broken into small pieces	Inadequate coolant flow, speed too low, feed too light
Long, straight	Speed too low or feed too light
Deep blue/purple	Speed too high, coolant flow too little
Not discolored	Speed too low
The hole	
Oversize	Okay
Rough finish	Okay
Off location, crooked, or bellmouthed	Starting holder too long, needs drill bushing, entering surface too rough or sloping, blade not properly sharpened, chip-splitters tracking, workpiece loose, burrs in Morse-taper spindle, blade not tight in holder
Step in hole	Blade was cocked, straightened out by itself
Large exit burr	Insufficient coolant at breakthrough, blade dull, radius worn on corner of blade, speed too high
The blade	
Edges chipped	Feed too light, feedrate varying, machine springing under thrust force, edges burned (annealed) during sharpening, chip-splitter grooves too deep, hesitant starting, drill walking, breakthrough at excessive feed, feed too light in work-hardening material, blade material too brittle, rake angle too high, blade hammered into holder, operation started on sharp edge of predrilled hole, blade cocked in holder, blade improperly sharpened
Edges burned	Speed too high, feed too light, flood coolant used, blade annealed in sharpening, dull blade used, running dry at breakthrough, blade improperly sharpened
Corner burned	Running dry on breakthrough, speed too high, blade material inadequate for workpiece material being drilled, margins worn undersize, inadequate coolant flow, running dull blade, soluble oil too diluted, improper sharpening
Tool broken	Chip-packing from too-light feed, high speed, or inadequate coolant flow; collapse of burned or chipped edge; rapid traverse into work; result of continuous edge-chipping, improper sharpening, or using too large a holder
Erratic tool life	Machine overloaded, variations in feedrate, variations in work-piece material, improper sharpening, blades run when dull, blades from different suppliers

Source: Universal Engineering Div, Houdaille Industries (Frankenmuth, Mich)

the tendency of the machine itself to spring back somewhat as breakthrough relaxes the heavy thrust of the operation. This tends to increase the effective feedrate of the last several revolutions of the drill.

To maintain the extra rigidity that may be required on breakthrough, as well as to improve the guiding accuracy of the tool throughout the drilling operation, Erickson Tool Co (Solon, Ohio) offers patented wear studs affixed to the holder by the blade retaining screws. These provide four-point contact within the hole ID in a manner somewhat similar to a double-margin twist drill.

Feeds and speeds at which spade drills are best used vary, of course, with both the tool material and the workpiece material. Blades are available in a variety of standard and premium high-speed-steel alloys, including cobalt-containing grades and with tungsten-carbide cutting edges. In general, speeds for standard-hss blades should be 20-30% lower than for twist drills under similar conditions, and speeds for cobalt-hss blades can be more nearly equal to those of twist drills.

Recommend heavy feeds

Feeds with hss blades, the experts stress, should be heavy. In a technical paper on production spade drilling presented several years ago, Frank Butrick recommended feeds ranging from 1.5 to 4 or 5 times as great as twist drills. He cited handbook values of 80 sfm and 0.016 ipr for drilling a 2-in. hole in 4150 steel with a twist drill. His own recommendations for drilling the hole with a spade drill were 70 sfm and "about 0.028 ipr, with 0.015 ipr being the recommended minimum and 0.050 ipr probably the practical maximum." At 0.028 ipr, he noted, "the spade drill will produce a given hole in 65.5% of the time required for a twist drill at the recommended speeds and feeds."

The heavy feeds, which admittedly result in somewhat rougher ID surface finishes, are necessary to produce easily flushed chips. Thinner chips produced by lighter feedrates have a greater tendency to tangle, pack the hole, and break the blade.

Because of the need for heavy and uniform feedrates to control chip formation, especially at the larger diameters, it is virtually mandatory to use power feeds in spade drilling. Thrust and torque requirements are roughly similar to those of twist drills, with advantages of the proportionally narrower web being offset by lower rake angles, other differences in cutting-edge geometry, and the requirement for heavy feeds. As an example, a 4-in. spade drill cutting mild steel at 60 sfm (57 rpm) and a feed of 0.020 ipr will require an axial thrust of about 8000 lb and a torque of about 1200 lb-in. Power required for such a cut is about 11 hp.

Rigid machine, rigid setup

Such values demand a rigid machine and a rigid setup. Lack of rigidity can lead to severe problems of springing as the drill enters the work and as it exits.

Inadequate thrust and horsepower capability in the machine can be overcome by producing the hole in two passes: first with a smaller drill (half to two-thirds the final diameter is recommended), then following it with a core blade. It's pointed out that, although this requires two operations, when they are performed at proper feedrates, this method generally gets the job done faster than using a larger tool at light feed and also avoids the difficulties that often accompany those lighter feedrates.

Another solution to the problem of limited thrust capacity—assuming adequate horsepower and spindle speeds are available—is the use of carbide-tipped spade-drill blades at higher spindle speeds and lower feeds. Such blades are available in a variety of designs, including core-blades with indexable carbide inserts (the latter from Butrick Mfg, which also reports that center-cutting spade-drill blades with indexable inserts are under development). Use of carbide-edged blades, however, does not sidestep the need for rigidity; if anything, it increases this requirement to avoid chatter and vibration problems that can quickly chip carbide cutting edges.

Aside from these general requirements, and the obvious requirement of sufficient travel to produce holes of the depth demanded, spade drilling does not require special machines. Spade-drilling operations are commonly performed on virtually all types of horizontal turning machines except for the smallest sizes, on horizontal-bar boring-milling-drilling machines, on the larger standard types of vertical drilling machines (including radials), and on both vertical and horizontal NC machining centers.

Spade-drill sizes

Series	Blade dia range	Blade dimensions T	A	B
A	1 – 1¼	3/16	¾	9/32
B	1 5/16 – 1½	9/32	1 1/16	9/32
C	1 9/16 – 2	5/16	1¼	7/16
D	2 1/16 – 2½	⅜	1¾	13/16
E	2 9/16 – 3	7/16	2 1/16	⅞
F	3 1/16 – 3½	½	2⅝	1
G	3 9/16 – 4	⅝	3 1/16	1⅛
H	4⅛ – 6+	11/16	3½	1¼
J*	4⅝ – 8+	11/16	3½	1¼
K*	7 – 9½+	11/16	3½	1¼
L*	9½ – 12+	11/16	3½	1¼
M**	12 – 15	11/16	3½	1¼

* Special
** Custom designed for each application
Does not include earlier Y-type holders

Source: Universal Engineering Div

Speeds for hss spade drills

Material	Speed, sfm
Soft cast iron	100-150
Medium-hard cast iron	70-100
Hard chilled cast iron	30-40
Brass, bronze	200-300
Aluminum	200-300
Magnesium	250-400
Alloy steel	20-30
Stainless steel	30-40
Automotive steel forgings	40-50
Tool steel, 1.2% C	50-60
Mild steel, 0.2% C	80-110
Steel, 0.4-0.5% C	70-80

Source: Erickson Tool Co (Solon, Ohio)

Indexable-insert drills

INDUSTRY'S NEWEST DRILLING TOOL is the indexable-insert drill for producing relatively shallow holes, which was first introduced commercially about five years ago by Metal Cutting Tools Inc (Rockford, Ill) and is now available in a variety of designs from at least half a dozen different sources, primarily manufacturers of carbide inserts.

In addition to providing the economic benefits of throwaway inserts—quick indexing of cutting edges without any need for regrinding and elimination of the need for replacement of the entire tool body—the use of these new drilling tools also provides the capability of producing holes at carbide speeds. This latter advantage is of major importance to users of numerically controlled lathes, the turrets of which generally bristle with carbide tooling at all stations. Early indexable-drill designs were thus suited for lathe applications; today's are suited for both fixed lathe-type use and rotating-tool-type use (as in a machining center). Most offer a choice of straight or Morse-taper shanks.

Currently available in diameters ranging from approximately ⅝ in. up to 3 in., indexable drills are essentially short-hole tools with maximum depth/diameter ratios limited to 2:1 to 3:1.

The different designs

As noted above, variations in design of indexable-insert drills are currently available. However, most are essentially two-fluted tools with two or three inserts fixed to the business end: one insert cuts to center, one cuts the ID of the hole, and, if a third insert is present (generally in the larger sizes), it serves to fill the cutting gap between the first two.

Various insert shapes are used in the products of different manufacturers. The original Metcut tool and now the Kendex/Metcut drill (Kennametal) uses square inserts, such as SNMM or SNMG styles, that are standard turning shapes. Sandvik and R. B. Tool Co use trigons. The Valenite drill mounts square inserts with chip-splitting grooves on their cutting edges and a round insert asymmetrically mounted but cutting to center. This firm uses a single diamond-shaped insert in smaller diameters. TRW Wendt-Sonis originally introduced its Mach 6 tool for lathe applications with a single, rather complex insert, which it offers as a standard product, but its recently introduced rotating version takes two inserts. Carboloy is about to introduce a line of indexable-insert drilling tools using a square insert at the tool's periphery and a triangular insert cutting to center.

The inserts on any tool may or may not be the same size, depending principally on the drill's working diameter, and they are mounted in positions and attitudes designed to counteract each other's lateral cutting forces and thus eliminate or at least minimize side loads on the drill. This is desirable because, like the boring tools they resemble, they are not guided by the hole being drilled and bushings are not used.

And one has helical flutes

Most designs incorporate straight flutes—an exception is R.B. Tool's Kub, which is helically fluted—and thus have no inherent chip-pulling action. The drill bodies, however, all provide for coolant delivery through the tool, which is effective in washing out the chips from the relatively shallow holes produced. Compressed air can be fed through the coolant passage to help evacuate the dust-

Carboloy

Soon to be introduced by Carboloy is indexable drill with triangle cutting to center, square at periphery

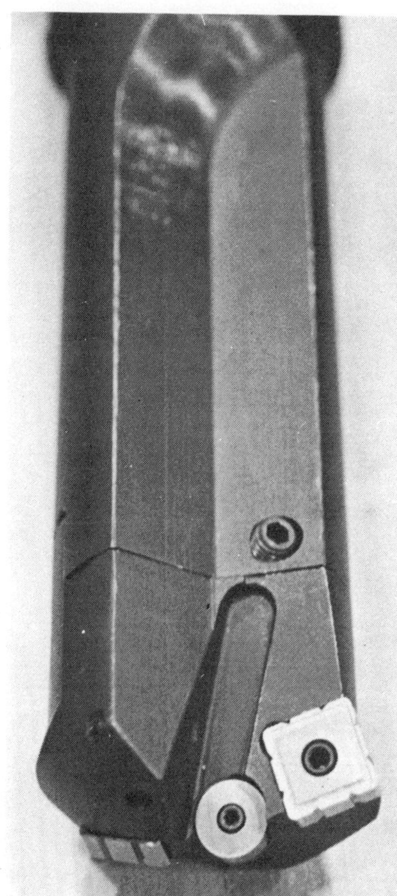

Using round at center and notched squares is Valenite indexable drill first shown at IMTS-78. Head is detachable

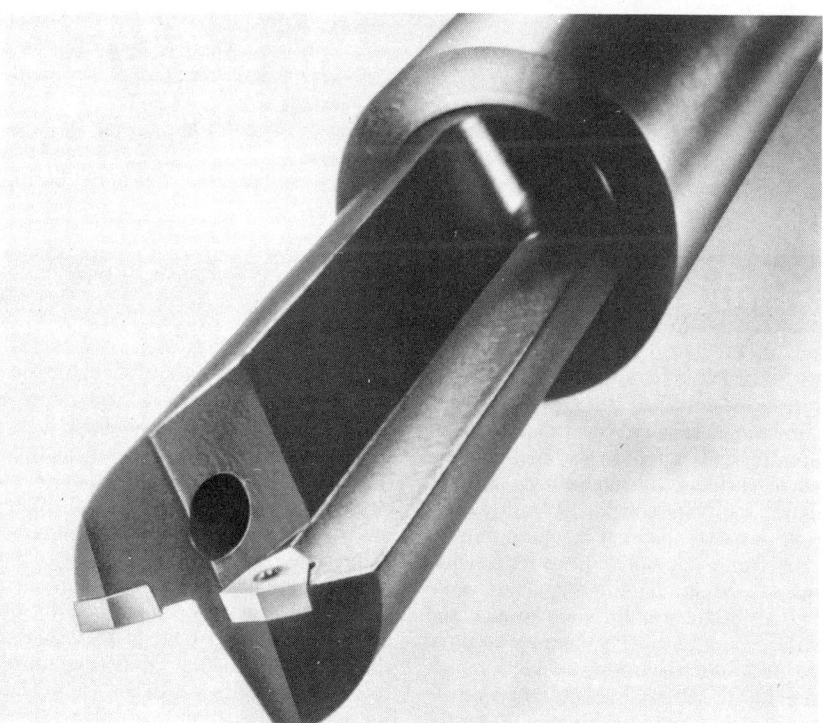

Sandvik

Trigon inserts are used at both center and periphery of Sandvik short-hole indexable drill. Inserts are dimpled on rake face for splitting chips

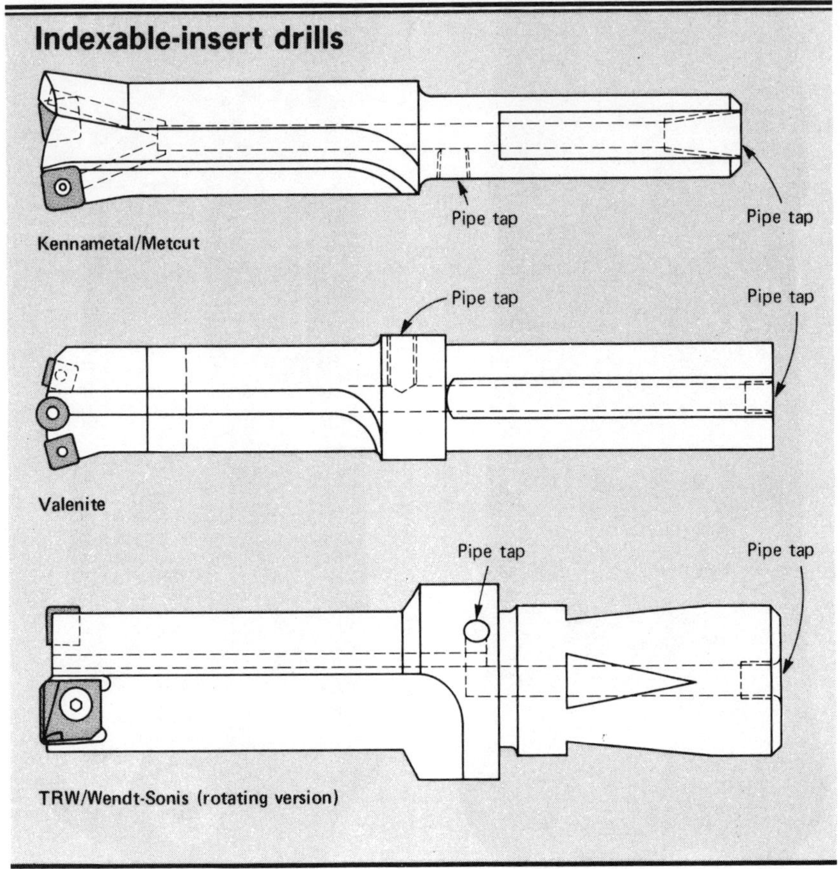

like chips produced in drilling cast iron.

One feature of indexable drills that is unique among drilling tools is their capability (in suitable machines) to perform boring operations in addition to their primary purpose. Surface quality of the holes produced by indexable drills is somewhat rough, certainly no better than that produced by any traditional drilling tool, but it is possible (with the tool mounted on an NC lathe's cross-sliding turret, for example) to step the tool radially outward and make a second pass—boring the hole to final diameter and providing a much improved ID surface. With some of the indexable tools, it is even possible to do this in a back-boring mode instead of first withdrawing the tool.

Indeed, Wendt-Sonis touted its original Mach 6 as a multifunction NC-lathe tool that was capable of drilling, boring, ID contouring, facing, and OD turning and contouring (the latter two operations being done after spindle rotation was reversed). Actual drilling diameter of the Wendt-Sonis tools is 0.030 in. under the nominal size to allow for the secondary boring operation.

Using the indexables

Some typical application recommendations for indexable drills: In a lathe, make sure that the drill and the work spindle are concentric within 0.005 in. (0.010 tir) or better and that longitudinal alignment is within 0.005 in. in 6 in. Similar alignment conditions should be met in rotating-tool applications. Setups should be as rigid as possible, which is always desirable but is especially so with carbide cutting edges. Coolant, which is not always necessary, should be at 10-40 psi minimum pressure in horizontal applications and higher in vertical setups.

Recommendations for speeds (starting conditions) are typically 300-500 sfm in free-machining steels and cast iron, 150-300 sfm in alloy steels and austenitic stainless, and 500-1000 sfm or more in aluminum. Feedrate recommendations offered by Kennametal: carbon steels, 0.006-0.008 ipr; resulfurized steels, 0.007-0.009 ipr; alloy and austenitic stainless steels, 0.004-0.008 ipr; cast irons, 0.010-0.015 ipr; and aluminum, 0.006-0.012 ipr.

Feed forces are no higher than those required by twist drills, and at least one company (Sandvik) claims that thrust requirements are as much as 40% lower than for twist drills. High cutting speeds, however, demand substantial spindle horsepower; drilling a 2-in. hole in 4140 steel at a cutting speed of about 300 sfm and a feed of 0.007 ipr requires a spindle power of about 18 hp.

Gundrills

If drilling is a family of processes, deep-hole drilling is a subfamily, itself consisting of a variety of techniques that are differentiated primarily by the method of cutting-fluid application and chip extraction. In common, the self-guiding tools used in these techniques share a remarkable capability to resist wandering at extreme depth-to-diameter ratios and to produce holes close to size and with generally excellent surface finishes. In addition, they provide high metal-removal rates that, in many cases, justify the use of "deep-hole" techniques for the production of short holes of one-diameter depth or even less. Although it is possible to adapt standard machine tools for these deep-hole processes, special machines are generally preferred.

The first of these deep-hole techniques is gundrilling. Its name indicates its origins and ancestry, but its applications today range far beyond the manufacture of firearms. The original gundrills were very likely half-round drills, themselves drilled axially with a coolant hole for the delivery of cutting fluid to the scene of the actual chipmaking. Modern gundrills typically consist of a crimped (to provide the single flute) alloy-steel-tubing shank with a solid-carbide or carbide-edged tip brazed or mechanically fixed to it. Guide pads following the cutting edge by about 90° and 180° are also standard.

Like the half-round drill, it is a single-lipped tool, and the major feature distinguishing the gundrill from previously discussed drilling tools is the delivery of coolant through the tool at extremely high pressures—typically from 300 psi to 1800 psi, depending on diameter—to force chips back down the flute. Successful application of a gundrill depends almost entirely on the formation of small chips that can be effectively evacuated by the flow of cutting fluid.

Standard gundrills are made in diameters from 0.078 in. (2 mm) to 2 in. or more, and specials have been made as small as 1 mm (0.39 in.). Depth-to-diameter ratios of 100:1 are considered "reasonable" by experts, and 200:1 is considered possible. At the other end of the scale, because of the rapid metal-removal rate of gundrilling and the excellence of hole quality produced, gundrills are often justified at depth ratios of less than 1:1.

What are some commonly held tolerances in gundrilling? Diameters under about 1/2 in. can be held to 0.0005-in. total tolerance, and, over 1/2 in., the tolerance shouldn't exceed 0.001 in. by much. According to one source, "roundness accuracies to 0.000,08 in. can be attained." Because of the burnishing effect of the guide pads, hole finishes of 10-20 μin. can be produced; as good as 5 μin., under favorable conditions.

What affects hole straightness?

Hole straightness is affected by a number of variables, such as diameter, depth, uniformity of workpiece material, condition of the machine, sharpness of the gundrill, feeds and speeds used, and the specific technique used (rotation of the tool, of the work, or both), but deviation should not exceed about 0.002 in. tir in a 4-in. depth at any diameter, and it can be held to 0.001 in. per ft.

Basic setup for a gundrilling operation, which is generally horizontal, requires a drill bushing very close to the work entry surface and may involve rotating either the work or the tool or both. Best concentricity and straightness are achieved with both the work and the tool rotating in opposite directions; the poorest is when only the drill is rotated,

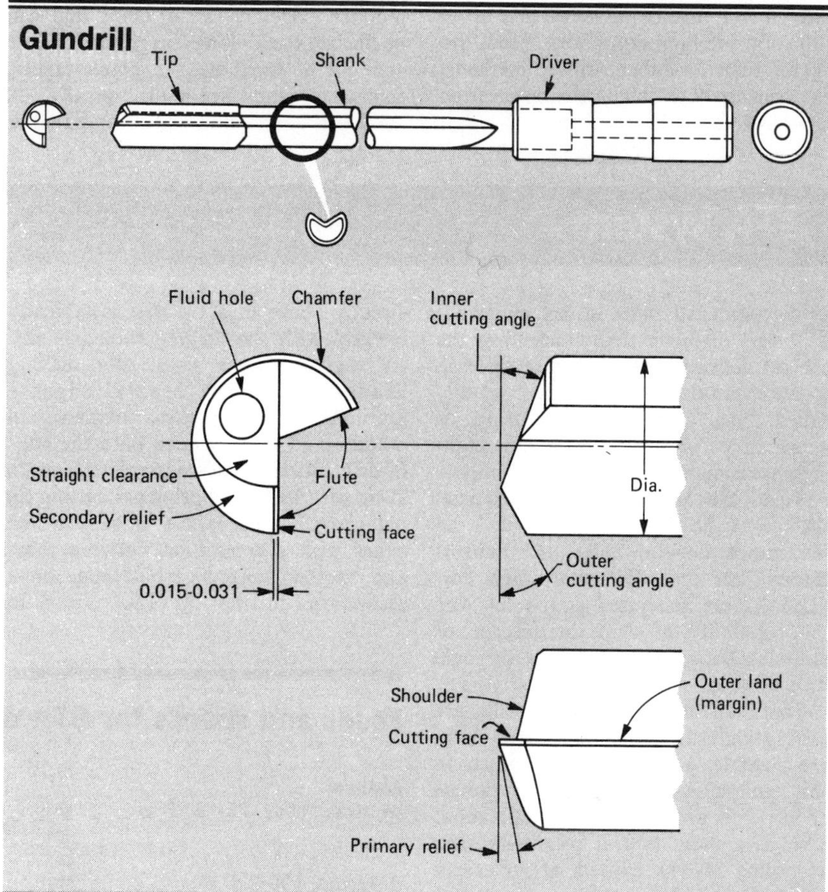

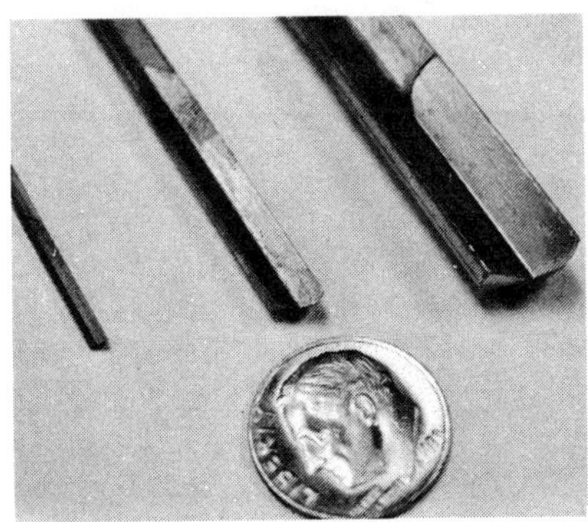

Smallest standard gundrill (left) is 2 mm dia, though smaller have been made. Center tool is 5 mm, right is 10 mm

Gundrilling feeds

(Carbon & alloy steel, under Bhn 300)

Dia, in.	Feeds, ipr Starting	Est. max
5/32 - 3/16	0.0002	0.0005
13/64 - 1/4	0.0004	0.0008
17/64 - 3/8	0.0005	0.0010
25/64 - 1/2	0.0006	0.0013
33/64 - 5/8	0.0007	0.0015
41/64 - 3/4	0.0008	0.0018
49/64 - 1 1/2	0.0010	0.003
1 1/2 - 2	0.002	0.004
2 - 2 1/2	0.0025	0.005

Source: Albion Corp (Ferndale, Mich)

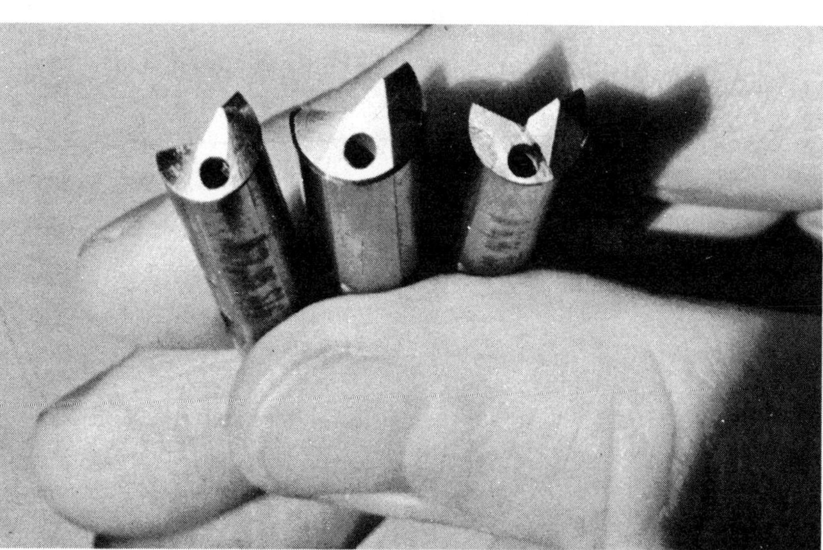

Three gundrill grinds: at left is conical clearance produced by rotating drill, center shows faceted equivalent, grind at right increases coolant flow to cutting edge

because of whipping of the long, slender shank at high rpm (antiwhip bushings are often used to control this effect).

With carbide cutting edges and wear pads, gundrilling speeds are high. Spindle speeds of 10,000 rpm and higher are not unheard of in gundrilling small holes in easily machined alloys. Cutting speeds as high as 300-500 sfm are sometimes used in free machining steels and up to 1000 sfm in aluminum alloys. Feeds are lighter than for other drilling methods, but, because of the high rpm, penetration rates can be in the range of 2-6 ipm (depending on workpiece material) at an optimum diameter of about ½ in. Penetration rates for both smaller and larger gundrills are lower, according to some experts. Axial-feed forces of about 500 lb per in. of diameter are required.

The combination of generally light feeds and special attention to the geometry of the cutting lip is designed to produce extremely small chips that can be flushed back along the gundrill's flute and out of the hole. As noted earlier, coolant pressures are high, typically 300 psi or so for gundrills of ¾-in. diameter and rising to as much as 1800 psi at diameters in the 0.080-in. range.

Straight cutting oils with EP (extreme pressure) additives containing sulfur or chlorine are commonly used. Flow rates are high—1 gpm for an 0.080-in. hole, 20 gpm for a ¾-in. hole. Temperature control and filtration are also important. Coolant-delivery hole through the tip of a small-diameter gundrill is necessarily small, and, to prevent the possibility of plugging it with the fines carried back by recirculated oil, filtration down to 10 μ is recommended for 0.080-in. gundrills.

BTA system

For drilling deep holes somewhat larger diameter than gundrilling, the coolant delivery technique is reversed in a European-developed method usually called "the BTA system" (Boring & Trepanning Assn). In this system high-pressure coolant is delivered externally—between the hole ID and the external surface of the tubular drill shank—and the return flow of chips and fluid is through the tool. Extremely high flow velocities are generated at the cut, and the possibility of chip interference or jamming between the tool and the hole ID is avoided.

Tool diameters as small as 7/16 in. are fairly standard—as small as about 3/16 in. are possible, according to specialists in this technique—and holes as large as about 12 in. are possible, although about 8 in. is a more typical maximum size, according to BTA experts at American Heller Corp.

Like a gundrill tip, a BTA drilling head presents an asymmetrical cutting edge (or edges) balanced by wear pads at approximately 90° and 180° behind the cutting edge (or main cutting edge) to provide guidance by the hole walls after the tool has entered the work through a guide bushing. Again like those of a gundrill, the wear pads on a BTA drilling tool also provide a burnishing effect on the hole wall.

The smallest tools provide a single cutting edge with a single chip mouth directly above in a tip that is typically integral with the drilling tube (shank). At slightly larger sizes, BTA drilling heads incorporate brazed edges—regrindable in some cases, throwaway in others—and are threaded onto the ends of drill tubes. In this size range (approx ¾-in. to 2½-in.), the principal cutting lip may have two separate brazed-carbide edges with a separation between them and another brazed carbide edge on a secondary cutting lip 180° away to remove material left between the cuts of the first two. Such a cutting head will have two chip mouths.

Still larger BTA drilling heads use three or more indexable carbide inserts clamped in a similar arrangement, with a primary and a secondary cutting lip and two chip mouths. In all cases, the primary lip performs most of the cutting and provides most of the cutting force, while the secondary lip contributes some balancing force so that the loads on the

Throwaway drillhead for BTA system has ¾-in. diameter, provides large chip mouth for internal coolant exhaust

Feeds and speeds for BTA drilling

Material		Hole diameter, in. ½	1	2	4	8
Aluminum alloy, 75-150 Bhn	sfm	800	800	800	800	800
	ipr	0.004	0.012	0.015	0.019	0.022
	ipm	25	36	23	14.5	8.4
Gray iron, 190-220 Bhn	sfm	250	250	250	250	250
	ipr	0.004	0.010	0.015	0.018	0.022
	ipm	7.5	9.5	7	4.2	2.6
Malleable iron, 200-255 Bhn	sfm	240	240	240	240	240
	ipr	0.002	0.005	0.010	0.011	0.014
	ipm	3.7	4.5	4.5	2.6	1.5
17-4 PH stainless, 150-200 Bhn	sfm	250	250	250	250	250
	ipr	0.0013	0.004	0.005	0.007	0.009
	ipm	2.5	3.8	2.4	1.7	1.1
1020 carbon steel, 175-225 Bhn	sfm	300	300	300	300	300
	ipr	0.002	0.005	0.006	0.007	0.009
	ipm	4.5	5.7	3.5	2.0	1.25
4140 steel, 175-225 Bhn	sfm	400	400	400	400	400
	ipr	0.002	0.005	0.006	0.007	0.009
	ipm	6	7.5	4.5	2.7	1.75
4340 steel, 250-300 Bhn	sfm	250	250	250	250	250
	ipr	0.002	0.004	0.0055	0.007	0.009
	ipm	3.7	3.8	2.6	1.7	1.1
M10 tool steel, 200-250 Bhn	sfm	200	200	200	200	200
	ipr	0.002	0.005	0.0055	0.007	0.009
	ipm	3	3.7	2.1	1.3	0.8
Inconel 600, 240-310 Bhn	sfm	80	80	80	80	80
	ipr	0.002	0.005	0.0065	0.007	0.009
	ipm	1.5	1.5	0.8	0.5	0.3
Zirconium, 140-280 Bhn	sfm	150	150	150	150	150
	ipr	0.0018	0.0045	0.008	0.012	0.016
	sfm	2.2	2.6	2.3	1.7	1.1

Source: American Heller Corp (Detroit)

wear pads (still approximately 90° and 180° behind the primary lip) do not become excessive.

In addition to avoiding potential chip-interference problems between the tool and the hole surface, the internal chip-extraction technique eliminates the need for a flute. Thus the drill tube is fully circular in cross-section, giving it a higher section modulus and considerably greater rigidity. This allows higher thrust forces to be used—up to double those of gundrilling—with resultant faster penetration rates at equal diameters. These greater stock-removal rates, however, may be gained at the expense of some of the precision hole tolerances and surface finishes that can be achieved with gundrilling.

A rule of thumb for coolant is a flow of 30 gpm per in. of diameter. Typical pressures are 800 psi in a 7/16-in. drilling operation, dropping to perhaps 100 psi in an 8-in. hole. Coolant viscosity should be 60-80 SSU at 100F—and a cooling system should be used, if necessary, to maintain cutting-fluid temperature at about that level. Alternately, coolant tanks should have a volume of at least 10 times the maximum flow per minute of which the coolant pump is capable.

Another similarity between gundrilling and BTA drilling is that either the work or the tool may be rotated, or both may be rotated in opposite directions (which again is the best technique when straightness requirements are critical).

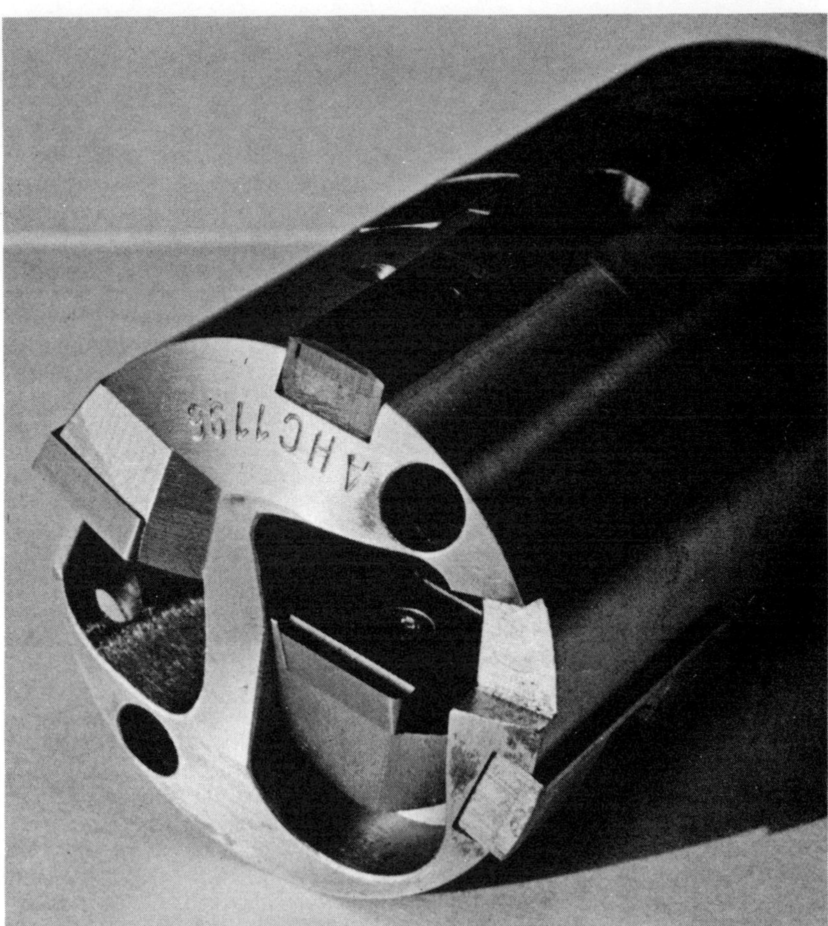

Divided cutting edges and dual chip mouths are incorporated in this 2¼-in. BTA drill (American Heller). Carbide inserts are indexable, wear pads replaceable

Ejector drilling

STILL ANOTHER TWIST in the application of cutting fluid and the ejection of chips forms the basis for a third deep-hole drilling technique—ejector drilling. Through an arrangement of concentric tubes, this system incorporates both an internal cutting fluid supply and an internal chip flow. The fluid is pumped between the drill tube and the inner tube to the drill head.

Most of the fluid is forced through holes in the drill head and cools and lubricates the support pads and cutting edges. The remainder is forced through venturi-shaped nozzles in the inner tube and is diverted back to the outlet. This creates a partial vacuum in the inner tube so that the fluid that has done the lubricating and cooling at the cut is sucked back into the inner tube together with the chips and is exhausted through the outlet.

Developed and patented by Sandvik, the ejector system is claimed to be the only deep-hole drilling system that can be fitted to existing machines because high-pressure seals between the work and a drill bushing are not necessary. In fact, the company points out that even the drill bushing can be eliminated in situations where a starting hole can be provided by conventional drilling from another turret station. Examples of machines on which ejector drilling operations are performed in production work include both turret and engine lathes (some with the workpiece rotating normally, others with the tool rotating and the work fixtured on the saddle) and horizontal boring machines as well as a variety of both standard and special deep-hole drilling machines.

Less coolant, lower pressure

Ejector drilling requires about a third less coolant volume than BTA drilling, about 10 gpm for a ¾-in. drill and perhaps 90 gpm at 6 in. Pressures are also lower, say 200 psi at ¾-in. diameter and dropping to about 100 psi at 6 in.

With the exception of accommodation for the differences in plumbing of the two systems, ejector drillheads are quite similar in design and cutting geometry to those used in the BTA system. Disposable, screw-on heads with brazed carbide edges and guide pads are available in diameters from 0.724 in. (18.4 mm) to 2.559 in. (65 mm). From the 65-mm size up to 180 mm (7.087 in.) and even larger specials, ejector drill heads are available with indexable carbide inserts.

For the ¾- to 2½-in. size range, cutting speeds range from, say, 100 sfm for some stainless-steel grades to about 400 sfm for the more free-cutting carbon steels and as high as 650 sfm for aluminum. In the same size range, feeds range from about 0.006 ipr up to about 0.012 ipr in steels, to about 0.016 ipr in cast iron, and as high as 0.028 or 0.030 ipr in aluminum. Starting conditions recommended for the larger indexable-insert drill heads call for feeds of about the same as BTA tooling.

[Continued]

Trepanning

ONE OF THE PRINCIPAL FEATURES of drilling, as noted earlier, is that the depth of cut is fixed at half the hole diameter when drilling from solid and reducing all of the material within the hole to chips. In larger diameters, this can demand huge amounts of spindle horsepower and high-tonnage axial thrust loads.

So why convert all of the material in a big hole to chips? This line of reasoning leads to the process of trepanning, in which an annular region is reduced to chips and a central core is removed in solid cylindrical form. Some truly impressive holes can be produced by this technique—with diameters well above 24 in. and depth-to-diameter ratios exceeding 100:1—but it is also applicable to holes of less than 2 in. dia. An additional advantage is that, because the trepanning tool does not cut to center, its cutting speed never drops to zero.

Pin-type gundrills, which may be obtained in diameters as small as 1/4 in., are arranged so that the uncut core passes back through the coolant hole in the gundrill tip, and chips are flushed back externally in the flute or flutes. In larger sizes of trepanning tools, coolant may be supplied either through the tool with external chip flushing or by the BTA system of external coolant delivery and internal chip removal.

With either coolant-delivery system, sufficient space must be allowed so that the core does not interfere with the flow of chips and coolant. In either case, of course, the system benefits from the fact that trepanning generates a smaller volume of chips for a given hole diameter than conventional drilling.

Depending largely on the diameter involved, trepanning heads may present either brazed or indexable cutting edges, which may be disposed in either single or divided arrays (the latter to provide improved balance of cutting forces, thus reducing the imbalance that must be supported by the wear pads).

Unique to trepanning are the problems posed by the solid core. Most obvious of these is the problem of removing it when the operation involves a blind hole. Special core-cropping tools are available from American Heller (Detroit) in diameters from 1 3/4 in. to 20 in., which incorporate a pivoting cutoff blade that is fed across the base of the core to produce a groove much like that in lathe cutoff.

Another problem, aggravated by hole depth, is sag of the core, which can create a pinching effect on the tooling and chip flow or an imbalance problem if the workpiece is being rotated.

A third problem can arise on breakout of the trepanning tool in a through-hole operation. The most advanced cutting edge is at approximately the center of the annulus being converted to chips, and, when this breaks through the end of the workpiece, a ring of material remains on the end of the core—much like a valve head on a valve stem. Unless the core is restrained, the heavy coolant pressure acting on this valve head tends to slam the core forward against the seal cup, from which it rebounds back into the cutting edge.

Coolant needs are similar to those of BTA drilling, except that the reference will be more to the area of the cutting annulus than ultimate hole diameter, which generally parallels reductions in end thrust and spindle power.

Trepanning is not a high-speed operation. Typical penetration rates are about 2 1/2 ipm for a 1 3/4-in. hole and about 3/4 ipm for a 12-in. one. Rather, the process is used because it minimizes the demands for spindle power and machine rigidity in larger diameters, thus lowering the overall costs of producing large, deep holes. Converting all the material within a 5-in. hole to chips might take 25 hp; trepanning it, about 12 hp. And machine-springing thrust forces are proportional to the width of the annulus, not the total hole diameter. ■

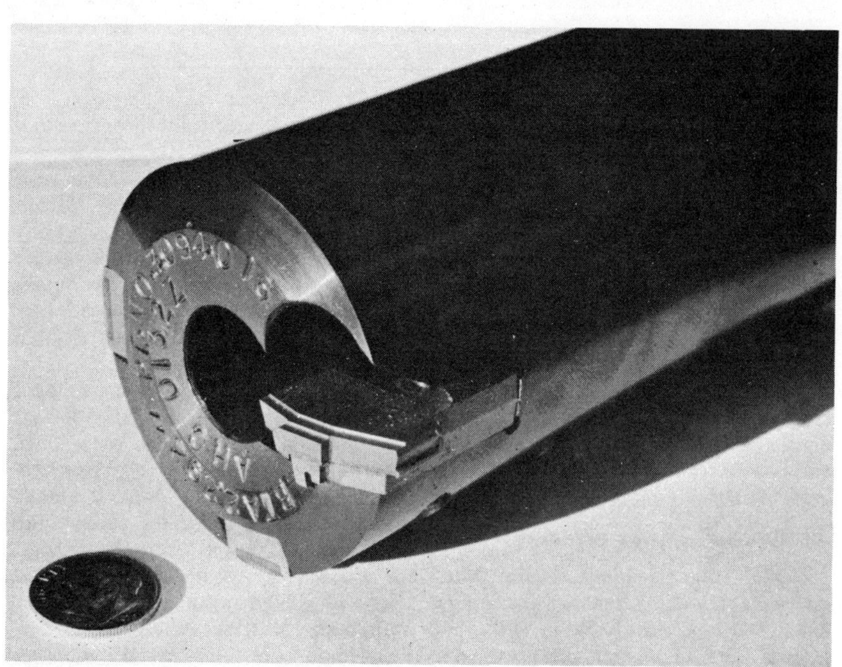

Trepanning tool cuts to 2.292 in. dia, leaves core of about same diameter as dime shown for size comparison. American Heller tool uses BTA internal exhaust

Acknowledgments

AM would like to thank the following organizations whose contributions in the form of specialized knowledge, helpful literature, and illustrations were particularly valuable in the preparation of this report:

Albion, American Heller, Balcrank Products, Bendix Industrial Tools, Burgmaster, Butrick Mfg, Carboloy Systems, Cleveland Twist Drill, DeHoff, Eldorado Tool, Erickson Tool, Giddings & Lewis, Guhring, Kennametal, Macor Inc, Madison Industries, Metal Cutting Tool Institute, Mohawk Tools, National Automatic Tool, National Jet Co, National Twist Drill & Tool, R&F Micro-Tool, Rockwell International, Sandvik, SIG Swiss Industrial Co, Society of Mfg Engineers, Sterling Precision, TBT Tiefbohrtechnik, TRW/Wendt-Sonis, Unitec National, Universal Engineering, US Drill Head, Valenite, Waukesha Cutting Tools, George Whalley Co, Zagar Inc.

Fundamentals of milling

It means more than making a lot of chips. Here's a brief review of the machines, tooling, and what happens at the cutting edge, plus an update on the latest technology

By Joseph Jablonowski, associate editor

The milling process, including machines, cutters, and their use, is one of the most important metalworking processes. Nine out of ten metalworking companies use it; it is most versatile, being able to carve huge workpieces or do trimming work on miniature ones; and it is one of the most efficient means of reducing a raw workpiece to a finished shape.

The term "milling" itself probably traces back to the Latin root *mola,* or grindstone. Early face-milling cutters, according to Edwin Battison, curator of the American Precision Museum in Windsor, Vt, closely resembled flour-milling grindstones of the beginning of the nineteenth century. Those circular stones crushed wheat on their face but, more important, featured radial grooves to permit the ground flour and chaff to escape centrifugally from the grind. Milling cutters' teeth looked a lot like those radial grooves, hence, the theory goes, the new metal-cutters were called "mills."

Milling machines in use, by SIC class

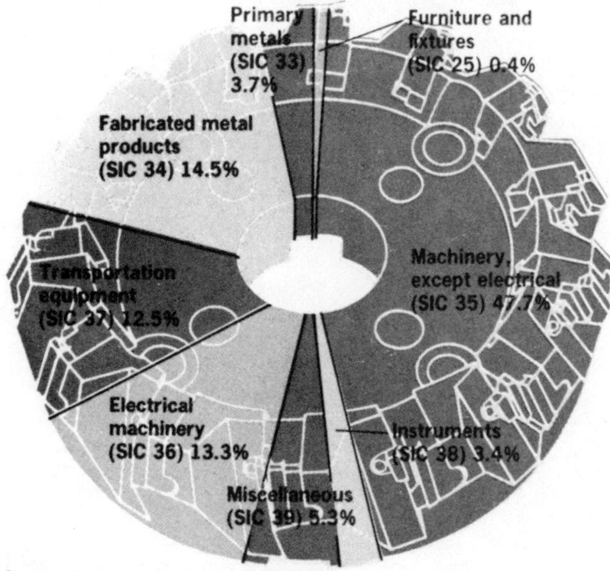

Source: American Machinist 11th Inventory

The first milling machine was made about 1818. It is now believed to have been built by Simeon North or one of his sons, and not by Eli Whitney as had long been believed. The 1818 machine had a head that had been taken from a lathe and had a wood cone pulley for a belt drive.

An important step in development took place at Harper's Ferry Armory, Va, between 1819 and 1826 and was the work of John Hall. He developed a series of precision machines, including three types of machines that were called cutting engines for straight cutting, curve cutting, and lever cutting. The straight-cutting machines turned out to be plain millers. The curve-cutting machines were profile-milling machines that had a rise and fall of the table in response to a pattern placed under the table. All we know about the lever-cutting machines is that they were similar to some forms of hand-millers.

Hall's technology spread rapidly to New England thanks to the number of visitors and workers traveling back and forth between Harper's Ferry and Middletown, Conn, where North had built an arms factory. An 1828 contract to North from the Ordnance Dept called for him to build 5000 Hall rifles, the parts of which were to be interchangeable with those produced in Virginia.

The Vermont gun manufacturer Robbins & Lawrence had already achieved a fair measure of success in making interchangeable rifle parts when, after an exhibition in London's Crystal Palace, the firm began to export milling machines. In 1855, the company shipped 157 machines to England's Enfield arms works; 75 of these were milling machines.

Early millers had a serious technical drawback: there was no convenient vertical adjustment. In effect, each designer seems to have approached the problem as though he were thinking of a lathe.

The first machine known to combine vertical adjustment and an adequate support for the spindle was made at Gay, Silver & Co about 1835. The headstock casting had the spindle on the bottom and served as an overarm, supporting the outer end. The headstock moved up and down on the vertical column and was adjustable with a hand crank. An idler pulley provided slack in the belt to permit the vertical head adjustment.

Frederick Howe served his apprenticeship at Gay, Silver and worked there until he went to the Robbins & Lawrence firm. Shortly after his arrival there in the mid-19th Century, he had built a plain milling machine. So far as vertical feed was concerned, this machine was a backward step, but, in all other respects, it was a major jump ahead—at least for the kind of work that confronted most armories. Howe improved this machine considerably in later models and, in 1850, developed an index milling machine. The work could be positioned in rotation in the horizontal plane and adjusted vertically by a leadscrew. The design lacked rigidity, however, something Howe corrected in his design of 1852—the design that was sold to the British for Enfield.

This milling machine was probably the design that Francis Pratt modified into the famous Lincoln miller. But the Lincoln miller was still a manufacturing machine: it was an excellent machine for performing the same operation over and over, but it totally lacked the kind of toolroom flexibility that the engine lathe was able to provide for turning operations.

The next step had to wait until 1862, when Joseph R. Brown developed a universal milling machine that was a true toolroom unit. By creating the knee-and-column arrangement, he brought the milling machine from the first crude idea of Simeon North to a machine of universal, industrial caliber.

Since those early beginnings, the development of milling machines and the tools that are used with them generally has taken off in two different directions: one aiming toward greater

Heaviest users of millers, all types

Industry	No. of machines
Misc machinery, except electrical	38,059
Special dies & tools, machine-tool accessories	37,308
Aircraft engines & parts	20,253
Special industries machinery	15,371
General industry machinery	12,369
Machine tools	11,070
Communication equipment	10,191
Screw-machine products	8,079
Misc manufacturing industries	7,468
Electrical components	7,381
Fabricated structural products	7,191
Ordnance & accessories	6,588
Cutlery, hand tools, hardware	6,350
Automotive parts	6,066
Office & computing machines	5,389

Source: *11th American Machinist Inventory*

versatility and the other pointed toward less flexible machines that do a single job at great efficiency.

Toolroom-oriented machines, such as those that evolved from Brown's universal, were influenced by the needs of those industries that used them for one-off manufacturing. Heavy-duty, production-oriented millers were the children of the transportation industries, especially automotive. The transportation-oriented Standard Industrial Classifications (SICs) still account for over 12% of the milling machines in use in the US, although, in many cases, the milling function in those industries has shifted to even more-specialized production units, such as the transfer line that combines milling functions with others, such as drilling, boring, etc.

Where the machines are

The transportation segment of industry, however, is not where the bulk of the US's milling machines are installed. Most of the units, some 137,500 of them, belong to companies in SIC 35 (nonelectrical machinery). Within that group, milling machines as a class of machine tools are found concentrated in the subcategory that includes the manufacture of machine tools, special dies and tools, and machine-tool accessories. That subcategory accounts for more than 48,000 milling machines of all types.

(Data on the number, age, and location of milling machines in the US come from the *11th American Machinist Inventory of Metalworking Equipment,* the latest such study providing complete information about all of metalworking. Similar, but more detailed studies on portions of the American metalworking "universe" were presented by this publication in 1976

The milling machine population, by type of machine

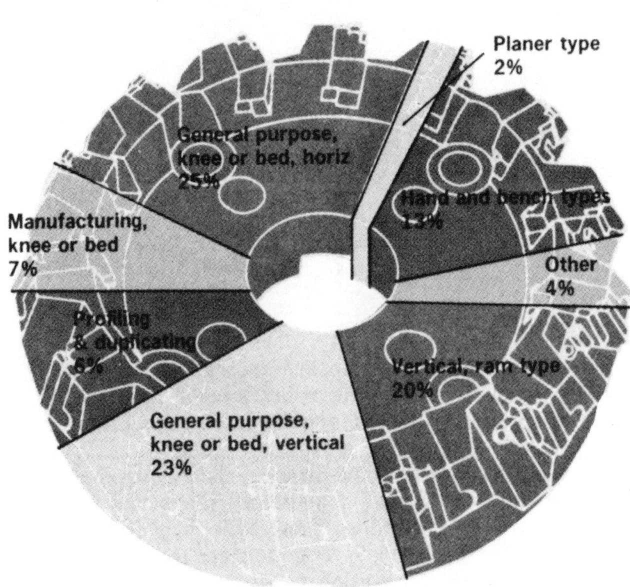

Source: American Machinist 11th Inventory

and 1977. In December, *American Machinist* will publish the third and final segment for that series of studies.)

The 11th Inventory, taken in 1973, showed that there were a total of 288,159 milling machines of all types used by US industries. Of that total, 38% were less than 10 years old (30% for turning machines of all types, 29% for drills, and 36% for grinders). Some 37% of all millers were between 10 and 20 years old (the same percentage held true for lathes, drills, and grinders). And 25% of the millers were 20 years old and over.

The Inventory further broke down the total population of milling machines into types of machines. It identified bench types and hand-feed types; vertical mills with rams; horizontal-spindle machines for general-purpose use; vertical-spindle machines for general purposes; what have come to be called "manufacturing-type" milling machines; planer types; profiling and duplicating machines, which include diesinking machines, skin mills, and spar mills; thread millers; and other special-function machines. (The proportion that each type of machine contributes to the total US milling-machine population is shown in a chart in this introductory section.)

Numerical control does not figure importantly in milling machines, according to the Inventory. Only 2%, or about 5000 machines (out of a total of around 288,000), were equipped with NC. The percentage of machines with NC becomes a little erratic when individual milling-machine types are analyzed: virtually none of the bench-type millers had sophisticated controls, as expected; however, as many as 8% of the machines in the category that included skin mills and spar mills were hooked up to NC units.

But, if you consider that much milling is performed on machines other than those that are called millers, numerical control indeed becomes a factor. Especially in the category of HBMs or "multifunction machines," whose detailed descriptions include operations like "drill-mill-bore, automatic toolchange," and "drill-mill-bore, indexing turret."

In combining those machines that do just milling operations with those that are multifunction in nature, the percentage of the combined total that is equipped with NC moves up to 5%. A much better percentage—considering that the survey accounts for all ages of machines—than that found among, say, lathes or drilling machines.

Shifts in new types of milling machines

"During recent years, many companies have purchased numerically controlled machining centers in place of milling machines," according to the most recent *Outlook for the US Machine Tool Market Through 1985,* published by the New York-based research firm of Frost & Sullivan. The analysts there note that "this trend is expected to continue."

Nonetheless, "modest" growth for strictly milling machines is seen by Frost & Sullivan through 1985. The firm places the estimated number of milling machines to be sold domestically in 1980 at 20,000, up 2000 from 1975; the prediction for 1985 is 22,000.

According to Commerce Dept figures, some 6600 milling machines were shipped in the first half of 1977 to domestic and overseas customers from American machine-tool builders. Annualized, that figure would exceed the 11,000-odd millers that were shipped in 1976, but it would not meet long-range expectations.

Part of the reason for that may be an even greater move toward NC machining centers than anticipated. Industry sentiment seems to bear that out. One New England builder that introduced a milling-machine-based CNC machining center at the International Machine Tool Show in Chicago a year and a half ago reports that its plant is having a difficult time keeping up with orders for that new unit. The same holds true generally with other machine-tool builders and those firms that specialize in retrofitting knee-and-column millers with controls and toolchangers.

But even if smaller, more dedicated milling machines are being supplanted by small machining centers with better flexibility, the milling function—also done on smaller machining centers—has grown.

After a brief discussion of the kinds of machines that have been traditionally classed as millers, this report will focus on the milling function itself—how chips are made, what kinds of tools make them, and the efficient ways to do the job.

Machines for milling

The shape and construction of milling machines vary greatly according to the purpose for which they are intended. There are literally hundreds of combinations of styles, sizes, and special designs: the National Machine Tool Builders' Assn *Directory* lists 25 major categories; a similar compilation by the West German machine-tool federation notes 37 varieties.

In order to streamline the discussion of types of milling machines, this report will identify nine distinctive designs, ranging generally from the most versatile machines to those of the most "dedicated" construction.

But no matter the particular style, milling machines share most if not all of the following design-component features:

■ Spindle. Power is transmitted from the machine's motor through a series of belts, gears, and/or clutches.

■ Base. The foundation of the machine and the location for the coolant reservoir. A trend away from use of castings toward welded-plate construction has been noticed recently.

■ Column. The main support for the spindle housing, it can contain the spindle motor.

■ Table. The workpiece is supported and clamped to the table with T-slots and fixtures. The table rests on the saddle and provides feed by traversing the workpiece longitudinally. Rotary motion is sometimes included in the table.

■ Saddle. In knee-and-column designs, the saddle rests on ways on the knee and positions the table transversely to the table's own movement.

■ Knee. A major structural support member that typically projects forward from the column, it is the mark of the widely used knee-and-column-type machine design. The knee moves on column ways to provide up and down motion for the saddle and table that rest on it.

■ Overarm. Projects forward from the top of the column and serves as the spindle carrier. In horizontal-spindle machines, the overarm usually carries an outrigger support for the nondriven end of the spindle arbor. In vertical-spindle machines, the overarm often contains a ram that permits adjustment of the spindle parallel to the motion of the saddle.

■ Arbor. An accurate shaft for holding, positioning, and driving the cutter body in horizontal-spindle machines. The full, overarm-outrigger-supported arbor is sometimes replaced with a stub arbor, which is used to mount a shell mill without outrigger support.

■ Ram. Attached to or part of the overarm, the ram moves the spindle or entire toolhead toward or away from the column. A feature usually found on vertical-spindle machines, the ram often can be swiveled about the central vertical axis of the column after turret clamps are released.

■ Quill. A feature of vertical-spindle machines, the quill houses the spindle and can be positioned up or down with hand or powered motion to provide rapid vertical positioning of the cutter or to provide feed for boring holes, drilling, or plunging an end mill.

The variations on the use and combination of each of these individual features are almost endless. Additionally, machine attachments individually can change a machine from one mode of operaton to an entirely different mode.

But, before the use of attachments—such as rotary tables, special arbors, universal milling attachments, etc—can be discussed, it is helpful to look at some of the more popular basic styles of milling machines in use today.

Hand miller

This small, low-power machine is a simple, basic design that usually provides feed of the rotating cutter into the workpiece by means of hand power applied through a lever. Distinguishing features of the hand miller usually include a horizontal spindle and provisions for quick loading and clamping of the workpiece.

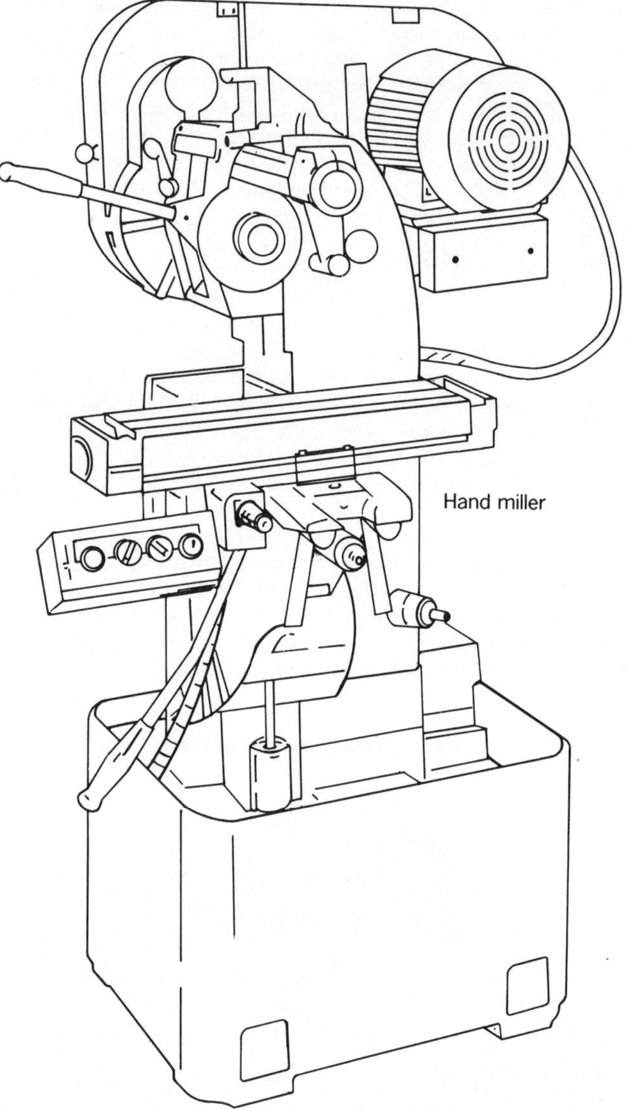
Hand miller

Hand millers may be stand-alone machines with their own floor-mounted beds, or they may be bench-mounted for light manufacturing. Occasionally, a powered feeding mechanism is provided, or the machine is inverted to place the milling cutter under the table with a portion of its cutting surface barely exposed through a hole or slot in the table.

Vertical-spindle, ram-type milling machines

"Bridgeport" is the name most usually associated with this very versatile type of machine, which has recently come to account for the greatest percentage of the US milling-machine "population." Indeed, most units in this type are patterned after the small Bridgeport design first built in the 1930s by Rudolph Bannow and Magnus Wahlstrom.

Typically, thc dcsign includes a knee-and-column support arrangement for the table, a design that dates back to Joseph R. Brown's universal milling machine of 1862. A saddle supporting the table provides in-and-out motions with respect to the column, and the motor is mounted on the toolhead, supported by the overarm.

The main distinguishing feature of the vertical-spindle ram-type miller is, of course, its ram. The ram, basically a sliding member of the machine's overarm, enables the spindle to move, via hand crank or power, in and out parallel to the movement of the saddle; that is, in the Y direction.

Additionally, ram-type vertical mills often incorporate other motion capabilities. In turret-ram styles, the entire ram overarm can pivot about the main upright axis of the column's backbone, thus positioning the spindle in an arc sweeping over the workpiece table to describe an arc.

Another contortion usually found in this type of miller is the capability to tilt the spindle axis away from the vertical, either by inclining the Z-axis left and right, tilting it forward and aft, or both. This is a function of the spindle-drive motor's proximity to the quill and spindle; straightforward belt drives do not operate through the base or column of the machine but rather are tilted with the spindle itself.

Machines with all of these motions contained in the overarm portion of the unit are generally built to operate at low

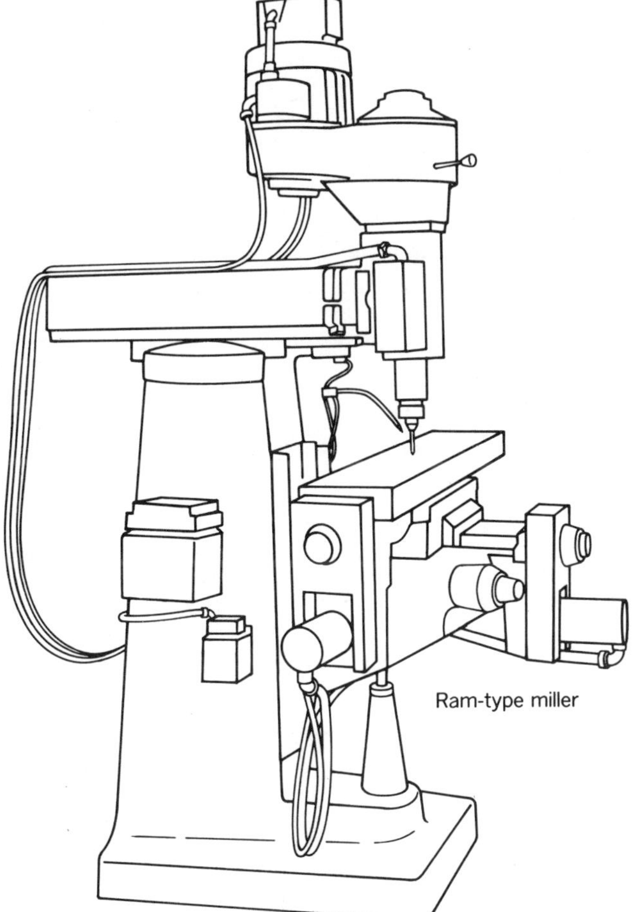

Ram-type miller

horsepower, with a fairly typical toolroom-oriented machine rated at 1-5 hp. End mills are the most common form of tooling, but formed cutters for slotting and dovetailing and shell-type cutters for cutting flat surfaces are also common. Additionally, machines of the vertical-spindle ram type also perform precision boring and tapping operations.

Vertical-spindle, knee-type milling machines

With heavier-duty construction and with higher-horsepower motors, vertical-spindle knee-type millers differ structurally from the ram styles: there is no provision for ram-like Y motion of the spindle carrier. Most other design details are similar, including the knee-and-column table support that provides flexibility in machining a wide variety of workpiece shapes.

As in ram-type vertical machines, the table holds the workpiece and moves left and right in the X-axis, which gives the primary feed motion for most milling. The table is mounted on the saddle, which provides the Y-axis motion, and the saddle, in turn, straddles the knee that elevates the work (Z′).

The knee slides up and down the column on ways and is partially supported with one or a pair of leadscrews, typically powered.

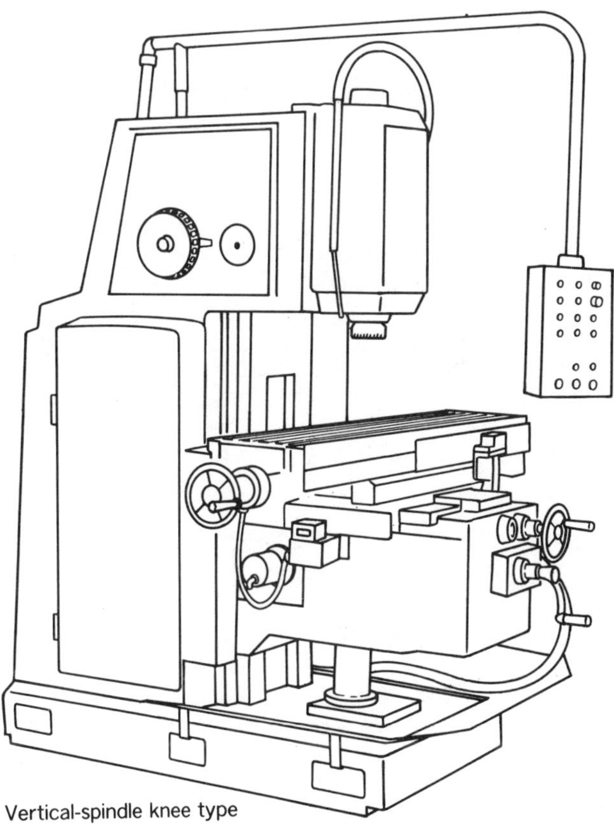

Vertical-spindle knee type

Universal versions of the vertical-spindle knee-type miller are common. In the universal version, addition of the capability for rotating the workpiece about the Z-axis, either through a rotary table or through a rotary vise mounted to the longitudinal T-slots in a plain table, permits angular cutting. A further extension of this principle, the omniversal design that's no longer current, rotated the workpiece and table about the Y-axis, in addition to providing the Z-axis rotation of a universal machine.

Bed-type vertical milling machines

Since all milling machines have beds, the term *bed-type* is inappropriate; a better description would be *fixed-bed* machines. These are characterized by the absence of the knee and, among vertical-spindle machines, are the most rigid class of construction and are usually built with the greatest horsepower capacities. Resting on rectangular beds that are single castings or are made up of weldments, the bed-type machines are able to withstand heavier loads on their work tables without deflections becoming a problem, and they can stand deeper cuts without causing tool chatter. As with other vertical-spindle machines, the table's right-to-left motion through the X-axis is the usual workpiece-feed method. [Continued]

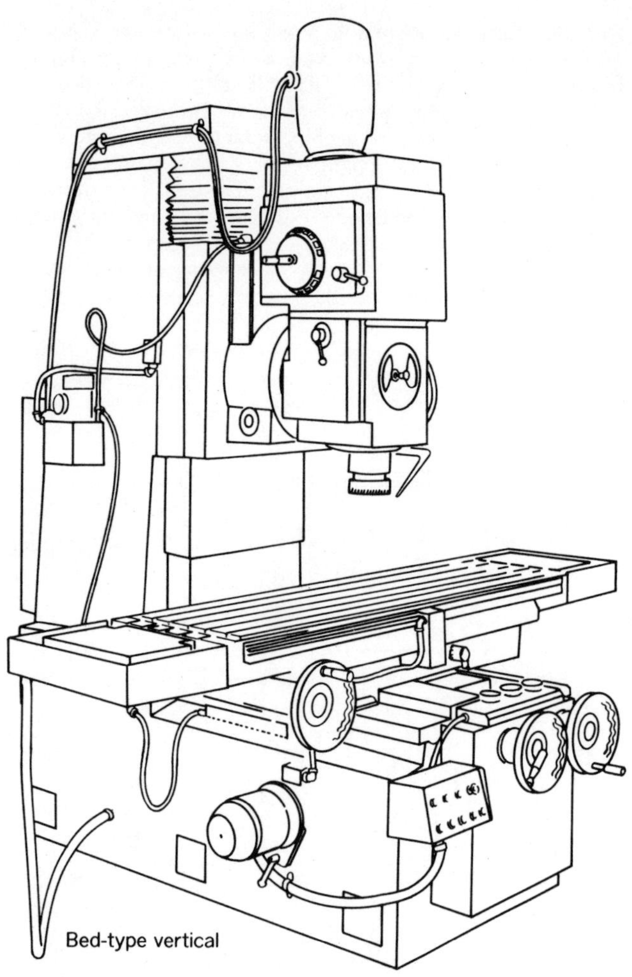
Bed-type vertical

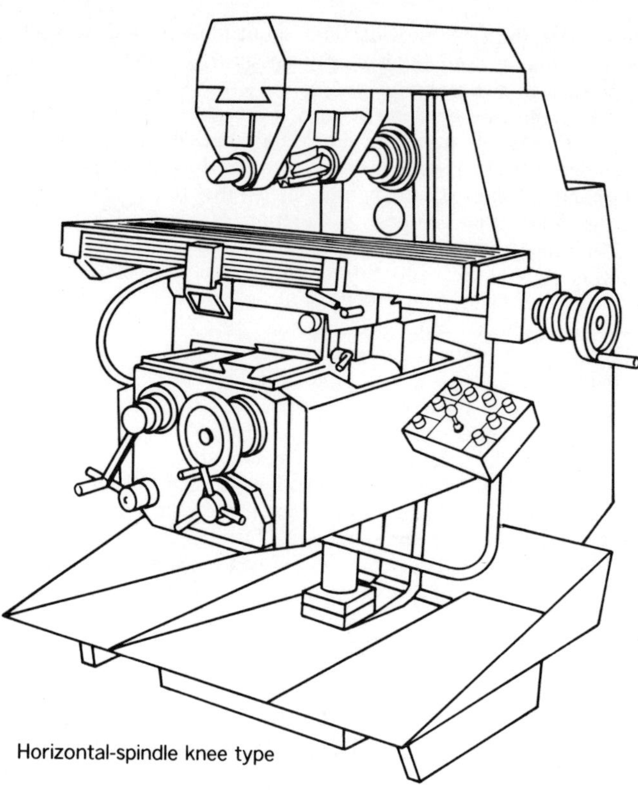
Horizontal-spindle knee type

The rise-and-fall miller is a specialized version of the fixed-bed design and incorporates a mechanism to raise and lower the spindle carrier through the Z-axis automatically in conjunction with the table travel. This synchronized movement is used in production applications to mill plane profiles in the same workpiece, to cut different heights on a single workpiece, or to mill different portions of ganged workpieces efficiently.

Another specialized version of the fixed-bed design swaps the table's X-axis linear motion for the circular motion of a rotary table. A continuously rotating table loaded with a number of fixtures brings workpieces sequentially past the spindle. Often, two spindles, one with a roughing cutter and the other one for finishing, are used.

Horizontal-spindle, knee-type millers

Although much depends on the intent of the individual machine-tool builder, horizontal-spindle millers often take heavier cuts than vertical-spindle varieties take. There are a couple of good reasons for this: with the spindle arbor supported at the column and at the end of the outrigger overarm (and sometimes in the middle by a third, center bearing), cutter rigidity is usually excellent; furthermore, with the motor and drive housed within the column, more room is available for heavy-duty equipment.

Horizontal-spindle millers with knee-and-column construction are sometimes called plain millers, which denotes that the table, typically long and narrow, is at a right angle to the spindle axis. Universal versions of horizontal-spindle machines incorporate a provision for rotating the table at angles up to about 30° away from perpendicular to the spindle axis, which permits cutting helical gears and threads; a dividing head is often added as standard equipment for this style.

The knee-type horizontal miller, of course, shares the knee-and-column construction with that kind of vertical miller. But, in many cases, it shares another vertical-miller feature: the vertical-spindle itself. An increasing number of horizontal universals are equipped with provisions for an added vertical spindle that mounts in the machine overarm when that construction member is not being used as an outrigger support for the horizontal main spindle.

Bed-type horizontals

In the bed-type horizontal machines, the spindle rigidity gained by generous support of a horizontal-axis cutter is combined with the table rigidity inherent in fixed-beds.

Often called *manufacturing-type* machines, fixed-bed horizontals indeed seem ideally suited for the limited versatility but high metal-removal rates typical of large-volume production. A good example of the use of one of these machines is in

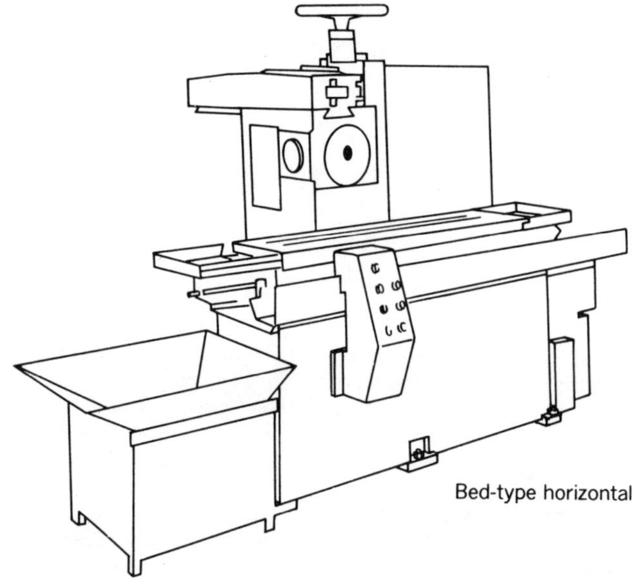
Bed-type horizontal

reciprocal milling, in which two fixtures, one at each end of the table, are installed and the operator unloads and loads one while the other presents the workpiece to the cutter.

This type of milling machine is most likely to be built in singlex, duplex, or triplex versions; the designations refer to the number of spindles employed. Dedicated, manufacturing-oriented machines are sometimes built with spindles located 90° or 180° apart, so's to carve multiple faces of a single workpiece in a single feed stroke.

Planer-type miller

Continuing in a progression toward greater horsepower and cutting capacity but with more specialization and less versatility is the planer-type mill, which shares its basic construction principle with that of the planer. In the planer, the cutting tool is held stationary by a large supporting rail supported by one or a pair of upright columns, while the workpiece, supported by the table, is reciprocated across the tool face. In the planer-type milling machine, the planing tool is replaced by a powered milling head; early versions of this class of machine indeed were the result of retrofitting powered heads onto existing planers.

Modern planer millers are characterized by their capacity to accept large workpieces. Many feature multiple milling heads (rail heads mount vertically on the cross rails that straddle the table; side heads have horizontal spindles that simultaneously machine, say, the sides of box castings). Feed is provided by the longitudinal motion of the table.

Openside planer mills are designed for ease of workpiece loading and unloading and have a single upright column that supports the cross rail. This is in contrast to more-conventional double-column (or "bridge-type") planer millers that can provide a high degree of structural rigidity.

Traveling-column milling machines

An alternate approach to machining a long, heavy workpiece is to mount it firmly in place on a nonmoving table and to traverse the milling cutter past it. The traveling-column machine does this by traversing the entire column along ways that run parallel to the length of the worktable.

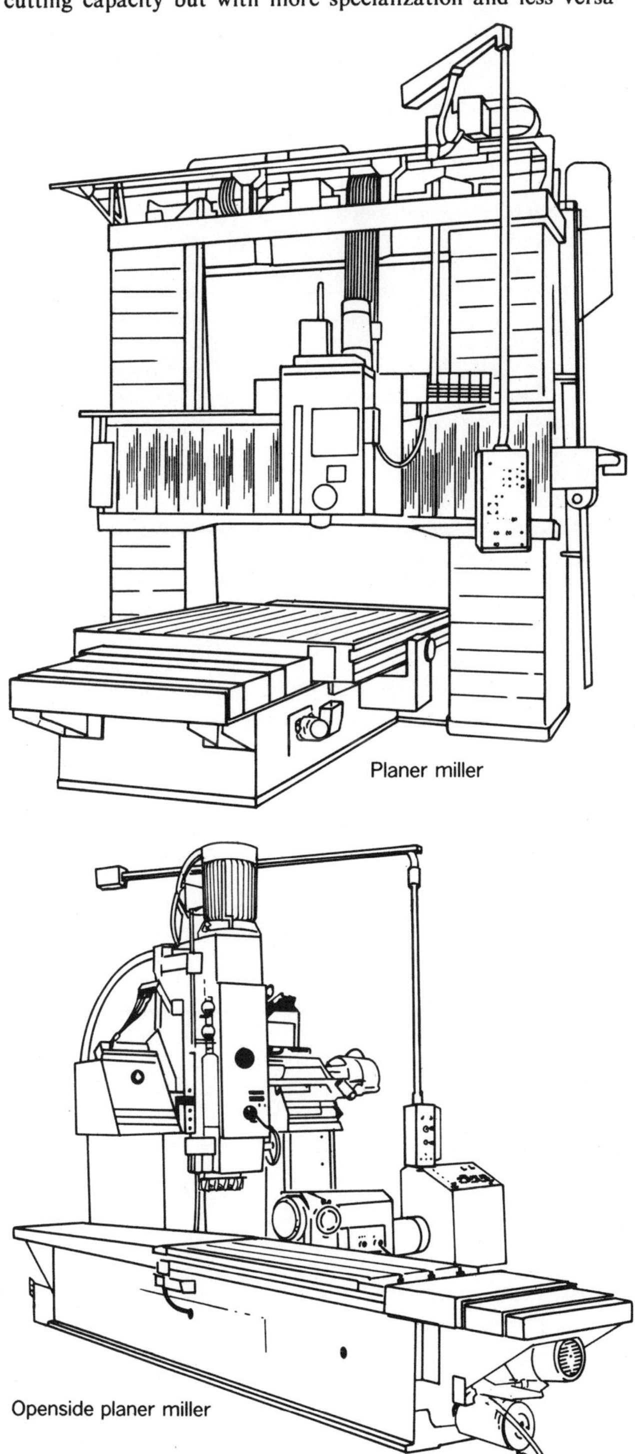
Planer miller

Openside planer miller

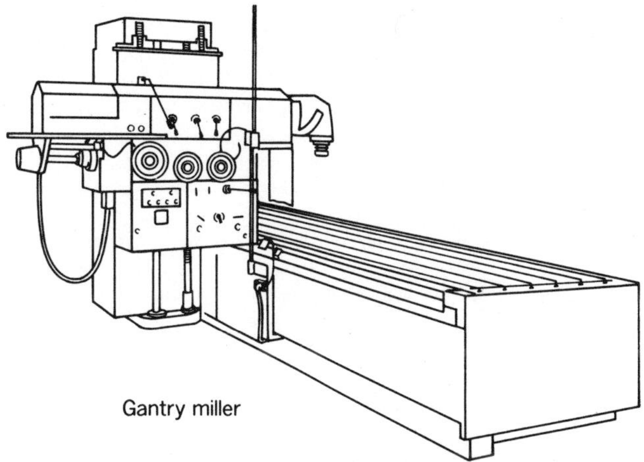
Gantry miller

Very specialized machines among gantry-type millers are variously known as skin mills or spar mills and are found in the aerospace industry, where they are used to profile wing skins and structural spars that are unusually long and require special airfoil-shaped contours. For these special-purpose production units, the rail-mounted vertical-spindle milling heads usually include additional axes of tilt in the A and the B directions to permit a full five axes of contouring capability.

In discussing spindle orientation in the larger special-mission behemoths like those found in the aerospace industry, where bed lengths are measured in tens of yards rather than in inches, it's important to realize that, for all practical purposes, the spindles can be regarded as vertical: they lack the outrigger support of the horizontal-spindle machines found in the shop.

A final type of miller might be the horizontal boring machine, not really a milling machine at all, but a unit that combines the functions of boring, drilling, reaming, and tapping with the ability to cut slots with end mills or shell mills. The HBM, next to a numerically controlled machining center, thus could be considered a milling machine.

There are two principal types of HBMs: table-type and floor-type. In the table-type unit, the base and column are connected, and the headstock moves up and down the column through the Y axis via ballscrew drive. In floor-type HBMs, the separate column traverses back and forth through X, while the table remains fixed. In practice, the size of table-type horizontal boring machines is denoted by spindle diameter; that of bed-type machines is denoted by table length.

Special accessories and attachments

Despite the fact that many milling machines are designed to do one particular job, a broad range of attachments can expand a machine's capability to make it competitive with another of an entirely different class of tool.

The **swivel-head vertical milling attachment** is often used to translate the horizontal alignment of the milling spindle to a vertical one. The attachment mounts to the column face of a horizontal-spindle machine and directs its spindle downward toward the workpiece table. Both fixed-spindle attachments and those equipped with quills that permit spindle feed in the downward direction are available.

Similar to the swivel-head attachment in that it mounts to the column is the **high-speed attachment** that boosts spindle speed for more-efficient use of smaller-diameter cutters. Often, high-speed attachments also provide angular redirection of the spindle axis.

Another common attachment that mounts to the face of the column on a horizontal-spindle machine is the **slotting attachment** that converts rotary spindle motion to a reciprocating motion of a few inches in stroke. The keyway-shaping accessory accepts a wide variety of single-point tools to cut slots, keyways, blind holes, and cavities beyond the capability of a rotary tool.

Special designs for special situations

Milling, it should be underlined, refers to a metal-removal process, not to any one machine. And while, over the years since North and Whitney, machine types have been developed to bring the process into best utility, many of the best milling applications are very specialized and call for the most unusual machines. A few examples prove the point:

At Babcock & Wilcox's Power Generation Group in Barberton, Ohio, process engineer Hal Misuraca had a problem. His company was building a large (roughly 30-ft-dia) cap for a nuclear pressure vessel. The cap had a cluster of 13-in.-ID standpipes projecting upward from it; each standpipe needed a large internal buttress thread to accept guide plugs for fuel rods.

Trouble was, threading had to be done after each pipe was welded precisely into position. Use of rotary tables or a VBM was eliminated because of the size of the workpiece.

The problem was solved when Misuraca enlisted the aid of a California firm that specialized in building planetary milling equipment. The company devised a stand-alone miller that mounted on a cylindrical column that had no base. Instead, the column was slipped into the standpipe bore adjacent to the standpipe to be internally threaded. The planetary-motion thread miller was able to complete the otherwise "impossible" job.

In a different application, this one at California's Long Beach Naval Shipyard, milling a structural component in place solved another manufacturing problem. The steam-driven aircraft-launch catapult on an aircraft carrier had a pair of guide rails each 240 ft long. When the ship was due for overhaul, launch specifications required that the rails' surfaces be held to close parallelism over their entire length.

A special base and column was designed to accommodate a 10-hp heavy-duty machining head from a Kansas-based builder of special machine tools. The unusual design permitted machining 9 in. below the base of the machine. End-milling tools worked the entire rail lengths, operating at varying spindle speeds from 50 rpm to 500 rpm.

Sometimes it's not an unconventional base or column that makes an application special. When Colt Firearms started producing the M-16 rifle at its Hartford, Conn, plant a dozen years ago, it turned to the same New Jersey-based special machine builder that had set up part of the production for the new weapon's predecessor, the M-14.

In setting up production on the workpieces—which typically include the upper and lower receivers, the bolt, its carrier, and the front sight—the strategy was to vary cutter speeds, feedrates, and dimensional movements by cams in the machines. Multiple-head machines were designed to surround a turntable on which the parts were fixtured; each cutter was identical to its neighbor, but each had its movements controlled by three or more circular cams.

And sometimes the best "milling" machine is not a milling machine at all. At India's National Aeronautical Lab in Bangalore, wings whose surfaces are mathematically developable were milled on a jig-borer employing tangential-milling techniques. In this process, the wing cross-section was approximated by a polygon that could be smoothed by hand finishing. The approximation was made so that each side of the polygon was a tangent to the airfoil shape.

The airfoil was defined by a finite set of points, derived from experiments, that were then joined smoothly by using spline approximation to achieve continuity of the first and second derivatives.

Splines and settings for the jig-borer were obtained on a computer: each setting of the jig-borer consisted of the cutter height and two revolutions of the turntable, one about the table axis and one about a fixed horizontal axis, so that the tangent plane coincided with the plane of milling.

Does the addition of a milling spindle on a lathe turn the turning machine into a miller? Of course not, but the application can't be overlooked. Vertical lathes have had side heads with milling spindles for decades. And the specialized turret on the "production center" from one New England NC-lathe builder includes a milling spindle—capable of up to 1500 rpm—on the end turret of what might be taken to be a slant-bed chucker at first glance. To use the profile-mill capabilities of the machine efficiently, the lathe operator or programmer must have a working knowledge of milling essentials, too.

The point is that milling is a metal-removal function, rather than just a description of a particular style of machine tool. Descriptions, at that, can be so narrow as to lose sight of what's actually happening: take one catalog title that defines a "horizontal-spindle, five-axis, double-headed, numerically controlled column-type profiler"; the job this machine does is milling!

Whether it means planetary-motion heads on base-less columns, machines that clamp directly onto a very large workpiece, multiheaded precision units, or applications of cutters on other machine tools, the function remains the same.

Rack-milling attachments can convert the horizontal spindle axis that is directed across a long worktable to one that is directed along the table. Used in conjunction with specially formed slotting cutters, the rack-milling attachment permits sequential machining of individual teeth on a long rack mounted along the table's main axis.

Attachments to the machine table can be just as important as those that work with the spindle. Almost without exception, milling machines require the use of **mechanical vises** to hold and clamp the workpiece. But which ones to use?

The plain vise with a sliding jaw forced by a rotating screw against a fixed jaw is the most common workholding mechanism for a milling machine. It is held in place on the machine's T-slotted table by keys and bolts, and it generally provides a safe and reliable, if uninspired, method of holding the workpiece. Add swivel capability to the base of the vise and the spindle/workpiece orientation is enhanced. The toolmaker's vise or universal vise further expands the versatility of individual workholding by enabling the workpiece to be tilted up and down in addition to being swiveled.

Suppliers of **magnetic chucks** have recently encouraged the use of these devices instead of more-conventional jaw-action styles. There are a number of variables in the equation; some workpiece materials and shapes will not work with chucks that hold workpieces via invisible flux lines. Besides the necessity for choosing a ferromagnetic workpiece material (excluding nonmagnetic austenitic stainlesses as well as the nonferrous metals) for this method, differences in the contact area, the cross-sectional shape of the workpiece, and the surface quality of the part are all factors that might limit full application of this style of workholding. Still, the relative ease of loading and unloading make magnetic chucks worthy of consideration.

Another popular on-table milling accessory, one that has been in use for some time, is the **dividing head.** Sometimes

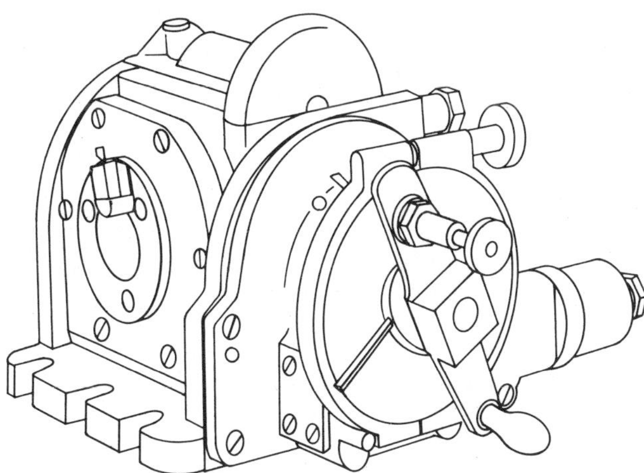

called an indexing center, it is a powered or hand-operated precision measuring device that moves and accurately locates a workpiece in a series of positions. It consists of a tailstock and a headstock, the latter provided with an index plate with holes or notches.

When manually operated, index equipment provides precise spacing between, say, teeth on a spur gear. Under power and equipped with followers attached to the machine feed, the equipment can be used to generate helixes and similar forms.

One of the most popular milling-machine accessories is the **rotary table,** perhaps even more popular than the indexing head. Provided with vertical spindle, the bolt-on table attachment can provide discrete or continuous indexing of the workpiece. Rotary tables with horizontal spindles are often paired with separate tailstocks and, in some cases, can replace index tables.

Trends in machine design

In just about every one of the preceeding descriptions of individual types of milling machines, the term "structural rigidity" plays an important part. For good reason: a less-than-rigid machine tool can't provide the dynamic stiffness needed to eliminate chattering of the tool.

Chatter is an important problem. To prevent that regular pattern of irregular cutting on the surface of the workpiece, a machine operator or programmer either has to lighten the cut—making less-than-full use of a beefy milling cutter—or take a second, finishing pass—time consuming, at best—or try to get use of a more heavy-duty machine.

"The challenge to structural designers comes when they have to cover huge spans without a lot of support members," says the manager of one machine-tool builder's mechanical and control development section. "In huge spar mills that need to cover large workpieces," he says, "the chatter problem can become acute indeed.

"One potential solution," the designer suggests, "is the use of tuned, damped absorbers." The concept of sympathetic vibration is shown by picking up a coffee cup and suspending it with a rubber band through its handle. When the hand moves up and down very slowly, the suspended cup remains almost stationary; similarly, when the hand holding the rubber band jitters in quick, short movements, the cup remains fairly still. However, at some regular hand motion midway between the slow rhythm and the quick jitter, the cup vibrates wildly.

By hanging a second, smaller weight from the cup, that midrange hand oscillation can be damped out. In much the same manner, newer machines from some builders contain structural appendages to their columns, gantries, and overarms that are precisely tuned to damp out vibrations in the range that causes tool chatter problems.

The concept, in which the smaller mass is excited by the tool vibrations while the larger mass remains still, is not a crutch for overcoming other design insufficiencies, designers maintain. Rather, it is a valid construction concept and one that will probably see increased use in millers, which, after all, machine with interrupted cuts.

Another trend in milling-machine design, of course, is precipitated by control technology. Numerical control probably has done more than anything to change the basic structure of the milling machine since Frederick Howe developed the index milling machine in 1850.

The shift from single-purpose milling machines to numerically controlled machining centers equipped with toolchangers is unmistakable, say experienced people in the machine-tool industry. (That shift, however, is difficult to analyze: in 1976, the last year for which complete Commerce Dept figures are available, the value of shipments for all machining centers moved up to exceed the value of shipments for all milling machines by about a third. Yet machining-center shipments represented only about a tenth the number of miller shipments. And what portion of the potential milling-machine customers replaced their mills with machining centers is difficult, if not impossible, to determine.)

Machining centers, after all, perform some milling functions. It is beyond the scope of this report to detail advances in machining centers; for more information, see Special Report 689, "NC machining centers," (AM—July '76). But it will be necessary to keep in mind that the following sections of this report—on chipmaking dynamics and milling cutters and their use—contain information that is applicable to machine tools other than dedicated millers.

Milling dynamics

Milling is, of course, a machining operation. In fact, in colloquial use, "to machine" often means to mill a particular workpiece.

More than that, milling is a *mechanical* machining operation, as opposed to, say, electrochemical machining. As such, certain mechanical principles that have to do with chip formation, cutter geometries, horsepower requirements, and the like apply to the process. Those aspects will be discussed in this section.

Milling generates a machined surface by removing—progressively—a certain amount of workpiece material. It involves relative motions between a cutter that rotates and a workpiece that also moves.

Motion can mean a number of different things. In some cases, the workpiece does not move at all but, rather, remains fixed while the rotating cutter is advanced across its face. In other cases, the workpiece moves in a straight line at a slow rate of feed while the rotating cutter remains in a fixed location. In still other instances, workpiece motion is rotary. In the specialized case of crankshaft milling, eccentric cranks are milled while the work is rotated about its main shaft and the cutter reciprocates in and out toward the eccentric; feed motions get complex, to say the least, and are often calculated with the aid of a minicomputer.

In all cases, the notable phenomenon is that each of a series of cutter teeth takes a single "bite" and does its share of the entire job. For that matter, sawing is a specialized form of milling.

Given the basics of a rotating, toothed cutter and relative motion between it and a workpiece, two different approaches to applying that rotating cutter can be used. Both have their special attributes. In **peripheral milling,** sometimes called "slab milling," the cutting teeth are located on the outside periphery of the cutter body and are most often *parallel* to the main cutter axis. Operations that are done with formed or shaped-profile cutters fall into this general class of milling.

When milling is done peripherally, it's generally done on a

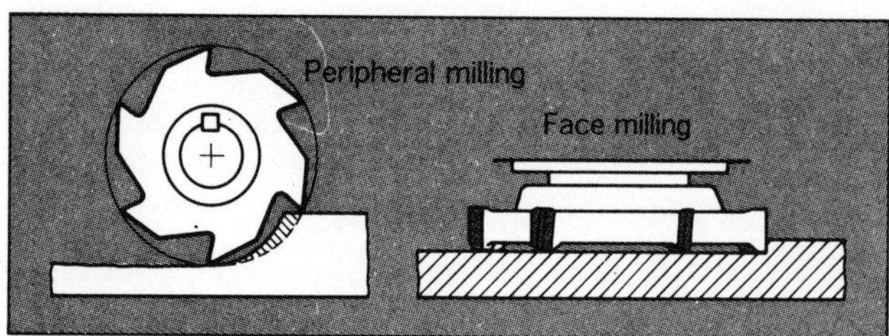

horizontal-axis milling machine, but it also can be done using the circumferential cutting edges of end mills in a vertical-axis machine, for example.

A distinction is often made between peripheral milling and **face milling,** but the differences are not all that great, except for the shape of the cutter. In face milling, which is generally the preferred method over peripheral milling, the cutter actually uses two different edges to achieve the finished surface.

Face-milling cutters hold their axis *perpendicular* to the surface of the workpiece and cut in two different ways: the periphery of the individual teeth remove most of the metal, while the face-cutting edges provide the finish of the surface that is being generated.

Other types of milling operations start from this basic distinction. In the *Machining Data Handbook*, published by Metcut Research Associates Inc, distinction is made between metal slitting, end milling, side and slot milling, slab milling, straddle milling, hollow milling, and gear milling. In many respects, those distinctions are based on the individual workpiece and the specific type of cutter, but they can all be classified as either peripheral- or face-milling operations.

There's another basic distinction to be made; this one has to do with the direction from which the cutter approaches the workpiece. The difference between up milling and down milling exists primarily in peripheral types of cuts. For facing types of cuts, a slightly different situation—actually a combination situation—exists.

Basically, the workpiece can be fed either with or against the direction of cutter rotation. When the cutter rotates in the feed direction, the technique is called **down milling,** or sometimes "climb milling" or "in-cut milling."

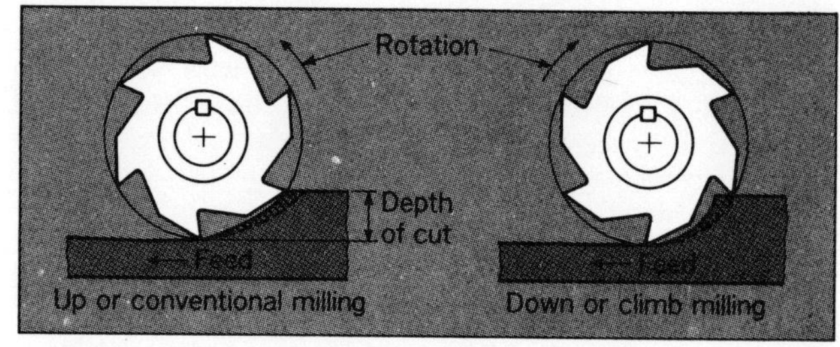

The entry of each tooth into the workpiece in down milling occurs when it is in a relatively slow portion of its path around the cutter rotational axis (with respect to the motion of the workpiece feed), compared with its relative speed when it exits the cut. It's because of this that in down or climb milling, the tooth takes a slightly bigger "bite" out of the metal at the entry point than at the exit point. The chips that are produced have tapered cross-sections, with the thickest side of that cross-section being the beginning of the cut. As expected, the forces involved make the cutter tend to be pushed up from the surface of the workpiece or to "climb" out of the cut, hence the alternate name for this technique.

The opposite combination of cutter rotation and workpiece feed occurs in **up milling,** sometimes called "conventional" or "out-cut" milling. The cutter rotates against the feed direction; each tooth starts from the portion of the work surface that its preceeding tooth has already machined.

The tooth's "bite" in up milling becomes increasingly larger as it works its way through the metal. That's because its speed—relative to that of the workpiece—diminishes.

Up or down or both?

The distinction between up milling and down milling is clear when the work is being done with peripheral-cutting tools. In

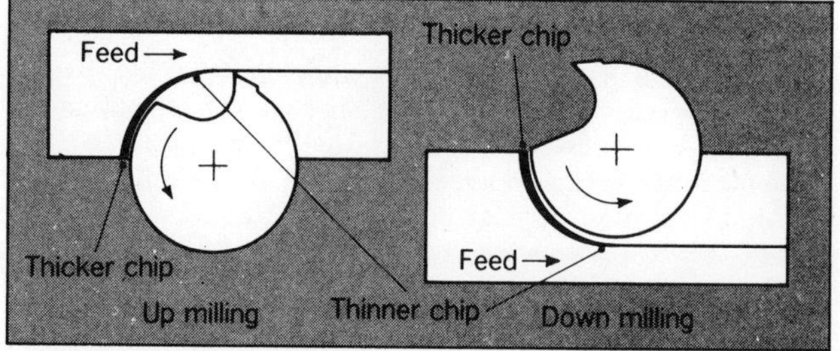

face milling, the up and down milling directions—depending on setup geometry—can occur at the same time.

If the swath of the cut lies across the cutter axis, as when the width of cut exceeds half the diameter of the milling cutter, each tooth will enter the cut in an up-milling mode and exit the cut as in down milling. Entry into the down milling mode would occur for an individual tooth when it crosses the feed

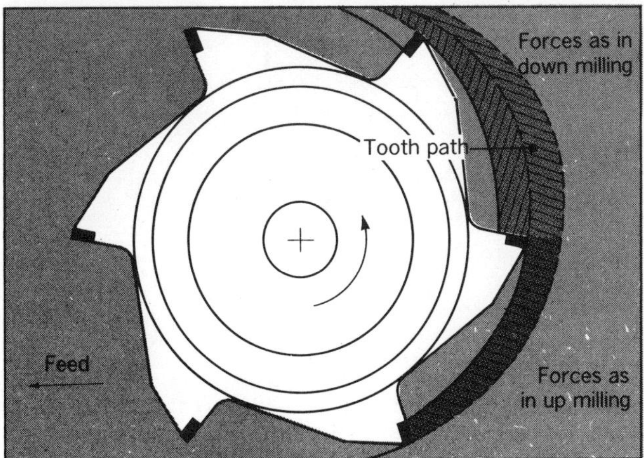

axis of the cutter. Entry into that second mode of cutting would almost be imperceptible, for it is a continuous motion.

In face milling across nearly the full width of the cutter, the choice between up milling or down milling cannot be made. But the situation can, and should, be avoided, especially in those extremes when the cut width is very close to the diameter of the face mill. The chip section at cutter entry in this case is very thin and will result in accelerated tooth wear from abrasion, as well as the likelihood that the chip will stick to or weld onto the booth, be carried around the noncutting side, and be recut.

Other problems caused by combined up and down milling include an unusual combination of tangential cutting forces that wind up acting in unison directly across the feed axis. This may cause workholding problems.

The simplest way to avoid combining both up milling and down milling when using a face cutter is to select a tool that is larger than the intended swath. Sometimes, as when pocketing with an end mill, combining the two modes is unavoidable.

When face-milling a narrower path, or when performing peripheral-milling operations, the choice between up cutting or down cutting can have an important effect on output, tool life, and surface finish.

As its alternate name implies, up milling or "conventional" milling is the way most jobs have been run over the years. As a practical matter, the cutting force opposes the feed force so that the feedrate remains fairly consistent, even if backlash exists in the spindle or the feed system of the machine.

Older cutter designs, especially those with sharper positive cutting angles, have less difficulty starting the cut under conditions of up milling; that is, starting with a zero-thickness chip. However, cutters with carbide cutting edges and neutral or negative cutting angles usually work better in down or climb cutting. The blunt cutting edges do not work well when they are forced to start cutting a zero-thickness chip; they work better taking a thick chip at the beginning of the cut and then easing their way out of the cut.

The direction, up or down, that milling is done will influence the direction in which chips are expelled from the cut. This can be a safety consideration.

Down milling is best performed on machines in good condition and when the work-moving slide is fed steadily. Otherwise, yielding of the work may cause a broken cutter, as well as damaged work and machine.

Down milling has certain advantages. The downward action of the milling cutter creates a force that tends to push the workpiece down into the vise or fixture. (Up milling, conversely, has the opposite tendency.) So, in down milling, simpler fixtures generally can be used.

Another benefit of down milling is a slightly lowered power consumption. And, with forces in the spindle opposing the supportive forces in the machine table, operation tends to be smoother. This, in some cases, will reduce the tendency to chatter.

Up milling, by the same token, has advantages of its own. For one, possible variation in the height of the workpiece stock would have little effect on the point at which the tooth enters the cut. (This can be important to tool life.) Up milling, furthermore, reduces or avoids the need for a backlash eliminator in the feed system.

Scale and abrasive dirt on the surface of the uncut work are never carried into the cut in up milling, and so there is less likelihood of abrasion damage to the tool. And, with cutting forces generally directed away from the cut, load on the spindle is reduced, and this results in longer spindle-bearing life.

Chips via interrupted cuts

Whether down milling or climb milling, face milling or peripheral milling, all operations have one thing in common: chips are formed discontinuously.

Most other metalcutting is done continuously; that is, while the cutting tool is working, each cutting edge is continually cutting and is subjected to relatively constant cutting forces. Examples of continuous operations include turning, boring, and drilling.

Milling is not a continuous metalcutting process. As a cutter rotates, each edge alternately enters and leaves the cut and is in the cut less than half the time. Entering and leaving the cut subjects the edge to impact loading, thermal cycling, and cyclical cutting forces. Interrupted cutting such as this is an important consideration in the design of cutter angles.

In the diagram at the top of the next page, only cyclical cutting forces are charted. The chart describes how forces change as a cutter rotates. As the blade enters the cut at (1), the forces build up rapidly (2) and peak as the blade crosses the direction of the feed (3). When the cutter leaves the cut, forces drop to zero immediately.

The fact that milling cuts are interrupted can have an effect on how the work should be done. One manufacturer of inserted-tooth face-milling cutters, for example, notes that a common cause of insert cracking is improper coolant application. Coolant streams, when misdirected or haphazardly aimed in the general direction of the workpiece/cutter interface, often cool the insert when it is out of the cut, rather than when it is working.

This sloppy procedure actually increases the temperature difference that the insert experiences in its normal operation. Temperatures build up along with the cutting forces that are shown in the diagram, and they drop off remarkably when the tooth moves around and starts "cutting air." When coolant is directed at the insert during this noncutting portion of its cycle, the fluid's evaporation absorbs heat from the cutter, further magnifying temperature differences. Thermal cracking can occur.

When the tooth or insert is in the cut, however, it shares characteristics with cutting tools used in other metalcutting operations, such as turning and boring. Chips are formed by the plowing and shearing of metal, and horsepower is converted into friction and heat. [Continued]

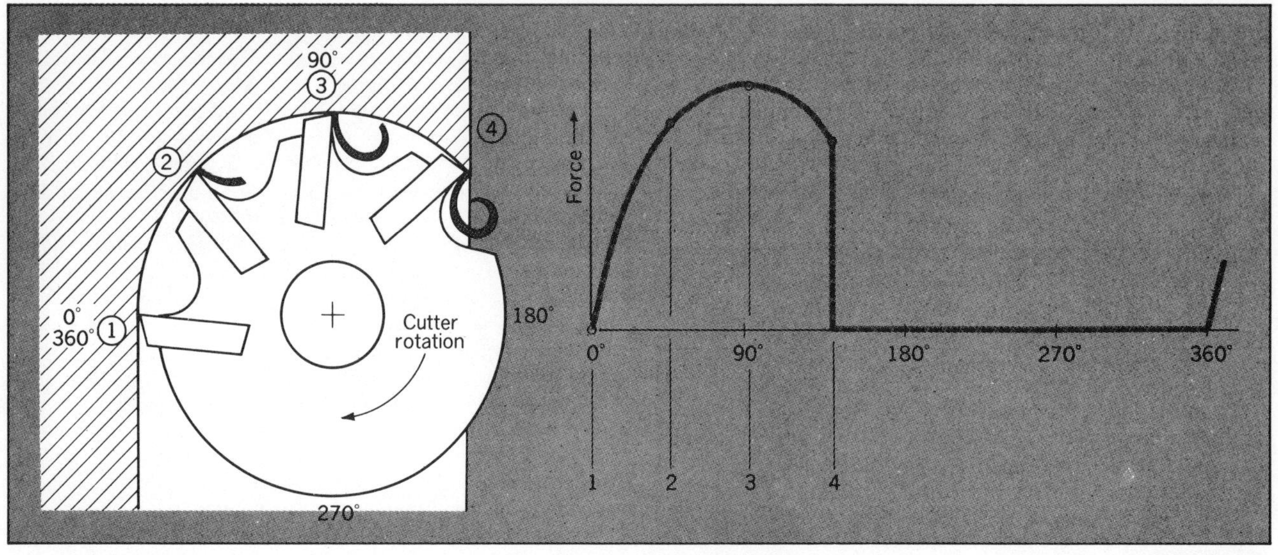

Chip formation involves a complex relationship between mechanical properties and chip structure, the relationship depending on plasticity conditions. Milling may be characterized by interrupted cuts, but, when the teeth *are* cutting, they form one of three different kinds of chips: discontinuous,

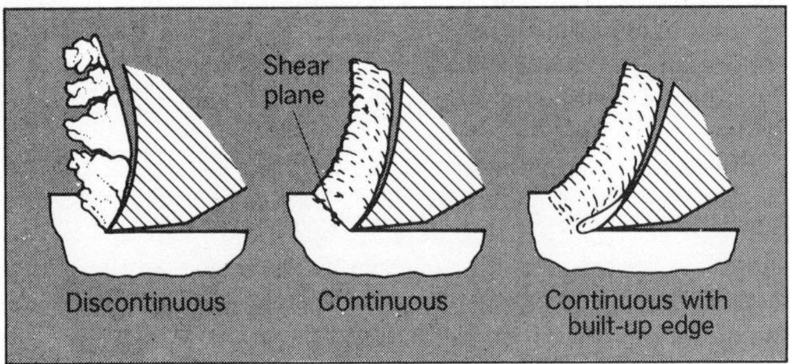

continuous, and continuous with a built-up edge. Of the three, the continuous chip is thought to be the most efficient and is diagrammed here, in exaggerated form, for detailed study.

As the tool engages the metal, a shearing process occurs along an area extending from the cutting edge to the workpiece surface ahead of the tool. This area, called the shear plane, makes a shear angle with the machined surface. The angle varies widely, depending on cutting conditions. The shearing process deforms the metal into layers, as shown in the model.

The total work expended in forming a chip is the sum of the work due to friction and the work required to shear the material. The majority of this work is converted into heat, which is concentrated in three cutting areas: the shear-plane area, which is the workpiece area located in front of the cutting edge; the area between the chip and the flute face, where the newly formed chip rubs against the leading edge of the tooth as it flows upward and away from the cut; and the area between the cutting edges and the workpiece, the location that lies behind the point of the tooth.

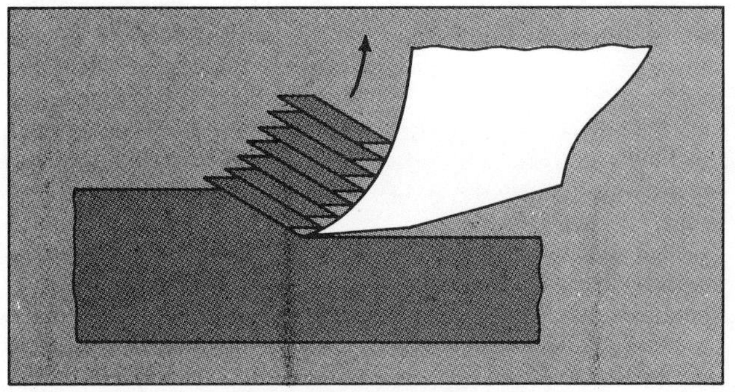

The heat developed in these three areas is a major consideration in tool failure as well as in efficient metal removal. Heat can be dissipated by a number of methods, including *proper* use of coolant, low-friction work and tool materials, good tool finish and sharpness of the cutting edge, and correct tool geometry.

What goes into chip formation

Both heat, with its negative characteristics, and metal shear, with its beneficial characteristics, are created by the power that's applied in the combined forces of cutter rotation and workpiece feed. That translates to the spindle horsepower and the table-drive horsepower. How the power moves from spindle and table to the chip is a function of the mechanical details of milling.

Let's look at cutter rotation first, since it is one of the most important factors in translating the machine's horsepower into cutting power.

Cutter rotation is measured in terms of speed, but which "speed?" There are two methods of measuring it. One is the rotational speed or **revolutions per minute.** This is simply a count of the number of times that a rotating object makes a complete revolution around the spindle axis in each minute. Machines are calibrated, of course, to deliver a specific rotational speed measured in rpm. But this has little utility as a measure of the dynamics of milling, for the peripheral *cutting* speed of the individual milling cutter varies with its diameter. Cutting speed is measured in **surface feet per minute** (sfm), or, in some cases, in meters per second (m/sec). Surface feet per minute is a measure of the actual velocity at which the tooth travels across the face of a stationary workpiece. It's analogous to the speed at which a simple planer or a shaper's cutting tool travels across a workpiece.

Calculation of the cutting speed depends on knowledge of the cutter's circumference. Since circumference is a product of π and the diameter of the cutter, the distance traveled (in feet) in one minute by a tooth rotating around the circumference of a cutter is equal to π times the cutter's diameter in feet times the revolutions per minute. For diameters measured in inches:

$$\text{sfm} = \frac{\pi D}{12} \times \frac{\text{rev}}{\text{min}}$$

This computes to sfm = rpm (D)/3.82; and, if we assume that $\pi/12 = 1/4$ (actually, it equals 0.2617), then, more conveniently:

$$\text{sfm} = \text{rpm} \times \text{diameter in inches}/4$$

This approximation is within 5% of being completely precise and is commonly used to quickly figure cutting speed, which is sometimes called V_c, or the velocity of the cutter. The more precise formula for cutting speed is $V_c = .262 \times D \times \text{rpm}$.

Just as the real cutting speed (sfm) is a function of the spindle rpm and the cutter diameter, the **feed** in milling is actually a combination of the machine's feedrate, the cutter's diameter, and the number of teeth on the cutter.

Milling implies cutting with multiple cutting edges, but it's the strength of a single one of those edges that limits the strength of the cutter. Also, the finish on the final product is related to the distance between each tooth's contact with the work.

The advance of the workpiece per unit of time (usually the table feedrate) is measured in simple feed or inches per minute (ipm). This is called f_m, or the feedrate of the machine. To obtain the tooth feedrate, called f_t, the machine feedrate is divided by the product of the spindle speed and the number of teeth on the cutter:

$$f_t = f_m/(n \times \text{rpm})$$

As a practical matter, the equation is usually worked backwards, in order to provide selected feed per cutter tooth (f_t). In that case, the conversion used is: $f_m = f_t \times n \times \text{rpm}$.

Setting the proper feedrate for the machine is one of the most crucial calculations for obtaining a particular finish. Generally, best finishes are obtained through low feedrates. But high production is directly related to higher feedrates. So there's a tradeoff.

Setting the machine feedrate further depends on the available horsepower of the machine, the machine's efficiency in

Terminology of formulas used

Various handbooks and company-sponsored users' literature employ different terminology to refer to the same measurement. Here is a comparison of terms that are widely used. Major sources of the terms listed here include Metcut's *Machining Data Handbook*; McGraw-Hill's *Manufacturing Process* and *Tool & Manufacturing Engineer's Handbook*; and Sandvik's Cutting Data Calculator *Handbook*, which uses "international" terminology. Symbols in parentheses are not used here.

Diameter of the cutter (in.)	D,(D_m, d)
Spindle speed (rpm)	rpm,(N, S)
Cutting speed (sfm)	V_c,(v)
Number of teeth	n, (z, N)
Feed per tooth (in.)	f_t, (f, sz, fpt)
Feed per revolution (in.)	f_r,(f′, sn)
Feedrate (of machine) (ipm)	f_m, (F, s′)
Metal-removal rate (in.^{3}/min)	Q, (V, wdf, CIM, R)
Depth of cut (in.)	d (a_r)
Time for cut (min)	t (T)
Power	hp (HP_m)
Efficiency (of spindle drive)	E
Machinability constant	K
Unit power (hp/cu-ipm)	P

Taking the recommended numbers . . . and using them

"If there's a Bible in this industry," says one machine-tool researcher, "it's the Red Book. No self-respecting shop could be without it." He refers to the *Machining Data Handbook*, first copyrighted in 1966 by Metcut Research Associates Inc in Cincinnati. It's a compilation of starting recommendations gathered from many industry sources, and it was made possible, in part, by funds provided under Defense Dept contracts. "Of course," continues the researcher, "no one should follow it to the letter. There's too much diversity among cutters, machines, even tooling materials to permit that."

That diversity does indeed exist, especially, as we shall see in the next section of this report, among different milling-cutter designs. According to one cutting-tool manufacturer's recommendations, face milling of medium steel (around a 220-320 Bn hardness) can take feeds anywhere from 0.005 in./tooth to 0.030 in./tooth and recommended cutting speeds from 40 sfm to 400 sfm, depending on the shape and material of the tool and whether the operation is a roughing or finishing one.

The best place to start, then, is the specific recommendation by the tool manufacturer for a specific tool design. Such information is readily available (photo, right). There are also convenient aids to breezing through the necessary calculations, including tables, special slide rules, and even a calculator with "canned" formulas.

Once again, though, the recommendations are starting points Experimentation—within limits—is encouraged.

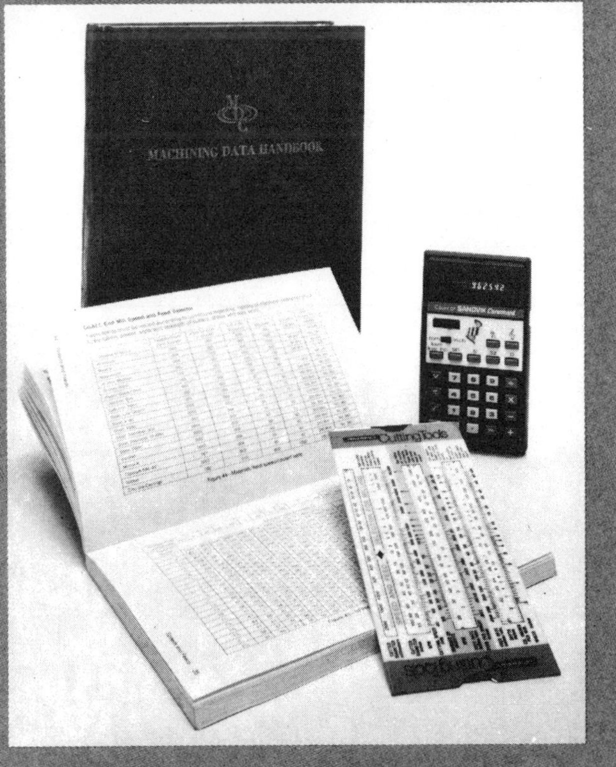

K factors for workpiece metals

Material	Constant
Steel, cold-drawn	1.0
forged and alloy	0.6-0.8
hard alloy	0.5
Stainless, free-machining	1.1
austenitic	0.7-0.82
Nickel	0.52
Titanium	0.75
Aluminum	2.25-4.0
Copper, annealed	0.85
Bronze & brass, medium	1.0-1.4
Cast iron, soft	1.35
medium	0.8-1.0
hard	0.6-0.8

Unit Power (P) for various metals

Material	Constant
Steel, cold-drawn	1.4
forged and alloy	1.6
hard alloy	2.0
Stainless, free-machining	1.0
austenitic	1.6
Nickel	1.9
Titanium	1.25-2.5
Aluminum	0.25
Copper, annealed	1.0
Bronze & brass, medium	0.4-0.5
Cast iron, soft	0.67
medium	1.0
hard	1.6

delivering that horsepower to the spindle, the kind of metal being milled, and, to some extent, the tooling material and the shape of the tooth or insert.

In order to determine the maximum feedrate that can be used for a given cut, the feed per tooth must first be calculated. That is given in this equation:

$$f_t = \frac{K \times hp \text{ (at cutter)}}{n \times rpm \times \text{cut width} \times d}$$

where K is an industry-applied constant machinability factor that varies for each type of workpiece material. The constant is derived experimentally by dividing the metal-removal rate (Q), which is measured in cubic inches per minute, by the horsepower measured at the spindle.

Q, the metal-removal rate, is the product of the width of the cut, the depth of the cut, and the machine feedrate.

Or: $Q = w \times d \times f_m$. The value figures in the calculation of horsepower required at the spindle.

To secure efficient metalcutting rates and to achieve an

Boost the feed or boost the speed: Two different approaches

What's the best combination of cutter-spindle speed and table feedrate for milling? To answer that question, another question first must be asked: What's meant by "best?" A high metal-removal rate might be the desired goal; so might metal-removal efficiency, with consumed horsepower entered into the calculation; other operations could require the best possible surface finish; and still another goal might be prolonging cutter life.

One cardinal rule has always been to use the highest feed (per tooth) that conditions will allow. But since feed is a function of cutter speed, the number of teeth, and the table feedrate, the question of which factor has the most significant effect is open to discussion.

Aiming at the most efficient use of energy expended in cutting, feedrate may be the most important parameter, according to Carl J. Oxford Jr. Writing in this publication, he has pointed out that energy consumption per part produced can be economically reduced by manipulating feedrates.

One of the most efective ways to remove metal efficiently is to increase the thickness of the chips, that is, increase the feed per cutting edge. This is energy-efficient because energy consumption increases less rapidly than the increase in feedrate.

In the normal range of feasible feeds, energy consumption varies as approximately the 0.7 to 0.8 power of the feed. This means that doubling the feed will increase the energy consumption by only about 70%. In terms of overall finished product, it translates into an energy saving of about 15% for each part produced.

But feedrate cannot be increased without limit. Since the cutting forces increase at about the same rate as energy consumption, a point will eventually be reached where either the cutting tool itself or the machine-tool drive will become overloaded.

Furthermore, increases in the feed generally cause some deterioration in the surface finish produced—whether this is a critical limitation depends on the type of operation (roughing or finishing) and the requirements of the workpiece. Sometimes chip-ejection problems are encountered, but, generally, the thicker chips resulting from increased feedrates break up more readily than thin ones do. Further, some machining setups are not rigid enough to tolerate the larger forces associated with increased feeds.

It is true, according to Oxford, that increases in either cutting speed or feed usually cause a reduction in tool life. But a trade-off comes into play here: speed increases cause a greater tool-life reduction than do corresponding boosts in feed. This fact is often the key to successful economic application of increased feed.

For example, if the feed can be doubled and the cutting speed cut in half, the result would be the maintenance of the same production rate, a large increase in tool life, and a reduction in energy consumption of about 10-15%. Alternately, a less-drastic speed reduction could yield the same tool life with an increased production rate while obtaining the same reduced energy consumption for each part that is produced.

Energy consumption, measured against a unit volume of material removed, is relatively unaffected by cutting speeds in the range normally feasible for high-speed-steel tools. If the feed per cutting edge is maintained at a constant value,

economical workpiece cost, proper use of available horsepower by a milling cutter becomes essential.

Horsepower calculations

A simple explanation of the power-requirement relationships can be made by comparing them to a balance scale. An increase or decrease in the metal-removal rate (Q) brings about a corresponding change in the required horsepower. This does not mean to imply that a balance must always be achieved; that would be the optimum condition. On tough materials, for example, the machine horsepower might far exceed the cut requirements, and it is the usual result when a larger machine is used for its greater strength and rigidity. Equally, the horsepower of a machine could be exceeded by as much as 25% or 50% for short durations without harm.

Each material has a narrow "cutting speed" and "feed per tooth" range, within which a milling cutter can achieve good life. Therefore, the depth and width of cut become the variables in meeting the horsepower conditions.

It's necessary to exercise a certain amount of judgment in applying power to a milling cutter. It would be unreasonable, for instance, to expect a ½-in.-dia end mill to handle a full 15 hp or to take a cut that is beyond the physical capabilities of the tool.

A simple means of determining the horsepower required at the spindle to mill at a specific metal-removal rate is given in this formula:

$$hp_s = Q \times P$$

where Q, again, is the metal-removal rate desired and P is Unit Power, which is a function of the specific metal being cut. Values for P, provided in a separate chart, are approximate, having been drawn from a number of industry sources. Consensus does not exist, because the figures are derived experimentally. Metcut Research Associates, for example, publishes two sets of values for P—one set obtained using a sharp tool and the other obtained using a dull tool. Experimentation within the shop is the best source of fixing these values.

The values and formulas for determining horsepower requirements, or the best use of available horsepower, remain the same for all types of milling on all types of machines. The same generally holds true for other basic distinctions discussed in this section. Peripheral milling is desirable in certain situations; in others, face milling is best. Similarly, up milling has advantages over down milling in individual instances.

But equally important to optimum machining is the body of information that deals specifically with cutter angles. As we shall see later in this report, the geometries of milling vary quite a bit, depending on the kind of cutter that is employed. Certain fundamental considerations in cutting with a face mill, for instance, do not apply when machining is done with a shell mill.

Similarly, recent advances in cutting-tool materials have affected their application. The newest ceramic-coated tungsten carbides, for example, are finding a niche in inserted-tooth-type cutter bodies, but they demand special handling in order to reach their maximum cutting potential. And not all milling should, or can, be done with insert-type cutters. The economics of cutting often justify the use of older-style solid-carbide or hss cutters.

horsepower requirements will increase directly with increases in speed, as will production rate, but the specific cutting energy per unit volume of material removed will remain substantially constant.

In all tests he conducted with regular and roughing end mills at National Twist Drill, Oxford found that as the feedrate is increased, specific cutting energy is reduced.

Explorations into another component of feed per tooth—the cutter's rotational speed—are being conducted by other researchers. By pushing spindle speeds far past those in conventional practice, engineers at Lockheed's California operation are charting what ultra-high cutting velocities can do to the milling function.

Lockheed's work started nearly two decades ago and included early experimental firing of simulated workpieces past a single-point cutting tool at velocities of 15,000-360,000 sfm. The experiments progressed through modification of two machine tools equipped with special milling spindles capable of very high rotational speeds (up to 100,000 rpm in one instance) and concomittant high cutting speeds.

Experiments on these actual machine tools in the milling of aluminum showed several findings: Increasing cutting speeds by 500% resulted in 300% improvements in cutting efficiency (measured in cu in. of metal removed per min per hp), regardless of the depth of cut. Revving up a ½-in. carbide end mill from 4000 rpm to 20,000 rpm reduced cutter side loads by 70% for a given chip load and depth of cut; furthermore, part deflection was reduced, allowing the milling of thinner sections. The higher the cutting speeds, the cooler the workpiece and the hotter the chips. In some tests, the chips were molten. And surface finishes produced by a two-flute, ¾-in.-dia carbide milling cutter operating at 18,000 rpm were in the range of 25-60 μin.

One of the conclusions drawn by the researchers was that the high-speed route could lead to cost reductions in the milling of aluminum alloys of 50% or more. It's important to note that although most of the tests in ultra-high-speed milling have been performed on aluminum workpieces, there's no reason, in the minds of researchers, that similar results could not be found with other metals.

Much of the ultra-high-speed milling work at Lockheed is now being done on a converted machining center equipped with a cartridge-type motorized spindle that provides speeds from 1800 to 18,000 rpm and delivers up to 25 hp. Its super-precision ball bearings are mist lubricated, and liquid cooling controls the heat in the spindle cartridge.

If the details of such experiments sound like they require very specialized equipment, it's true. Spindle runout on a machine that Lockheed is testing in a production application is less than 0.0001 in. The cutter used is a solid-carbide end mill, selected for its high stiffness and designed to facilitate ejection of chips.

Commercially viable ultra-high-speed milling spindles lie in the future, still the concept of increasing productivity by boosting spindle speed holds much promise.

Cranking up the feedrate of the machine tool also effectively increases productivity, and, although the results may not be as dramatic as super-fast machining, the technique's employment is available to more machinists.

Milling cutters and applications

If the machines used for milling are in themselves diverse, the cutters for doing specific milling jobs are even more varied. One manufacturer of cutting tools lists close to 100 different styles of milling cutters in its catalog. Furthermore, as we've seen, although there are certain fundamentals in milling dynamics that apply to all chip-forming operations, the forces, geometries, and applications can vary drastically, depending on the type of cutter that is used.

This section of our report will discuss what has generally become accepted to be the four basic styles of cutters. These include "plain" and side-milling cutters, "shell" mills, insert-type "face" mills, and "end" mills. In addition, the tooling materials and, in particular, the inserts that can be common to a number of different styles of cutters will be explored.

It's here in the discussion of individual types of cutters that the distinction, made in the last section, between peripheral milling and face milling becomes most apparent. Actually, this is an artificial distinction, because all milling cutters cut on their periphery—the outside of their cylindrical shape. In addition, some also cut on their faces; these include certain applications of shell mills and end mills, as well as face mills.

Plain cutters mill simple planes

Perhaps the most basic type of milling cutter, the "plain" mill, by definition, has teeth *only* on its circumference. Designed in a simple cylindrical shape, these cutters usually have through holes for locating them on the arbor of a horizontal-spindle milling machine.

A keyway slot is the most common form of transmitting driving energy to the cutter, although some designers are experimenting with polygon-shaped holes for mounting the cutter on a polygon-shaped arbor.

For mounting a key-slotted arbor type cutter, accepted practice dictates the use of a key long enough to enter the shims and spacers on both sides of the cutter. Cutters, collars, and shims should be cleaned before assembly onto the arbor. Dirty or scored collars and cutters that are assembled onto the arbor will cause the cutter to jump out of alignment.

Accepted practice also calls for use of a dial indicator to check against both radial and axial runout. Many horizontal-spindle, overarm-type milling machines are equipped with a special bracket to hold the indicator for checking. In other machines, the indicator may be clamped in the workholding fixture.

Runout and alignment problems when the cutter is mounted on an arbor can be minimized by using less common shank-type cutters instead of arbor-type tools. Although usually limited to small-diameter tools, shank-type plain cutters (not to be confused with end mills) have a straight or tapered shank that mounts in an adapter or directly to the machine tool's spindle taper. Straight-shank cutters can be positioned in a spring collet but more often are fastened with one or more screws that fit against flats on the shank. Taper-shank cutters have self-holding tapers. These fit standardized toolholding collets, such as Morse, Brown & Sharpe, or NMTB.

Whether they are arbor-type or shank-type, plain milling cutters are intended to cut straight surfaces. The teeth on them can either be designed to lie straight across the periphery of the cutter or to lie at a helical angle. A helix angle of about 25° is described as low; 45° is medium; and 52° is called a high helix. Helixes can be either right-hand or left-hand: a cutter viewed from one end, with flutes that twist away from the observer in a clockwise direction, has right-hand helix; a cutter with flutes that twist away in a counterclockwise direction has a left-hand helix. Some narrow cutters alternate succeeding teeth between left- and right-hand axial rake angles.

The helixes aid in chip clearance and in shearing away the chip. Whether they are left- or right-hand makes no difference

in actual cutting of chips, but rather it becomes important in directing cutting forces in toward the machine. That's because right and left handed also describe the rotation of the milling cutter.

The hand of rotation, sometimes called the hand of cut, describes the direction the tool rotates in relation to the machine. When viewed toward the machine spindle and rotary motion is clockwise, the cutter is operating with "right-hand rotation." If rotary motion is counterclockwise, the cutter is operating with "left-hand rotation."

So, in a plain cutter designed for milling surfaces where the width of the cut is narrower than the width of the cutter, left-hand helix angles are generally combined with right-hand cutter rotation. This keeps the pressure or thrust against the spindle bearing and helps prevent chatter.

Helix angles and the direction of rotation are only two of the features common to most milling cutters that are best described in relation to simple "plain" mills. Basically, there are two major sets of angles: radial and axial. The helix angles described above fall under axial angles; radial angles are those in the same plane as that in which the radius is measured (the flat surface in many cutters).

The **land** width, the flat section on the outer periphery of the tool located after the cutting edge, varies from about 1/64 in. for small-diameter cutters to 1/16 in. or more for large-diameter cutters. Sometimes, a narrow cylindrical margin is aft of the cutting edge. The relieved surface is then provided back of the margin.

To keep the tool from rubbing on the workpiece, the peripheral **relief angle** is formed between the land and a tangent to the cutter's outside circle. Following this feature is the peripheral **clearance angle,** the one measured between the surface aft of the land and a tangent of the cutter's outside circle that passes through the cutting edge. This radial clearance angle includes the relief angle.

The part of a tooth behind the cutting edge may be a smooth curve, or it may be made up of the land and one or more lines built up from compound angles. Some cutters have secondary and even tertiary clearance angles.

The way relief angles are designed into the cutter brings up another basic distinction between tools. **Profile-relieved cutters** are those on which the relief is obtained and sharpening done by grinding a narrow land back of the cutting edge.

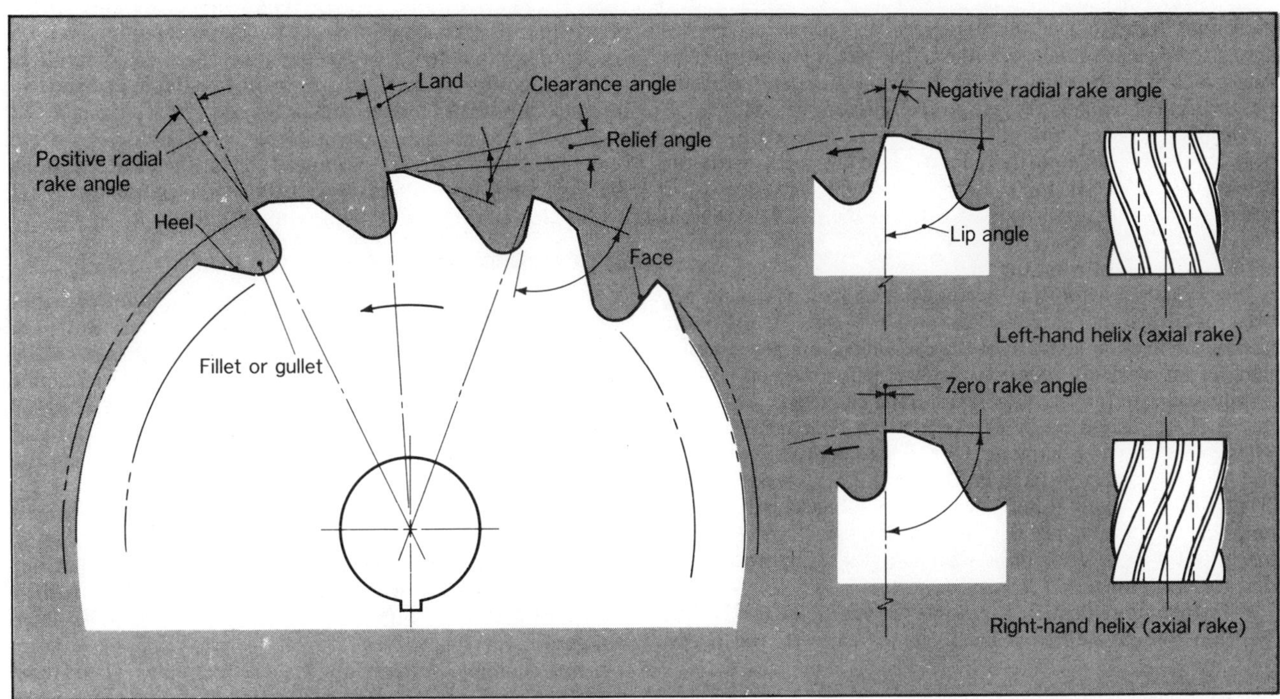

Profile-relieved cutters can be made to cut straight surfaces, or they may be designed to leave behind a contoured or irregular surface. These latter cutters are sometimes called form cutters, but this terminology can confuse them with form-relieved cutters.

When a curved relief in back of the cutting edge is produced by a cam-actuated tool or grinding wheel, the milling cutter is called a **form-relieved cutter.** The distinguishing mark of these cutters is that they keep the same form throughout the length of each tooth. To sharpen a form-relieved cutter, it's necessary to grind only the front of each tooth—into the gash. Since the diameter decreases slightly with each grind, the cutter has to be reset in the machine before it is used again.

Of the two categories, solid profile-relieved cutters are usually lowest in first cost and, therefore, may be the best for short runs or toolroom use. Form-relieved cutters are generally higher in first cost, but the expense of sharpening them is considerably less, especially when a complicated shape is involved.

The term "gash" refers to the angle formed by the last clearance plane (the heel) with the leading edge of the next cutting edge (the face). The gash angle helps determine (along with the pitch) the size of the flute, or chip space, as well as the rake angles that form the cutting-edge geometry.

The rake angles do the cutting

Rake is defined as the angular relationship between the tooth face (or the tangent to the tooth face at a given point) and a given reference line. In plain milling cutters, it's the **radial rake angle** (see illustration at top of this page) that cuts the metal. When the cutting edge of the tooth face leads the rest of the face into the cut, the radial rake angle is *positive*; the angle is *negative* when the cutting edge tilts back from the tooth face; and it's zero when the tooth face is in the same plane as the cutter's radius. Another angle sometimes referred to, especially in cutter regrinding, is the lip angle: the included angle of metal between the face of the tooth and the surface of the land. These two surfaces form the cutting edge.

Plain milling cutters are designed with more-or-less standardized values for axial rake angles (and helix angles) and radial rake angles, depending on their tooling material and the kind of workpiece material they are intended for. For hss cutters, axial rakes are usually in the vicinity of 10-15° for cutting most steels, as high as 25° for cutters intended for machining aluminum, magnesium, copper alloys, and as low as 0° for such workpieces as molybdenum alloys. Carbide cutters, which are wear-resistant but prone to cracking, are often built with lower axial-rake angles: a high of perhaps 20° for aluminum but 0° to −10° for most steels and even more negative for milling tungsten alloys.

Similarly, plain milling cutters made of hss usually have radial-rake angles of 5-12° for steel workpieces and as high as 20° for biting into materials like cast iron and aluminum. But carbide plain mills feature positive radial rakes only for tough aluminum and similar alloys and negative rakes (down to −10°) for most steels.

The next measurable feature of the plain milling cutter is actually a count: **pitch** counts the number of teeth. Described as teeth per inch (tpi), the pitch of a cutter is seldom used in feedrate calculations. Rather, the number of teeth in the overall circumference of the cutter (the value "n") is plugged into formulas, along with cutter diameter.

Still, pitch—or, more accurately, tooth spacing—is an important parameter in milling, for it affects the way chips are expelled from the cutter and away from the point of cut. Pitch can vary widely in plain milling cutters according to use. One heavy user of cutters has standardized recommendations: a maximum of 3 tpi for cutting steel, 5 tpi for milling cast iron, and a coarse ½-2 tpi for cutting aluminum. The recommendations come from practical experience in feeds and speeds: if too fine a pitch is used in milling aluminum, for instance, the teeth will clog, because of the high speeds and large chips. Conversely, wide spacing for harder materials would result in damage to an hss cutter.

The number of teeth on a cutter also affects the final surface finish of the part. Consider a single tooth's path as a plain milling cutter is fed past a flat workpiece. As the tooth takes an individual "bite" out of the workpiece, the cutter also moves across some parts of the work where that bite is not taken. This leaves a cusp on the part, and the height of the cusp represents the theoretical irregularity left on the finished surface.

Very few practical milling cutters have a single tooth, and so it is interesting to note the effect of adding teeth upon the generated surface. If, for example, three additional teeth are added, each extra tooth would cut into the cusp left behind. The height of the cusp, in addition to its length, would be reduced, thus improving the finish.

The calculation for the theoretical height of the cusp has been published in National Twist Drill and Tool's *Metal Cuttings*. The exact solution of the cycloidic geometry involved requires an unwieldy equation, but the approximation is sufficiently accurate for any practical milling situation:

$$\text{Cusp height} = D/4 - (\pi \times f_t/\pi \times D \pm n\, f_t)^2$$

where D is the cutter diameter, f_t is the feed per tooth in inches, and n is the number of teeth in the cutter. In the formula, plus or minus is used to signify that the + sign is used for up milling and the − sign is used for down milling. That's because the tooth path for down milling has a greater curvature than that for up milling.

A further simplification is possible for very light feed where the feed per revolution is small compared with the cutter diameter. Here, since the tooth paths approach circular arcs, the difference in cusp heights generated by climb and conventional (down and up) milling becomes negligible. The simplified relationship:

$$\text{Cusp height} = \tfrac{1}{4}\,(f_t^2/D)$$

Intuitively, one would expect that the longer sweep of a large-diameter cutter would produce a smaller generated surface-cusp height than a small-diameter cutter would produce. According to the equations, this is so: the cusp height is seen to be approximately inversely proportional to the cutter diameter.

This means that, if all other factors are held constant, the height of the cusp can be cut in half by doubling the cutter's diameter.

This discussion makes a strong case for the use of a large number of teeth to produce a small cusp height, thus a smooth surface. But it also makes a strong assumption: that the cutter has very little runout—the cumulative inaccuracies in the cutter, the arbor, and the milling machine.

If a single tooth is higher than the rest, it will dig more deeply into the workpiece and obliterate the contribution made by adjacent teeth. In a four-tooth cutter, this means that a single tooth generates about half of the finished surface and that irregularities are slightly greater than the amount the single tooth is higher than the other three teeth.

But, if a single high tooth is bad for surface finish, cutter runout, with its rotation of the cutter about an axis slightly offset from its true axis, produces the same condition. With a true-running four-tooth cutter, such an offset could result in one high tooth, a diametrically opposite low tooth, and the two intermediate teeth on approximately the correct diameter but slightly off true angular spacing. Here, as in the case of a single high tooth, one tooth creates most of the finished surface, while the low tooth contributes very little. An increase in runout could eliminate the effect of intermediate teeth.

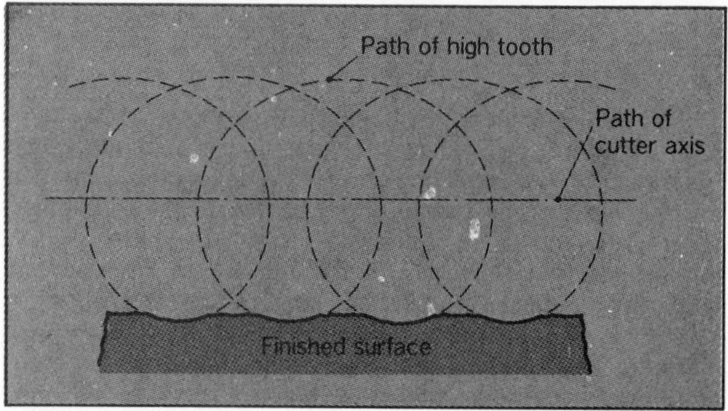

Some 'plain' cutters aren't plain at all

It has been useful so far to limit the discussion of cutter design to what are called "plain" milling cutters. But most milling cutters that fit into this category are anything *but* plain! The economies of designing and building milling cutters dictate that the most complex shapes and special forms to be placed into milling cutters be ground into solid-steel tool bodies that have the least intricacies in their construction. (As always, there are exceptions.)

So it's within this class of milling cutter that the most complicated shapes are to be found. These job-tailored tools, especially those with form-relieved tooth design, may start like plain milling cutters, with teeth projecting radially from their rotational axis, but after that all similarity between them disappears.

Convex-milling cutters are a good example. Their teeth curve outward on the circumferential surface to form the contour of a semicircle. They are designed to mill a concave surface of a circular contour equal to half a circle or less. So a convex-milling cutter, since it performs some cutting in a plane that's not exactly on its outer circumference, is the first step toward a **side-milling cutter.**

A side-milling cutter is one that has cutting edges on both its cylindrical surface and on its face (or end). These face edges also have relief angles, clearance angles, rake angles, and lands; and these features are measured similarly to those in the simplest plain cutter.

With the capability to align cutting edges in a number of different directions, "plain" milling cutters are built to handle complicated shapes in a single pass. The designs are almost endless: special cutters mill precise tooth forms into hacksaw blades on a production basis; others cut distinctive "Christmas-tree" shaped pockets into the base of turbine blades; narrow side-milling cutters, often called slitters, can have side chip clearance for deep slots or plunge cuts where many teeth are buried in the work; and some complicated combination cutters gang together profile-relieved and form-relieved portions onto the same cutter.

One special category of side-milling cutters is known as **shell mills.** These versatile tools have teeth along both their axial and radial sides and are made hollow in order to mount onto an adapter, similar to a centering plug, and the drive is obtained by a key on the adapter engaging a slot on the back of the cutter. The cutter is locked in place with a retaining bolt or screw, and the whole assembly is mounted into the spindle of either a vertical or horizontal milling machine. A typical shell mill is shown below.

High-speed steel or carbide-tipped shell mills typically are made in diameters from 1¼ to 6 in. with corresponding face widths of 1-2½ in.

Milling the face of a workpiece

When using a shell mill to cut across the flat surface of a workpiece, what the machinist is doing is face milling. This goes back to a basic distinction, made early in the Milling Dynamics section of this report, between peripheral milling and face milling. Actually, a little of both types is involved, because, as the end teeth in a vertically mounted cutter dig down into the workpiece, the teeth on the cylindrical surface cut in the direction of the workpiece feed.

Face mills are those specialized milling cutters suited to do this operation best. They are designed to machine a surface that is parallel to the face of the cutter (or at right angles to the axis of the cutter). The machined surface per cutter pass may equal the whole width of the face of the cutter (the cutter's diameter) or any fraction of the whole diameter. Generally, the less cutter width used, the better.

A feature that sets face mills apart from plain and shell cutters has to do with the design of the teeth and the cutter body. Whereas plain milling cutters have the cutting edges designed right into the body of the cutter, virtually all face mills use inserted blades that are physically separate from the body of the cutter itself.

The inserted portions involved can be one of two types: resharpenable blades or replaceable inserts. The first type (drawing, bottom of this page), often called "grind-type" blades, are held in place by a system of wedges and individual clamps, which can be loosened to permit the blades to be repositioned for sharpening. The second type (upper drawing, this page) also has individual clamps for the inserts, but the cutting inserts themselves are smaller, can be made of a number of tooling materials, can be indexed, and are designed to be thrown out after they are worn.

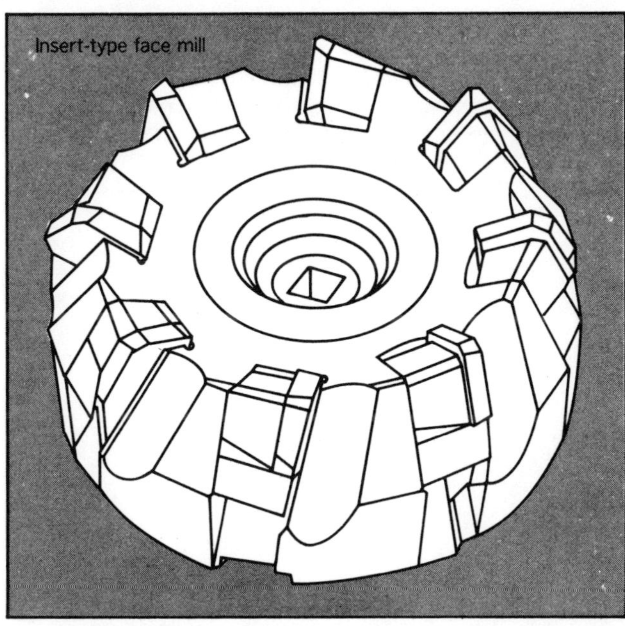
Insert-type face mill

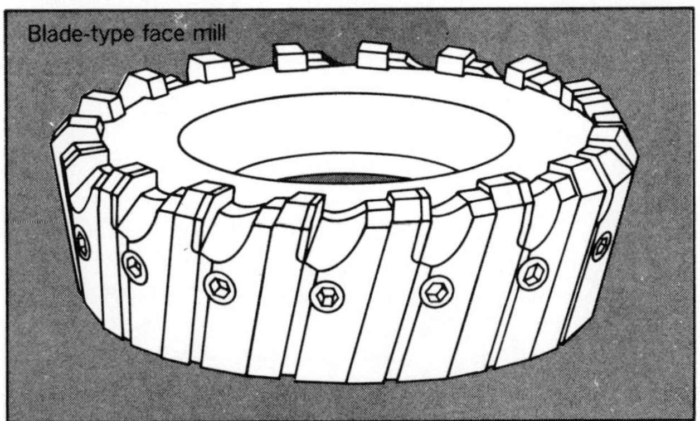
Blade-type face mill

The geometries for both grind-type-blade face mills and insert-type face mills are similar. Three major angles are involved in the cutting operation: the radial rake, the axial rake, and the face-cutting edge angle.

Axial and radial cutting angles determine both the cutting energy applied to the shear zone and the way the chip is thrown from the cutting zone. **Axial rake** is measured by comparing the peripheral edge of the blade or insert with the spindle axis. "Sharp" teeth are positive; "blunt" ones are negative. **Radial-rake** angles are measured along the bottom edge of the blade or insert and are compared with the radius of the round cutter body. Once again, sharp angles are positive, and, when the cutting edge leans ahead of the cut, the radial-rake angle is denoted negative.

Of the two, radial rake has the most effect on horsepower, and axial rake affects the chip forming and thrust on the spindle. In general, the rake angles should be as sharp as the combination of workpiece material and cutter edge material will allow. Three combinations of axial and radial rake angles are the most popular: double negative, double positive, and positive-axial negative-radial. (The illustration shows at top of next page a double-positive cutter on the left and a double-negative cutter on the right.)

Double-negative cutters offer strong edges and can be used to mill hard steels effectively. The chip formed with this geometry resembles a clock spring and can become trapped in the cutter's gullets. For this reason, soft, ductile materials can cause chip jamming and should be avoided. Guidelines dictate the use of double-negative cutters when inserts are subject to heavy impact stresses, cast iron or hard materials are machined, ceramic tooling is used, or the machines used have poor axial bearings.

Double-positive cutters offer the best cutting action and are capable of more metal removal per horsepower than any other geometry provides. Clearance must be provided on the inserts themselves (negative rake angles form their own clearance), which means that positive indexable inserts have only half the number of cutting edges as negative-rake inserts. This can be misleading, however, in terms of economy. True, the positive insert may not be flipped over to provide additional cutting edges, but, in many cases, it will outperform a negative-rake insert.

General guidelines for choosing double-positive tool geometry in a face mill include situations when the workpiece is unstable, for machines with limited horsepower, and for workpiece materials with which negative rakes produce edge buildup (stainless, aluminum, soft steels, and heat-resistant alloys).

Positive-axial, negative-radial cutters are often used for rough, heavy milling operations. The positive-negative geometry directs the chips away from the cutter—a distinct advantage for heavy depths of cut. The positive-axial, negative-radial combination of rake angles is basic to Ingersoll's patented Shear-Clear geometry, which adds a bevel to the external side of a regrindable-blade insert. Chips, especially springy ones thrown from a steel workpiece, depart the cutting area as spirals pointing outward.

Bevel, the angle at which the main cutting side of the insert engages the workpiece, is sometimes called the cut-entering

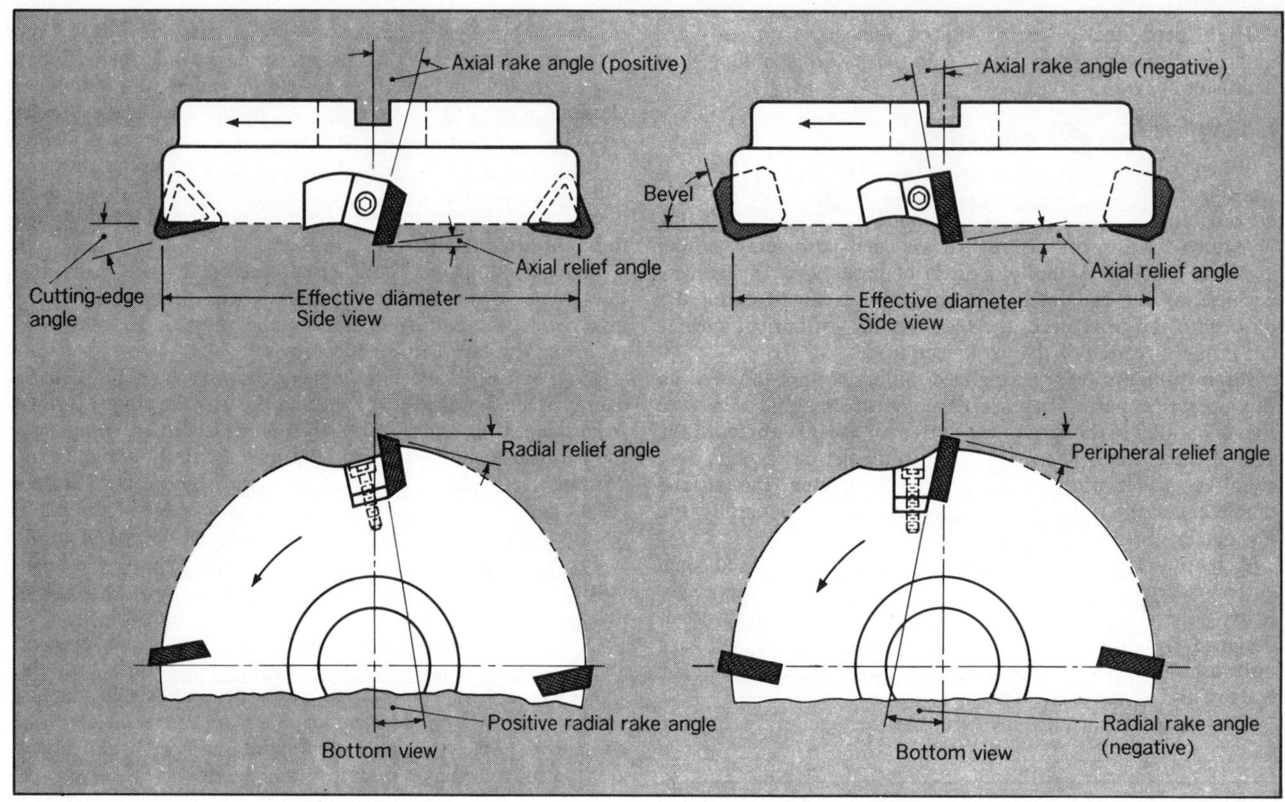

angle (not to be confused with the attitude at which the cutter enters the leading edge [box, next page]) and is measured from the machined surface to the surface generated by the main cutting edge. When the outside of the peripheral cutting edge is perpendicular to the machined surface, the entering angle is 90°; when the peripheral edge leans out toward the workpiece, the entering angle decreases.

The size of the entering angle affects axial and radial forces on the cutter, as well as the thickness of the chip. For a given feed per tooth, a vertical orientation between the peripheral cutting edge and the workpiece surface results in the maximum chip thickness. As that angle decreases (toward a minimum of 45°), chip thickness also decreases. Both the axial and radial forces increase with a smaller entering angle. The axial load is affected most, but milling machines are capable of withstanding high axial forces. The real effect will be on unstable workpieces and on inadequate clamping.

Corner angle affects strength

For high (90°) entering angles, the weak corner angle limits the tool, and edge breakout is common in materials like cast iron. For medium entering angles (75-60°), fairly large portions of the available cutting lengths are used, roughing is generally done with entering angles in this vicinity. Small (about 45°) entering angles cut down on edge breakout in milling cast iron; these are used in milling very hard materials when power and stability are sufficient.

Face mills are sometimes used for cutting shoulders, and the entering angle is predetermined by the required angle of the shoulder. The typical shoulder-angle requirement is 90°, impossible to achieve with a square insert because some clearance is required to ensure that most of the bottom of the insert does not scrape the workpiece (except for finish milling when this is desirable). Right-angle shoulders are usually cut with triangular or diamond-shaped inserts, which are mounted so that their outside cutting edges are at 90° to the workpiece surface.

Thus far, all axial angles in face milling have been described with the assumption that the axis of the cutter is normal to the face of the finished workpiece. This is seldom the case.

Whenever possible, milling machines are set up so that the spindle is tilted very slightly in the direction of feed. The tilt—just a degree or so—is enough to ensure that the rear (noncut-

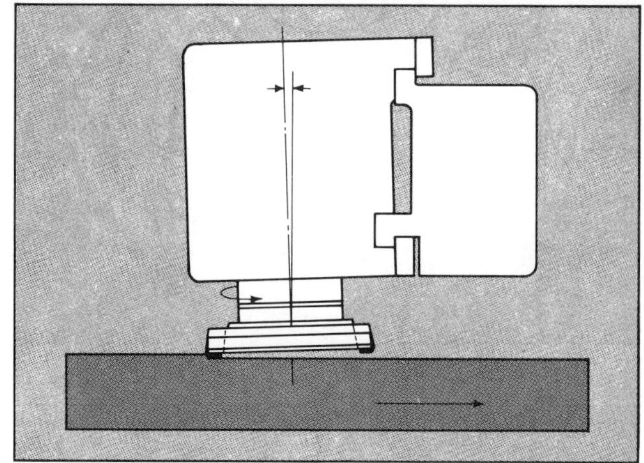

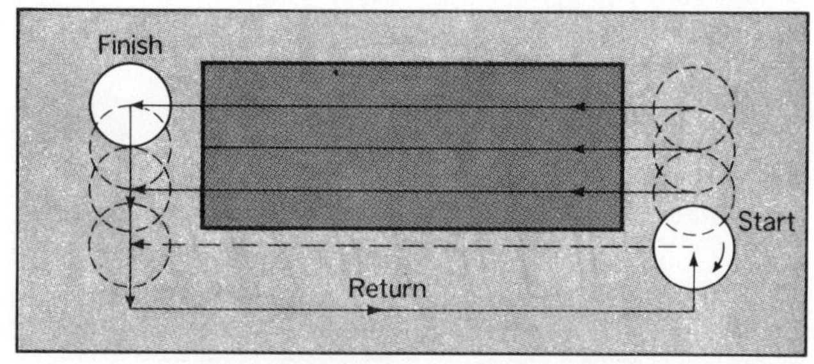

ting portion) of the rotating spindle is lifted off the surface of the workpiece and does not scratch it or carry small chips around the cutter's rotary path.

For good finishes, it's not enough just for the cutter to clear the milled surface. Once again, the geometry of the insert or blade comes into play.

Finishing face mills cut with two portions of the cutter surface: the circumferential periphery removes most of the metal, and the flat face of the tooth shaves the finish.

Some face-milling cutter bodies have additional teeth located on their bottoms but placed within the outer circumference; these teeth are for finishing purposes only. For these internal teeth and for those on the outer periphery that also generate the finished surface, the flatness of the portion in contact with the face of the work-piece is important. But, as always, there are exceptions.

One finishing-cutter design calls for slightly elliptical bottoms to be ground onto each grind-type blade. The very large elliptical radius ensures that there are no corners to contact the surface and scratch it. The elliptical-bottom blades do not produce quite as good a finish as those with flat bottoms; but then they do not have to be aligned quite as critically.

Operator technique is also important in generating fine finishes. Rules to keep in mind include keeping surface speed high; always cutting in the same direction, returning the cutter to the starting end of the machine when two or more passes are required to cover the workpiece; and never changing depth-of-cut settings between passes since, in practice, it is impossible to return to precisely the same setting for the next pass.

So face-milling cutters, while they share certain nomenclature with plain types, differ remarkably in their application. One feature, in closing, remains the same as for plain mills: that of "hand." Face mills are either right- or left-handed,

Entering with the right attitude

Although many machine operators leave the angle of entry to fall where it may, the attitude of the cutting edge when it enters the cut can have a great effect on tooth breakage. Keith Smith, of Greenleaf Corp, points out that this is especially true in high-positive-radial-rake face mills that use carbide, ceramic, or diamond inserts.

What is the angle of entry? If we assume, for illustration purposes, a theoretical milling cutter with zero positive, zero negative rake angles, then whenever the cutter's center is outside the workpiece, entry angle of the tooth will be positive. That is to say that the initial impact between the insert and the workpiece will occur at the weakest point of the insert, at its edge. This is shown at the left in the first top-view drawing.

If, on the other hand, the same workpiece is repositioned under the cutter so that the spindle axis is inside the workpiece edge—as in the right-hand side of the same drawing—the cutter's angle of entry will become negative. What has happened is that, in the first portion of the cut, up milling—or a situation resembling it—has been achieved.

In this second situation, the initial impact on the tooth or insert will occur away from the relatively weak tip. Insert breakage is less likely to occur, especially with positive-geometry cutters.

These factors have an implication for cutter selection: if the cutter is too large for the intended path of cut, there's more likelihood for entry angle to be positive (and potentially damaging.

Real complications of this theory start to occur when the cutter enters a cut or leaves it. In the right-hand side of the second drawing, the conditions of the entering cutter's teeth are much different from those of the exiting cutter on the left-hand side of the same drawing.

As the entering cutter proceeds further into the workpiece, each succeeding initial bite taken by the same insert becomes progressively more "negative." Once the cutter has moved completely into the workpiece, there is no angle of entry to worry about, except in those situations in climb milling when the insert starts cutting at a side edge of the work.

As the cutter exits the workpiece, it will start to break through the far edge so as to create two "islands." Again, the angle of entry will vary as the cutter removes these two islands, cutting first on one and then crossing the gap to re-enter the other.

Still further complications occur when the workpiece face contains holes or cavities. Here the cutter's entry angle can be changing constantly as each pocket is succesively encountered, cut, and then exited by the cutter. The situation is very different from face-milling a workpiece with a continuous surface.

For cutting intricate parts like these, fine-pitch cutters for placing the maximum number of blades into the cut are best. This tactic minimizes the variation between entry angles for one blade or insert to the next.

Negative-radial-rake cutters or negative-positive-rake cutters with lead angles also help to offset this angle-of-entry effect.

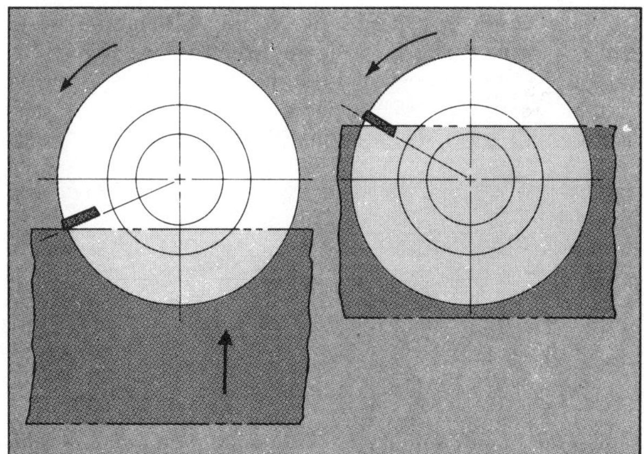

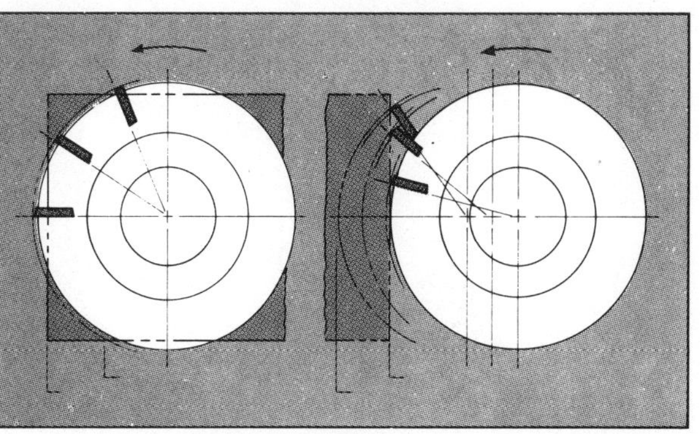

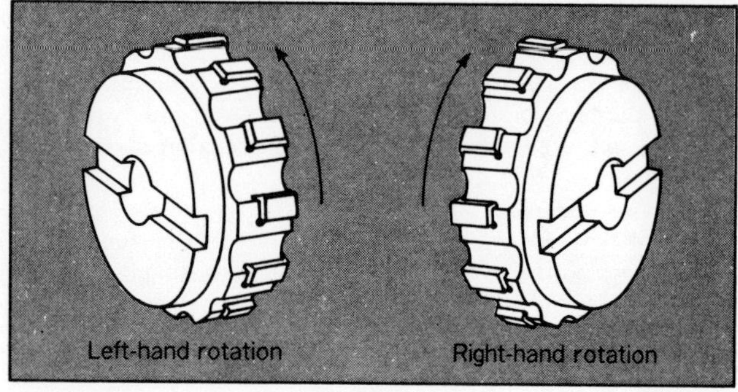

depending on the way the spindle rotates. Sometimes axial rake angles, like those for plain cutters, also are denoted in right and left hand.

End mills: combination tools

The last of four major classes of milling cutters, end mills are best described as tailored to suit a combination of types of cut. Whereas most plain and shell cutters are built to cut best on their outer periphery and face mills are designed for wide flat workpiece faces, end mills cut on their ends as well as on their periphery. Typical operations for these longer fluted cutters include facing, slotting, profiling, plunge cutting, and cavity cutting, for which plunging and slotting operations are combined.

The helical flute is the characteristic of end mills. Prior to the development of the design—most often attributed to Carl Bergstrom of Weldon Tool—end mills usually had straight flutes or, at best, flutes with very slight helixes. In heavy side-milling applications, these tools had a tendency to chatter and break teeth. In straight-fluted-end-mill applications, the load on each cutting edge builds up almost instantaneously, causing deflection, springback, and, thus, chatter. On the other hand, when long helixes are used, the chip load is applied to the entire flute length in a progressive sliding action, similar to that of a snowplow with its blade angled off to one side.

But the thrust of a helical tooth has a tendency to pull a tapered shank out of the machine spindle. To overcome this, many end mills use a straight shank with a single flat ground

into its side. A locking screw holds the cutter in place and drives it; this straight-shank design also permits double-ended tools to be designed to allow quick changeability and economy of materials.

To be absolutely precise, end mills as a class are just a specialized version of shank-type helical side-milling cutters, and the geometry and nomenclature of the various parts of end mills indeed follow those of plain mills.

Like plain mills, end mills can have right- or left-hand helixes to their flutes, the gooves that are milled or ground into the body's circumference. Peripheral teeth may have positive, negative, or zero rake. The cutting edges are followed by lands, areas that are ground back from the cutting diameter to a relief to allow the tooth to clear the cut. The relief may be flat, concave, or eccentric, and different companies make varying claims for the effectiveness, edge support, and regrindability of each. Following the relief angle, the clearance angle in side teeth is the angle of the heel, and it allows coolant to enter and chips to clear.

For side teeth, a distinction should be made between the flute length and the length of cut; the two are not equal. The side teeth themselves are formed by the intersection of the flute's hook or rake and the land.

End teeth on an end mill also perform cutting operations. An end gash is cut into the body of the mill to provide the end cutting edge. In addition, the end of the cutter is not flat but rather has a dish or a concavity to provide center relief for each end-cutting tooth, and the mill moves laterally across the workpiece.

The style of these mills affects the function for which they can be used. The two-fluted end mill can plunge-cut like a drill. Three-fluted ones have one end tooth cutting all the way to the center and also can plunge cut. Regular four- and six-fluted mills have none of their end teeth extending through to the central axis of the mill and generally are not intended for drilling-type operations or cutting through to center. Four-flute center-cutting end mills, however, are available; these have secondary or even tertiary gashes at the tool's central axis. But they are costly and difficult to sharpen, and because of this they are seldom used.

Regrinding of end mills is done on a tool-and-cutter grinder and can be done by the user or by an outside contractor. In regrinding, one manufacturer recommends that a maximum of only 12-14% of the original diameter be removed before the cutter is discarded. This is so for a number of reasons. A progressive reduction in diameter reduces the tool's cross-sectional area, permitting greater deflection under load. Furthermore, with each regrind and reduction in diameter, there is a subsequent and significant reduction in hook angle

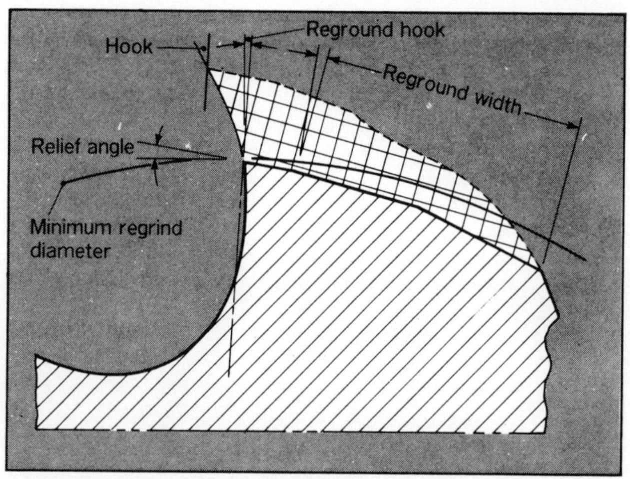

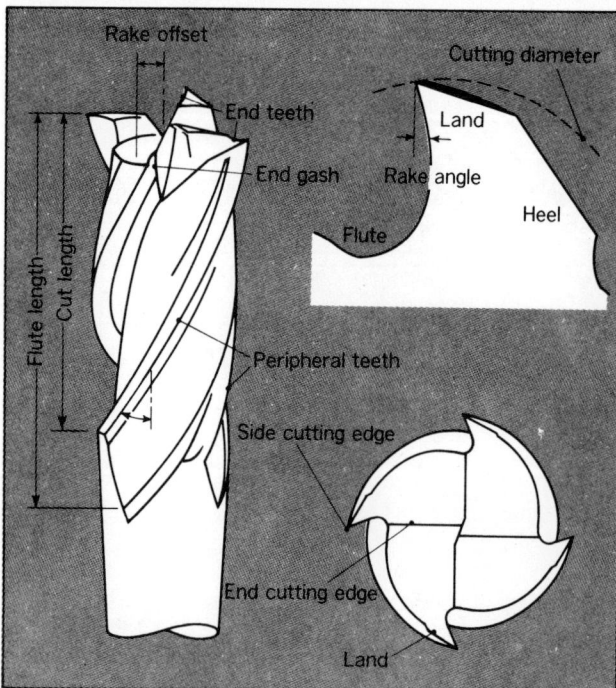

(radial rake). Flute depth also is reduced in regrinding, and there may sometimes be insufficient chip clearance in a too-often-resharpened cutter. Lastly, as the diameter is reduced, the cross-sectional profile causes a rapid increase in the width of the secondary land, thereby increasing grinding time.

Deflection is an important problem. It can be due to too heavy a feed, to a cutter that has been resharpened too often so that its diameter (and mass) has been reduced, or to the development of a wear land, a rubbing spot on the relieved portion of the tooth, as the tooth becomes worn. Deflection can cause leaning keyways, chatter, and tool breakage. By holding tool length to the minimum that is required, by slowing the feed, and by using stiffer, solid-carbide—rather than hss—cutters, deflection can be minimized.

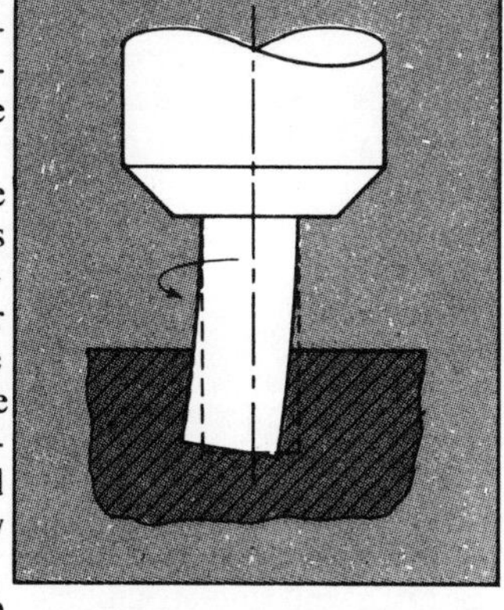

Special types of end mills are designed to do certain jobs best. **Roughing** end mills—also called hogging cutters or chipbreaker cutter—have grooves or scallops around the body that are either semicircular or sinusoidal in shape and have an advantage in that they produce well broken-up chips.

Roughing end mills absorb less side load than those with smooth helical teeth, and they have a heavier cross-section. Stock removal, because of these factors, is often up to three times that of the usual end mill, and these designs very often can be made to benefit from increased machine feedrate (see box on pages SR-14 and SR-15). The finish obtainable with roughing end mills is, as might be guessed, worse than that of regular cutters but, depending on workpiece requirements, may be acceptable.

Roughing end mills are form-relieved tools and are resharpened by grinding only the flute. With each resharpening, the gap between teeth becomes wider, and this can actually improve their performance on materials like aluminum.

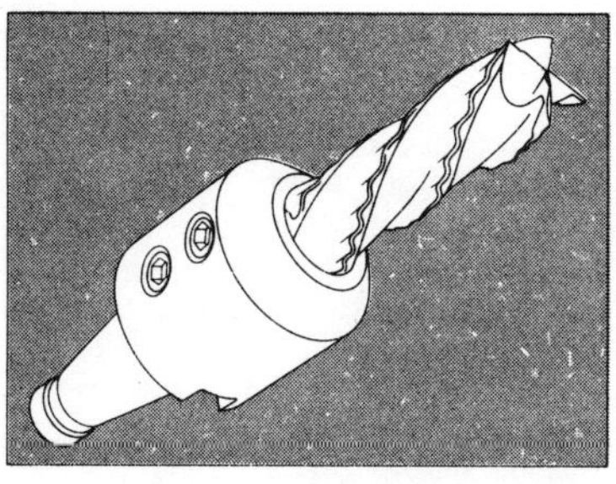

Roughing end mills, like the one shown here with a sinusoidal wave pattern on the cutting edge, are different from regular end mills in the way they bite the workpiece. In addition, end mills—like plain milling cutters—are designed for specifically different workpiece-shape requirements. Ball-end styles are popular for duplicating milling machines: These have a certain fixed radius built right into the end teeth. Other specialized styles, not true end mills, are made to cut T-slots or dovetails. This is usually a two-step operation in which a regular tool cuts a starting slot and the specially profiled end mill starts from one side of the workpiece to remove the internal shape, with only its straight shank protruding from the previously cut slot.

Other specialized end mills use straight flutes for cutting slots and keyways, and still others employ individual inserted blades and even indexable inserts.

What the teeth are made of, and why

Tool-material selection for those end mills that use replaceable inserts is similar to that for face mills that use them. But most end mills—and most plain and shell mills and some face mills, for that matter—are made of cast-alloy tool material or high-speed steel.

Cast alloys exhibit a uniform red hardness and keep edge hardness at temperatures up to 1500F. Tools of these materials are usually recommended for use with cutting speeds between 100 and 200 sfm.

High-speed steel is a tool steel that can maintain cutting hardness at elevated temperatures, because it includes one of two alloying elements, tungsten or molybdenum, or a combination of both. All hss also contains carbon, vanadium, and chromium, and some grades also include cobalt as an alloying element.

As with most metals, steels that are harder and more heavily alloyed tend to be more brittle and prone to cracking. Also, since there are about 30 different grades of hss currently in use, it's best to follow manufacturers' recommendations for specific types of cutters for the job at hand.

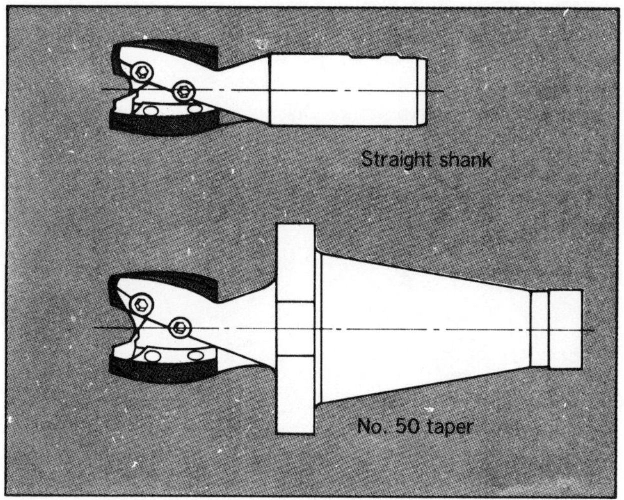

Even harder cutting-tool materials are the **carbides.** Capable of retaining their hardness at even higher temperatures than hss, these tungsten-carbide-based materials are generally classified into eight grades—C1 through C8—with the first four having the better abrasion resistance and the last four capable of withstanding cratering and deformation. (ISO grades—15 grades in three groups of Blue, Yellow, and Red—are analogous but not similar. See Special Report 686 (AM—Apr'76,p67), "World directory of carbides," for full specifications and listings.)

Milling cutters may be solid carbide, generally limited to small plain cutters and end mills, or they may be tipped with carbide, in which case the carbide is either brazed onto the tooth of the cutter or is used as a solid blade or insert that is clamped into position within the cutter body.

For milling cutters that use inserts, there are a number of other choices. **Coated carbides** have thin (perhaps 3-10 μm) coatings of titanium carbide, titanium nitride, hafnium nitride, or similar substances that are vapor-deposited on solid-tungsten-carbide bases. Advantages of the coated grades include better resistance to flank wear and cratering, plus the ability to work at higher feeds to build up cutting temperatures and avoid edge buildup.

Certain grades of **polycrystalline diamonds** lend themselves to the rigors of interrupted cuts. Used for milling aluminum, the diamond-tool blanks are often brazed to steel inserts and ground to shape with diamond wheels. Tool life can be hundreds of times greater than for carbide, according to some field tests.

[Continued]

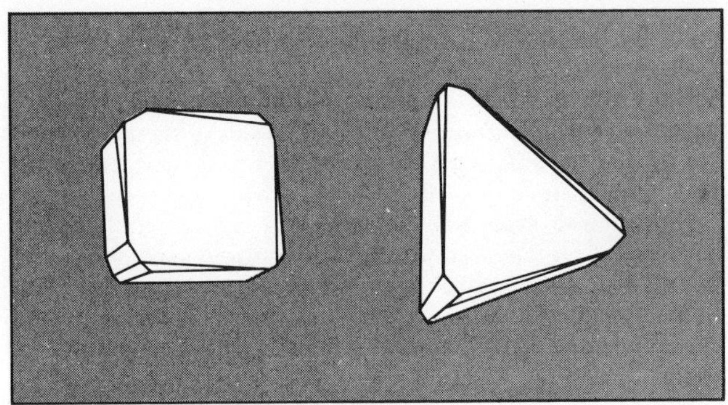

Ceramic inserts have been used with varying degrees of success for some time; as far back as the late 1950s, ceramic tools milled steels at 2000 sfm under laboratory conditions. But ceramic inserts generally require very rigid machines to prevent fracturing. One of the latest developments is the appearance of **ceramic-coated carbide** inserts. These fairly recent tools have performance characteristics better than those

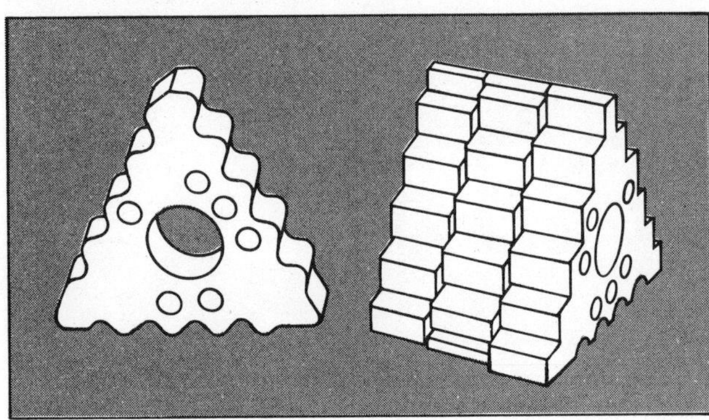

of regular carbide and more forgiving than those of solid ceramic.

Whatever the specific tooth material, cutters that use inserts, particularly indexable ones, are growing in popularity. Especially with NC machines, for which the cutter has to be preset and "qualified," changing and setting the tool faces of a reusable cutter body can provide a substantial saving.

The shape of the insert itself can be a major point of contention between different manufacturers. Although square and triangular are the two basic shapes, milling-cutter bodies are also designed to accept hexagons, parallelograms, and

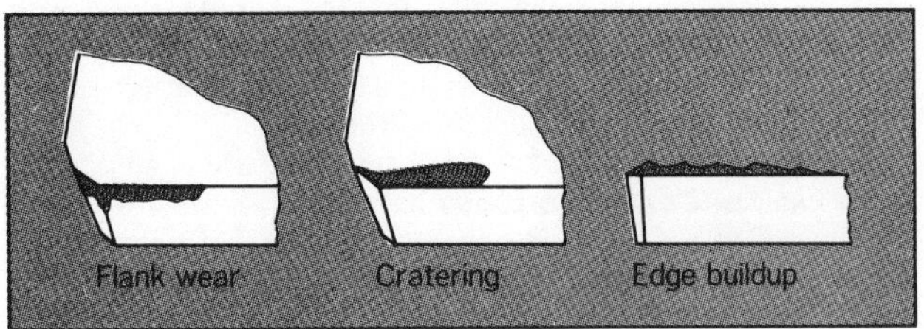

round (sometimes rotating) inserts. Serrations are sometimes provided on the cutting edges of inserts; these are analogous to the roughing end mills with ridges or sinusoidal wave patterns described above. In rough-cutting operation, each ridge takes a small bite out of the workpiece; succeeding ridges, slightly out of phase with each other, proceed through the cut, removing more metal until the cut is complete.

Insert orientation also can be different. Some milling cutters present the insert edgewise to the cut in order to place more bulk behind the cutting edge. Those from another company use inserts shaped more like cubes than wafers; the idea, again, is to put maximum strength behind the cutting edge.

For insert-type cutters, even the way in which the tool bits are held varies from one manufacturer to another. Some use insert nests; others directly clamp the insert bit with individual wedges and setscrews; others have a spring-actuated lever system that can be released quickly via a small prybar; and still others use self-centering screws that mount through holes in the inserts themselves. Replacement of inserts can be done while the cutter is mounted in the spindle or is off the machine and in a toolroom. One design parts the cutter in two: the main body stays on the machine while a demountable shell goes to the toolroom.

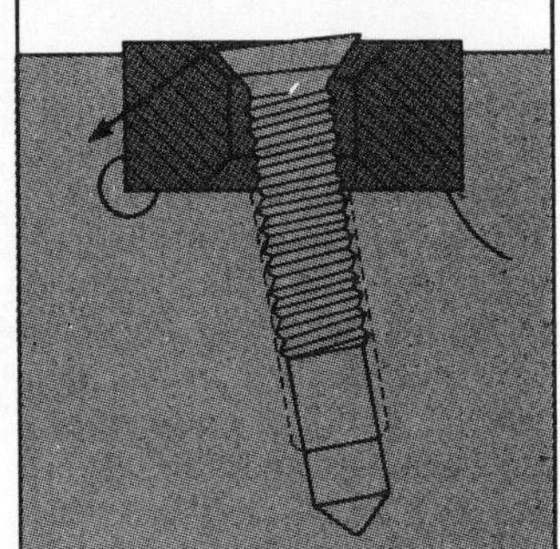

It's the tooling, then, that can have a profound effect on the very profitability of a milling job. The selection of a basic type of machine, of course, is crucial, as we've seen in the first sections of this report. But the fact remains that most milling in the US is done on machines—be they milling machines or machining centers—that are fairly flexible: The basic production units can usually accommodate a very wide variety of workpieces, and they provide a wide enough range of feeds and speeds for a number of cutter types.

In day-to-day practice, milling productivity is usually determined when the operator or engineer selects the cutter that will do the job and the parameters within which it will operate. It's here that money is made, or lost. ■

Acknowledgements:

Many companies and associations have been very helpful with information for this report. Of particular help were the New England Research Applications Center, which provided computerized information retrieval, and Sandvik, in whose laboratory AM photographed the cover illustration. In addition, the editors would like especially to thank Bendix, Boston Digital, Bridgeport Machines, Cincinnati Milacron, DoAll, Ex-Cell-O, GE Carboloy, Greenleaf, Harrisville Tool, Hoglund Tri-Ordinate, Illinois/Eclipse, Ingersoll Cutting Tool, Jones & Lamson, Kennametal, Master Machine Tools, Metal Cutting Tool Institute, Metcut Research Associates, National Twist Drill, W.H. Nichols, Society of Manufacturing Engineers, VR/Wesson and Weldon Tool.

Fundamentals of Boring

As one of the basic metalcutting operations, boring is performed by a great many machine tools. For certain machines, though, boring is a primary purpose. This Report analyzes the fundamentals of the operation, the machines that perform it, and the tools, equipment, and accessories they use

By Richard T Berg
associate editor

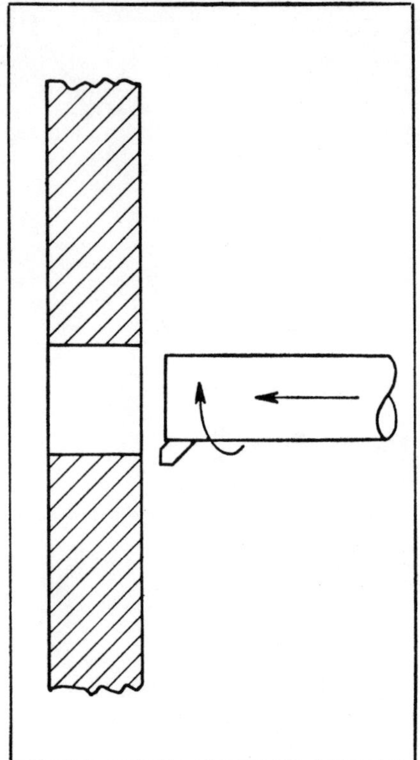

Horizontal boring machine rotates cutter and feeds it into the work

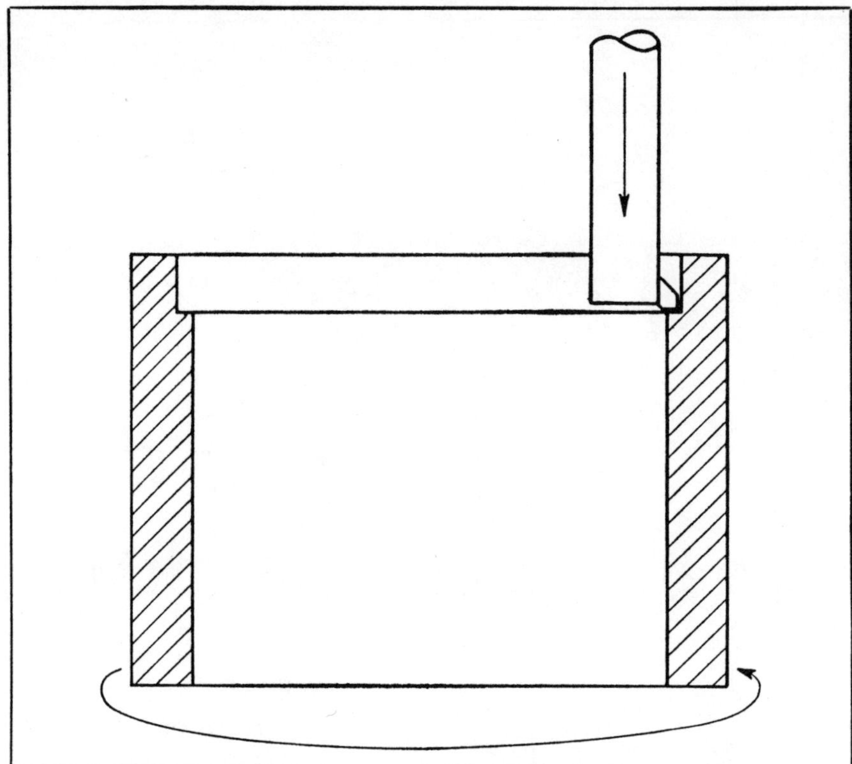

Vertical boring mill rotates the work about a vertical axis and feeds the non-rotating cutter downward into the work for boring

What is boring?

Boring, in the broadest sense of the term, is an operation than can be performed by almost any machine tool with a rotating spindle and a feed motion, and most of them (excluding grinders) do perform boring of one sort or another.

But boring is also a specialized operation and often involves requirements that are best met by machines designed specifically for boring. This specialization has proceeded along several broad, general lines, and the requirements include such varied factors as precision, workpiece size, versatility, and high productivity.

Therefore, the machines known as boring machines are quite a mixed group. In fact, about the only thing they all have in common is that their primary purpose, or at least one of their primary purposes, is to perform boring.

First, though, let's analyze boring in general, so that all boring is included. Then we can zero in on the operations that are performed on the various boring machines themselves.

Boring can be defined as the machining of an internal surface under three conditions, all of which must be met:

(1) with a single-point cutting tool,

(2) with either the cutting tool or the workpiece rotating about the axis of the surface being machined, and

(3) with a feed relationship between the workpiece and the tool in a direction parallel to the surface being machined.

Actually, it is not uncommon to see a two-cutter boring tool, but the essential factor is that these tools could still do their work, even if they had only one cutting edge. Drilling, reaming, and tapping do not meet this requirement and are not included in boring. Milling is out because the cutter does not rotate about the axis of the surface being machined.

It should be noted that boring is not employed to *create* a hole. Its main purpose is to open up a drilled, cored, or fabricated hole and machine it to the desired diameter and surface finish.

Precision was not mentioned in the definition of boring because it is not a basic requirement. However, in common usage, boring usually implies work within fairly close limits—accurate diameters, accurate locations, good finish, and most important, true, straight holes. In fact, this is often the reason why boring is performed instead of drilling.

In the range of hole sizes that can be finished by either drilling or boring, drilling is more economical, and boring is done only if the quality of the product demands it.

Many types of boring

Starting from these general considerations, however, boring takes off in a great many directions. As indicated in the definition, the cutting tool may rotate, or the work may rotate. The cutting tool may feed into the work, or the work may feed into the tool. And the axis may be horizontal, or vertical, or any other angle.

No single machine, of course, can take off in *all* these directions, not even a boring machine. As a group, however, they can and do cover the full range of boring indicated, and they can perform many other operations, too.

Any machine that rotates either the tool or the work has no difficulty with drilling or reaming, for example, and most of these can also perform facing in one form or another. Turning and grooving are also possible, and some boring machines perform milling as a routine operation.

Among these "extra" jobs, some are more easily performed when the cutter rotates and some when the work rotates. On the basis of extra jobs then, rather than boring itself, there are two general types of boring machine, and the distinction can be important in selecting the best one for a particular job.

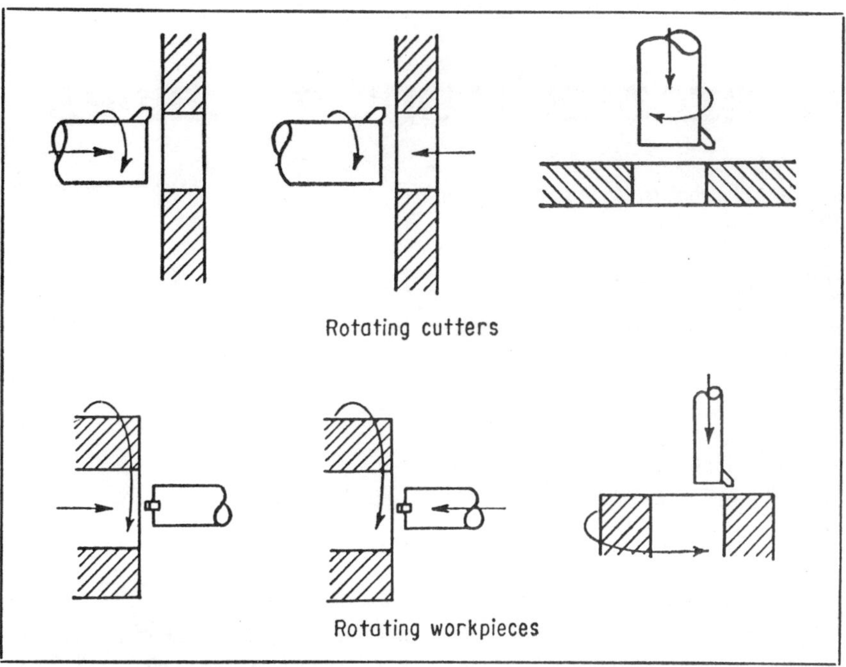

Precision production boring machines come in a variety of arrangements, with either the work or the tool rotating and with either one being fed

Major machine types

In addition, the various boring machines can be classified in four groups, according to the way they are constructed and according to their special purposes. Each of these will be discussed in detail in the sections that follow, but here is a quick summary.

Horizontal boring machines have a horizontal spindle that rotates the cutting tool and advances it into the workpiece, which is clamped to a horizontal table. They are built along the same lines as a horizontal milling machine, except for the added feature of a power-fed spindle.

One of the major virtues of these machines is that they can bore a number of different holes and do a variety of other jobs at different locations on the workpiece without changing the setup. In normal application, they are employed for medium- to small-lot production.

Vertical boring mills have a horizontal faceplate or chuck that rotates the workpiece about a vertical axis. For boring, the tool (nonrotating) is advanced into the work by an overhead ram on a crossrail.

This type of machine can bore extremely large diameters, and even for smaller parts that could also be bored by a lathe, for example, the vertical boring mill sometimes has an advantage in that heavy or awkward workpieces are easier to set up on a horizontal faceplate. Production runs may range from one-of-a-kind work through continuous production.

Jig borers traditionally have a vertical spindle that rotates the tool and advances it into the workpiece, which is clamped to a horizontal worktable. The arrangement is similar to a vertical milling machine, except that the jig borer usually has a power-fed spindle. Another type of jig borer has a horizontal spindle and the same general lines as the HBM.

But the distinctive feature of jig borers is that they have carried the usually high accuracy of boring to an even higher level, largely because they are designed and built toward that end. Intended originally as toolroom machines, they have also been put to work on production jobs where the tolerances on hole location and size are extremely close.

Lot sizes in toolroom work are normally quite small, and this still constitutes the major application for jig borers. When they are used for production jobs, the quality of the work is usually more important than the quantity.

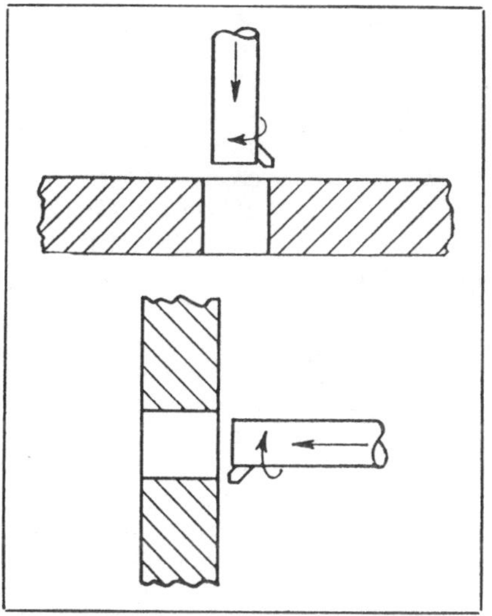

Jig borers have either vertical or horizontal spindle for rotating tool

Precision production boring machines do not have one specific form. In fact, just about every possible arrangement of workpiece and cutter can be found in this classification.

The common purpose of most such machines, however, is to perform precision boring, and other finishing operations, in medium- to high-volume production. Some of them are designed as general-purpose machines, but more often they are built to machine a single part or a group of parts requiring only minor changes in setup.

Lathes? Lathes do perform boring, of course. However, because lathes, by themselves, are a major category of machine tools, and because they cannot be treated fully under the subject of boring, they are not covered in this Report.

Even among boring machines, a 16-page report cannot describe in detail all the variations and "specials" that have been developed. Instead, its purpose is to analyze each of the basic machines, its method of operation, and its special abilities, i e, what it is, how it works, and what it's good for.

Horizontal boring machines

The characteristic features of the horizontal boring machine are:

- a horizontal spindle that rotates the cutting tool.
- a horizontal work surface on which the workpiece is mounted.
- power feed for the spindle so that it can advance the cutting tool into the work for boring.
- power-fed motion between the work and the spindle in at least the two axes perpendicular to the spindle axis for milling.

An additional feature found on some, but not all, HBMs is a support for the outboard end of a line-boring bar.

But these machines are also called "horizontal boring, milling, and drilling machines" because of the variety of work they can do. A major advantage is that they can do these different jobs at various locations on the work without changing the basic setup.

Working in this manner, they can maintain accurate relationships between the areas that are machined. This is especially helpful in machining complex castings or plates with many interrelated dimensions.

Lot sizes range all the way from single-piece jobs to medium-sized production runs, and special tooling is available for efficient high-volume production, as well. However, at the higher levels, custom-built equipment such as multi-spindle machines or transfer machines may be used, even for jobs normally done on the HBM.

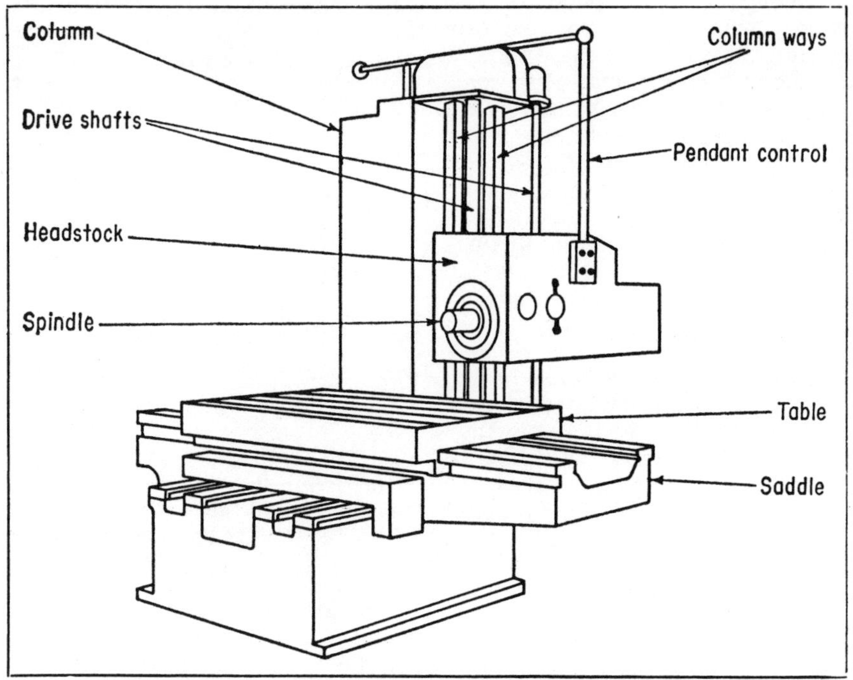

Table-type horizontal boring machine can move the workpiece in two axes and can move the headstock vertically in addition to feeding the spindle

Three basic types

The three major types of horizontal boring machines are distinguished by the way they perform milling operations and by the way they locate for boring. That is, they are distinguished by the way in which relative motion (other than spindle feed) is established between the work and the spindle.

Table-type machines can feed the workpiece in two horizontal axes because the table is mounted on a saddle which in turn is mounted on ways. The column is fixed, but the head can be fed vertically on the column, and the spindle, of course, feeds horizontally.

Because both the spindle and the table can move parallel to the axis of the spindle, the table can be moved in close to extend the "reach" of the spindle, or the work can be fed toward the tool in order to maintain a constant amount of spindle overhang.

Planer-type machines have the same characteristics, except that there is no saddle, and the table feeds only longitudinally (perpendicular to the spindle axis). However, with this feed motion and power feed for the spindle and the head, all three major axes are covered, and some planer-type machines retain the other feed possibility by having the column mounted on ways at right angles to the table ways.

Floor-type machines have a power-fed head and spindle like the others, but the work is mounted on a stationary work surface called a floor plate. For longitudinal motion, the column is mounted on a runway so the spindle can move past the work.

This type is intended for large or exceptionally heavy parts that would be difficult to machine accurately on a moving table. On some floor-type machines, the column can also move *across* the runway to increase the reach of the spindle or to reduce overhang.

In general, horizontal boring machines tend to be rather large. The spindles on standard units range from about 3 inches up to 10 inches in diameter, and this dimension identifies the size of the machine. Power ratings range from 15 to 75 hp.

Large working space

The size of the working area is also generous. On a small machine, the table might measure 2½ by 3 ft, and on a medium-sized machine, it might be 5 by 12 ft. A really large floor-type machine can have a 70-ft runway and still be a standard machine.

In most horizontal boring machines, the relative location of the spindle and the work is established by power traverse, and in some cases the positioning is done automatically, with gage stops. Numerical control is also available for most HBMs and is especially useful for complex workpieces that require accurate location of a number of different holes.

Considering boring alone, and

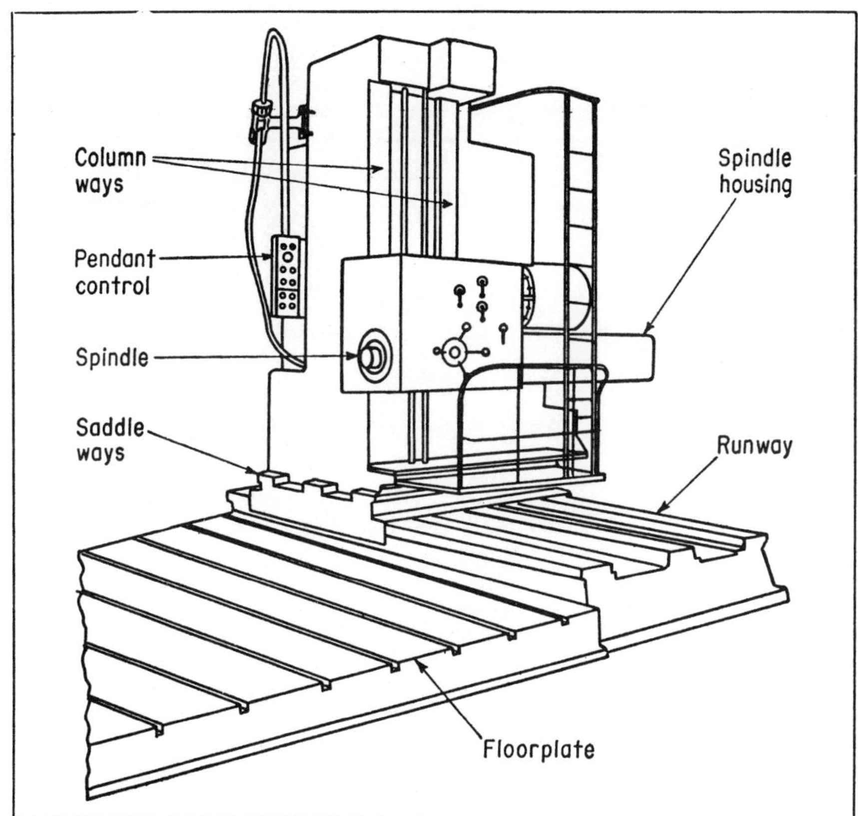

Floor-type horizontal boring machine operates with fixed workpiece. The column moves along runway and can also move across it on some machines

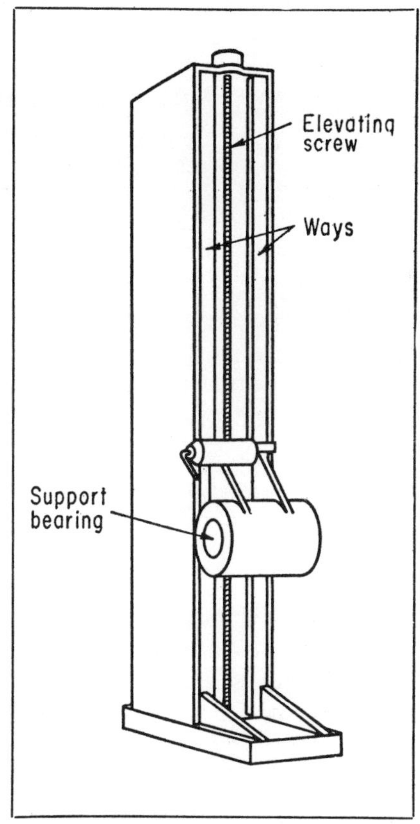

End support is aligned with spindle and supports outboard end of line bar

excluding milling and the other operations an HBM can perform, there are two general classifications, line boring and stub boring. Boring heads are closely related to stub-boring bars, but we will discuss them later in connection with the "additional" operations.

Line boring vs stub boring

Line boring is done with a bar that extends through the workpiece and is supported at one or more points in addition to its connection with the spindle. Often, it has a piloted end that is inserted in the machine's outboard support.

The main advantage of line boring is that it eliminates, or greatly reduces the effect of overhang. Therefore, when the ratio of bar length to bar diameter is high, or when boring must be done with unbalanced cutters, line boring can hold closer tolerances on size, alignment, and finish than stub boring can.

Stub boring is done with a bar that is supported only at the spindle. For most jobs, however, this gives adequate support, especially if the work is positioned close to the spindle. Furthermore, when there is no through hole in the workpiece, stub boring is the only choice, without special fixturing.

However, even though there are certain jobs that one bar can do better than the other, the similarities between them are more important than the differences. For most boring, the two types are approximately equal, and in general, they both use the same kinds of cutting tool.

One of the basic cutting tools for horizontal boring is the flycutter, which is discussed in the Tooling section of the Special Report. Flycutters can be mounted in either line bars or stub bars.

Another basic cutter is the pencil-type boring tool. This is mounted in the end of an adjustable head and represents a form of stub boring. As pointed out in the Tooling section, both types of cutter are available with micrometer adjustment devices.

In addition, a number of other tools have been developed to improve the efficiency of horizontal boring. One of the more popular of these is the block-type tool, which is inserted in a slot in a boring bar and locked in position with a pin of one kind or another.

The big advantage is that a block tool can be removed from the bar and replaced later to bore the same diameter hole. In the meantime, other block tools can be installed in the bar for other purposes. Thus the same boring bar can do a series of operations, and the cutters do not have to be adjusted for size each time they are installed in the bar.

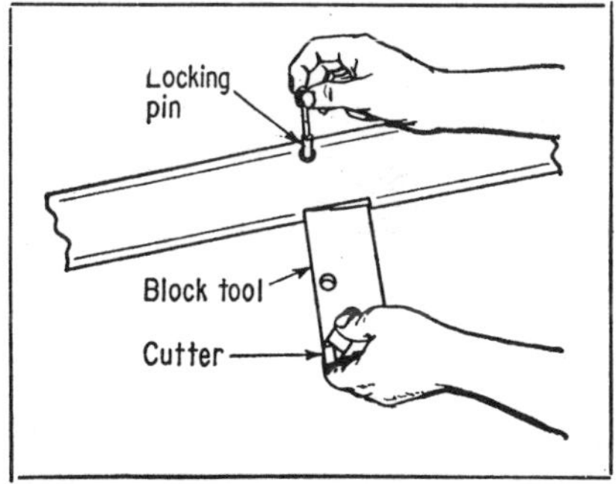

Block tool is set or ground to size and then locked in slot in bar

These block tools are available in a number of different styles for use with either throwaway carbide inserts or HSS cutting tools. Most of them can be adjusted radially, to allow for tool wear and grinding or to suit any size within the adjustment range.

Some are designed to carry one cutting tool; others carry two, one at each end of the block. The single-cutter blocks are usually easier to change from one bore size to another and therefore offer flexibility for short runs.

On the other hand, two-cutter blocks have the advantage in production runs and in making heavy cuts. In addition to the fact that metal is being removed by two cutters simultaneously, the cutting forces are balanced, and a more accurate cut can be made.

For a complete analysis of block-type boring tools and other presettable boring tools, see Special Report No. 524, "The Revolution in Tooling, Part III" (**AM**–June 25 '62, p73).

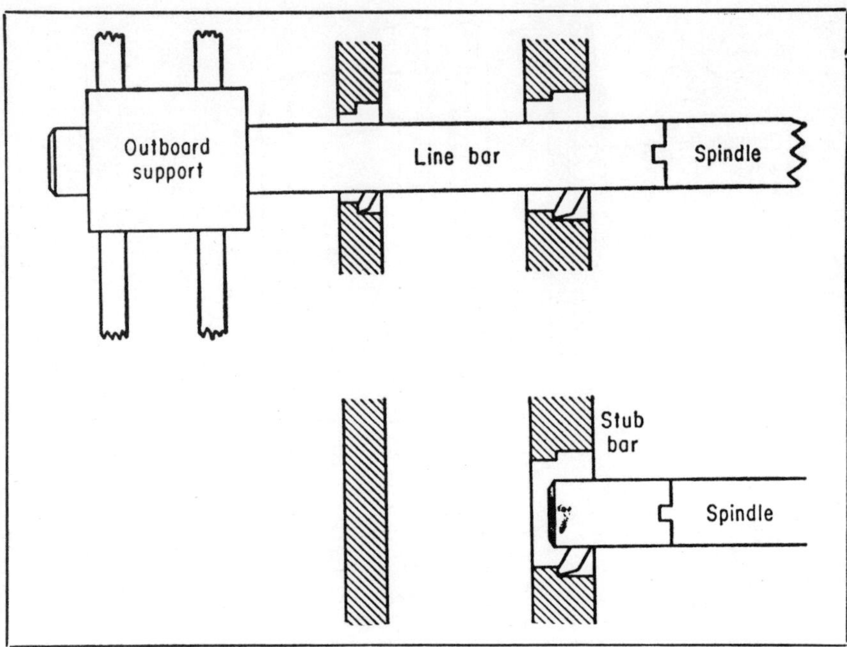

Line boring bar extends through workpiece and is usually supported at outer end; stub bar must be used when there is no through hole in workpiece

Combined operations

Cluster tools employ any or all types of boring tools, mounted as a single unit, to perform a number of operations simultaneously. All the cutting edges, of course, must be set to a common reference in order to maintain accurate size relationships. When high production quantities can justify the additional tool cost and setup time, cluster tools can save a great deal of operating time.

The tools discussed up to this point might be called the "straight boring tools." Even though some of them have fairly sophisticated adjustment devices, the tools are fed straight into the work once they are set for the correct diameter.

Operations that can be performed this way are boring, counterboring, chamfering, plunge facing, and plunge backfacing. Most of these tools are limited to a working diameter of 12 to 14 in.

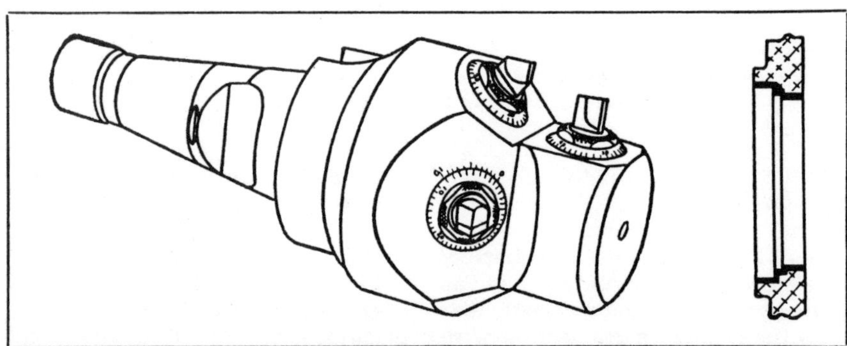

Cluster tool can perform a number of boring, facing, chamfering operations in a single pass but requires careful positioning of every cutting edge

Tooling for additional jobs

But horizontal boring has a much greater scope than this would indicate—both in size and versatility. For larger diameters, several types of boring rings and boring heads and extensions are available, with standard diameters ranging up as high as 60 inches. These can even be incorporated into a cluster tool.

Equally important, though, is the fact that some of these boring heads have a gear or cam arrangement that can feed the cutting tool radially to a preset stop. Or the feed mechanism can be disengaged and the cross slide locked in place so the unit can be fed straight into the work.

With these heads, the horizontal boring machine is capable of a number of operations that would only be possible on a machine that rotates the workpiece, such as a vertical boring mill or a lathe. These operations are grooving, turning, and feed facing (feeding across a face or backface as distinguished from plunge facing).

Other special tools and refinements permit taper boring, spherical boring, and even tracer-controlled contour boring. Therefore, when we add the fact that horizontal boring machines duplicate the capabilities of a horizontal milling machine, and when we consider that the available accessories include right-angle heads, swivel heads, and rotary tables, it becomes apparent that the HBM is a very versatile machine tool.

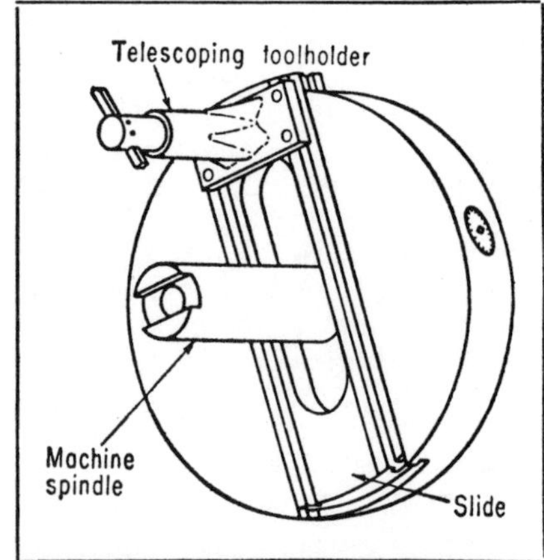

Cross-feed head extends scope of horizontal boring machine because it can feed tool radially for grooving and facing. In addition, with the telescoping toolholder, the tool can be positioned for long turning cuts and back-facing as well

Vertical boring mills

The characteristic features of the vertical boring mill are:

- a horizontal table or faceplate that rotates the workpiece about a vertical axis.
- an overhead ram, mounted on a crossrail, that can feed a non-rotating tool either vertically or horizontally into the work.

Additional features found on many vertical boring mills are:

- a swivel-mounted ram that can feed the tool diagonally.
- a side head that can feed either horizontally or vertically.

In operation, the basic vertical boring mill works much like a lathe turned up on end. In addition to boring, it can perform all the typical lathe operations including turning, facing, chamfering, and internal and external grooving.

In fact, these machines are sometimes called **vertical boring and turning machines**, and when the ram is equipped with an indexing turret, the name **vertical turret lathe** is frequently used, especially for the smaller units.

(At the end of this section we will attempt to define the difference between a VTL and a VBM. For the present, however, we will consider the VTL as a *type* of vertical boring mill.)

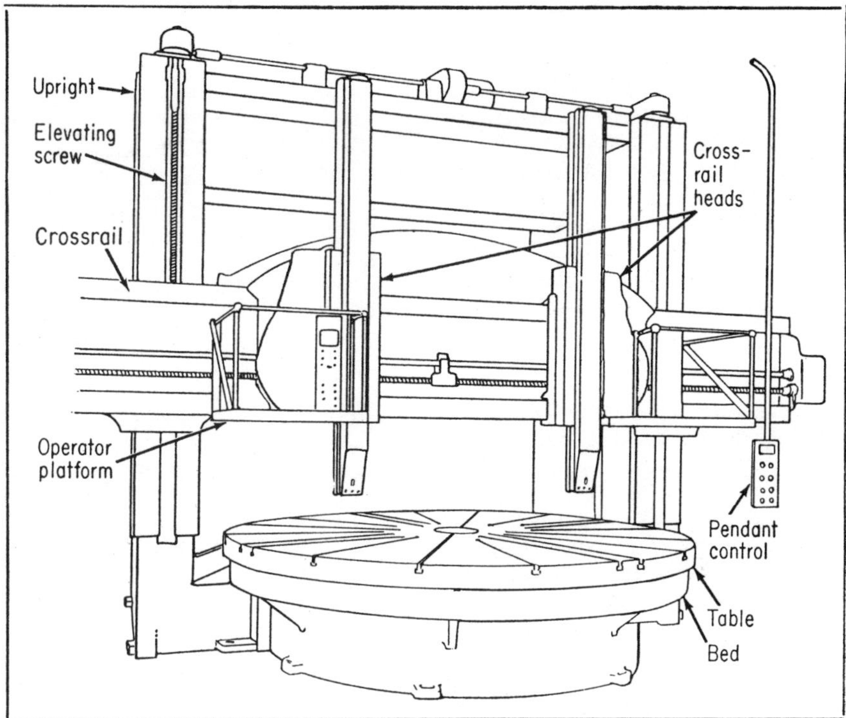

Vertical boring mill can rotate very large workpieces for boring and a variety of other operations. The heads swivel and have independent control

For large, heavy parts

Compared with a lathe, the main advantage of the vertical boring mill is that heavy or awkward workpieces are easier to set up on a horizontal faceplate than on a vertical faceplate. There's also an advantage in the fact that the VBM can machine much larger workpieces than the lathe can.

In comparing the vertical boring mill with the horizontal boring machine, we have a comparison of opposites, and not just because one is vertical and the other is horizontal. The real difference, the one that affects the kinds of work done by each machine, is that one rotates the workpiece and the other rotates the tool.

As far as the basic operation of through-boring is concerned, the two techniques are approximately equal. But there is a great differ-
However, there is a great difference in the additional operations that are often performed along with boring.

Milling is a good example. This is an important extra job for the horizontal machine and an easy one to perform because the cutter rotates. The VBM can't do milling at all, though it can approximate face milling with a straight cut across the top of the work.

On the other hand, facing, turning, and grooving are "naturals" for the VBM because the work rotates, and the tool can be fed almost at will. The horizontal can do these jobs only with special heads or accessories.

The "ideal" workpiece

All things considered, the "ideal" workpiece for a vertical boring mill would be cylindrical in shape, with a number of concentric diameters (grooves, steps, etc) both inside and outside, but it would not have any flat surfaces except for the top and bottom.

Size and weight are also important sometimes, but a part doesn't have to be large and heavy to be suitable for a vertical boring mill. Machine sizes are designated by table diameter, and these range from 2 ft to 24 ft in standard machines. Horsepower ratings range from 30 hp to over 100 hp.

As noted earlier, most vertical boring mills have certain refinements in addition to the rotating table and the overhead ram. These accessories, some of them optional and some of them standard, can contribute a great deal to the machine's versatility and productivity

The swivel ram, for example, adds taper boring and taper turning to the basic operations, and it's not unusual to see two such leads, each under separate control.

The indexing tool turret reduces non-cutting time because the tools needed for a job can be mounted in sequence in the turret. Some machines have one turret head and one swivel ram and sometimes the turret head itself can swivel.

The side head is a standard option on most VBMs, and some machines can even have two of them. A side head operates in very much the same manner as the main head and may also be equipped with a tool turret, but its position at the side gives it easier access to the outside areas of some workpieces.

For machining additional shapes that are beyond the normal capabilities of the VBM, several accessories are available. Among them is a variety of tracer attachments for contour machining. Of course, contours can also be machined under numerical control, which is available for most VBMs.

In addition, geared taper attachments are sometimes needed because most swivel rams are limited to 30° or 45° either side of the vertical position. Thread cutting and drum scoring are also possible with special gearing.

Combined operations

A significant factor in the productivity of vertical boring mills is that two or more heads can be operated simultaneously. Therefore the machines have a good deal of metal-removal capacity—more than an unaided operator can usually manage efficiently.

Simultaneous cuts *can* be made manually, provided the cuts are fairly long, but they are seldom attempted in production without full automatic control of the machine cycle. This can be achieved with mechanical cycle control, punched-tape control, or magnetic-tape control.

The cutting tools and toolholders employed in vertical boring depend to a large degree on whether the operation is being performed manually or automatically.

For manual operations, the tools are usually conventional turning and boring tools clamped in toolposts and boring bars. Disposable inserts are popular, but there is seldom a need for micrometer-adjustable tools. It is easier to set or adjust a cut by moving the head along the crossrail than by changing the position of the cutter in its mounting.

For automatically controlled operations, however, preset and quick-change tooling is almost essential because each cutting edge is part of the complete machining system. When the tools are preset to size, and when they can be changed quickly, downtime is held to a minimum.

For more on preset and quick-change tooling for vertical boring mills, see Special Report No. 550, "The revolution in tooling: Part 5" (**AM**—Apr 13 '64, p165).

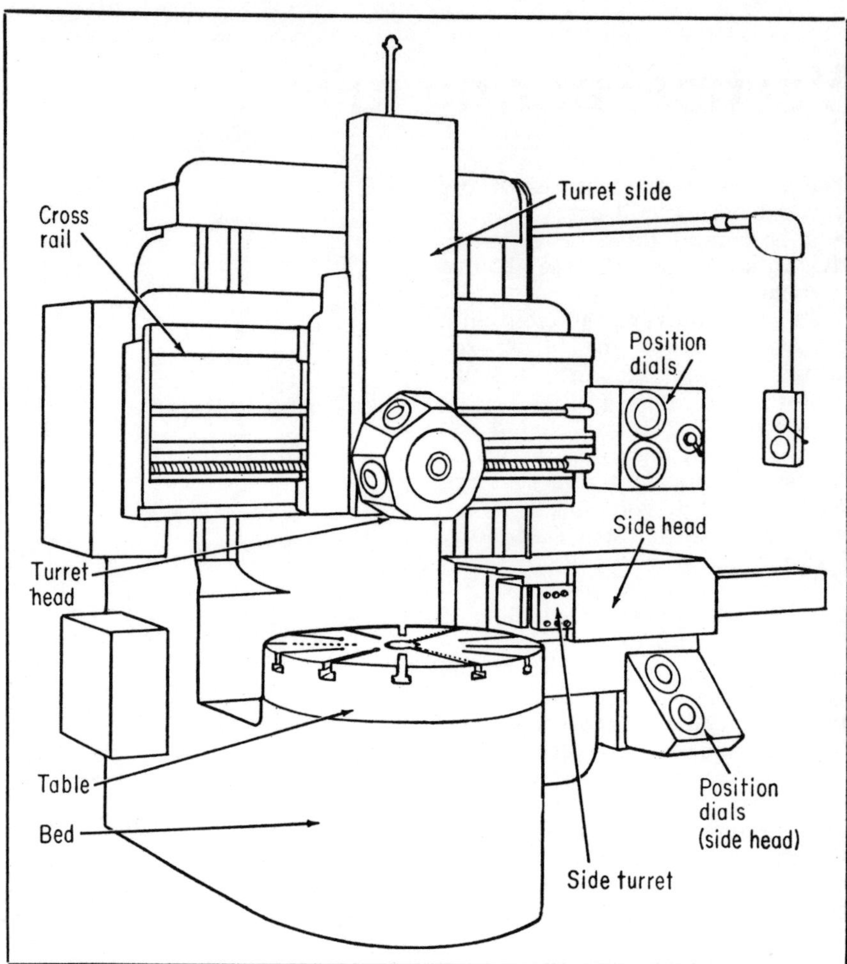

Small vertical boring mill is often called vertical turret lathe, especially when assigned to automatically controlled high-volume operations

VBM or VTL?

Everything that has been said about vertical boring mills in this section could also be said about vertical turret lathes. What is the difference, then, between a VBM and a VTL?

Actually, the difference goes beyond tooling, accessories, and controls; it's more a question of the kind of work performed, especially the volume of work.

The VTL is a machine intended for high production rates. It usually works on jobs that justify a good deal of planning to make sure that non-cutting time is held to a minimum and that simultaneous cuts are made whenever possible. Therefore, most VTLs have automatic control and at least two turreted heads, main and side.

On the other hand, a machine can have this same lineup of equipment and still be considered a vertical boring mill rather than a VTL. This is particularly true of the larger machines because they are more likely to do short-run jobs.

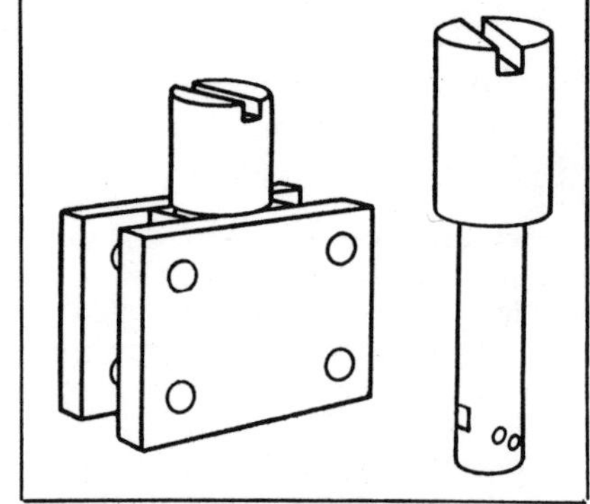

Toolpost and boring bar are typical

By the same token, the smaller machines are more likely to be equipped and *used* as VTLs because the workpieces within their range tend to be produced in higher quantities. As to the machines that are neither large nor small, we can only say that the name of the machine depends on how it is operated.

Jig borers

The characteristic features of jig borers are:

■ a precision spindle that rotates the tool, with power feed to advance it into the workpiece for boring.

■ a built-in precision measuring system for positioning the workpiece with respect to the spindle.

■ ACCURACY. The accuracy of positioning and squareness is guaranteed for many of these machines and ranges from 0.0002 to 0.000,05 in. (200 to 50 "millionths").

But there are two major types of jig borer. One type started out as a toolroom machine and has gradually worked its way into production operations that involve a high degree of precision. The other type was intended, right from the start, for the precision production operations as well as the toolroom type of work.

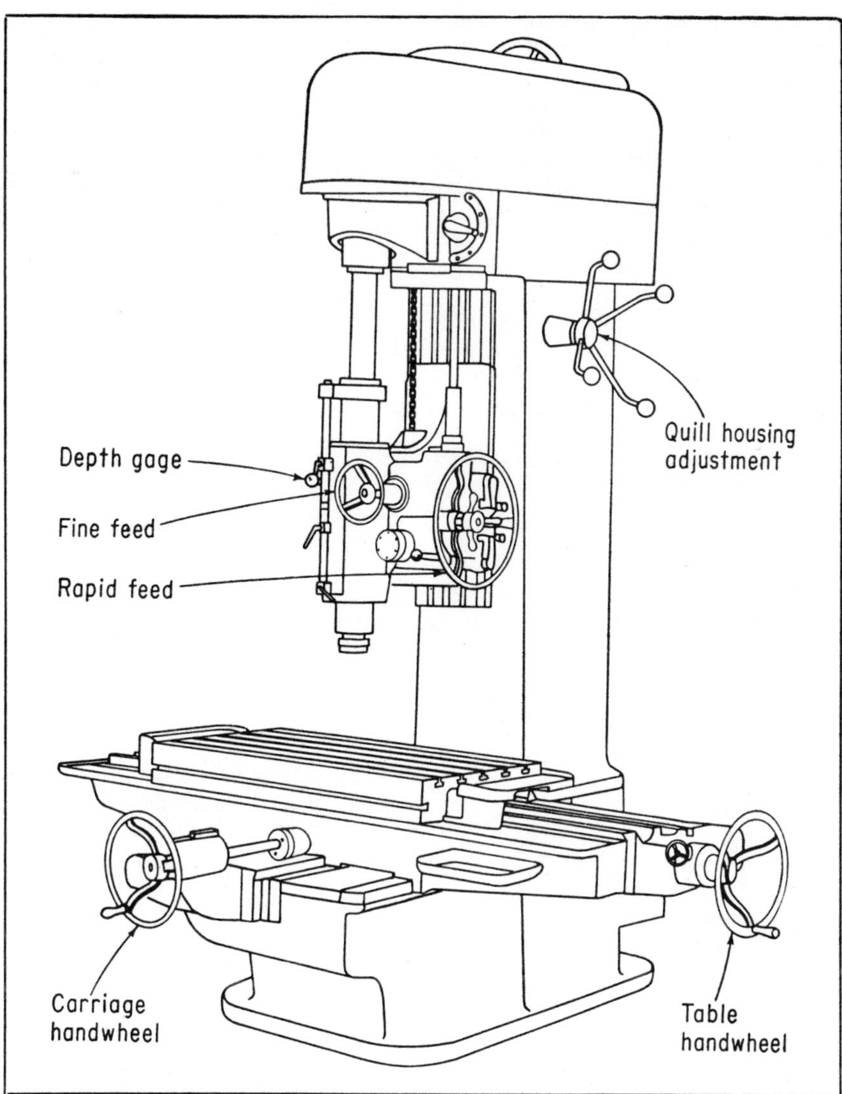

Traditional jig borer has vertical spindle and a horizontal table that moves workpiece in two axes for accurate hole location and boring

Traditional jig borer

The traditional toolroom jig borer usually has a vertical spindle and a horizontal table. On most of them, the table is moved longitudinally and transversely to position the workpiece. One important exception to this, however, is a machine whose spindle head is mounted on a crossrail between two vertical columns. The table moves longitudinally, and the spindle head moves transversely.

The primary purpose of the traditional jig borers is to locate and bore holes to very high standards of accuracy. They can also perform a certain amount of milling when necessary, but the typical milling jobs would be light cleanup cuts or end-milling the details in jigs or fixtures or other tools.

They are not intended for heavy milling operations. This is indicated by their low power ratings (spindle motors range from 1 to 4 hp) and by the fact that they are rarely equipped with power feed for the table, though some have rapid traverse for positioning.

Jig boring and milling

The other type of jig borer is definitely intended for heavy and light milling, as well as boring, so we will identify it with the name "jig boring and milling machine." Machines of this type always have a selection of table feeds, and their motors range from 5 to 20 hp.

Some jig boring and milling machines have a vertical spindle and the same general construction as the traditional jig borers. Others have a horizontal spindle and are built along the same lines as a horizontal boring machine, except that they are built to higher standards of accuracy.

As might be expected, there is a difference in the range of table sizes for the two types of machine. The traditional jig borers range from about ½ ft by 1½ ft to 3 ft by 6 ft. The jig boring and milling

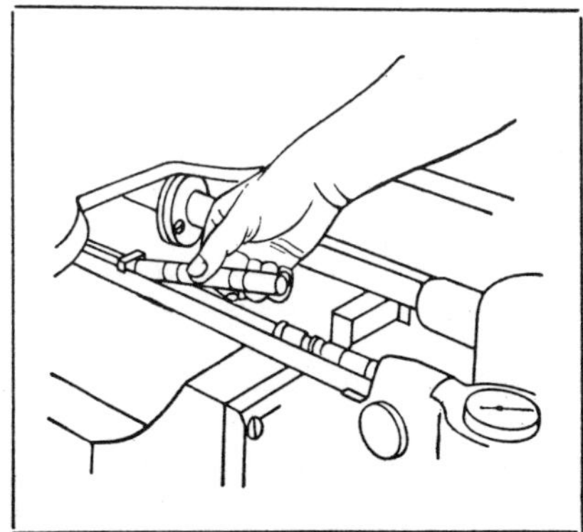

End measure system for jig borer uses gage rods, inside vernier micrometer and dial indicator to locate table and carriage for precision boring

machines range from 2 ft by 3 ft to 4 ft by 12 ft.

Another point of difference among jig borers, of both types, is in the measuring systems they employ. There are at least three major systems, and there are variations within each system.

The leadscrew measuring system uses either precision leadscrews or leadscrews whose errors are compensated for by automatic correctors mounted in the machine. Direct readings of table position are made to 0.0001 or 0.000,05 in. on a dial and vernier.

The end measure system uses gage rods (end measures) to establish the table's position with respect to the reference line in each coordinate. Both manual and automatic end measure systems operate to 0.0001 in.

Optical measuring systems have a standard scale attached to each moving member and an optical system that magnifies and projects the lines from this scale onto a screen at the operator's station. When the operator aligns the index on the screen with the projected image of the line, he can read the position of the table, either on a micrometer head or a dial, to 0.0001 or 0.000,05 in.

In addition, numerical control is available for many of these machines, and it's not unusual for the numerical control system to be given complete responsibility for positioning.

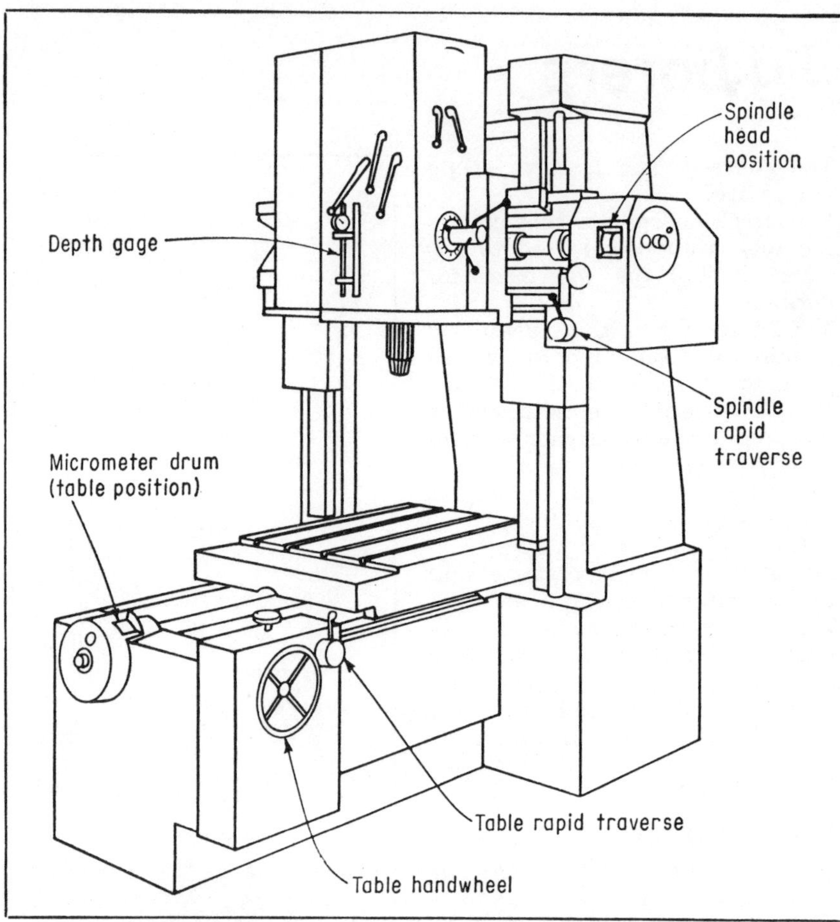

A variation on traditional jig borer moves spindle head on crossrail. Horizontal-spindle jig borer looks like table-type horizontal boring machine

Accuracy is the goal

Regardless of these differences, however, all jig borers have the same goal of high accuracy in locating and boring, and performing other machining as well. In essence, a jig borer is a precision measuring machine that can also perform machining operations.

When a jig borer, of either type, is assigned to a repetitive production operation, the main thing it can offer is accuracy. In other respects, the operation is performed much as it would be on another machine.

In the toolroom, though, or in the model shop, a fully equipped jig borer offers a great deal of versatility in addition to accuracy. In general, it is operated approximately as follows:

After a workpiece is placed on the table and aligned with the coordinate measuring system, a specific reference location on the piece, such as the center of a hole, is "picked up" with the spindle. From this location, other locations can be established by precisely measured movements along the coordinates. With a rotary table and/or a sine plate, angular displacements can also be made.

Operating in this manner, the jig borer practically eliminates the tedious surface-plate layouts and the locating templets or fixtures that would be needed for other machines. And because it is at least as accurate as most shop inspection equipment, it is often trusted to inspect its own work after the machining is completed.

Tools for locating

Among the tools used in jig boring, one of the most important is the dial indicator and feeler that are mounted in the spindle for aligning the workpiece and picking up locations. Additional tools that help in picking up locations are edge finders, line finders, and proving bars.

For centering the spindle over reference points that are difficult to locate by other means, a locating microscope with a tapered shank is often employed.

The basic cutting tools and toolholders for boring are described in the tooling section of this Report. The same descriptions apply to jig boring, except that the equipment used in jig boring is usually capable of a higher degree of precision than normal.

In addition to cutting tools, a jig borer usually needs a variety of accessories to help in setting up and manipulating a workpiece for precision machining without fixtures. Basic setup accessories include precision scraped parallels, angle plates, and vises.

Rotary tables are useful for spacing holes in a circle and for aligning small workpieces, and some jig borers have a built-in rotary table. Tilting tables and sine plates position the work for oblique holes, and in conjunction with a rotating table, they can establish compound angles.

Precision production boring machines

The characteristic features of precision production boring machines are:

- precision: They are designed and built to hold tolerances of less than 0.001 in. with good surface finish.
- production: They are usually intended for high rates of production and for semiautomatic or automatic operation.

However, it is impossible to classify them by any single type of construction or by any single mode of operation. The rotating member may be the tool, or it may be the work; and the tool (whether rotating or non-rotating) may be fed into the work or the work into the tool. And the spindle axis can be horizontal or vertical or any other angle.

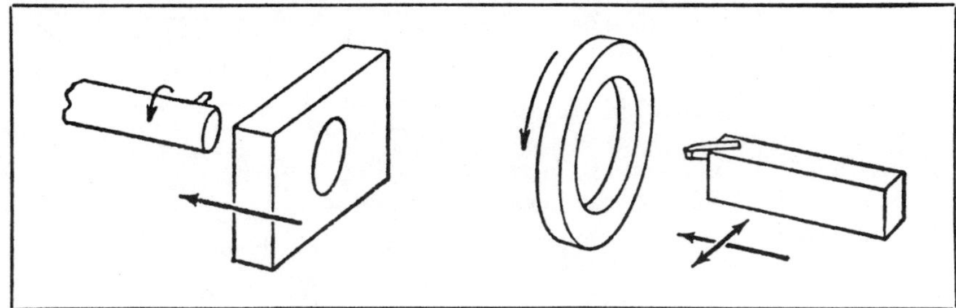

Normal arrangement in standard precision production boring machines is for the non-rotating member to be fed into the rotating member

Some machines are standard

Fortunately, the situation is not quite this complex in practice. On most standard precision production boring machines, the rotating member is fixed, and the non-rotating member is moved for feeding and traversing.

An exception to this rule is that, on the large transfer machines and way-type machines, the normal arrangement is just the reverse. There, the rotating member (the tool) is usually advanced while the non-rotating member (the workholder) remains stationary. However, the following discussion will be largely devoted to standard precision production boring machines and will start out with the horizontal types.

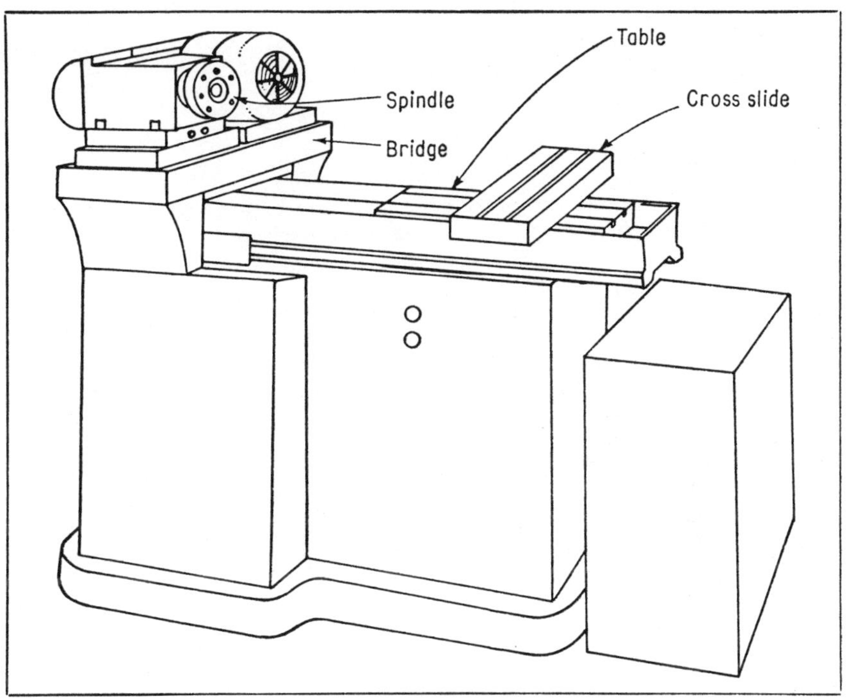

Simplest precision production boring machine has single spindle at one end only. The table feeds longitudinally, and cross slide is an optional extra

Most such machines are designed so they can be operated either with rotating tools or with a rotating workpiece. Therefore, in workplanning, it is not a question of which type of machine is needed for a particular job, but rather, which mode of operation is better suited to it.

In the normal arrangement of these machines, the spindle is mounted in a fixed position at one end, usually the left end. A table, mounted on ways, can be fed or traversed toward the spindle, and a cross slide can be mounted on the table for feed motions or lateral indexing at right angles to the spindle axis.

As to the general run of sizes, a small unit will have a mounting surface about one ft square and a table travel of about ½ ft. Larger units may have a table pad as large as 12 sq ft and a travel of several feet.

Usually, the table is driven by hydraulic power or by a pneumatic-hydraulic combination, and stepless adjustment of feed rates is normal. On many machines, a variety of automatic cycles can be set up to suit the job requirements and can include such elements as rapid traverse, feed, cross feed, dwell, and return. Cam-controlled machines have special features and are described on the next page.

A double-end machine is practically the same as a single-end machine, except that spindles are mounted at both ends, and the table moves between them. In both types, single-end and double-end, multiple spindles are often employed to increase production.

Rotate the tool or the work?

The basic distinction in all boring is whether the tool rotates or the work rotates. Other types of boring machine are stuck with one mode of operation or the other, but in precision production boring, there is a choice.

For any job, this choice is influenced by the particular operations that have to be performed and also by the size, shape, and balance of the workpiece. For a rough rule of thumb, complex machining is more easily performed by rotating the work, but if the workpiece is awkward or unbalanced, it's easier to rotate the tool.

If the job is a simple one involving only straight boring and possibly facing and chamfering, and if the workpiece can be rotated easily, the two modes of operation are very nearly equal.

Complex part; rotate tool

However, if the workpiece has an irregular shape, the job would probably be done with a rotating cutter. If the workpiece has more than one hole to be bored, there's no doubt about it. Multiple-spindle heads can be set up to bore a great many parallel holes in one pass, whereas the rotating-work type of operation would require a different setup for each hole.

This, by the way, is the kind of job that would be done on a horizontal boring machine if production quantities were not high enough to justify the tooling expense for production boring.

Complex cuts; rotate part

On the other hand, if the job requires grooving, tapering, or contouring, there's an advantage in rotating the workpiece, assuming it's the right size and shape to be rotated. Rotating tools *can* perform these operations, but they have to be mounted on special heads with internal gearing or cams to move the tool radially.

In contrast, a non-rotating tool can be moved about a rotating workpiece almost at will. Grooving involves simply an in-feed and a cross-feed to the correct depth, and for tapering, the axis of the work is set at an angle to the table travel. Contouring can be done by direct mechanical tracing, or a cam-controlled machine can be used for this purpose.

Cam-controlled machines are the same as the others we have been discussing, except that the motions of the table and the cross slide are each controlled by a cam. The cams are coordinated so that a single-point cutting tool can be moved through any desired path to generate radiuses, tapers, or other shapes in addition to boring.

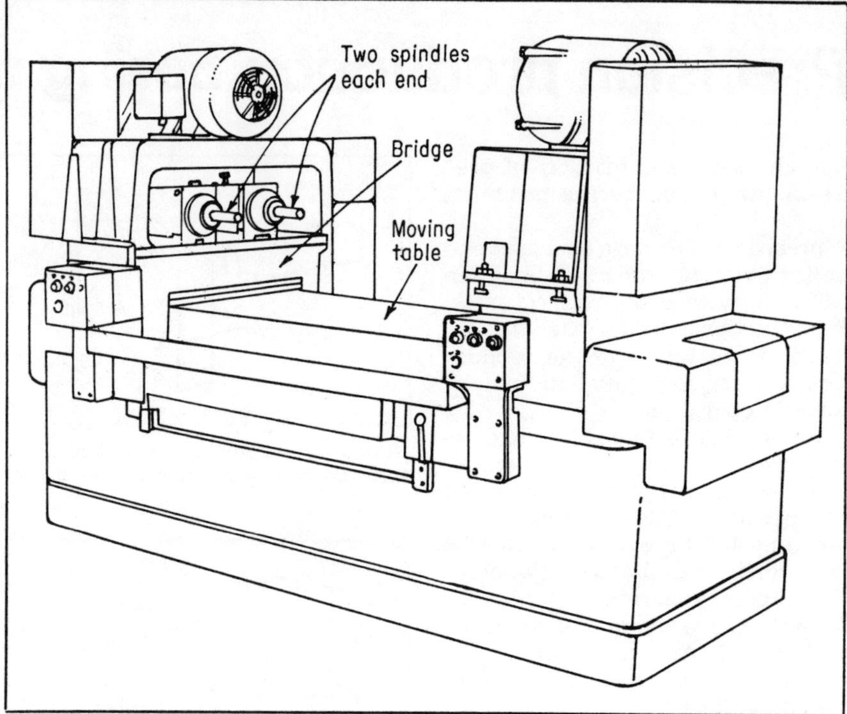

Double-end, two-spindle machine for high production can machine both sides of two workpieces in single setup or can do other combined operations

One advantage is that specially ground form tools are not needed, and another is that fewer tools need to be set for any given part. However, all this is true only if the workpiece can be rotated. If the work has to be fixtured on the table and machined with a rotating cutter, the cam-controlled machine is on a par with the others.

Vertical machines

Although all the above comments have dealt with horizontal operation, most of them would also apply to vertical precision boring. The same options of rotating either the tool or the work are available, and the feed can be hydraulic or mechanical (cam controlled).

Among the advantages claimed for vertical operation are that heavy workpieces are easier to load and unload, that the tools are in a more convenient position for changing and adjusting, and that a vertical machine takes up less floorspace than a horizontal.

These machines are popular for heavy workpieces that can be rotated on a vertical spindle while the tools are fed from above by a carrier. In this application they resemble vertical boring mills, except that accuracy is held to closer limits, and workpiece sizes are usually a good deal smaller than they would be on a VBM.

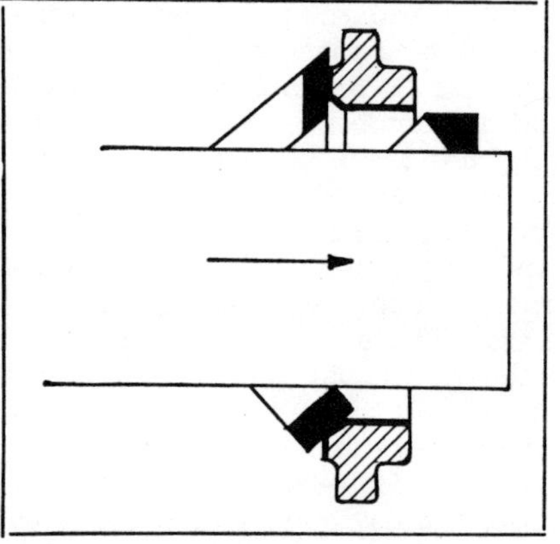
Boring bar with clustered tools cuts three surfaces in a single pass

Automatic production lathes are not included in the precision production boring category, even though there is a strong family resemblance. As a general rule, the automatic lathes are not intended to work at the high accuracy level of precision boring.

Another factor that distinguishes these machines is that automatic lathes normally have two or more toolslides that may feed into the work along different paths. In precision boring machines, all the tools are almost invariably mount-

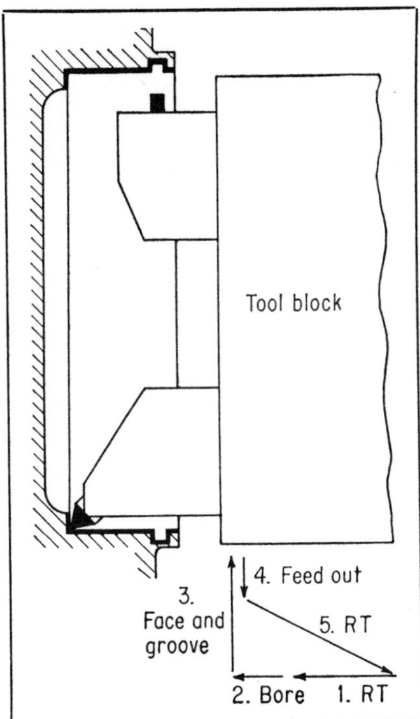

Non-rotating tools feed in and then across for boring, facing, grooving

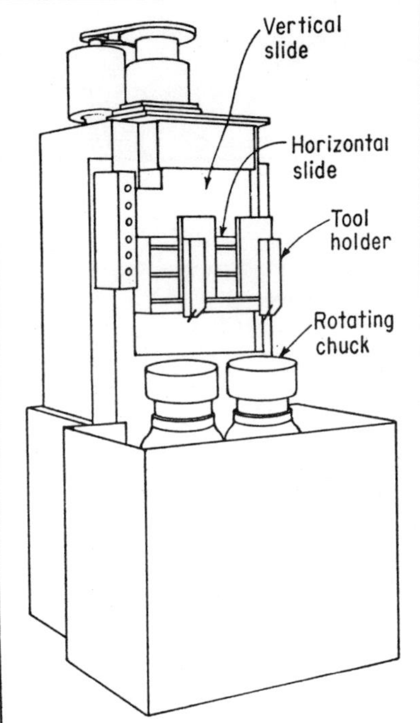

Vertical precision borer is easy to load and unload with heavy parts

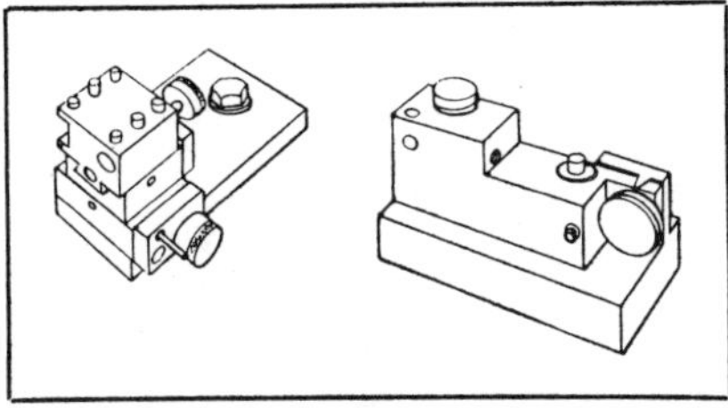

Holders for non-rotating tools have precise adjustment for tool position

ed on a single carrier—the spindle, the table, or the cross slide.

This unit-mounting maintains a fixed relationship between the cutting edges throughout the operation, which helps to hold the desired accuracy on interrelated dimensions. In addition, precision devices for setting and adjusting cutting tools are found more often on the boring machines than on the automatic lathes.

Planning for production

But precision is not the only goal of these machines. High productivity is usually just as important, and a great deal of ingenuity is often exercised in planning the tooling for a job to achieve the maximum possible output.

The objective should always be to complete the job in one pass or, at least, to combine as many cuts in one pass as possible. In addition to reducing operating time, this also holds related dimensions within close limits.

When the cutting tools are rotated, a large number of them can be mounted in a cluster on the same holder to perform a variety of operations such as boring, plunge facing and chamfering in a straight plunge cut. Moreover, a cross-feeding head can be mounted with this cluster for grooving and back-facing.

Frequently, too, the tools for roughing, semi-finishing, and finishing can be mounted on the same holder so as to follow one after the other.

Non-rotating tools can also be clustered, of course, by mounting them in their proper relationship on the table or cross slide. As noted earlier, the cross slide gives this mode of operation an additional flexibility because it adds another feed direction.

It stands to reason that when a number of cutting tools are mounted on the same carrier, there should be some means for making fine adjustments. Otherwise, it would be difficult to establish and maintain the correct relationships between cutting edges.

For this purpose, cartridge-type tools with micrometer adjustment are gaining popularity, both in rotating and non-rotating tools. In addition, non-rotating tools are often mounted in micrometer-adjustable toolposts, which can even be equipped for automatic adjustment to compensate for tool wear.

Another aid to accuracy and productivity is the principle of preset tooling, which is partly based on adjustable tools such as these. A detailed discussion of the tools and how they are preset for precision boring is contained in two Special Reports in the series, "The Revolution in Tooling," Special Report No. 524 (**AM**—June 25 '62, p73) and Special Report No. 555 (**AM**—July 20 '64, p77).

Additional productivity

All the above tooling techniques for high production rates can be put to good use, even in a single-spindle setup. For even greater production, additional spindles may be employed.

Multi-spindle operation has been mentioned earlier in connection with multi-hole workpieces, but for workpieces that have only one hole to be machined, two or more spindles placed side by side can multiply the output by doing the same operation simultaneously. Or they might be set up to do a series of operations in sequence.

Beyond this level, there is the added potential of the double-ended machines. A typical job would be the machining of both the front and back of a hole in a single setup. Of course, these machines can also be operated on a multi-spindle basis, so there's an almost endless number of possible job combinations.

Efficient short runs, too

On the other hand, we should make it clear that high productivity does not necessarily mean continuous operation or even extremely large lot sizes. By applying the principles of preset and quick-change tooling, job setups can be changed fairly quickly, and these machines can work efficiently on medium-sized production quantities as well.

With universal fixtures, they can even do the short-run or single-piece jobs that do not warrant special fixtures and tooling. These jobs, however, are not really typical of the precision production boring machine, which in this case, would be substituting for a horizontal boring mill or a jig borer.

Tools for boring

The least common denominator in boring tools is that they all remove metal by making chips. There are different styles of cutter, different types of holder, and different techniques for making adjustments, but all these serve only to put the cutting edges in the correct location and support them while they make chips.

Another common factor is that every tool has a side cutting edge and an end cutting edge. These terms are oriented to the length and width of the tool shank and are part of the standard nomenclature for single-point cutting tools (ASA B5.36-1957).

To keep the terms straight, it's easiest to think of a tool whose shank is clamped at 90° to the axis of the spindle. It's fairly clear that such a tool would cut with its side cutting edge when boring and with its end cutting edge when facing.

This situation is not altered by the fact that many boring tools are designed to be mounted at other angles; they still present the same cutting edges to the work. In the following discussion we will consider only the requirements for boring, so the side cutting edge will always be the working edge.

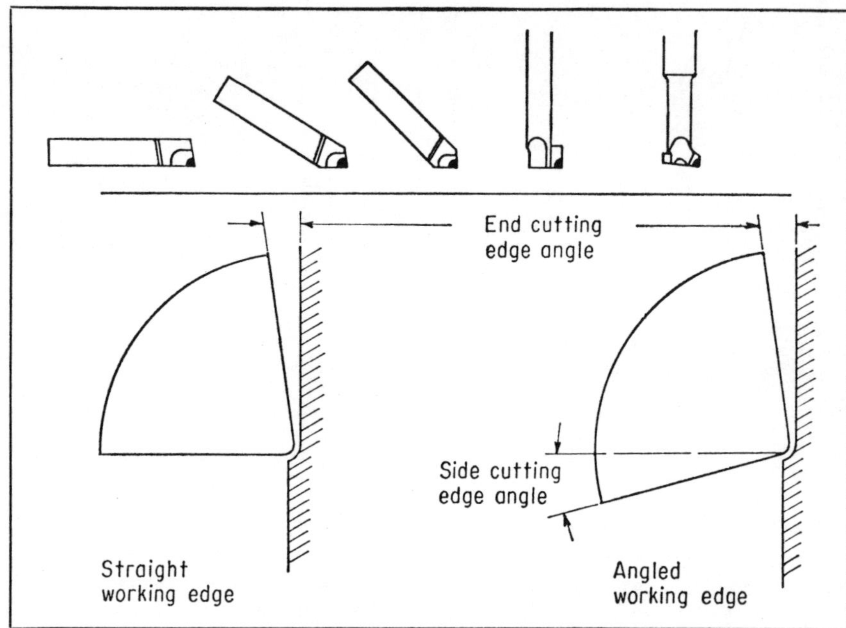

Top view of cutting tip in position for boring is not influenced by shape or type of shank, but different jobs call for different edge angles

Top view for cutting edges

Considering the top view of the tip then, the side cutting edge may have an angle (often called the lead angle) or it may be straight (perpendicular to the axis of the bore).

The angled side cutting edge (10° to 30°) is recommended for heavy cuts because it spreads out the chip load and stabilizes the cutter, but this type of tool can be used only on cuts that go clear through the bore. A straight side cutting edge is required for boring to a shoulder, and this shape also produces a better finish on the bore.

The end cutting edge usually has enough angle to keep it from dragging when the other edge is cutting. On many commercial carbide-tipped tools the angle is 8°, but a larger angle may be needed for soft materials and a smaller angle for hard or tough materials.

Between the two cutting edges there is usually a nose radius, though sometimes this corner has a chamfer instead. Either treatment strengthens the corner and avoids chipping in roughing cuts. For finishing cuts, the size of the radius is a function of the feed rate and the surface finish desired.

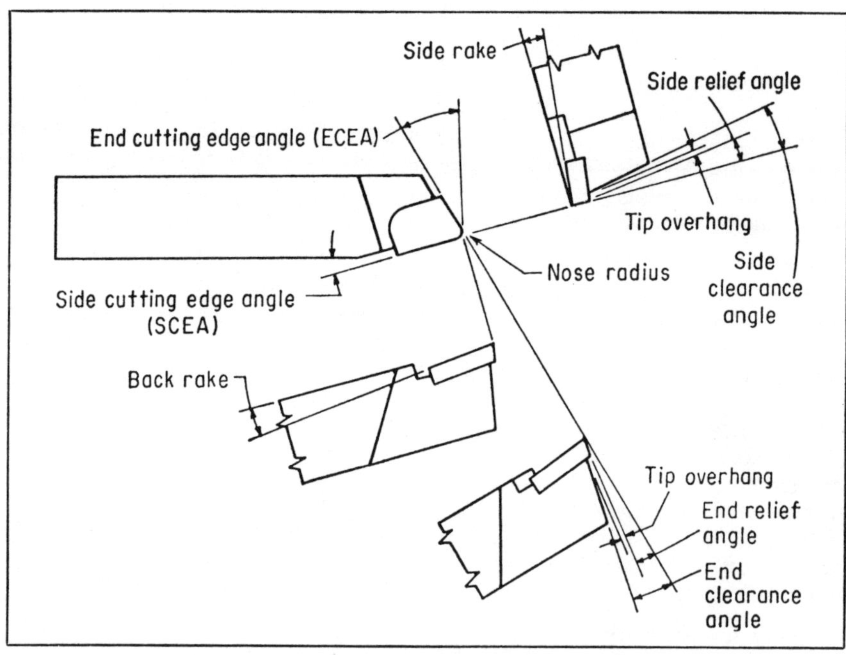

Nomenclature diagram is adapted from ASA B5.36-1957, "Carbide Blanks and Cutting Tools," and shows typical single-point, carbide-tipped tool

Looking now at the end projection in the nomenclature diagram, we have the working profile of the tool when boring. The side relief angle (sometimes called front relief angle) is usually 5° or 7°, with a clearance ange of 10°.

The side rake angle (sometimes called axial rake) depends very much upon the material being machined. A positive side rake of 5° or 6° is suitable for general purposes, but for soft materials like aluminum or copper, a positive rake of 15° to 20° is better. On the other hand, negative rakes of -5° to -15° are required for very hard, tough materials, scaly work, and for interrupted cuts.

The side projection shows the profile of the tool around the end cutting edge. All the angles in this view are greatly influenced by the position of the tool with respect to the radial center of the bore and by the diameter of the bore as well.

Effect of bore size

For example, if the tool is set slightly above center, as it usually is, the surface to be machined approaches the edge from behind, as it were. This means that, if the tool is ground to a neutral back rake, its *effective* back rake will be negative. The amount of change depends on how small the diameter is and on how far the tool is placed above center.

To counteract this effect, general-purpose boring tools are often ground to a positive 12° back rake. In certain applications, this angle may have to be altered, of course, depending on the diameter of the bore and the material being cut. And for some of the tougher materials, the effective back rake should be negative to help break chips.

End relief and end clearance are usually considered together in boring tools, and here again, the bore diameter and position of the tool exert a controlling influence. A clearance of 10° will suit most applications, but this may have to be increased for small diameters to keep the bottom of the shank from rubbing on the work.

Two basic cutting tools

As indicated by the miniature sketches on the opposite page, the tool shank may be any one of various shapes. It is possible, however, to identify two general types of tool, based on the way the shank is mounted: those that are mounted parallel with the spindle axis and those that are mounted at some other angle.

The parallel-mounted rotating tools are usually called pencil-type boring tools. Because of the fact that their shank is parallel with the spindle, these tools are not adjusted for different bore diameters by shifting their position in their mounting. Instead, they are mounted in adjustable heads.

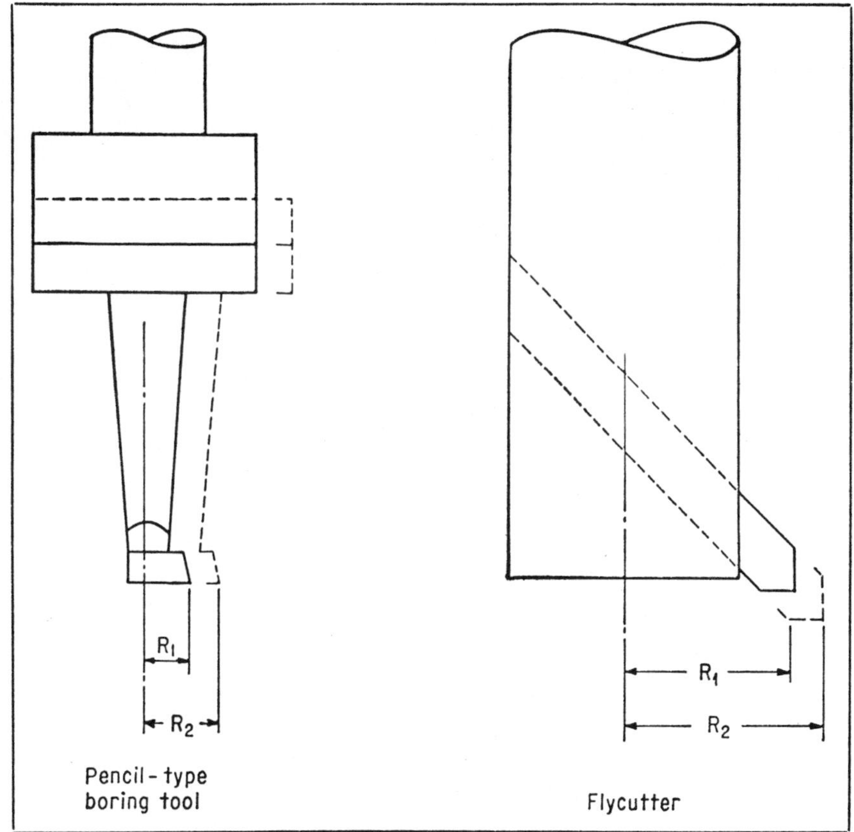

Method of adjustment determines type of tool: parallel-mounted tools need adjustable head; angular-mounted tools are adjusted within their mounting

When holes smaller than about ½ inch have to be bored, the pencil-type tool is usually the easiest way to do the job and sometimes the only way. They are made for bores as small as 0.020 in., but the larger pencil-type tools have a working range up to about a 6-inch diameter.

These tools are made either with a plain cutting tip or with a form-relieved tip. The plain tip is resharpened by grinding the side and end to the desired angles. The form-relived tip is re-sharpened by grinding the top only because the cutting-edge angles and the clearance angles are "built in."

The angular-mounted tools are a much larger group and a much more varied group. In operation, the main thing they have in common is that their cutting diameter can be adjusted by moving the tool within its mounting slot. In many applications, this is not the only type of adjustment employed, but it is always possible.

The simplest form of angular-mounted tool is the flycutter, which is clamped in a slot with setscrews and adjusted by loosening the screws and moving the tool. For more accurate adjustment, especially when this is the only type of adjustment being used, there is usually an adjusting screw at the end of the mounting slot.

Precision flycutters

Furthermore, the adjusting-screw idea has been refined in a number of devices for making precise adjustments in the flycutter's position. These have a dial that permits direct reading of the amount of adjustment, usually in increments of 0.001 or 0.0001 in. The minimum bore diameter is about 1 inch.

A thorough discussion of these devices can be found in Special Report No. 524, "The revolution in tooling, Part 3" (**AM**—June 25 '62, p73). In fact the Report covers just about all the available equipment for accurate setting (and presetting) of boring tools and for adjusting them when necessary. Therefore, we will only describe them briefly here.

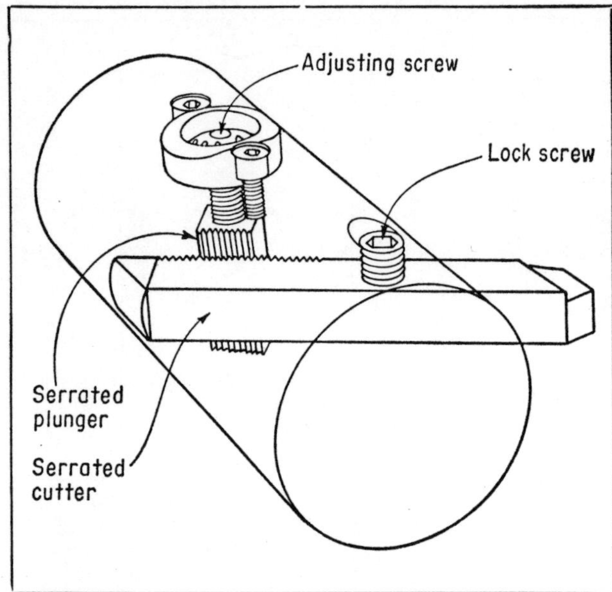

Precision flycutter in boring bar has adjusting screw and dial with graduations in units of 0.0001 in.

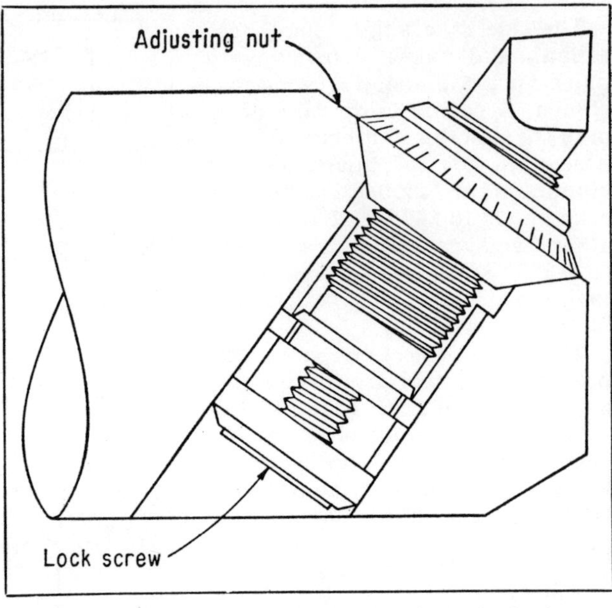

Cartridge-type tool can be preset away from the boring bar and reads to 0.0001 in. with vernier

Cartridge-type flycutters are another in the general family of micrometer-adjustable angular tools. In addition, however, they can often be preset away from the boring bar and can quickly replace a worn cutter that needs resharpening. Minimum bore diameter is ½ inch.

Block-type boring tools are also in the angular-mounted group. They may be screw adjustable, micrometer adjustable, or ground to size, and they have a working range of ¾ inch to 17 inches. These are discussed under horizontal boring machines in this Report and also in Report No. 524.

Types of mounting

All the above cutters can operate either as rotating tools or as non-rotating tools, but for the sake of simplicity, we will assume for the present that they are rotating. That means they must be attached to the spindle in some way.

For this purpose, they are mounted either in rigid boring bars or in adjustable boring heads. As a general rule, the more refined angular-mounted tools are mounted in the rigid bars because the tools have their own adjustment.

Rigid boring bars are either line bars or stub bars, and the distinction between them is explained under horizontal boring machines in this Report. The boring bars used in most other machines are stub bars (supported only at the spindle). Therefore, for high-quality work, the bar should be as large as possible for the hole being bored, and the overhang should be kept to a minimum.

Adjustable boring heads are made in several different types. Some have a slide that can be moved by a micrometer screw; others have an eccentric mounting between the spindle and the tool; and still others have a wedge-type device to adjust the cutter.

An adjustable head is essential, of course, for the pencil-type boring tools because they have no adjustment in themselves. In addition, many of these heads are designed to mount flycutters, in which case the simpler flycutters are satisfactory. Practically all such heads have a micrometer type of reading, graduated in increments of 0.001 in. or less.

Cross-feed heads

For feeding a rotating tool across a face or counterbore or into a groove, a variety of cross-feeding heads are available. These usually have a slide that is moved radially by internal gears, cams, or levers.

Cross-feeding heads for jig borers and horizontal boring machines can usually be clamped in a fixed position for boring or turning operations, and they often have a micrometer-type adjustment. The crossfeed heads for precision production boring machines are hydraulically or mechanically operated in such a way that they can be synchronized with the other parts of the automatic cycle. ■

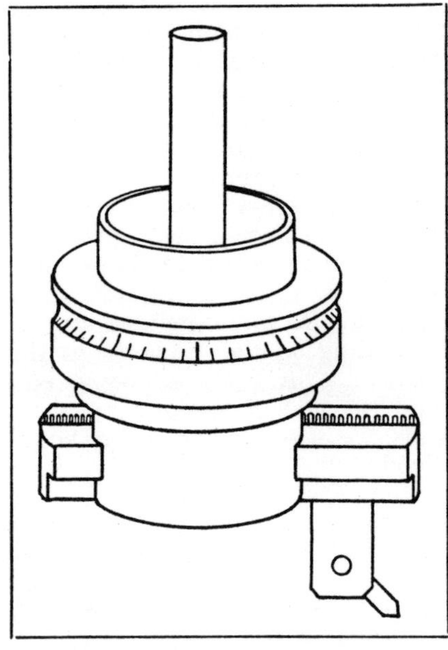

Facing head feeds cutter radially

Fundamentals of Turning, 1

Why are there so many kinds of lathes? This Report, the first of two parts, analyzes the differences and explains the basic operation of each of the major categories

By Richard T Berg, associate editor

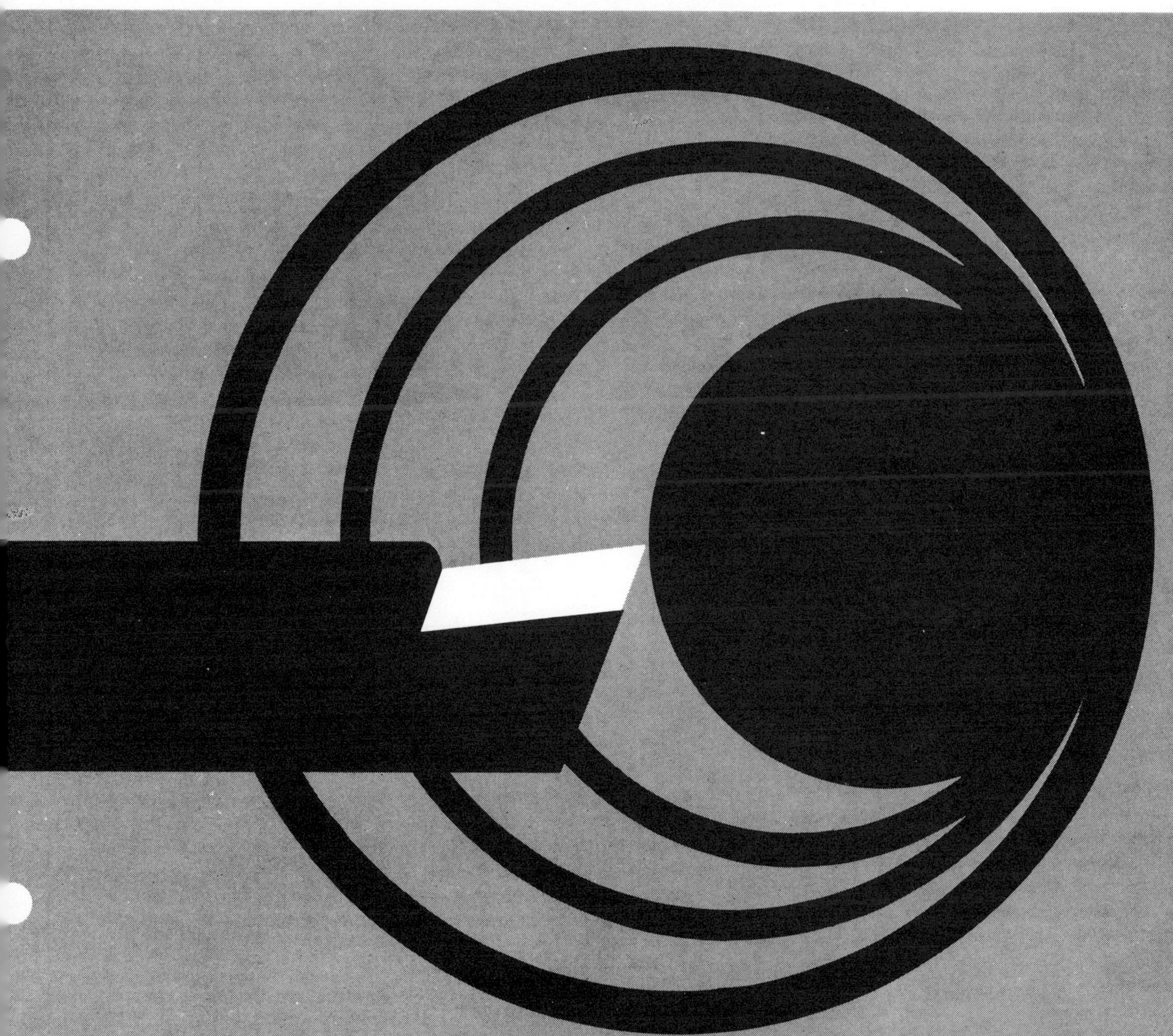

What is turning?

Turning is the machining operation that produces cylindrical parts. In its basic form, it can be defined as the machining of an external surface:

- with the workpiece rotating,
- with a single-point cutting tool, and
- with the cutting tool feeding parallel to the axis of the workpiece and at a distance that will remove the outer surface of the work.

Taper turning is practically the same, except that the cutter path is at an angle to the work axis. Similarly, in contour turning, the distance of the cutter from the work axis is varied to produce the desired shape.

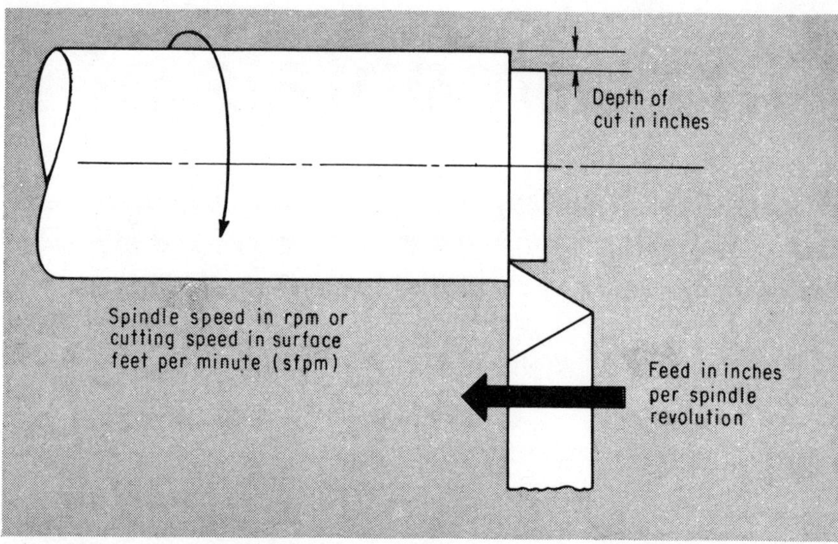

In the basic turning operation, and in most other machining operations, speed, feed, and depth of cut are the major cutting factors controlled at the machine

Even though a single-point tool is specified, this does not exclude multiple-tool setups, which are often employed in turning. In such setups, each tool operates independently as a single-point cutter.

Two exceptions, however, must be made to the basic definition. One is that some types of boring machines perform turning by rotating the cutter about a non-rotating workpiece. The other is that the Swiss-type automatic screw machine feeds the work longitudinally past the cutters. However, these are special cases.

Adjustable cutting factors

The three primary factors in any basic turning operation are speed, feed, and depth of cut. Other factors such as kind of material and type of tool have a large influence, of course, but these three are the ones the operator can change by adjusting the controls, right at the machine.

Speed always refers to the spindle and the workpiece. When it is stated in revolutions per minute (rpm) it tells their **rotating speed.** But the important figure for a particular turning operation is the **surface speed,** or the speed at which the workpiece material is moving past the cutting tool.

This figure is simply the product of the rotating speed times the circumference (in feet) of the workpiece before the cut is started. It is expressed in surface feet per minute (sfpm), and it refers only to the workpiece. Every different diameter on a workpiece will have a different cutting speed, even though the rotating speed remains the same.

Feed always refers to the cutting tool, and it is the rate at which the tool advances along its cutting path. On most power-fed lathes, the feed rate is directly related to the spindle speed and is expressed in inches (of tool advance) per revolution (of the spindle), or ipr. The figure, by the way, is usually much less than an inch and is shown as a decimal amount.

Some machines, however, have a hydraulic feed that is completely independent of spindle rotation. In this case, the rate of feed is expressed in inches per minute (ipm).

Depth of cut is practically self-explanatory. It is the thickness of the layer being removed from the workpiece or the distance from the uncut surface of the work to the cut surface, expressed in inches. It is important to note, though, that the diameter of the workpiece is reduced by two times the depth of cut because this layer is being removed from both sides of the work.

The basic turning machine

The lathe, of course, is the basic turning machine. It can perform a good many other operations, besides, but for the moment, let's consider turning alone, and let's look at it in the light of the basic definition. What are the rock-bottom requirements of a turning machine?

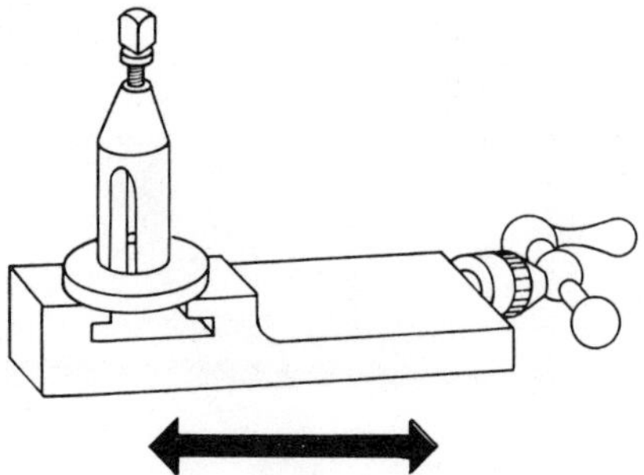

Simple screw-fed slide can carry single-point tool for turning

The first point in the definition concerns the work-handling end of a lathe; it must have some means of holding and rotating the workpiece. This is done with the spindle, which is mounted in the headstock. A workholding device such as a chuck, a collet, or a faceplate is mounted on the nose of the spindle, and the whole assembly is made to rotate by power, usually from an electric motor, applied to the spindle by belts or gears or both.

The second and third points of the definition are concerned with the other major part of a lathe, the tool-handling part. There must be a sturdy mounting for the tool so it can withstand the cutting forces, and there must be a mechanical means for moving it through the desired cutting path – parallel to the work axis for straight turning. In addition, the distance of the cutting edge from the axis of the work must be adjustable so that different diameters can be turned.

Putting the parts together

In the simplest possible turning machine, the tool-handling part would have a toolpost that could clamp the tool at the correct angle and at the same height as the centerline of the work. The toolpost, in turn, would be clamped to a T-slot in a slide that could be moved longitudinally with a hand screw feed.

For the crosswise adjustment needed for different diameters, the toolpost could be loosened and moved as necessary and then reclamped to the T-slot. However, in the interest of accuracy and convenience (and progress) we'll say that the whole previous assembly is mounted on another slide that can be fed crosswise.

With the addition of the cross slide, though, the machine can do more than just turning operations. By cross-feeding it can make facing cuts on the end of the work, and with the appropriate tools, it can be fed into the side of the work for forming, necking, and cutoff operations.

Of course, it can also do similar operations *inside* a workpiece, where the counterparts of turning and necking are boring and recessing, respectively. Furthermore, if we say that the top slide can be swivelled and locked at any desired angle, the machine can also do taper turning and boring.

Does almost everything

Therefore, this very simple turning machine can do just about all the basic operations expected of a lathe. As a matter of fact, there are a good many small but highly respected lathes in operation today that are very little more complex than this one.

But if this is all a lathe needs in order to perform its work, why are there so many different kinds, and why are some of them so much more complex? Why couldn't they all simply be bigger or smaller, according to the kinds of workpiece they are expected to machine?

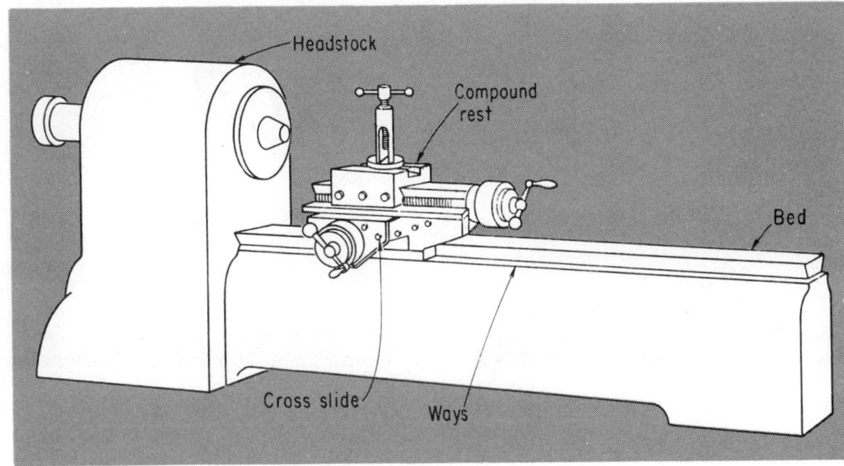

Basic turning machine can rotate the work and feed the tool longitudinally for turning and can perform other operations by feeding transversely

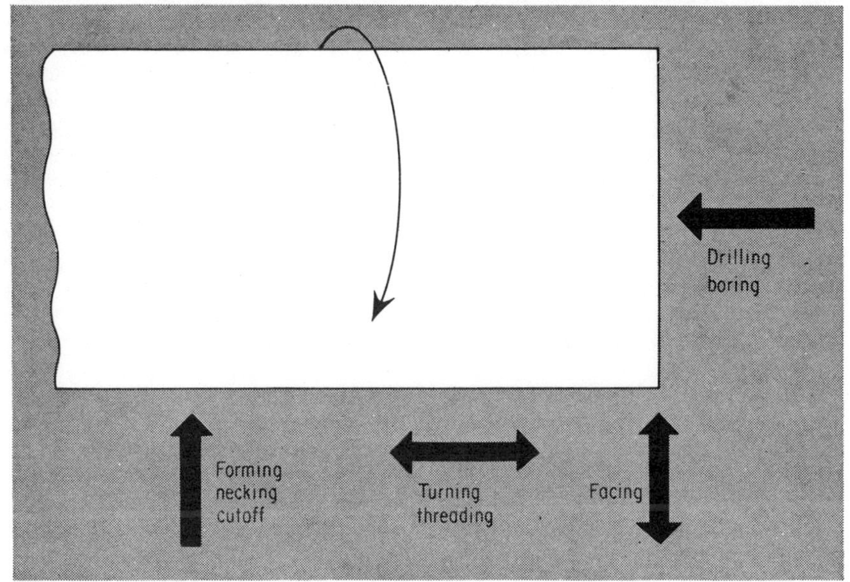

Lathe operations go by different names, depending on what direction the tool is fed and on what portion of the rotating workpiece is being machined

Actually, they could be. Oh, for long workpieces, a lathe would need a tailstock so the work could be supported between centers; and for large, heavy work, it would need power feed because of the heavier cuts that are likely to be taken. These additions will increase the versatility of our basic lathe, but they won't make it a different *kind* of lathe.

In order to increase its efficiency, we can also add a choice of spindle speeds. Then the operator can set the speed for each cut at the rate that will give an optimum tool life (either for lowest cost or for highest output). With one of the more advanced controls, he can even preset it for the next cut.

Along with this we can add a gearbox so that the tool can be fed through the cut at any of a number of different rates. This will permit high rates of feed for roughing cuts and low rates for finishing cuts when a good surface finish is required.

About the only standard component we can still add is the equipment needed for precision threading operations, so let's give it a leadscrew and a chasing dial.

At this point we have a lathe that is thoroughly versatile. It can do any job that a lathe ought to do, and it is easy to set up and change from one job to another. (It comes pretty close to being an engine lathe, which is more fully described in the following section, but for the present we will con-

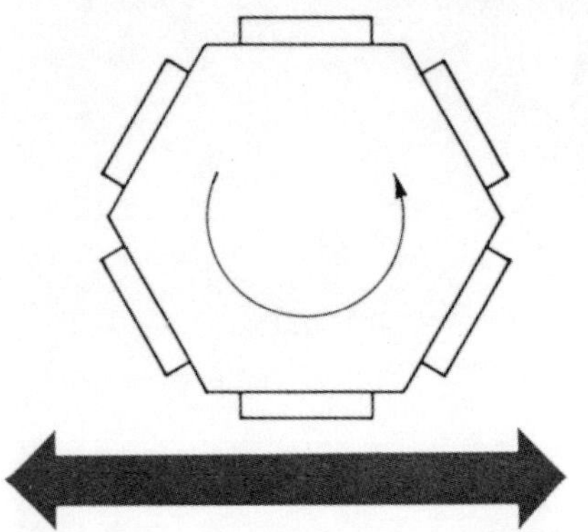

Indexing-tool principle keeps tools ready to perform their work in turn

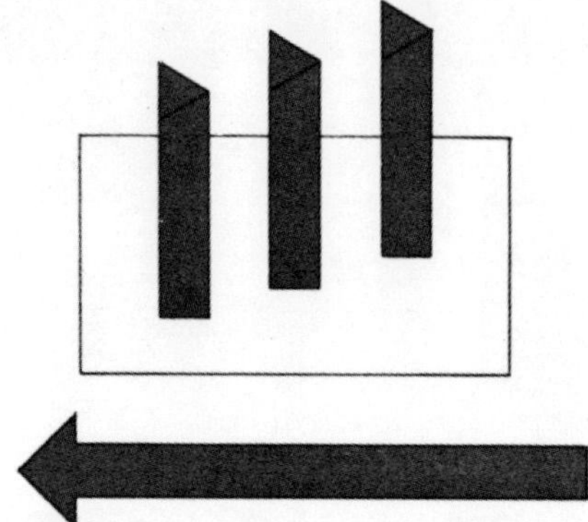

Massed-tool principle cuts multiple surfaces of work at the same time

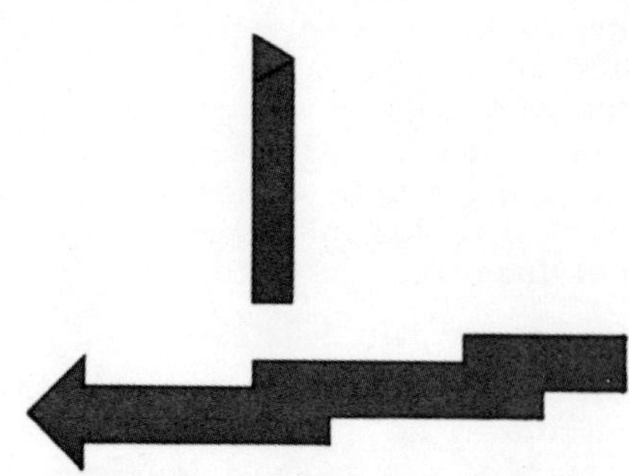

Guided-path principle moves one tool as needed to make a series of cuts

tinue to call it 'the basic lathe.')

We have also made this lathe about as efficient as we can without adding special accessories. However, on a production job involving intricate parts and a large number of cuts, the machine will be idle a good part of the time, just because of the way it is operated.

For each diameter that must be turned, the operator has to adjust the position of the tool. Furthermore, if he changes from one operation to another (from turning to necking, say), he has to remove one tool from the toolpost and replace it with another.

And of course, that's why there are a number of production-oriented accessories. That's also the reason why there are different kinds of lathes. All of them are aimed at higher output and lower cost per piece, and each one has its own way of going about it.

Tool handling is the key

There are certain fundamental improvements that can be made in the tool-handling part of a lathe, even in the basic lathe, with the help of accessories. However, in the different kinds of lathes, one or another of these improvements has been carried far beyond the level at which it can be applied to the basic lathe. In fact, the whole design of the machine is often based on its tooling.

Therefore, by noting how the cutting tools are mounted and manipulated, we can sort the production lathes into fairly distinct groups. Within each group, there are various types and various levels of automaticity, but there are only three major kinds, based on the following tooling principles:

- *Mount all the tools needed for a job on a turret so that each one can be quickly indexed into the correct position to take its cut.* This principle does not reduce the actual cutting time, but it makes a considerable reduction in the time between cuts. Therefore production and efficiency are increased because the machine is cutting for a larger percentage of the time.
- *Mount all the tools needed for a job in such a way that they all cut at the same time and so that the job is completed in a single pass.* This principle improves productivity two ways. Ideally, the cutting time is divided by the number of tools employed, and there is no time between cuts.
- *Make one tool do the whole job by guiding it through whatever path is necessary for machining all the surfaces from one end of the piece to the other.* This principle eliminates the time between cuts because the tool moves directly from one cut to the next. In addition, it can machine curved, and irregular surfaces.

Of course, no one kind of lathe has a patent on any of these principles. However, each of the major groups specializes in one principle or another.

But we should point out immediately that these principles might better be called 'ideals' or 'goals'. In most cases, one principle does not have the ability to machine all surfaces, especially on a complex workpiece. Therefore, even though each kind of lathe is based on one principle, their designs often include certain features borrowed from the other principles, to round out their capabilities and also to increase their efficiency.

In addition, there are so many accessories for various purposes that it's fair to say that almost any lathe can do almost every lathe operation. The result is that the lines of distinction between the different kinds are very blurred at times.

Therefore, the only way to see how and why each kind of machine is different is to consider the distinctive features and abilities of standard machines in each group, stripped of accessories and gadgets for the purpose of observation. Once the group has been defined, important accessories will be mentioned separately, but they cannot be considered distinctive features if they are used by more than one kind of lathe.

Each group specializes

As mentioned earlier, each group is characterized by the fact that it specializes in one principle of tool manipulation. Furthermore, because of the way these principles have been applied in actual machines, it turns out that there is a characteristic tool path for each group. If we know what kind of path the tool moves through when being fed into the work, on a standard machine, we can tell what group the machine belongs to.

It also turns out that the cutting path of each group tends to make it particularly well suited for some jobs and not so well suited for certain others. Here again, accessories can often fill the gap, but they tend to complicate the setup.

Therefore, in selecting a machine for a job, especially a job involving a good many parts, it is best to select one that has a 'natural ability' for the work to be done. And the cutting path of each kind of lathe is an important clue to the kinds of work it can do most easily.

In the remainder of this Report, therefore, the cutting path will be used as the major distinction between different kinds of lathes.

Kinds of lathes

Basic, general-purpose	Production: Indexing-tool lathes	Production: Massed-tool lathes	Production: Guided-path lathes
Hand-fed machines			
Hand-fed basic lathe Bench lathe Hi-speed precision lathe Speed lathe	Hand-fed turret lathe 2nd operation machine Hand screw machine	(None)	(None)
Power-fed but manually operated machines			
Power-fed basic lathe Engine lathe Toolroom lathe Gap lathe T-lathe Facing lathe	Ram-type turret lathe Saddle-type turret lathe Fixed center Cross feed Precision chucking mach Hand screw machine (Vertical turret lathe is listed under Boring)	(None)	Tracer-controlled lathe w/manual start & stop
Automatic-cycle machines			
(None)	Automatic turret lathe Horizontal turret Vertical turret Single-spdl automatic Bar machine Chucking machine Automatic screw mach Turret type Sliding-head type Multiple-spdl automatic Bar machine Chucking machine Vertical chucking mach	Multiple-slide automatic Single-spindle automatic Single-spindle chucker	Automatic tracer lathe Copying lathe Duplicating lathe NC lathe Engine-lathe type Turret-lathe type Cam-controlled lathe Single-point threader

The section on each group will start with a diagram showing the cutting paths for that group.

Other important factors such as type of power supply and type of control over speed and feed may be *typical* of a certain kind of lathe, but they can usually be employed on other kinds, as well. The type of control over the cutting path does identify certain lathes, especially in the guided-path group, but even here, it is the cutting path that determines the group and the control that identifies the type of lathe within the group.

The degree of automaticity is not a difference of kind, but it is an important difference of type, and it is distinct enough to mark off three different levels of lathes.

The least automatic machines are those on which the tool is fed into the work by turning a handwheel, usually referred to as hand-fed lathes. Next are the machines that are power-fed but manually operated, or manually controlled in the case of powered rapid traverse. These will be called simply powerfed lathes.

In the top category are the automatic-cycle lathes. No distinction will be made between single-cycle automatic and continuous automatic because most machines that are capable of completing a single automatic cycle are also capable of continuous operation, when they are loaded automatically.

The table showing kinds of lathes and degrees of automaticity is intended to show most of the common names in use today, plus a few that have been invented because the common names do not adequately identify the machine.

Many of the names in the table represent different types of machines, each with its own individual features. However, there are at least a few instances where two or more names stand for the same machine.

This is notably true of the massed-tool lathes. All three names shown can apply to a single machine. And some names are used so loosely in the shop that they have to be shown here under more than one heading—names such as chucker, for example.

There are many additional machines that can be classified as lathes, but most of them serve in highly specialized applications. Instead of attempting to cover all these 'specials,' the report will concentrate on the broader groups that have more general use.

The following pages will cover the basic, general-purpose lathes, the massed-tool lathes, and the guided-path lathes. Part 2 of the Report, which will follow in a later issue, will cover the indexing-tool lathes, along with a discussion of tools for turning.

Basic, general-purpose lathes

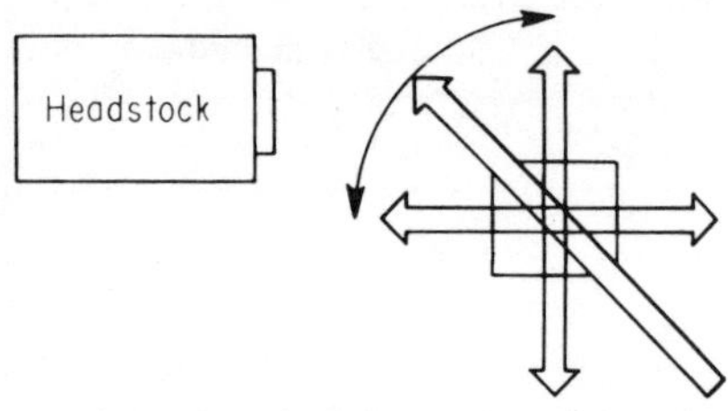

The distinguishing features of the basic, general-purpose lathe are:

- Cutting is performed by one-single-point cutting tool at a time.
- The tool can be fed in any direction.

In these lathes the emphasis is on versatility. They can do any and all lathe jobs, including some that no other machines can handle, and because they operate with one cutting tool at a time, they are easy to set up. They are at their best in situations where many different parts are produced in small quantities.

On the other hand, in their basic form, they are not particularly well suited for high-volume production. With accessories, of course, they can do many such jobs efficiently, but only when the workpieces are simple and require very few cuts.

Both hand-fed and power-fed lathes are included in this group, with the hand-fed machines at the low end of the size range.

Hand-fed basic lathes

There are many names for these lathes, and probably the one heard most often is 'bench lathe,' simply because it's small enough to be mounted on a workbench. Other names, however, say a little more about the equipment and its application: high-speed precision lathe, instrument lathe, and speed lathe.

They *are* small machines, with a maximum swing ranging from four inches to about 9 inches and center-to-center distances of no more than 24 inches, some of them much less. And they are intended for small workpieces, which usually involve light cuts over short distances. Therefore power feed is not necessary.

Even more important, though, is the fact that hand feed is essential for many of the delicate, high-precision jobs these lathes do, so the operator can feel the action of the cutting tool going through the workpiece material.

Small-diameter workpieces, of course, must be given a fairly high rotating speed in order to attain a suitable cutting speed, and that's why speed is emphasized in some of the names for these lathes. Top speeds range from about 3000 to 5000 rpm, which is not really fantastic, but it's somewhat higher than the speeds on larger lathes.

Speeds are adjusted by various different methods, depending on the manufacturer of the lathe. Some have multiple-step pulleys, others have a variable pulley arrangement, and still others have adjustable speed d-c motors.

The tool-handling part of these lathes is very simple and is substantially the same as the 'simplest possible turning machine' described on page 81. Transverse tool motions are provided by a cross slide clamped directly to the bedways wherever needed, and a swivel-mounted compound slide provides tool motions in all other directions.

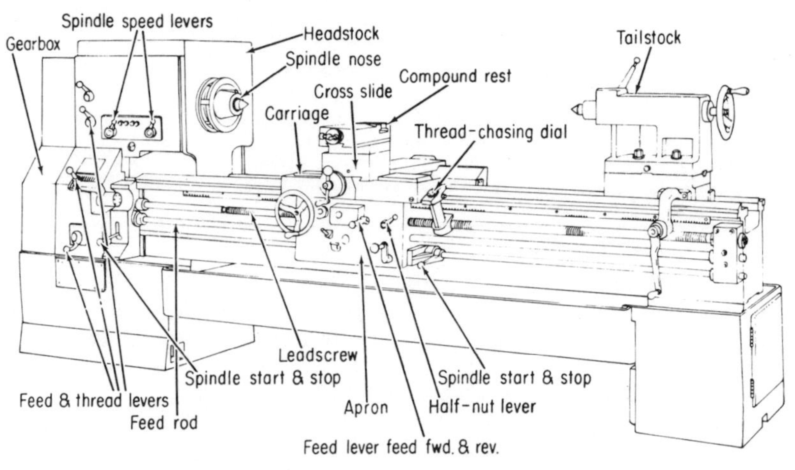

Standard engine lathe, one of the power-fed basic lathes, is a versatile machine that is easy to set up and operate and is intended for short-run jobs

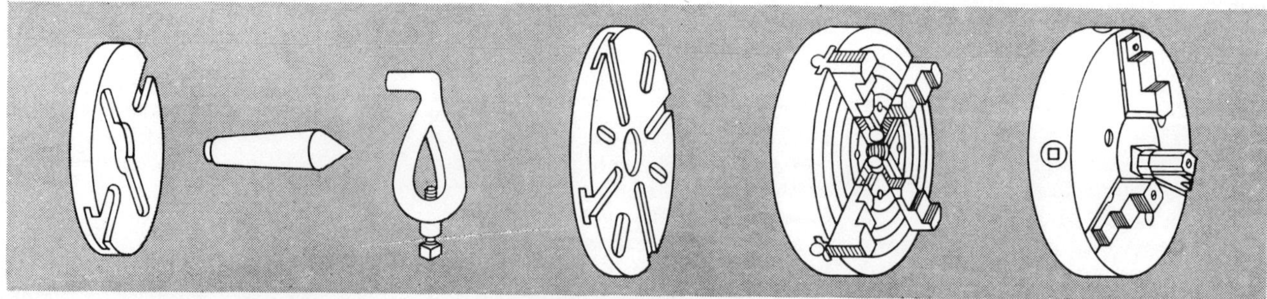

Various types of work holders include driving plate, center, and dog (at left), which are used together for between-centers work. Others are faceplate and three- and four-jaw chucks

Both slides are usually operated by handwheels or ball cranks connected to feed screws. However, lever-operated slides with a rack-and-pinion arrangement are also available for speeding production applications.

The tailstock, like the cross slide, is clamped directly to the bedways where needed. As on any lathe, it can support one end of a long workpiece, or, if the workpiece is mounted at the spindle only, the tailstock can be used for feeding drills, reamers, or threading tools longitudinally.

Speed lathe, a variation

A variation on this lathe is a machine intended for short parts that do not have to be supported between centers. It does not have a tailstock, and it has a shorter bed, no more than about 10 inches.

Usually called a 'speed lathe,' it can serve a variety of purposes. With the same cross-slide-and-compound unit as the lathe described earlier, it does most of the same jobs, except that it does them on shorter workpieces. Or it may have a double-tool cross slide with a tool mounted at each end so the operator can feed one way and then the other to perform two cross-feeding operations quickly.

Power-fed basic lathes

The major machines in this category are engine lathes and toolroom lathes. Actually, the toolroom lathe *is* an engine lathe, built to somewhat closer accuracy, as set forth in American Standard Accuracy of Engine and Toolroom Lathes (ASA B5.16-1962), published by The American Society of Mechanical Engineers.

For example, one of the 25 accuracy tests specifies that the spindle center runout for any size toolroom lathe must be within 0.0003 in. (tir). On an engine lathe with a swing of 12 to 18 inches, it must be within 0.0005 in.

Mechanically, the two machines are practically the same, and the following discussion of engine lathes also applies to toolroom lathes. Similarly, there are other machines that are based on the engine lathe, but certain features such as size or arrangement have been adapted to a specific kind of work or operating procedure.

The gap lathe, for example, has a cut-out section in the bed immediately in front of the headstock so

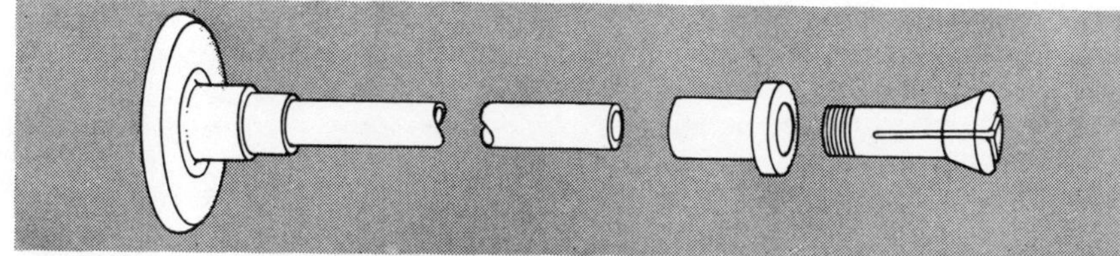

Collet (right) is seated against adapter by collet drawbar (left) that extends through spindle. Collet holds small-diameter work of uniform size

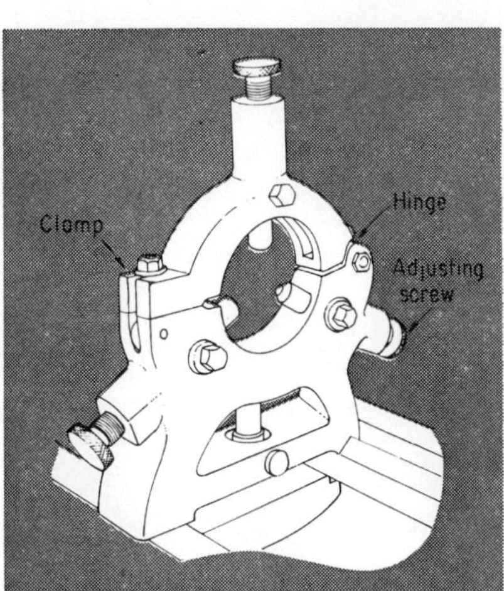

Steady rest is mounted on ways when needed, to support long, slender work

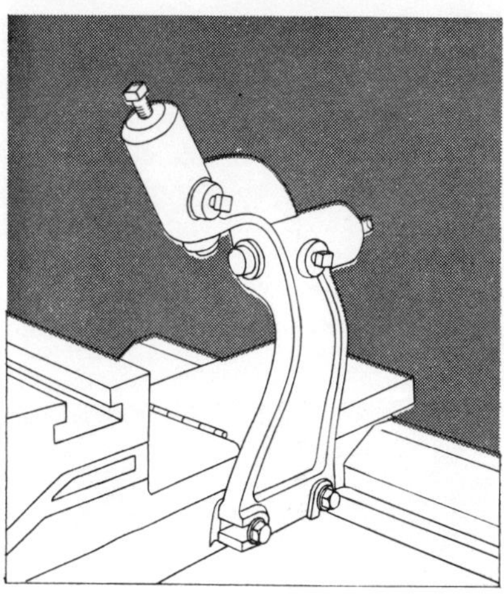

Follow rest is attached to carriage and provides moving support for the work

it can swing a larger workpiece. T-lathes (also called right-angle lathes and facing lathes) have a bed set at 90° to the spindle axis instead of the usual parallel arrangement so they can machine large disks where facing cuts predominate.

In size, the power-fed basic lathes pick up about where the hand-fed machines leave off. In fact there are even some engine lathes small enough to be called bench lathes. At the other end of the scale are the giants that have as much as a 9 or 10-ft swing over the ways and can be ordered to almost any length. In spite of their large size, however, these machines are seldom powered by more than 50 to 75 hp.

The obvious difference between all these lathes and the hand-fed lathes is in the power-operated mechanism that feeds the cutting tool. Another point is that engine lathes normally have a geared headstock with a selection of as many as 36 spindle speeds that are set by lever, and on some machines they can be preset for the next cut. Adjustable d-c motor drive is also available.

In addition, the engine lathe is further distinguished by the fact that it has two *kinds* of feed and the fact that the feed rates are always directly related to the rotating speed of the spindle. One feed is for the regular machining operations, and the other feed is for single-point threading. This ability to do single-point threading makes the engine lathe different, not only from hand-fed lathes, but from almost all other lathes.

Power for both kinds of feed is taken from the spindle and conveyed by a gear train (the 'end gears') to the quick-change gearbox, where the ratio of tool advance to spindle speed can be selected. A choice of 32 to 60 or more feeds and threads is not unusual.

From the gearbox, the power is conveyed to the carriage apron by a leadscrew and a feed rod. If regular machining is to be performed, it is then converted into longitudinal or transverse tool feed by gearing in the apron. Or if single-point threading is to be done, the 'half nut' or 'split nut' is engaged directly with the leadscrew.

An alternate arrangement found on many engine lathes is a combination leadscrew and feed rod with a keyway cut in the leadscrew to deliver feeding power.

The carriage consists of three sliding components for tool feeding, which is one more than the hand-fed basic lathe usually has. The additional part is the carriage, which is mounted directly on the bedways and is moved longitudinally, either by power feed or manually with a handwheel.

The cross slide and swivelling compound, mounted on the carriage, are the same as on the smaller lathes, except that the cross slide is power-fed. On some larger machines, even the compound is power-fed, and sometimes these machines also have powered rapid traverse for the carriage and slide.

Another convenience on the medium and large machines is a spindle control lever at the apron. This adds another rod parallel to the leadscrew and feed rod; and still another one is sometimes added, especially on the toolroom lathes, to provide leadscrew reverse at the apron for threading jobs.

Single-point threading is really a specialized turning operation. The tool is fed longitudinally, as it is in turning or boring, but the rate of feed (expressed in threads per inch) is usually much faster, and the tool cuts a helical groove instead of a relatively smooth cylindrical surface.

Successive cuts are made through this same groove, each one at a greater depth. The shape of the groove is determined by the shape of the tool, which is ground to cut threads conforming with one or another of the various threading standards.

Because it takes a good many cuts to reach the required thread depth, this operation is fairly slow. Much faster threading can be done with taps and dies, but single-point threading does offer a number of advantages.

One is that it makes a more accurate thread. Another is that it can do this threading over any length and anywhere on the workpiece, not just on the end. Furthermore, the tool cost is quite low, which is an important factor in the short-run jobs normally done on an engine lathe.

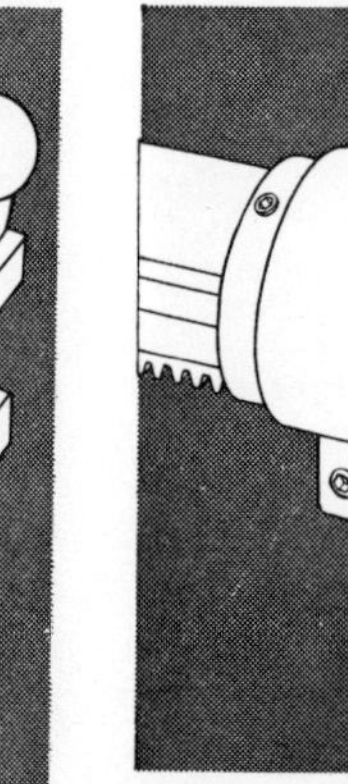

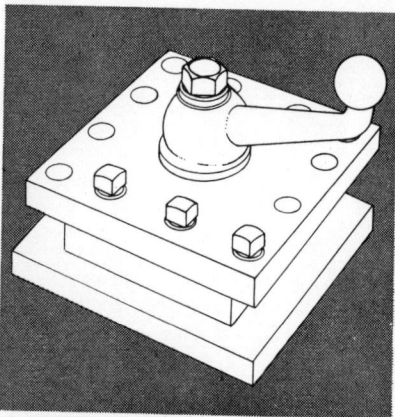

Toolpost turret holds tools ready for work and reduces tool-change time

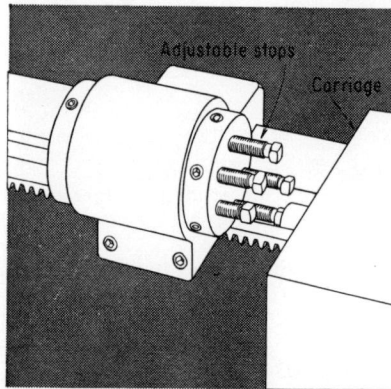

Carriage is stopped at end of cut or positioned at start of facing cuts

Accessories help do the job

Accessories for the engine lathe are many and varied, and among those more frequently used are the steady rest and the follow rest. The steady rest is clamped at any desired position on the bed to support long workpieces and keep them from springing or whipping while being machined. The follow rest is attached to the carriage and supports the work at a point opposite the cutting tool.

Many engine lathes also have taper attachments for machining the tapers that are too long for the compound slide. With a radius-swinging attachment, internal and external spherical surfaces can be machined.

Carriage stops are an important aid to both speed and accuracy, and similar stops are also available for the cross slide.

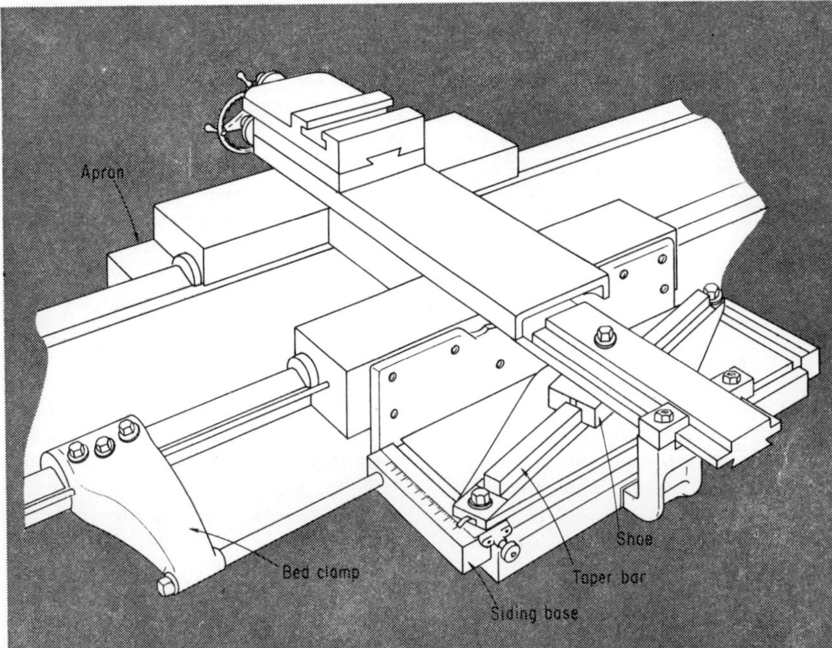

Taper attachment moves cross slide transversely when carriage moves, but only if the bed clamp is fastened. Taper bar is set for the angle to be machined

Production aids

So far, we have discussed the engine lathe as a basic, general-purpose lathe, but it frequently borrows production ideas from the other kinds of lathes. One of these is the four-way turret or square turret, which is mounted on the compound in place of the single toolpost and keeps four tools handy for immediate use.

Beyond this point, the productivity of an engine lathe can be further increased with a rear toolpost for the cross slide, a hexagon turret attachment to replace the tailstock, a tracer attachment, and various others. However, these tools will be covered in detail in the sections devoted to turret lathes and tracer lathes.

Nevertheless, we cannot leave the engine lathe without citing additional evidence of its complete versatility. There *are* accessories that permit it to do both grinding and milling operations.

Multiple-slide automatic lathes

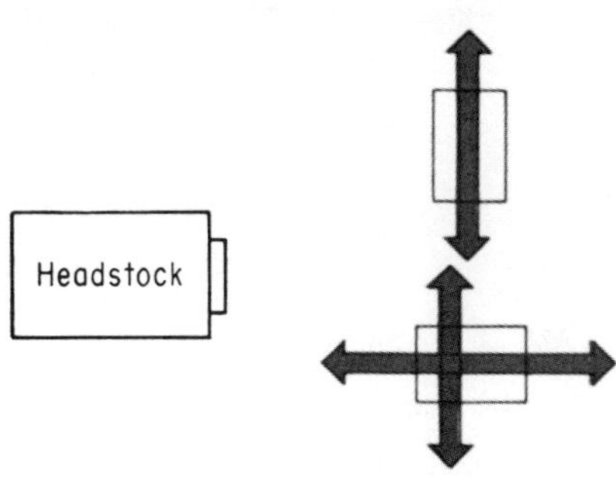

The distinguishing features of the multiple-slide automatic lathe are:

- Cutting is performed by massed-tool arrangements.
- Two or more independently fed toolblocks can move along different paths and feed into the work simultaneously, each one making a single pass.

The tools are not *always* mountin multiples, nor are the toolblocks always fed into the work simultaneously. However, if a machine is capable of these things, it belongs in this group.

Here, the primary goal is high-speed production. With a number of tools cutting at the same time and with no time lost for tool indexing or tool positioning, these machines certainly rank among the most productive lathes.

Setting up all those tools, however, tends to be fairly complicated and time-consuming. Actually, this depends on how complex the work-piece is, but by and large, these lathes are best employed for medium to long production runs.

Too fast for hand operation

There are no hand-fed or hand-operated machines in this group, mainly because it takes a good deal of power to feed a multiple-tool arrangement and because there's just too much happening to rely on manual operation, even with power feed. Therefore, the operation is strictly automatic, either single-cycle automatic or full automatic when parts are loaded and unloaded mechanically.

These machines go by a variety of different names. One of the more popular is 'single-spindle automatic lathe,' which is often shortened to 'automatic lathe' or simply 'automatic.' Another, for models not equipped with a tailstock, is 'single-spindle chucking machine' or 'chucker.'

These shorter names are suitable, once you know you are talking about a particular *kind* of automatic lathe or a particular kind of chucker, and they will be used in this section, but the longer name at the head of the section is intended to identify these machines better in the whole field of lathes where nicknames sometimes lead to confusion.

Possibly one of the reasons why the popular names are somewhat indefinite is that the machines included in this category are a varied group. Some are horizontal while others are vertical; some have level toolslides while other have sloping toolslides; some are screw fed, others are fed by a hydraulic or air-hydraulic system, and still others are fed entirely by cams.

The basic similarities

However, instead of being distracted by the differences within the group, let's start by looking at the basic similiarities. For the moment we will overlook some of the special capabilities so that we can dig down to these basic factors that are the same for all.

As noted earlier, the automatic lathe is distinguished from other lathes primarily in the way tools are mounted, the way they are arranged about the work, the way they are fed into the work, and of course, in the fact that the operation is done in an automatic cycle.

The cutting tools are usually mounted in toolblocks that can carry a number of individual single-point tools. The tools are positioned so that most of the cutting is done simultaneously, and this is cne of the major advantages of the automatic lathe.

Moreover, this type of machine always has two or more such toolblocks, each mounted independently on its own slide, or carriage, or other moving support. Their motion is synchronized, but each one can be set for its own travel path and distance, and the feed rates are usually set individually.

(This, by the way, is the only point of distinction between the automatic lathe and some of the precision production boring machines that have similar toolblocks, but without independent motion.)

Each toolblock has own job

In the basic arrangement, the toolblock at the front of the lathe feeds toward the headstock to perform turning operations. It can also be fed crosswise, and cutting may be performed during this motion, but more often than not, it serves the purpose of putting the toolblock in position for turning.

The other toolblock is located at the rear of the lathe and feeds across (toward the front) to make facing, grooving, and forming cuts.

Thus the automatic lathe is well suited to workpieces that have a large number of diameters, regardless of whether the diameters are of increasing or decreasing size or mixed sizes. The total cutting time is only the length of time required

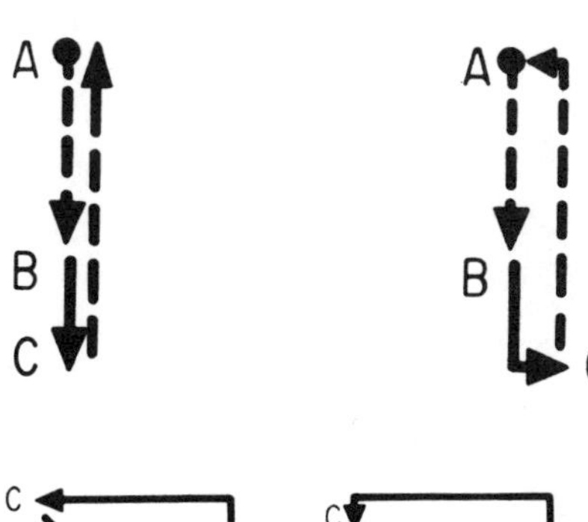

Typical cutting path of rear toolblock (left) is straight in and out and may have tool relief. Front toolblock is basically for turning and boring and may follow various paths to and from actual cutting pass (below). Rapid traverse is in color, feed in black

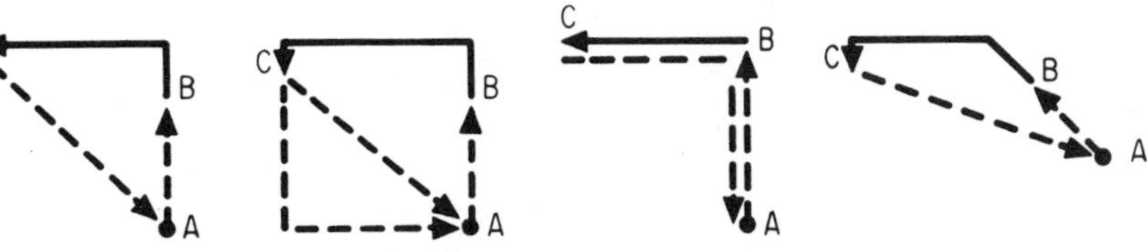

for the longest cut; all the others are 'free', as far as cutting time is concerned.

Another good job for this machine is turning long workpieces, even if the diameter is the same throughout. For this, a series of tools can be set on an extended front carriage so that each one cuts a portion of the total length during a short feed stroke. Furthermore, roughing and finishing tools can be mounted in tandem and both operations completed in a single pass, one cutting tool following right behind the other.

Drilling and boring can also be performed with appropriate mountings on the front toolblock. It is even possible to combine boring and turning in a single operation if the workpiece permits.

The operating details

In the simple multi-diameter operation shown in the sketch, the front toolblock leaves its stop position and cross-traverses, changing to feed rate just before it reaches the work. It enters the work and feeds to depth and then feeds longitudinally toward the headstock, turning all three diameters simultaneously.

At the end of the feed stroke, the tools return to the stop position. On some machines they retrace their path; on others they complete a square pattern; and on still others they return by the most direct path.

Meanwhile, the rear toolblock has advanced to the work under rapid traverse and has made the facing cuts at feed rate. To keep this example simple, there is no tool relief on this cross slide. Therefore when the tools withdraw under rapid traverse, they will leave a spiral tool mark on each face. However, a number of devices are available that can relieve the tools, i e, move them a slight distance away from the work at the end of the cut so that no withdrawal marks are made.

Multi-diameter turning and facing jobs such as this are typical of the work that can be done by any automatic lathe with its basic equipment and the necessary cutting tools. However, the machines are by no means limited to straight turning and facing. With special cams and feed devices and with swivel toolblocks and extra toolslides they can be made to cut a great many different angular and shaped surfaces. These will be discussed later.

Setups can be complex

Because the average job calls for a good many cutting tools, and all of them must be set to interrelated dimensions, the automatic lathe normally works on medium to long production runs. Simple jobs involving only a few tools can be set up more quickly, of course, and such jobs can often be put on an automatic lathe advantageously, even for short runs.

However, it is the more complex jobs and the longer runs that really pay off. And here, both setup time and tool-change time can be held to a minimum with up-to-date tooling such as the preset and quick-change tool techniques discussed in Special Report No. 555 (**AM**—July 20 '64, p77).

Simple, one-pass cycle

Actually, the basic automatic cycle itself is fairly simple. The tools advance at rapid traverse, make their cuts at feed rate, and return to the starting point at rapid traverse, controlled either by trip dogs and stops or by cams.

All these machines have a number of speeds and feeds available. However the spindle speed, once selected, normally stays the same throughout the cycle. For special requirements, speed changes are available as an optional extra, but almost all standard machines have single-speed headstocks.

This means, of course, that, when different diameters are being turned simultaneously, the speed for the job must be a compromise. The tools working on the smaller diameters will have to cut at less than their optimum speed in order

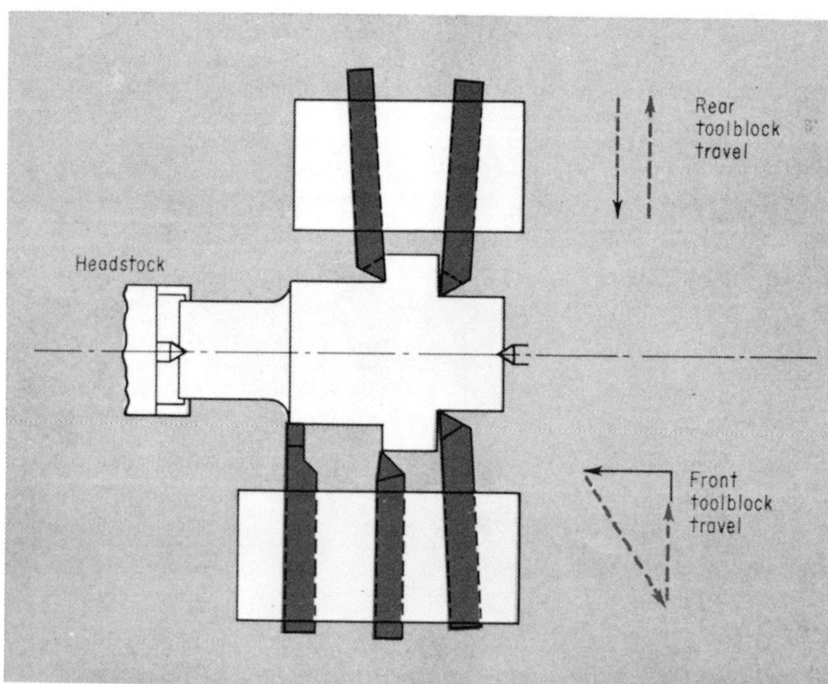

Multiple-step workpieces are machined in a single pass on automatic lathe with tools cutting simultaneously to turn out parts at high production rate

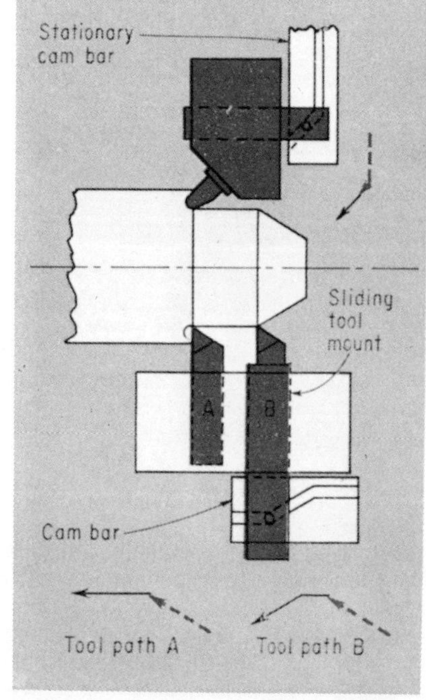

Slotted cams can move one or more tools through special cutting paths

to avoid burning up the tools working on the larger diameters. However, as both large and small diameters are being machined at the same time, the increased metal removal more than offsets the effect of slower speds.

Feed changes are somewhat more common, with finer feeds being used to produce a better surface finish on certain areas of the work. As noted earlier, feed rates for the different toolblocks are often set independently. In addition, feed rates can sometimes be changed under load, especially on the hydraulic-feed machines, but this is usually an optional feature, not included in the machine's standard equipment.

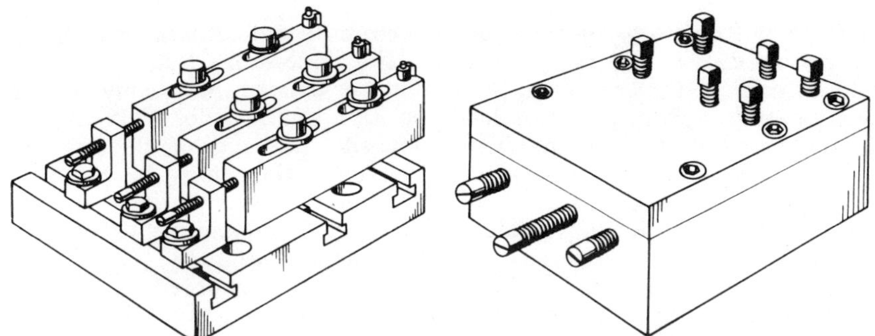

Tools are mounted either on universal tool support (left) or solid toolblock (right) that can be preset from master gages to reduce time spent in setup

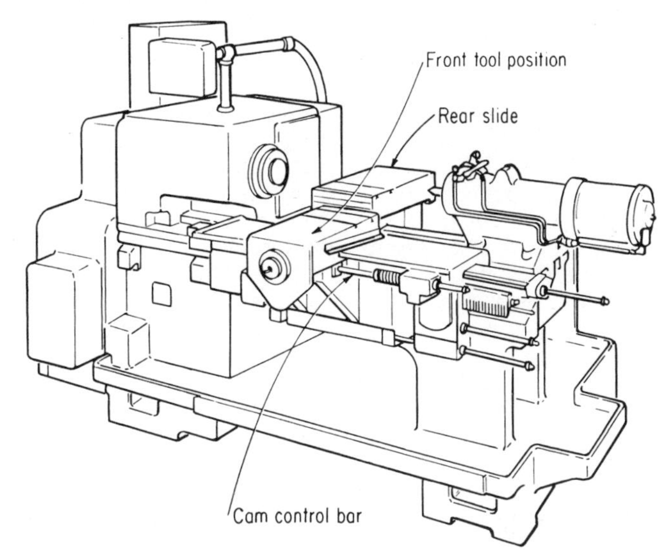

Multiple-slide automatic lathe cross-feeds the front toolblock with cam as carriage moves toward headstock. Rear toolblock moves straight in and out

Pickoff gears change speed

Therefore, with fairly simple automatic cycles and with production runs in the medium to long category, there is no need for complex gearing arrangements in the headstock. Speeds are changed only when setting up a new job, so convenience is not a major factor, and the changes are usually made with pairs of pick-off gears.

Horsepower ratings tend to be on the generous side because it takes a lot of power to keep the workpiece rotating when a number of carbide tools are cutting simultaneously. For example, a machine with a turning diameter of 10 inches is very likely to have a 15 to 20 hp motor and may even go as high as a rating of 40 hp.

This doesn't mean that a machine of smaller horsepower cannot be called an automatic lathe; some are intended for smaller workpieces. However, in order to perform multiple-tool operations, they usually have a good deal more power than a comparable machine that cuts with a single tool.

Up to this point we have been discussing the similarities in automatic lathes. There are many differences, too—differences in method of operation and even in the kinds of work that can be performed .

Two general groups

However, it is possible to identify two broad, general groups. The machines in one group are designed to support the workpiece between centers, and those in the other group support the work only at the spindle end, in a chuck, a collet, or a special workholding fixture.

Of these two, the between-centers type might be called the 'complete' multiple-slide automatic lathe because it can work either way. It can support long workpieces between centers, or it can hold shorter pieces at the spindle and perform end-working operations such as drilling and boring, in addition to the usual turning and cross-feed operations.

The other machine, which is often called the automatic chucking lathe, or simply 'chucker', came into being because of the large number of applications that do not require two-ended support for the workpiece. Another factor is that, without the tailstock, the machine usually takes up less floorspace in the shop.

In addition to these fairly obvious differences between the two types, it is interesting to note that they also tend to have different methods of operation. Feeds, for example, are almost always mechanically driven on the center-type machines, which have various combinations of screws, gears, and cams, depending on the manufacturer. Almost all the chuckers, though, are fed hydraulically.

Another point of difference is in the way the two groups have solved the problem of machining angles, curves, and irregular cuts, even though the basic feed paths of these machines are straight longitudinal and transverse lines.

Cams for special paths

Many of the center-type machines use straight cam bars to transmit power to their crossfeed motions, so it is not unnatural for them to apply the cam principle in different ways to change the shape of the feed path. Thus, a rear toolblock, which would normally feed straight across, can be mounted on a slide and be deflected into an angular path or even a curved path by a slotted cam.

A somewhat trickier arrangement

is to mount only one of the tools in any group so that it can slide, and to control the motion of this one tool with a cam. With this setup on the front carriage, the cam-controlled tool might turn a taper or a special curved shape while all the other tools are turning straight diameters.

Most of the chuckers, as noted eariler, have hydraulic feeds, and cam control is seldom seen on these machines. Instead, many of them have swivel-mounted rear tool slides that can be set for angular or tapered cuts, and all of them offer tracing attachments for the more difficult curves and irregular shapes.

Variety in machine shapes

Another difference between the two groups is in the general structure of the machines, but here, the variety is all in the chucker group. Most machines in the between-centers group look pretty much like lathes, but a chucker may or may not look like a lathe.

Most of them do have horizontal spindles, but there are also some with vertical spindles. Both can do the same kinds of work, but the vertical machines boast an added advantage in being able to handle heavy or bulky pieces that would be difficult to mount on the spindle of a horizontal machine. Two-spindle models are offered in both horizontal and vertical machines.

The operator's station at a chucker may be in the conventional position at the side, or it may be opposite the end of the spindle. The tool slides may be mounted horizontally or at an angle or even vertically. And another variation is that sometimes both (or all) the toolslides are mounted on a platen that rapid traverses them longitudinally into and away from their working position and may also provide an additional feed motion.

Accessories increase scope

Accessories for automatic lathes include a wide variety of devices intended either to increase the range of operations performed or to increase the production rate or both. Probably the most prevalent accessory is an additional tool slide that puts additional tools to work and may feed them in from different directions.

This extra toolslide is often mounted above the work on the headstock. An extra front carriage on the between-centers machines can split a long job and reduce machining time by about a half. And for special applications, an additional cutter head may be inserted in the bore of the spindle to do back boring or back facing while the conventional tools are working on the front.

Tool relief at the end of a cut, to avoid dragging the cutting edge back over the finished surface, is sometimes provided in the basic cycle of the machine. If not, and if tool relief is necessary, hydraulic, pneumatic, or cam-operated devices are usually available.

Another useful device is a 'speeder', sometimes used for drilling and boring so that these small-diameter end-working tools will not be penalized by a slow spindle speed that might be required by work being performed on larger diameters of the workpiece.

The speeder also sees a good deal of application in threading operations. Other threading techniques for automatic lathes employ thread rollers or self-opening die heads and collapsible taps.

Indexing tool turrets are even available for certain machines, so that two or more passes can be taken at the work. However, this requires a modification of the usual one-pass automatic cycle.

As might be expected in connection with machines that are designed for high-speed production, almost every manufacturer offers mechanical loading and unloading devices. Some are only to help the operator mount a heavy workpiece on the machine, but others may do the whole job, transferring the workpiece from a previous machine, loading it, and later unloading it and passing it along.

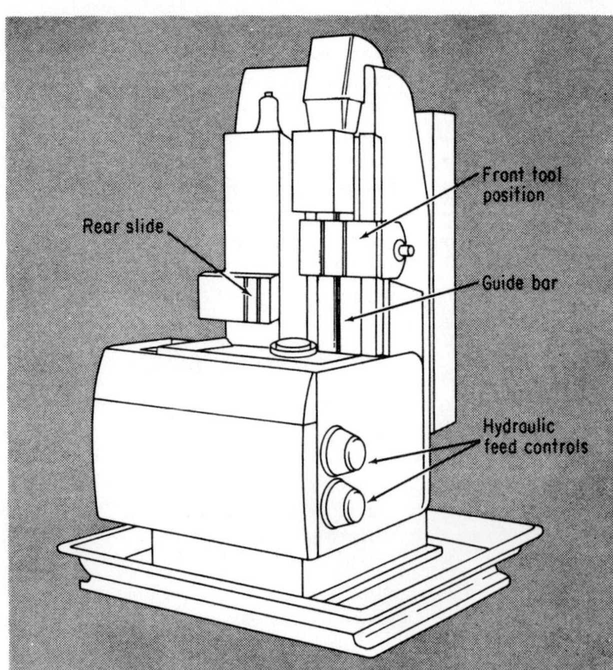

Vertical automatic lathe has same slides as on horizontal model but is intended for work on heavy or bulky parts

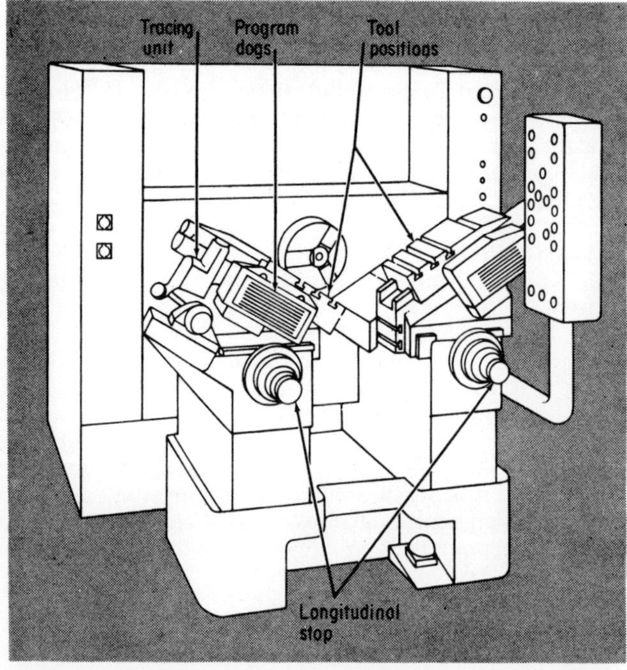

Automatic chucking lathe has hydraulically fed slides plus tracing unit for machining curves and irregular contours

Guided-path lathes

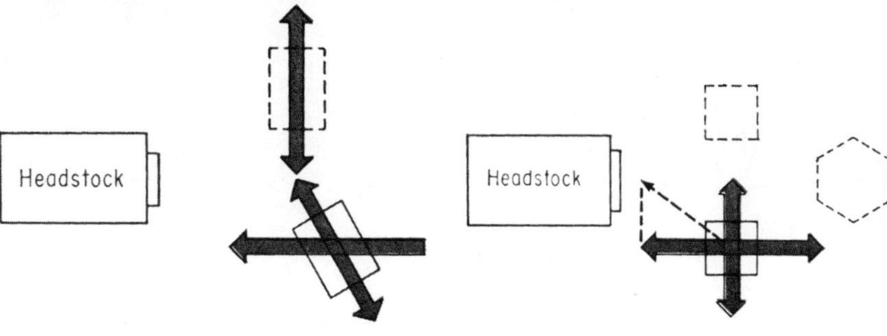

The distinguishing features of the guided-path lathe are:

- The majority of cutting is performed by a single-point tool.
- The tool can be moved through any desired cutting path by simultaneous feed in two axes; the rate and direction of feed in one axis (or both) are controlled to achieve the desired path.

Additional tools are often used, simultaneously or sequentially, but the major advantage of these lathes is their ability to machine complex workpieces, including contoured surfaces, with simple tooling and with no lost time or motion between cuts. Therefore, many difficult jobs can be done at high production rates.

Tool setup is easy, but preparing the control medium (either a templet or a punched tape) is somewhat more time-consuming. However, this preparation can often be justified by medium to short production runs, and the machines are especially well suited to short runs that are repeated occasionally.

The major machines in this group are tracer-controlled lathes and numerically controlled lathes. Also included are cam-controlled lathes and automatic single-point threading lathes.

Hand feed is still practiced in some templet-following applications, but most templet work is now being done by tracer control, either on power-fed but manually operated machines or on automatic-cycle machines. All the other machines in this group are automatic.

Power-fed, guided-path lathes

Tracer control, as an added accessory, and even as a built-in feature of a lathe, very often falls into the manually operated category. The tracing operation itself is automatic, but the operator has to position the tracing unit for each cut and has to return it to the starting position for the next cut.

In most such applications the tracer is used on a lathe that is already identifiable as one of the standard kinds. The tracer increases its capabilities and efficiency but does not make it a different kind of lathe. As there is no detailed discussion of tracer control in the other sections of this Special Report, their general application will be covered here.

In most contour tracing systems for turning, the single-point cutting tool is mounted on a slide that is rigidly connected to a tracing head during the cutting operation. These two are a unit and always travel together.

On the tracing head itself is a sensitive stylus that controls the source of power for the toolslide and therefore controls its own motion as well as that of the cutting tool. Thus, when the stylus is set to follow the contours of a templet, it moves the cutting tool through a parallel path, cutting the same contours in the workpiece.

These systems are called either single-axis or two-axis tracers, depending on whether the tracing unit controls crossfeed only or controls both cross and longitudinal feed in a turning operation.

Single-axis tracers

In the single-axis system, longitudinal feed is constant and is not under the control of the tracer The carriage, carrying the cross slide, cutting tool and tracing unit, moves steadily toward the headstock. Meanwhile, the tracer, which controls the feed and direction of the cross slide, moves it in and out as necessary to keep the stylus against the templet.

The usual arrangement for a single-axis system is to mount the cross slide at an angle to the work, frequently either 60° or 45° to the axis of the spindle. With this setup, the tracer can move 'out' fast enough to machine a 90° shoulder while the carriage continues a constant longitudinal feed.

A disadvantage, though, is that the cross slide cannot feed 'in' fast enough to machine a 90° shoulder between decreasing diameters in the direction of tracing. However, these inaccessible shoulders can be reached if the piece is reversed in a second operation, and many such pieces require a second operation, anyway, if they are to be machined all over.

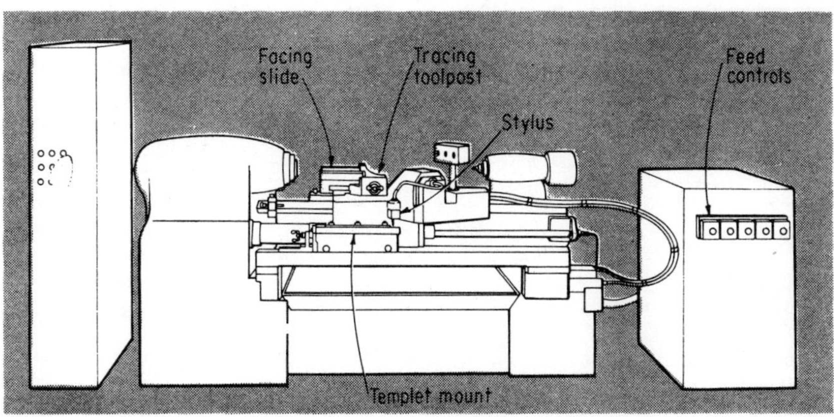

Automatic tracing lathe guides tool through cutting path for rough and finish passes and has a rear facing slide that feeds straight in for cleanup cuts

Single-axis systems are also available with a swivel slide instead of one that is fixed at either 60° or 45°. These offer the advantage that the slide can be set at the best angle for the particular contours to be machined, but they still cannot machine two opposing 90° shoulders.

Two-axis tracers—360°

In contrast with this, the two-axis tracing systems *can* perform this operation, and other complex tracing jobs as well, because the tracing unit controls the rate and direction of both the longitudinal and transverse feeds. These are sometimes called 360° systems because they can trace around a complete circle in either direction.

The only limitation of this system is imposed by the cutting tool itself, and this can be overcome by techniques that use a special templet to bring one tool and then another against the workpiece.

180° tracers—1½ axis?

Between the single-axis and the two-axis systems, there are what might be called the 1½-axis systems, though they are usually called 180° systems. This type of tracing unit has control over the rate and direction of one axis (the 'tracing axis') and only the rate of the other (the 'feeding axis').

Thus, in a turning operation, the cross slide would be the tracing axis, and the longitudinal direction would be the feeding axis, just as in single-axis tracing when the cross slide is set at 90°. However, in the 180° system, longitudinal feed can be controlled, all the way down to zero, to coordinate with the motion in the tracing axis.

The steeper the contour, the slower the longitudinal feed will be, and this produces a substantially constant rate of tool feed across the surface being cut. The longitudinal feed cannot be reversed, but it can stop and wait while the cross feed machines a 90° shoulder.

Additional flexibility is given to the 180° system by the fact that either axis can be selected as the tracing axis (travel in both directions), and the other axis becomes the feeding axis (travel in one direction only).

All these tracer systems have a major advantage in being able to machine shapes that are difficult to machine on other lathes. In addition, they use inexpensive single-point tools, and they can move the tool through a whole series of cuts, both straight and contoured, with no lost time between cuts and with no opportunity for operator error.

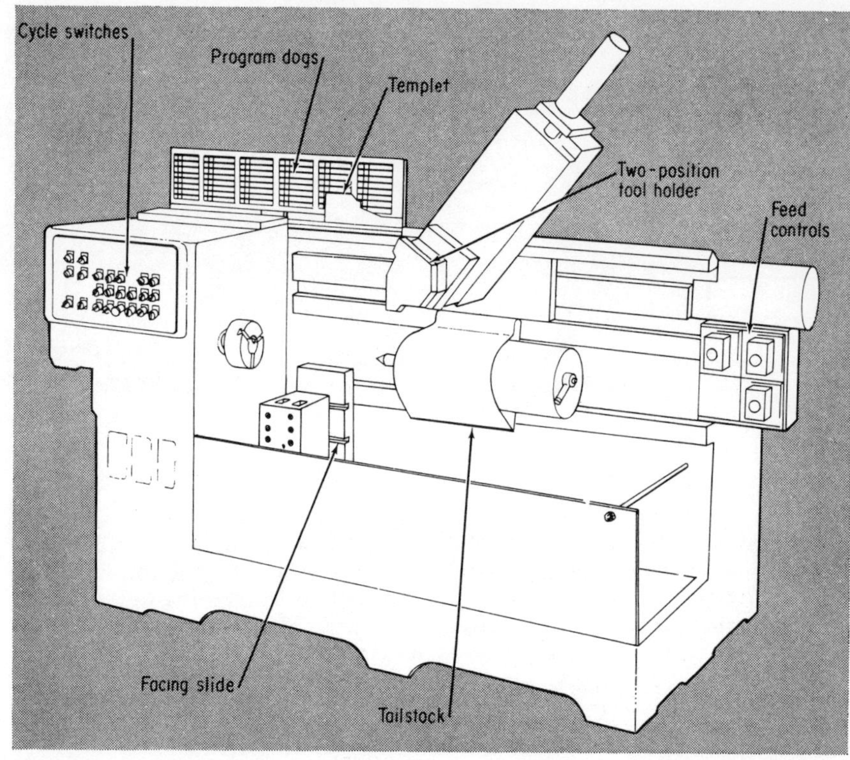

Tracing unit is mounted on upper ways on this automatic tracing lathe, and the work rotates clockwise so operator can observe action of tracing tool

Because the dimensions of the part are built into the templet, and because the stylus operates within close limits of accuracy (usually within ±0.001 in.), one piece will be practically identical with the next, except for tool wear. This means that complex parts can be inspected quickly, just by checking the first and last cuts.

It is very difficult, and somewhat dangerous, to generalize about setup costs. Once the templet is made, setup is very fast, of course. However, the cost of preparing the templet must be included in any honest appraisal. Even so, on suitable parts, the templet cost can usually be recovered within a day's work.

Additional deails on tracer application are contained in Special Report No. 553 on 'Contour Tracing' (**AM**—June 8 '64, p81).

Automatic guided-path lathes

In automatic tracing lathes, or copying lathes, the tracer principle has led to a distinct kind of lathe in which the tracing system is usually combined with a separate, independent cross slide so that the workpiece can be completely machined in an automatic cycle. The cycle may even take the tracing tool through a series of passes for roughing and finishing cuts.

Some of these machines have the conventional engine-lathe layout, but in many instances the shape of the machine is quite different. Some, for example, have a sloping bed with the tracer unit mounted either above or below the work.

Others have a vertical box structure with upper and lower ways tied into a unit by the headstock at one end and a column at the other end. On these, the tracing unit is mounted on the upper ways, and a cross slide or turning slide is mounted on the lower ways.

The tracing systems on most of these machines are single-axis systems. The carriage moves longitudinally at a constant rate, and a tracing slide, at a fixed angle to the work, feeds the tool in or out as directed by a stylus that follows either a flat templet or a master part. Like the single-axis tracing attachments, they can copy to a 90° shoulder on increasing diameters, but they have a blind spot on decreasing diameters.

And that, of course, is one of the

reasons for having a separate, independent cross slide. A number of tools can be mounted on this slide to come in and clean up the cuts the tracer wasn't able to reach. Or it can carry roughing tools and come in before the tracing pass starts so as to save wear and tear on the tracing tool.

Not all automatic tracing lathes, however, have the single-axis tracer. At least one has a 180° tracing unit, which can machine more complex contours and has more flexibility for contour facing in addition to turning and boring. Unlike the machines with single-axis tracers, this one is intended primarily for chucking operations. Nevertheless, it also employs independent slides, either for roughing work or simply to put more tools on the job.

The typical cycle

In a typical cycle on either type of automatic tracing lathe, the tool and stylus are rapid-traversed to the templet where the feed rate is started and the tracing unit takes charge. The machining is done as in conventional tracing, but when the end-of-cut stop is reached, the tool and stylus are returned to the load-unload position, again at rapid traverse.

Meanwhile, at a preset time during the cycle, the facing slide comes in to perform its operations, moving at traverse rate except for the actual machining portion of its in-and-out stroke.

Therefore the operation is very similar to that of a multiple-slide automatic lathe equipped with a tracing attachment. The difference is only a matter of emphasis. Each kind of machine is based on its own tool-manipulating principle but has borrowed the other principle to round out its capabilities and improve its efficiency.

Can make multiple passes

Many tracing lathes also offer an additional pass so that semi-finishing and finishing are done automatically, one right after the other. One type of machine works with two templets and automatically shifts from the roughing templet to the finishing templet while the tool is moving into position for the second pass. A number of other machines have a two-position indexing turret carrying a separate tool for each pass.

Some machines also offer additional roughing passes for jobs involving heavy stock removal. These passes come before the final two passes and are essentially straight cuts across the work at successively greater depths to machine the general shape of the product.

Controls for automatic tracing lathes tend to be rather sophisticated. Spindle-speed and tool-feed rates can usually be set for each pass, and on some machines these rates can also be changed during the cut. Thus, cutting speeds can be held near the optimum rate, even when the diameter of the cut changes, and fine feeds, which slow up the operation, can be used only on surfaces that require a good finish.

Maximum swing over the ways ranges from 12 to 23 inches, and swing over the slides ranges from 8 to 21 inches. However, to accommodate multiple-tool cutting, most machines are powered with a generous 30 to 40 hp.

Numerically controlled lathes

Numerically controlled lathes can be compared almost directly with the two-axis tracer lathes, at least up to a point. Both move the cutting tool through the desired path by controlling the rate and direction of feed in two axes simultaneously.

The main difference, of course, is in how the commands originate. In the tracer lathe, they come from a stylus in contact with a templet. In the NC lathe they come from a punched tape. As far as the individual cutting tool is concerned, it can be guided through exactly the same paths with either control.

However, because all NC lathes have access to a number of cutting

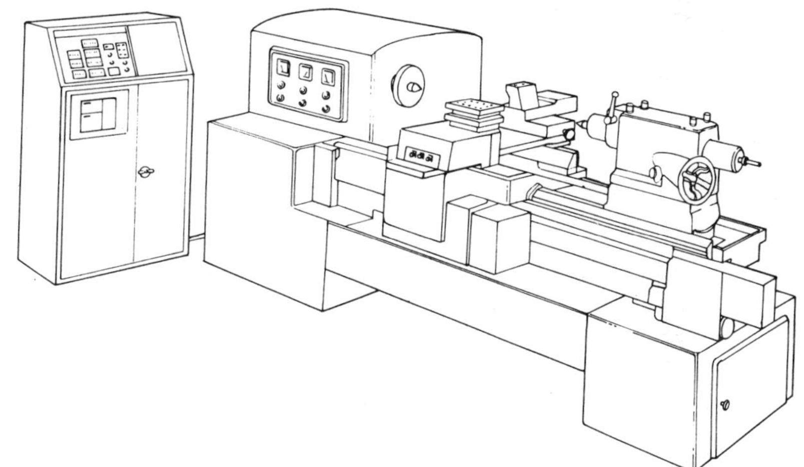

Numerically controlled lathe makes contoured and straight cuts from indexable turret on cross slide, may also have toolpost or turret at rear of slide

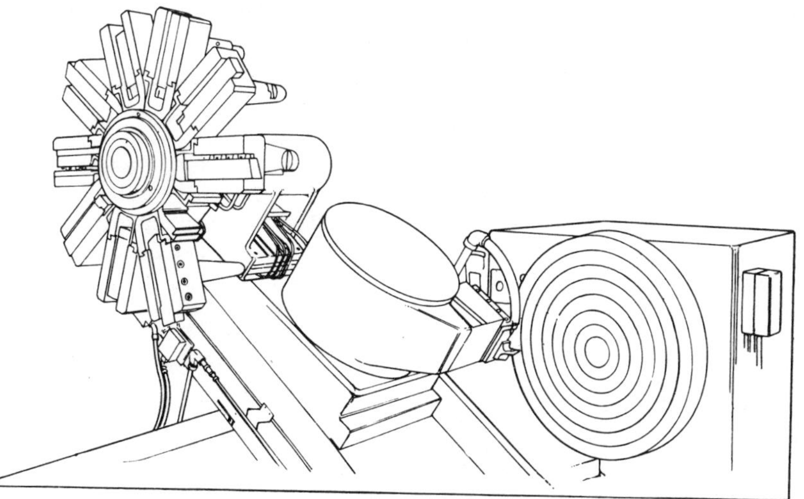

Tool changer for NC lathe can carry 10 tools. On tape command, the toolpost (center) rotates 180°, and tool magazine (left) exchanges one tool for another

tools during the machining cycle, they tend to be somewhat more versatile. For example, they can do boring, facing, turning, and back-facing, all in one operation, using the appropriate tool for each surface to be machined. Contouring, of course, can be included in any phase of the operation.

Another feature of the NC lathes is that all auxiliary functions (speed and feed changes, spindle direction, turret index, coolant on and off) are all controlled by the tape. The tracer-controlled lathes must have additional controls for these functions.

In general appearance, most numerically controlled lathes look very much like engine lathes or turret lathes, with the exception that the familiar handwheels and levers are largely absent from the machine itself. These have been moved to a control panel where they appear in the form of knobs and switches.

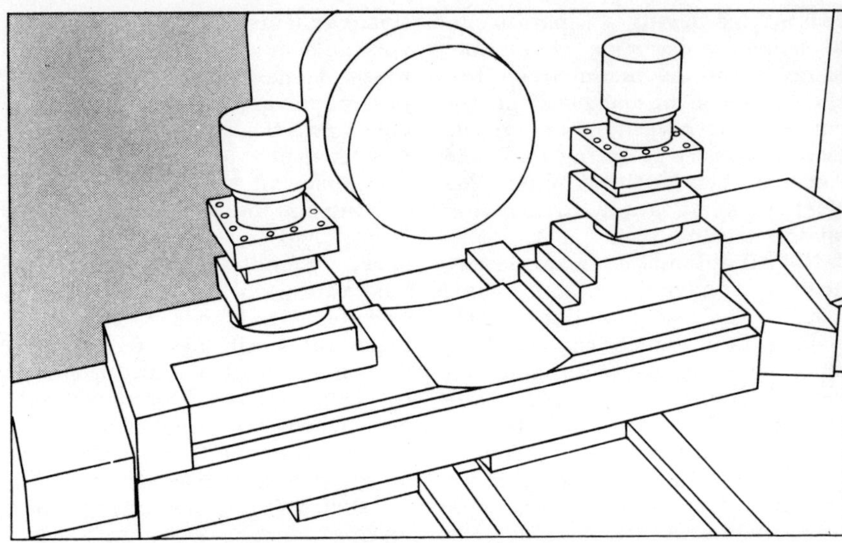

Cross slide of NC turret lathe moves in two axes and has two square turrets, each carrying a unit that supplies coolant to only the tool doing the cutting

Tool arrangement

On both the engine-lathe type and the turret-lathe type, contouring is done from a square, indexable turret that is mounted on a cross slide and carriage to give it the necessary two axes of motion. An additional square turret is often mounted at the rear end of the cross slide to bring other tools into play, and the turret-lathe type also has a hex turret, mounted on the bedways in line with the spindle.

There is one NC lathe, however, that is different. Instead of mounting the tools on a turret, this one carries as many as 10 of them in an automatic tool changer that removes a tool from the toolpost and replaces it with the one the tape calls for next.

In size, the NC lathes cover a wider range than the automatic tracer lathes do, and they also tend to be larger. Maximum swing over the slide ranges from 12 to 36 inches, and power ratings range from 15 hp all the way up to 75 hp. Direct-current motors are frequently employed on these machines to provide a large number of spindle speeds with small incremental steps.

Contouring with cams

Cam control, with flat circular cams controlling the feed rate and direction in two axes, can guide a cutting tool through approximately the same paths as those generated by the two-axis tracer control and by numerical control.

This type of control is found on both horizontal and vertical machines that can perform turning operations. However, because they usually carry a number of tools on a single cam-controlled tool mount and because their operation closely resembles that of precision production boring machines, they have been discussed in 'Fundamentals of Boring' (**AM**—Sept 12 '64, p91)

Single point threading

Automatic single-point threading lathes are not actually guided-path lathes as defined earlier, because threading is not a contouring operation. However, these machines do guide a single-point tool through a series of automatic passes to generate a desired form.

In the threading operation, the cutting itself is much the same as the single-point threading done on engine lathes. The cutting tool makes successive longitudinal passes at greater and greater depth to cut a helical groove whose shape is determined by the shape of the tool. The number of threads per inch is determined by the ratio of tool-feed rate to spindle speed.

However, in the automatic threading lathe, the entire operation is automatic and can be performed at high production rates. Once the operator has installed the workpiece and pressed the start button, the machine does the rest, advancing the tool a predetermined amount for each pass, making finer cuts for the finish passes, and automatically stopping after a certain number of passes.

On one type of machine for this work, the longitudinal feed and rapid return are driven by a reversing leadscrew, and the operation is controlled hydraulically. On the other, the whole operation is driven and controlled by cams.

With inexpensive single-point tools, these machines can produce precision threads of practically any kind. The machines come in a range of sizes and can do either internal or external threading, with external thread diameters ranging up to 17 inches.

In addition, the machines can be equipped with a facing slide and a tracer attachment, and these can even be included in the automatic cycle. Therefore, some workpieces can be completely finish machined and threaded in one operation. ■

Fundamentals of Turning, 2

Concluding the analysis of the four major groups of lathes, this Report covers the lathes in the Indexing-tool group, from hand-fed turret lathes to multiple-spindle automatics

Synopsis

Part 1 of 'Fundamentals of Turning (**AM**—Apr 26 '65, p79) analyzed the basic operation of turning and described the simplest possible machine that can perform this operation. It then explained how the amazing variety of modern lathes can be sorted into four major groups, depending on the tooling principles employed and on the cutting paths through which the tools move.

This was followed by a detailed analysis of three of the major groups: General-purpose lathes, Massed-tool lathes, and Guided-path lathes. The fourth and largest group, the Indexing-tool lathes, will be covered in the following pages.

The basic operation

As defined in Part 1, turning, in its most basic form, is the machining of an external cylindrical surface:

- With the workpiece rotating,
- with a single-point tool, and
- with the tool feeding parallel to the axis of the workpiece and at a distance that will remove the outer surface of the work.

A very simple machine can perform this operation, of course, and with only a few additional components it can do all the other normal lathe operations as well: taper turning, boring, and the cross-feeding operations such as facing, forming, necking and cutoff.

Beyond this, the addition of a tailstock and adjustable speeds and feeds and a few other refinements brings us to the engine lathe, which is the modern general-purpose lathe. It is a thoroughly versatile machine, but the basic engine lathe is only moderately productive.

Three types of improvement

Therefore, a number of production-oriented accessories have been developed to improve the tool-handling part of an engine lathe. These accessories are based on three fundamental principles—the indexing-tool principle, the massed-tool principle, and the guided-path principle—which are symbolized in the sketches at right.

But these same principles have been employed even more effectively by other kinds of lathes that have been specifically designed to emphasize one principle or another.

This, then, is the basis for the four broad groups of lathes—the general-purpose group and three production-oriented groups. The groups are further identified by their characteristic cutting paths, the path the tool moves through when being fed into the work.

Of course no individual lathe concentrates on one tooling principle to the exclusion of the others, but if we look at the basic standard machine without the accessories that round out its capabilities, we can usually determine which group it belongs to.

Still, there are certain hybrid units that combine the features of two or more groups in a way that defies classification. Most of these are quite specialized, however, and it is the purpose of this Report to cover the broader groups that have more general use.

By Richard T Berg, associate editor

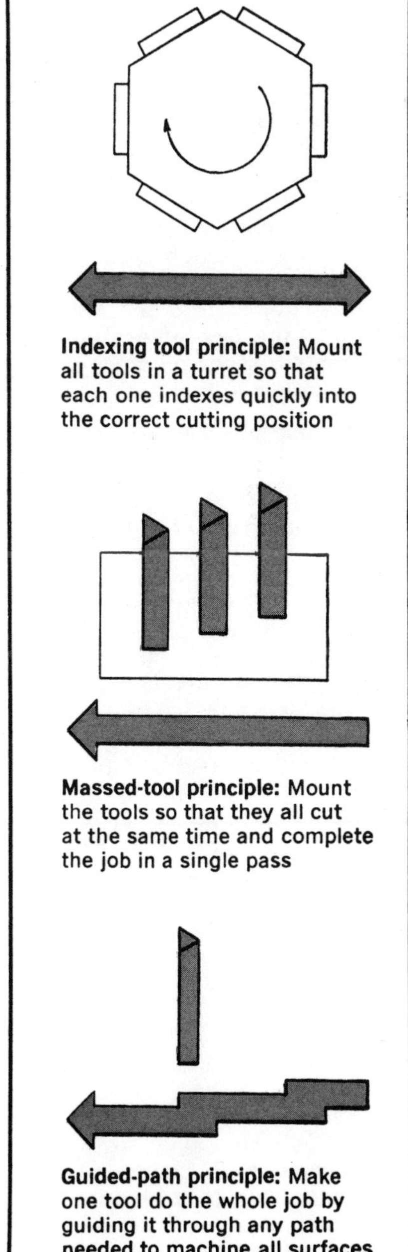

Indexing tool principle: Mount all tools in a turret so that each one indexes quickly into the correct cutting position

Massed-tool principle: Mount the tools so that they all cut at the same time and complete the job in a single pass

Guided-path principle: Make one tool do the whole job by guiding it through any path needed to machine all surfaces

Indexing-tool lathes

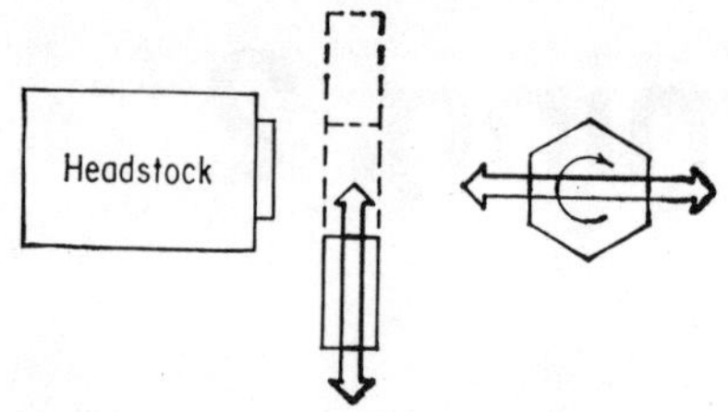

Simple, basic tool path applies to all lathes in indexing-tool group

The distinguishing features of the indexing-tool lathes are:

- Cutting is performed by a succession of tools that can be indexed, one after the other, into cutting position.
- The indexable tool mount (the turret) moves longitudinally to feed its cutting tools into the end of the workpiece, and most machines also have one or more cross slides that move transversely to feed other tools into the side of the work; the cross-feeding tools are mounted either on a toolblock or on another turret.

Largest group of lathes

This is the largest and most diverse group of lathes, and it extends through the full range of automaticity, including a number of variations and combinations at each level. (Even the multiple-spindle automatics are included, if it is understood that 'tool indexing' is relative in this instance, because the workpieces are indexed instead of the tools.)

Along with the fact that the major portion of the tooling is mounted opposite the end of the workpiece, it stands to reason that the machines in this group do not have tailstocks for supporting long workpieces between centers. This imposes some limitation on the length-to-diameter ratio of workpieces they can machine.

On the other hand, they are particularly well suited to workpieces that require internal machining, especially when a series of cuts must be made over the same surface, as in drilling, reaming, and tapping, for example. And the same can be said for external surfaces, which can be rough and finish turned with successive tools.

This leads to the fact that even the simplest machines in this group have a productive advantage over the general-purpose engine lathe because the turret-mounted tools are always ready to cut when indexed into working position. And simultaneous cuts can often be made with the two tool mounts.

The basic tool-path diagram for the indexing-tool lathes is shown at left, and it applies generally to all the machines in the group. However, the different types of lathes within the group have a number of variations on the same basic pattern. These are minor variations, but they will be shown on the following pages to help explain the different machines.

The major categories in the remainder of this Report will be based largely on ascending degrees of automaticity. Thus it will cover hand-fed indexing lathes, power-fed (but manually controlled) indexing lathes, and two general types of automatic machines—those that are sequentially controlled and those that have a timed cycle, which is usually controlled by cams. Numerically controlled lathes have been discussed in Part I of this Report.

Hand-fed indexing lathes

The simplest indexing lathes are the turret lathes that are fed manually, by means of either a lever or a handwheel. As on most lathes, the work-handling part of the machine is at the left, and the tool-handling parts are mounted on the bedways, which extend from the headstock to the right-hand end of the machine.

The major tool-handling part is, of course, the turret, which moves longitudinally, and there is also a cross slide that moves transversely. The tool-path diagram for these machines is the same as the basic diagram for all indexing lathes.

On hand-fed machines the turret is usually either round or hexagonal in shape and has six tool stations, with a hole bored at each station to accept the shanks of standard tools and holders. When a station is in working position, the center of this hole is in line with the axis of the spindle.

All hand-fed turret lathes are ram-type machines. That is, the turret is mounted on a ram, and the ram in turn is mounted on a base that is clamped to the bedways at a fixed location for any particular operation. Only the ram moves as the cutting tools are fed, into the work.

On most such machines the turret indexes automatically from one station to the next when the op-

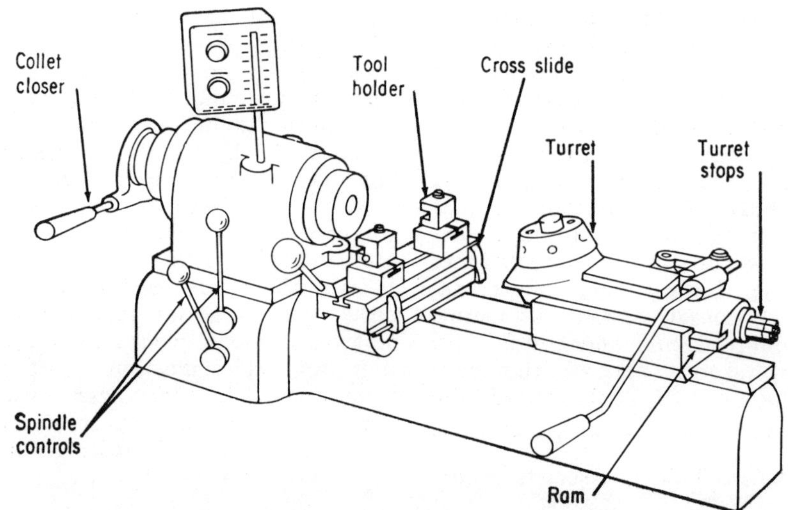

Hand-fed turret lathe is a ram-type machine that is operated by levers or handwheels for the machining of small workpieces at high production rates

erator moves the ram to the fully retracted position (all the way to the right). At the same time, the stop drum on the right-hand end of the ram is also indexed to the next position. This device carries six stop screws (one for each turret station) that can be adjusted to stop the feed motion of each tool at the correct point.

The cross slide is mounted on the bedways between the turret and the headstock. Its base, like the base of the turret, is clamped at a fixed position during any operation. In a typical operation, the cross slide carries two toolblocks, one at the rear and one at the front, but the front toolblock can also be replaced with a square turret or even a swivelling compound slide for making diagonal cuts.

Thus, these machines have eight tool mounts. The six on the turret are normally used for single-point boring and turning tools and also for drills, reamers, taps and dies, while the two on the cross slide are for facing, form-cutting, and cutoff tools.

Simplicity is a strong point

One of the advantages of the hand-fed turret lathes is their simplicity of operation. Once they are set up, there is very little chance to make dimensional errors. The turret always moves through the same longitudinal path; the cross slide always moves through the same transverse path; and the end point of each pass is established by stops.

In addition, some of these machines have automatic headstock controls that can be preset for the desired spindle speed and direction for each turret station. This feature will be described more fully under power-fed turret lathes.

Hand-fed lathes are small

Hand-fed turret lathes tend to be relatively small machines. On most of them, the swing over the cross slide is no more than about 5½ inches, and collet capacities are 1⅛ inch and less. Maximum travel of both the turret and the cross slide is in the neighborhood of 4 inches.

Operating either as chuckers or as bar machines, these lathes are intended for machining small parts—often precision parts—at high production rates. In work such as this, involving delicate workpieces and tools, there is a definite advantage in hand-feed; it lets the operator 'feel' the cutting action. At the same time, however, he must be careful to feed the cutting tool at a uniform rate, if surface finish is an important factor.

In addition to these machines, there are also larger hand-fed turret lathes, some of them running to about twice the sizes mentioned above. These are usually intended for matching nonferrous parts, which involve low cutting forces. Work such as this can be done at fairly high rates, even on these relatively simple machines.

Power-fed indexing lathes

In contrast with the hand-fed turret lathes, all of which are ram-type machines, the power-fed turret lathes include three general types, depending on how the turret is supported and fed:

Ram-type turret lathes. Except for the addition of power-feed, the turret mounting is the same as on the hand-fed machines. The turret is mounted on a movable ram, which is supported by a base clamped to the bedways.

Saddle-type turret lathes with fixed center. On these, the whole turret-support assembly is mounted on the bedways and moves longitudinally along the ways.

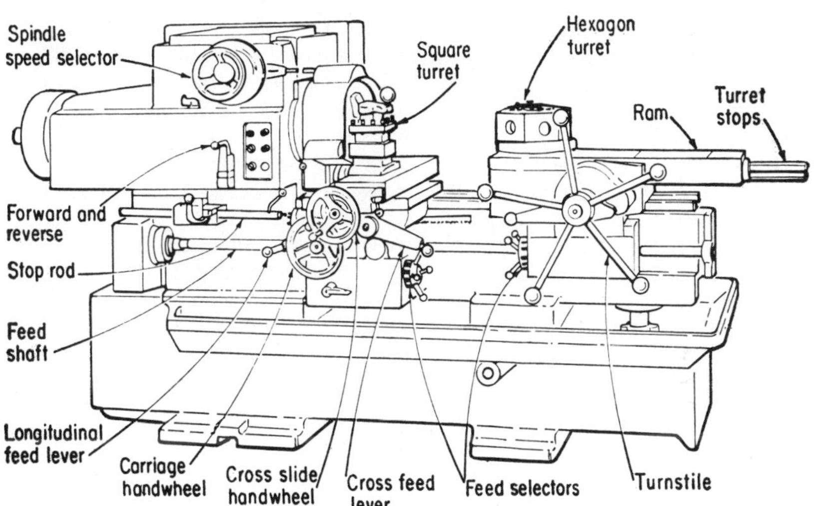

Power-fed turret lathes are most versatile of indexing-tool group. On ram-type machine here, turret support is clamped to ways, and only the ram is fed

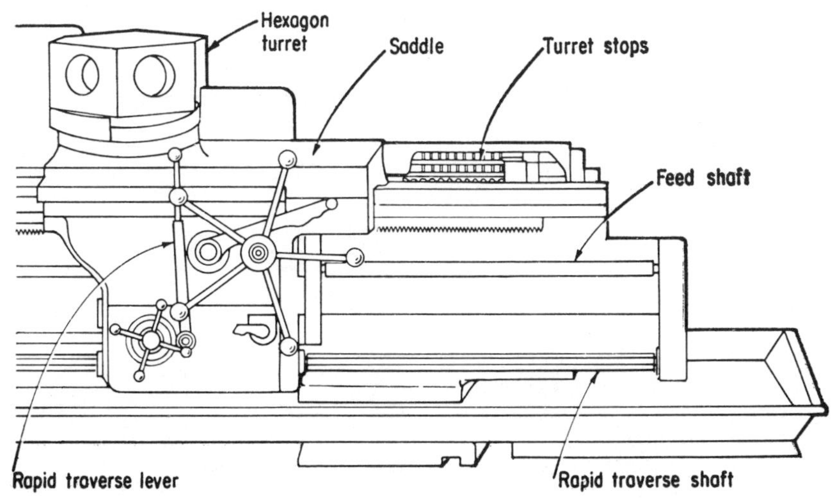

Saddle-type turret lathe (right-hand end shown here) has saddle that moves on ways to feed turret toward headstock. Powered rapid traverse is a usual feature

Saddle-type turret lathes with cross-feeding turret. Again, the whole turret-support assembly can move longitudinally along the ways. In addition, however, the turret support includes a set of ways at 90° to the bedways, so the turret can be fed crosswise as well as longitudinally.

Power for the feed motions on these machines comes from the spindle and is conveyed by 'end gears' to a feed rod. This rod extends along the front of the lathe, passing through the 'apron' where it delivers the feeding power.

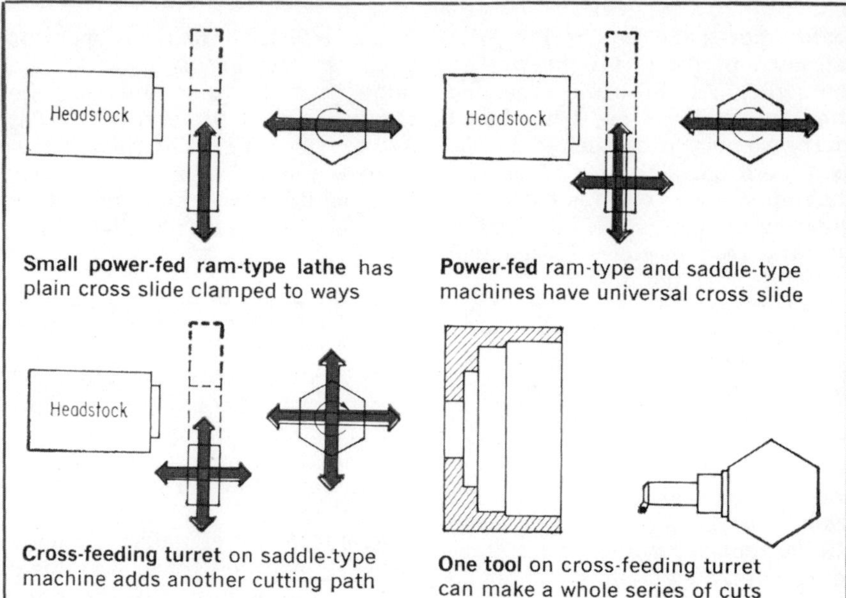

Small power-fed ram-type lathe has plain cross slide clamped to ways

Power-fed ram-type and saddle-type machines have universal cross slide

Cross-feeding turret on saddle-type machine adds another cutting path

One tool on cross-feeding turret can make a whole series of cuts

Not all tools are power-fed

All the saddle-type machines have full power feed for the turret and cross slide. The ram-type machines, too, have power feed for the hexagon turret, but some of the smaller ones have hand-fed cross slides just like those on the hand-fed machines. Called 'plain cross slides,' they are clamped to the ways during setup, and they move crosswise only.

Even on the larger ram-type machines, the power feed for the hexagon turret operates in one direction only—toward the headstock—and the operator retracts it manually. However, these larger machines do have power-fed cross slides, and the feed motion here is both forward and reverse, either longitudinally or crosswise.

In this respect, the larger ram machines are like the saddle-type turret lathes, and when the cross slide has these four directions of power feed, the machine is called a 'universal' turret lathe. Therefore, most of the ram-type machines and all of the saddle-type are universal turret lathes.

This is an important factor because it distinguishes these machines from almost all other indexing-tool lathes. On the hand-fed machines described earlier and on the automatic machines to be discussed later, the cross slide feeds crosswise only, and all operations requiring longitudinal feed (turning, drilling, boring) must be performed from the hexagon turret.

Therefore, the universal turret lathes, with their additional cutting path for the cross slide, have greater freedom in tooling arrangements. They can also do a greater variety of work with simple tools.

This is a big advantage on short-run jobs, especially those that would otherwise require special tooling. Even between-centers work is possible, but in this instance the turret lathe would only be substituting for an engine lathe.

Going a step beyond the universal turret lathe is the saddle-type turret lathe with a cross-feeding turret. This machine is universal in two senses because both the turret and the cross slide can be fed in four directions.

The advantages of the cross-feeding turret are substantially the same as those of the universal cross slide. Both add to the versatility of a turret lathe, and both permit much simpler tooling than is required for the machines that are restricted to longitudinal feed for the hex turret and transverse feed for the cross slide.

However, even though the four-way feed on these machines is an advantage for short-run jobs and highly complex workpieces, it is usually a slower way to operate, and it should not ordinarily be employed for production jobs.

After all, the turret lathe is a production lathe, not a toolroom lathe. One of the main benefits of a good turret-lathe setup is that each tool is ready to make its cut as soon as it is indexed into working position. If the operator has to stop to reposition the tool mount (either cross-feeding turret or cross slide) for depth of cut before each pass, much of the natural advantage of the turret lathe is being thrown away.

Ram vs saddle

In general, the ram-type machines tend to be smaller than the saddle-type machines. They have a maximum swing (over the ways) ranging up to 22 inches and a bar capacity up to 4½ inches.

Compared with this, the saddle-

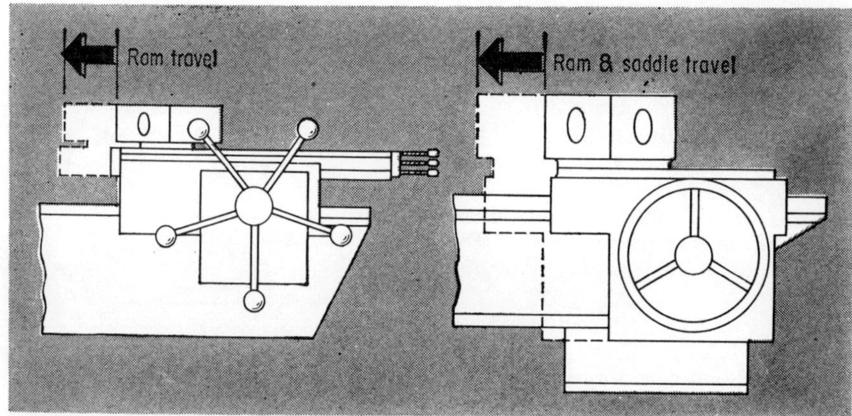

Ram-type turret (left) is light and easily manipulated for short, fast cuts. Saddle-type turret has constant tool overhang and is better for longer cuts

type lathes have swings ranging from 16 to 40 inches and bar capacities from 2½ to 6 inches and even 12 inches, though the larger-sized machines are more likely to machine chuck-mounted work than bar stock.

A more significant difference, though, is in the length of travel of the hexagon turret. The maximum for the ram-type machine is about 14 inches, because the turret base is clamped to the ways, and the feed motion is in the ram.

On the saddle-type machines, however, the whole turret support travels along the bedways. Therefore, machines of this type may have a longitudinal turret travel of about 2½ to 8 feet. This difference in construction accounts for differences in application.

Advantages for each type

For small (or short) workpieces, the ram-type machine has an advantage in that the turret base can be clamped at a point that requires minimum travel. And the low inertia of the turret-and-ram assembly permits rapid manipulation with least exertion.

On the other hand, heavy cuts should be avoided on the ram-type machine because the tool moves farther from its support as the ram advances. The saddle-type machines do not have this problem because the amount of tool overhang remains constant.

In addition to having greater rigidity, the saddle machines usually have more horsepower, too. Maximums range from 20 to 75 hp, compared with a top of 30 on the ram-type machines. And because of their greater size, they almost invariably have powered rapid traverse.

On routine production jobs, the work the operator does is about the same on either type of machine, once the setup is made.

Making the typical cut

For each cut made from the hexagon turret, he traverses the tool to a position close to the work and then engages the power feed. At the end of the feed stroke, the length stop disengages the power feed, and the operator retracts the tool, moving the turret to the right until it automatically indexes to the next station. Then he adjusts speed and feed for the next cut, if necessary, and repeats the same procedure.

On universal machines, the cross slide also has automatic stops, so the operator usually doesn't have to pay strict attention to a cut after he has engaged the power feed. Therefore, in an efficient operation, he should be able to make some of the cuts simultaneously—when the hexagon turrret is machining a long bore, for example.

On some machines, he can preset the spindle speed for the next cut while the present cut is in progress. Then all he has to do is push a button or flip a lever when he wants the next speed.

And on some of the smaller, ram-type machines, which are often set up for a rapid series of short cuts, there is an automatic control for the headstock. Most of the conditions for each cut are preset into the control, and when the turret is indexed to a new station, the control may start or stop the spindle or change its speed or direction. On some machines, the feed rate is also controlled this way, and on others, the same control is exercised over the cross slide.

Accessories

Most production threading on turret lathes is done with taps and dies, but there are several kinds of leadscrew attachments that help produce more accurate threads. The most common is the leader-and-follower system.

The leader is a sleeve, 6 to 12 inches long, with a precision thread on its outer surface. This sleeve is clamped on the feed rod, and the follower, which is a mating split-nut device, is mounted on the apron.

The system is used for single-point threading or for 'leading' a tap or die into the work at the correct rate. The operation is quite similar to the leadscrew work done on an engine lathe, except that the screw is much shorter, and a different leader and follower must be installed for threads of a different pitch.

As an alternative, however, some turret lathes can be equipped with a gearbox that usually offers four different pitches with any leader and follower. And some machines can be factory-equipped with a full-length leadscrew and quick-change gearbox like those on engine lathes.

Taper attachments are a standard accessory for turning and boring angular surfaces with the cross

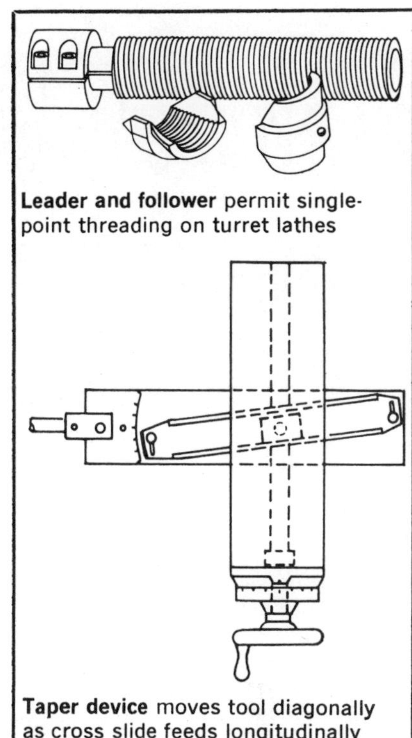

Leader and follower permit single-point threading on turret lathes

Taper device moves tool diagonally as cross slide feeds longitudinally

slide, and they are also available for the cross-feeding hexagon turrets. For tapers that are too steep for a taper attachment, a swivel compound slide can be mounted on the cross slide of most machines.

Tracer attachments for machining curved or irregular surfaces are another standard turret-lathe accessory. Some are mounted on the cross slide, and others are mounted on a cross-feeding hex turret. Still others are mounted on one station of the hex turret and are indexed just as the other cutting tools are.

Precision chucking machines

The machines that go by the name 'precision chucking machines' have turret-mounted cutting tools, and they also have power feed. Therefore they can be classified as power-fed turret lathes, but there are a goood many differences that justify a separate discussion.

One of the distinctive features is that there is usually no cross slide; both cross and longitudinal motions are performed by the turret. Thus the turret has the same variety of cutting paths as does the cross-feeding turret on a saddle-type turret lathe.

Because of this, the same advantages apply. Simple tooling can be used for turning, boring, and cross-feeding cuts, and a series of cuts

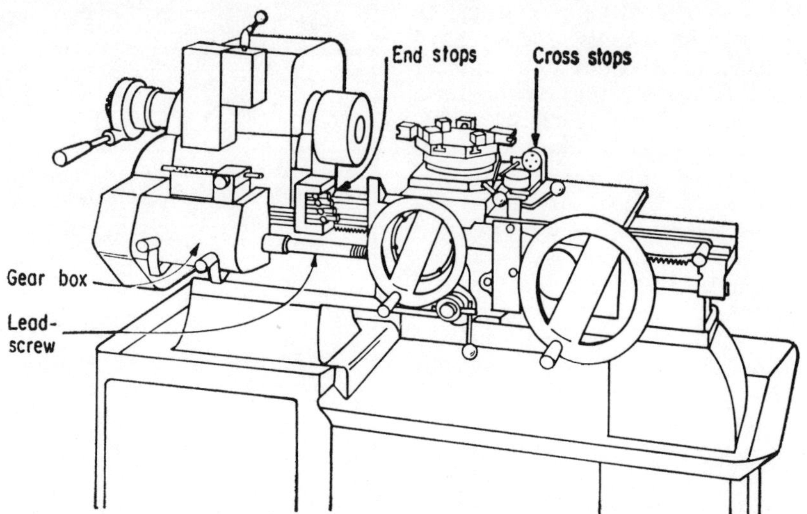

Precision chucking machine has cross-feed turret but no cross slide. All cuts are made from the turret, which has longitudinal and transverse power feed

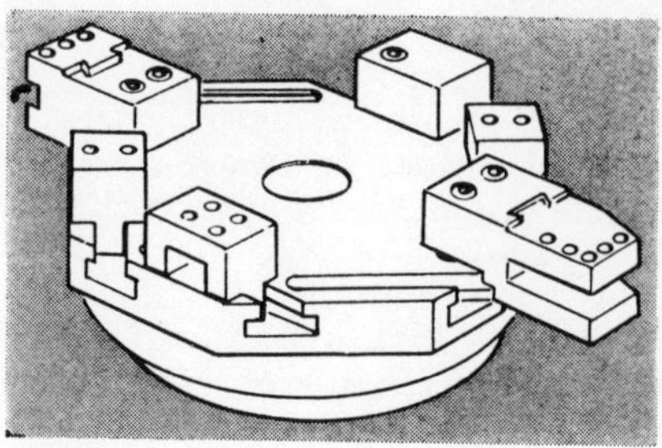

Eight-sided turret is often used on precision chucking machines

can be made with one tool (stepped diameters, for example).

But the precision chucking machines are a good deal smaller than the saddle-type turret lathes. Their maximum swing is about 13 inches; collet capacity is 1-1/16 inch; and they usually have 2 to 3 hp, with spindle speeds ranging up to 3000 or 4000 rpm.

Another difference is in the method of operation. On saddle-type turret lathes, the four-way feed is something extra; it adds to the versatility of the basic machine, but it is usually reserved for short-run jobs or particularly complex parts.

On the precision chucking machines, however, all cutting is done from the turret. This means that the crosswise position of the turret must be adjusted for the depth of longitudinal cuts, and longitudinal position must be adjusted for the depth of cross-feed cuts. It follows, then, that the precision chucking machines cannot match the productivity of a properly set-up turret lathe.

However, it would be a mistake to compare them with turret lathes only. In fact, they can be compared more directly with a small toolroom lathe, if the toolroom lathe has a square turret in place of the compound slide. (Most of the chuckers have eight-station turrets, and the tools are mounted in horizontal T-slots on top of the turret instead of on the vertical faces, as on most other lathes.)

Thus the main differences are that the chucker has more tool positions; it has a unified tooling system; and it makes more use of travel stops when performing repetitive operations.

Both the toolroom lathe and the precision chucking machine place a strong emphasis on precision, and they are designed and built to do close-tolerance work. Both of them also do single-point threading, and some of the chuckers can do it automatically.

In summary, then, the precision chucking machine has collected several advantages from different machines and offers them in a combined package. It claims greater productivity than a toolroom lathe, greater accuracy than the small turret lathes, and faster spindle speeds than the larger turret lathes.

Therefore, its best applications are in close-tolerance work on fairly small parts and in fairly small lots that wouldn't justify the cost of tooling up a fully automatic machine.

Automatic indexing lathes

There are a number of ways to achieve automatic operation, as will be seen in the following pages, but the basic factor is found in the controls. On automatic lathes, the controls make it possible for the machine to go through a complete operating cycle, or a whole series of cycles, without attention from the operator.

In the non-automatic turret lathes discussed perviously, there is already one important element of automaticity in that the machines have automatic length stops, which disengage the power feed when the end of the stroke is reached.

However, the operation always stops at that point. The operator has to be there to withdraw the tool and prepare for the next cut.

In the automatic machines, the essential feature is that the operation does not stop at the end of each stroke. Instead, the control directs the machine to withdraw the tools, index, and start the next cut, using the appropriate speed and feed for each cut and rapid traverse when not cutting.

One general type of automatic lathe operates approximately as this suggests. The length stops not only disengage the feed but also send a signal to the control, reporting that a particular phase of the operation is complete.

This automatically triggers the start of the next phase, and the operation continues this way from one phase to the next. In the following pages, these will be called 'sequential automatics.'

The other general type of automatic lathe has a control system based on cams and gears that establish a fixed amount of time for each cycle. These will be called 'timed-cycle automatics.'

These are not common shop names, of course, but they help make a major distinction. Each type of machine requires a different approach to operation planning and setup, even though the two groups have substantially the same tool paths and do approximately the same kinds of work.

Sequential automatics

Most of the automatic lathes with sequential control are members of the turret lathe family. In addition, the vertical, multiple-spindle machines also operate with this type of control, but the application is somewhat different.

On the single-spindle machines, there is a set of controls that determines the functions (the things that happen) and the speeds and feeds (the rates at which they happen) for each turret station when that station is in working position.

The control system may be electro-mechanical, with trip dogs and limit switches and a function drum that indexes in coordination with the turret; or it may be an electrical or electronic system having a control panel with a complete set of selector switches and dials for each turret station. In either case, the completion of any stroke by the turret triggers the next phase of the operation.

Two distinct types

But the single-spindle machines are also separated into two general types, depending on how the turret is mounted. The 'saddle-type automatic turret lathes' are quite similar to the saddle-type lathes described earlier.

The turret moves along the bedways and is usually mounted in a horizontal plane. There are usually six tool-mounting faces, and the one in working position is centered on the axis of the spindle.

The other type of single-spindle machine has a turret mounted in a vertical plane on a shaft that extends from the headstock. It has five or six tool-mounting faces, which are parallel with the centerline of the spindle but offset from that line.

These machines are often called 'single-spindle automatic bar or chucking machines.' However, they are also called 'open-ended turret lathes,' so we will use the simpler name this time.

In spite of this difference in turret arrangement, the tool paths of the two types are very similar. On both, the turret moves longitudinally for drilling, reaming, boring, turning and threading.

The distinction is that, in moving longitudinally, the saddle-type turret moves toward the end of the work, and the open-ended turret moves alongside the work. Therefore, considering the factor of tool-overhang, the open-ended machines can claim an advantage in long turning operations on large diameters, whereas, the saddle-type machines can claim an advantage in boring deep holes.

However, in the normal run of jobs that do not involve exceptionally long passes and where both turning and boring are done on the same workpiece, these factors do not make a great difference. With suitable tooling, each can perform a full range of work.

Cross-feeding operations are performed by cross slides, of course, either a bridge-type cross slide with a toolblock on each end or two independent cross slides,

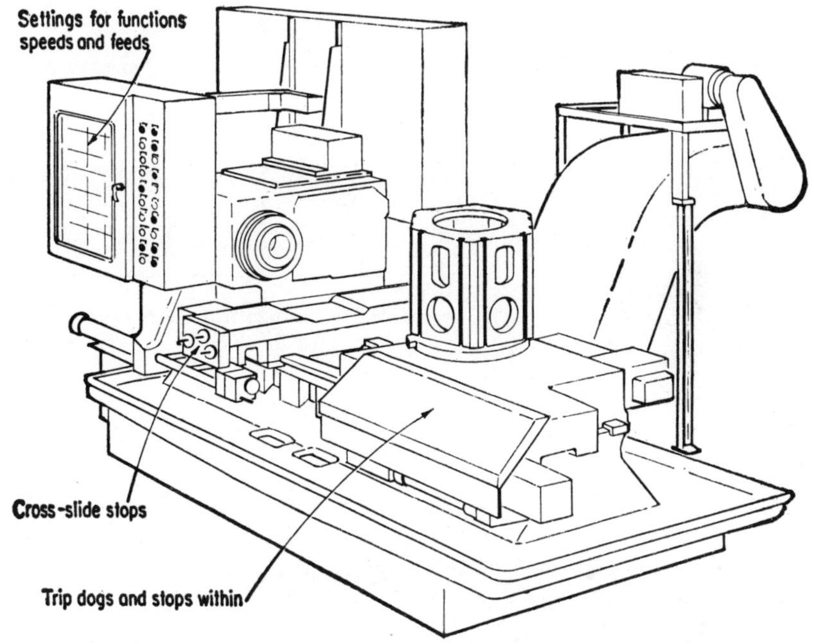

Large automatic turret lathe has saddle-type turret and bridge-type cross slide. This unit is also equipped with a conveyor to remove chips produced

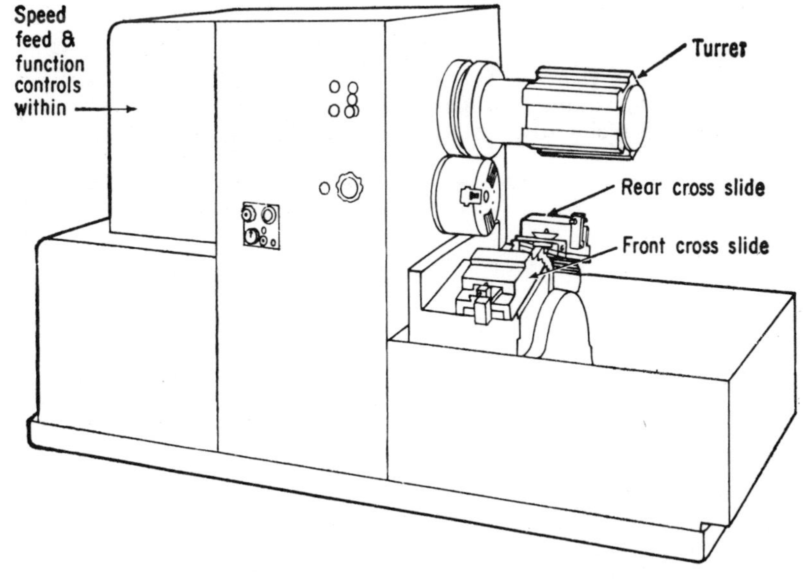

Open-ended turret lathe has turret mounted on shaft extending from headstock. Independent cross slides may operate simultaneously or at different times

depending on the design of the machine. (Bar machines may have a third slide for cutting off.)

On most machines the function control can be set so the cross slides operate with any turret face, feeding either simultaneously with the turret or before or after the turret advances. But on all these machines, the cross slides move crosswise only.

On some machines a cross slide can be set at an angle, but it must remain at the same angle for the whole job. For the more intricate cuts such as recessing, the motions of the turret and a cross slide are often coordinated.

This is done with a 'slide tool' mounted on the turret and a 'pusher' mounted on the cross slide. The turret advances to cutting position and dwells there while a 'late cross slide' advances to push the cutter to the required depth. The same principle is used for cutting tapers and some contours.

In addition, a variety of other special tools and accessories are available for spherical boring, angular facing, and tracing. Threading is usually done with collapsing taps and self-opening dies.

Therefore, these automatic machines are capable of performing substantially the full range of production work done on non-automatic turret lathes. One exception is that the non-automatic machines tend to go to larger sizes.

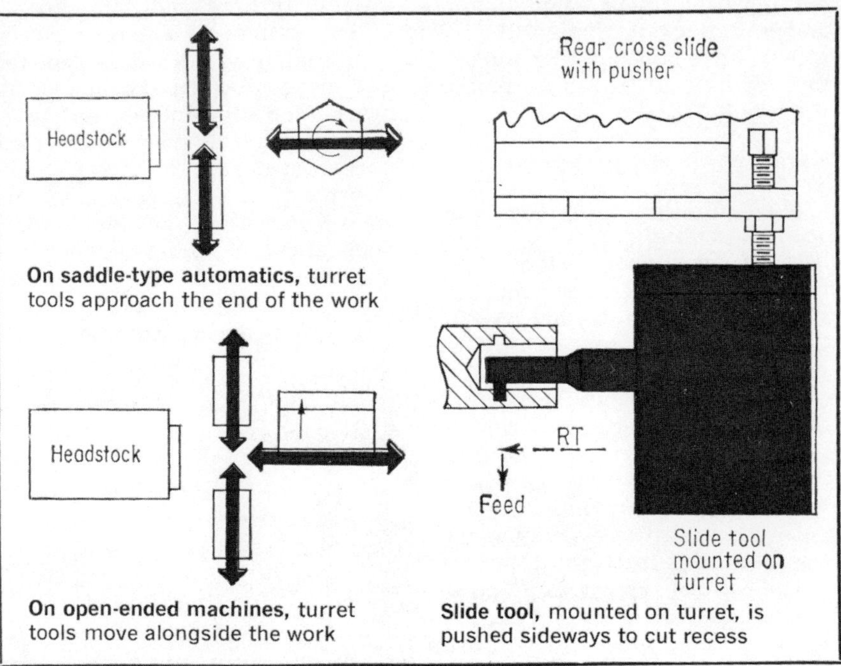

On saddle-type automatics, turret tools approach the end of the work

On open-ended machines, turret tools move alongside the work

Slide tool, mounted on turret, is pushed sideways to cut recess

Added setup pays off

Setup is somewhat more complicated on the automatic machines, largely because all the controls have to be set for every pass before the operation can begin. Once the setup is aproved, though, the controls do not have to be touched again.

Machine cost for the automatics is higher, of course, because the control system adds a large item of expense. However, productivity is also a good deal higher, and the output is more uniform because the human element has been eliminated.

As to sizes, the saddle-type automatic machines are generally larger and have longer working strokes than the open-ended machines. Thus the saddle-type machines have a swing over the cross slide ranging from 12 to 24 inches. The cross-slide strokes range from 5 to 9 inches each side of center, with maximum turret strokes from 12 to 30 inches. Power ratings are 15 to 60 hp.

In comparison, the open-ended machines have swings of 6 to 19 inches, cross-slide stroke of 1 to 6 inches, and turret stroke of 3 to 11 inches. Power ratings range from 7½ to 40 hp.

Other machines

Another machine with sequential control is the automatic precision chucking machine, which is a saddle-type lathe with a cross-feeding turret but no cross slide. All cutting is done from the turret, just as on the non-automatic precision chucking machines described earlier. However, on the automatic machines there is no pause between cuts to adjust the turret position for the next cut.

On the vertical, multiple-spindle machines there is a modified sequential control. These machines have a number of vertical work-holding spindles (6, 8, 10, or 12) mounted in a carrier that indexes about a vertical column. Each working station has an overhead tool mount that does its assigned portion of the complete operation every time the carrier indexes. Therefore, as there is a workpiece at every station, each indexing produces a completed piece.

The tool mounts do not index, of course. Each one makes the same pass on successive workpieces, and they have individual electro-mechanical controls for rapid traverse, feed rate, length of cut, and rapid withdrawal. Thus, they may or may not finish their cuts together, but the spindle carrier cannot index until all tool mounts are fully retracted.

Timed-cycle automatics

The machines in this category, whether single- or multiple-spindle, are often called 'screw machines,' because they were originally designed for machining screws and other small threaded parts. The term still refers to the high-speed production of small parts, but they may or may not be threaded.

Actually, the multiple-spindle machines are usually called 'multiple-spindle bar or chucking machines,' or simply 'multiple spindle automatics.' The term 'screw machine' is more often used for the single-spindle machines, but, as there are two distinct types, we will call them the 'turret-type screw machine' and the 'sliding-headstock screw machine.'

Of course, the factor that brings all these machines together under one heading is that they employ the same general type of control system—the time-cycle control.

Here, the total time for the operation is fixed in advance, and the precise moment when each phase begins and ends is controlled by cams, which also provide the driving force for all sliding members. The cams rotate in unison, and one revolution of the camshaft (or shafts) completes the cycle.

On most other automatic lathes, the familiar factors of speed, feed, and length of cut are set directly on the machine. On the timed-cycle machines, spindle speed is set in the same way (usually with pick-off gears), but the length of cut is established by the cams, and there is no machine setting for rate of feed.

Instead, the rate of feed for any cut is established by two other factors—the shape of the cam and the speed ratio between the spindle and the camshaft. Here's how it works.

First, the length of the cut and the desired rate of feed are converted into revolutions required to complete the cut: length of cut plus a safety factor (in.) ÷ rate of feed (ipr) = revolutions required.

On single-spindle machines, the revolutions required for all cutting and non-cutting functions (less allowances for overlap) are added together to find the total revolutions required for the operation. On multiple-spindle machines, the only figure needed is the number of revolutions for the longest cut.

In either case, the total number of revolutions is the important factor. Once this is known, another pair of pick-off gears (not the ones for spindle speed) can be installed between the camshaft and its drive so that the camshaft will make one revolution while the spindle makes its total required revolutions, as found above.

But the rotating speeds of the spindle and the camshaft only set the general framework and establish the total time for the cycle. Within this framework, the shape of each cam determines when a tool (or group of tools) will start to advance and how fast and how far it will advance.

This type of control requires more planning and preparation for an operation than the sequential type of control does. However, the machine itself can be a good deal simpler, and less expensive, because the control is largely in the cams instead of being built into the machine.

The principle is applied in two different ways, however, as will be explained in the following sections. The main distinction is that the single-spindle machines make a series of cuts throughout the cycle, and the multiple-spindle machines make all their cuts simultaneously, in one part of the cycle.

Timed-cycle: single spindle

The two types of single-spindle automatic screw machines have completely different cutting paths, but they both use the timed-cycle

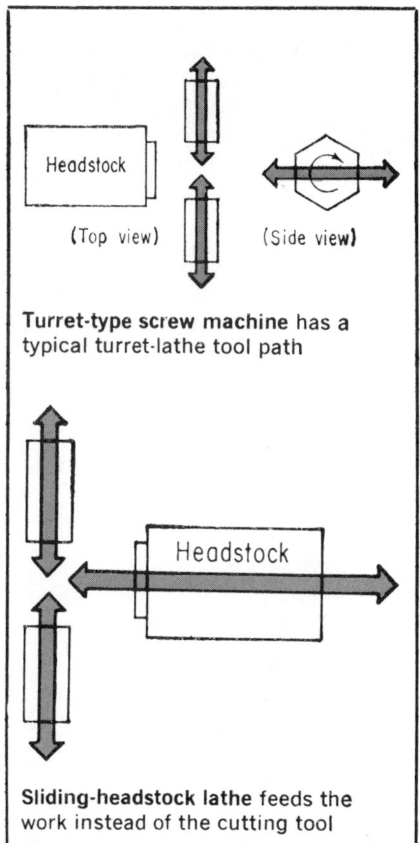

Turret-type screw machine has a typical turret-lathe tool path

Sliding-headstock lathe feeds the work instead of the cutting tool

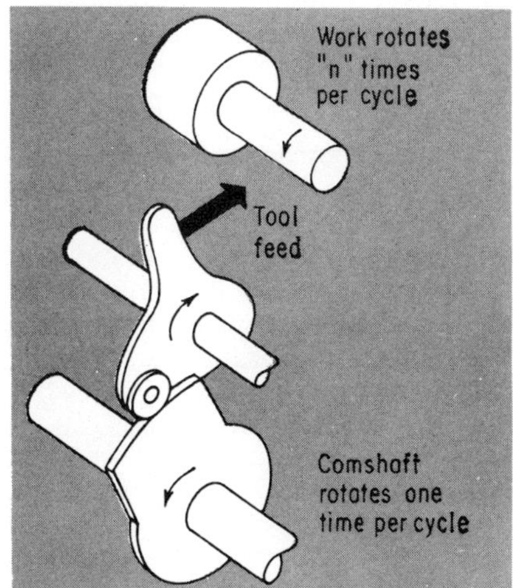

Total cycle time is determined by gears that drive camshaft; cam drives lever that feeds the tool

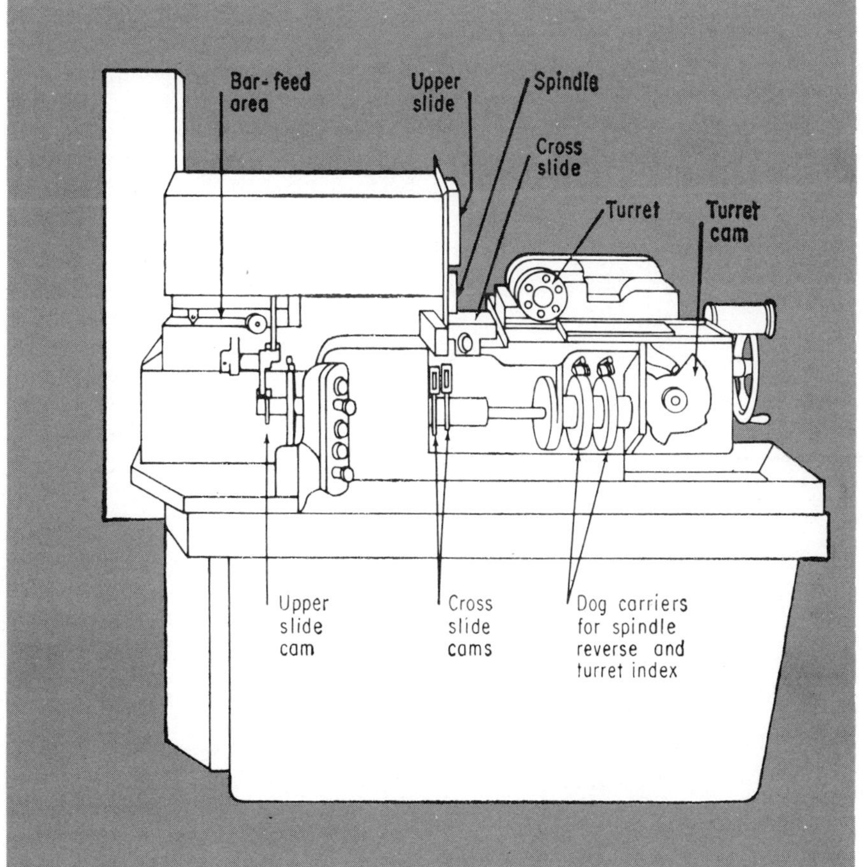

Turret-type screw machine has turret mounted in a vertical plane. Turret cam drives turret through successive passes; other cams drive the cross slides

control, and they apply it in approximately the same way.

These machines work on only one piece at a time, but they work on it with a good many different tools. There are a number of cross-slide tools, and the control must also provide for a series of longitudinal motions.

Obviously, these motions cannot all take place at the same time. They must be scheduled at different times during the cycle in order to complete the required cuts without interference.

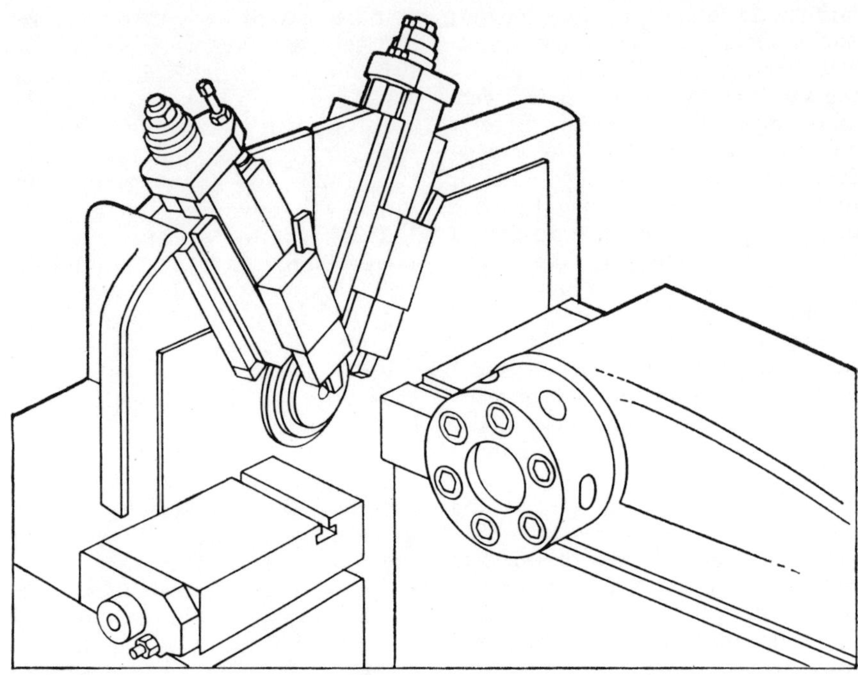

On single-spindle automatic, the workpiece is surrounded by cutting tools, as on this turret-type screw machine. Tools must take turns to make cuts

Constant-speed camshaft

Therefore, the camshaft of a single-spindle machine usually rotates at a fixed rate of speed throughout the cycle, and the cams that control the various motions are designed so that each motion starts and stops at a suitable time.

For example, a cross-slide cam should reach its high point (the end of the cutting stroke) at a time when the tool and holder will not collide with other tools.

This is one of the advantages of cam control. The clearances for closely timed and intricate tool movements can be calculated in advance. However, these calculations are a job for an expert, and the cams often have to be designed and cut for the individual job.

On the other hand, all the setup man has to do is install the cams and proper gearing and adjust the tools to cut to the correct depth. Furthermore, as the feed rates are established by the shapes of the various cams, the optimum rate can be used, even with constant camshaft rotation.

Two screw-machine types

Up to this point, the turret-type screw machines and the sliding-headstock screw machines have a good deal in common, and they also use similar accessories, which will be covered later. However, the machines are so different in their methods of operation that they will have to be treated separately.

The turret-type machines operate with almost the same cutting paths as the automatic saddle-type lathes discussed earlier, even though the turret in this case is usually round and is mounted in a vertical plane parallel with the spindle axis.

A series of end-working tools, usually 6 or 8, is mounted on the turret, which feeds toward the headstock. In addition, there are always two independent cross slides, plus one or two additional slides mounted on the headstock.

Maximum bar capacities for these machines range from ¼ inch to about 2 inches, and maximum turret travel is from 1½ inch to 3½ inches. Spindle speeds are as high as 9000 rpm on the smaller machines and power ratings range from 3 to 7½ hp. Equipped as chucking machines, they can carry chucks up to about 4½ inches.

Other type feeds the work

In contrast with these machines, the sliding-headstock lathes (called Swiss-type automatics) work on an entirely different principle. For turning operations, they feed the *work* longitudinally, and the cutting is done by a cross-slide tool positioned in the path of the moving workpiece.

As it is fed, the work is supported by a close-fitting guide bushing in a stationary support. And because all the cutting takes place immediately in front of this bushing, long slender workpieces can be machined with very little error caused by deflection.

For cross-feeding cuts such as forming and cutoff, the headstock can be made to dwell at any point while the cut is being made.

The toolholders (usually 5) are mounted radially on the same support as the bushing, and the cutting path for all these tools is essentially straight in and out. However, by combining the motion of a cutting tool with the motion of the headstock, curves and tapers can also be generated, and intricate shapes can be machined with relatively simple tools.

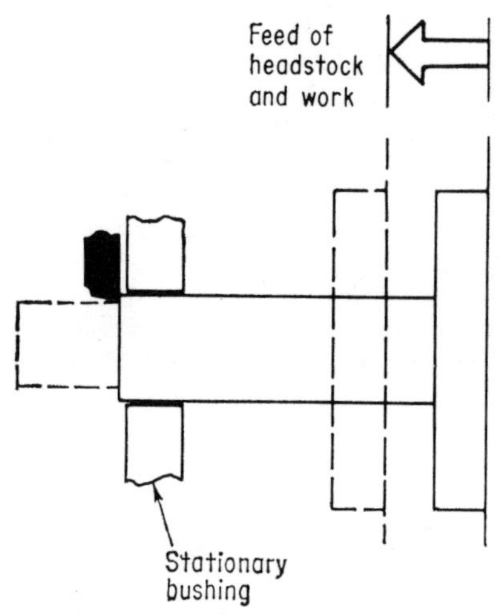

On sliding-headstock machine, the headstock is at right-hand end

End-working operations such as drilling, reaming, boring, and threading are performed with accessories mounted on the bed opposite the headstock. These accessories may have rotating or non-rotating spindles.

A typical sliding-headstock

lathe with a maximum collet capacity of ½ inch may have a standard feed capacity of about 4 inches. Spindle speeds for such a machine are 7000 to 8000 rpm. The smallest machines have a maximum collet capacity of about 5/32 inch and the largest, about 1¼ inch. Power ratings range from 2 to 5 hp.

Accessories

Both types of screw machines offer an impressive array of accessories. Some of these merely extend the capabilities of the machine in normal lathe operations (with the workpiece rotating), but many of them perform work that is not normally done on a lathe.

Among the accessories for lathe-type operations are attachments for contour turning, single-point threading, high-speed drilling, and eccentric turning.

Among the accessories for 'extra' operations, one group is based on the principle of stopping the spindle and then feeding a rotating cutter radially to perform cross drilling, milling or tapping. The spindle can also be equipped with an indexing device so that two cross-drilled holes, for example, can be aligned at 90°.

The other group works with a pick-off arm that transfers the cut-off workpiece to an auxiliary work center on the same machine. There the cut-off end can be drilled, burred, slotted, milled, or threaded while the next piece is being machined from barstock.

Timed cycle: multiple spindle

As noted earlier, the multiple-spindle automatics make all their cuts simultaneously. They can do this because they have much more working space per tool than the single-spindle machines do.

There may be 4, 5, 6, or 8 spindels, but each spindle is served by only one standard tool mount for end-working tools, and there is no more than one cross slide for each spindle. Thus the tool-path diagram is very simple, but there is one for every station.

The 'end slide' has a tool-mounting surface opposite each working spindle, and on most machines the end-working tools on this slide all move together and travel the same distance. The cross-slide tools are usually individually mounted and individually cammed.

The spindles are mounted in an indexable carrier, and each workpiece is machined progressivly at successive tool stations until the operation is completed in the final station. As there is a workpiece in each spindle, and as all the cuts are made simultaneously, a finished piece is produced every time the carrier indexes. This is one cycle, or one revolution of the camshafts.

Because all the cuts can be made at the same time, the cycle is usually divided into two parts: the cutting portion and the non-cutting portion or 'idle time.'

During the idle time, the cam-

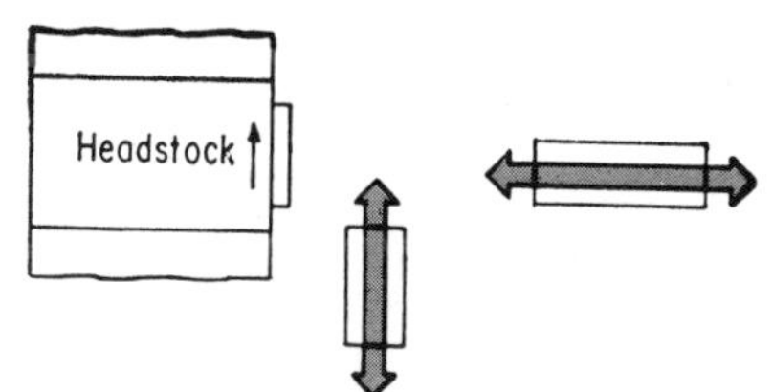

All spindles on multiple-spindle automatic have the same tool path

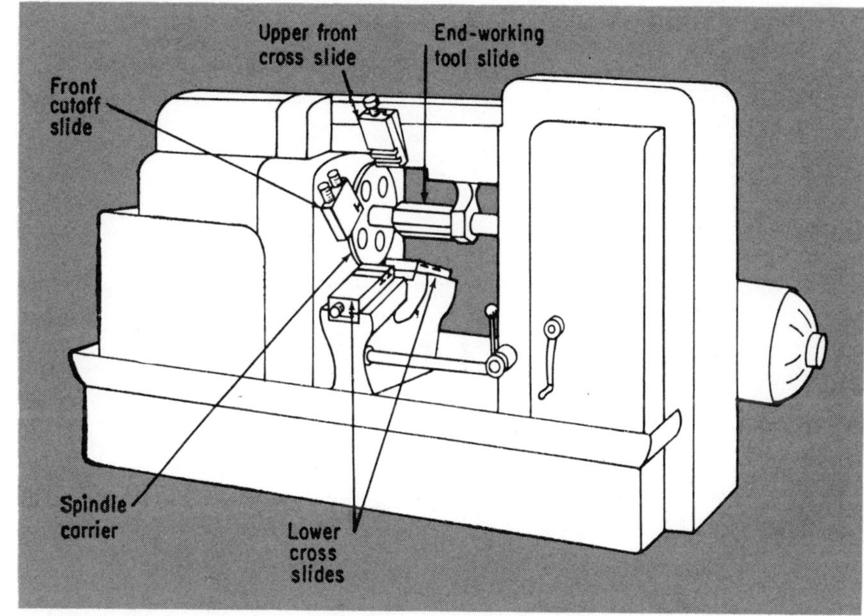

Multiple-spindle automatic makes all cuts simultaneously and then performs the non-cutting functions (tool withdrawal, index, bar feed) at high speed

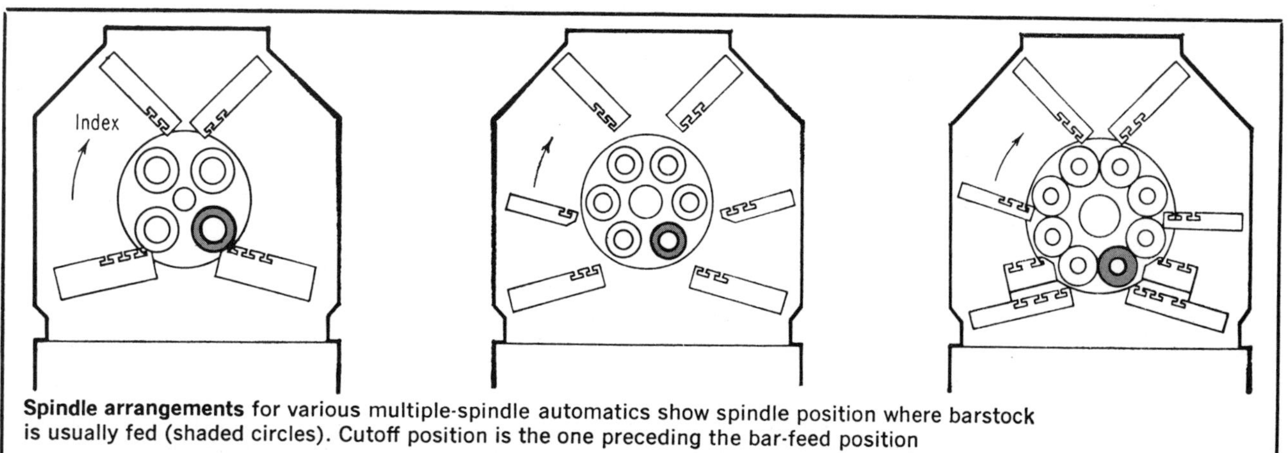

Spindle arrangements for various multiple-spindle automatics show spindle position where barstock is usually fed (shaded circles). Cutoff position is the one preceding the bar-feed position

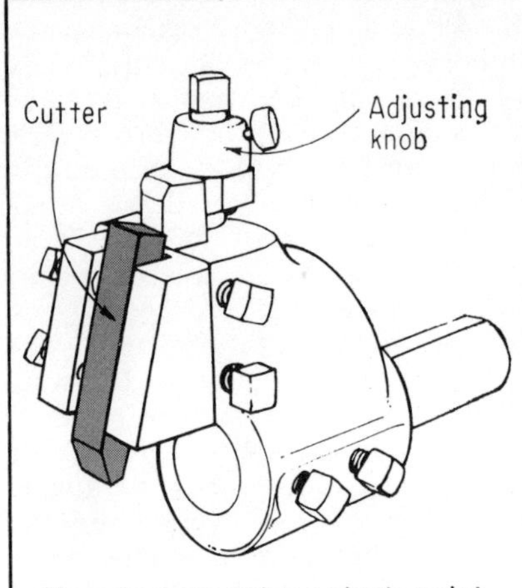

Knee turner carries a single-point cutter and a boring bar or a drill

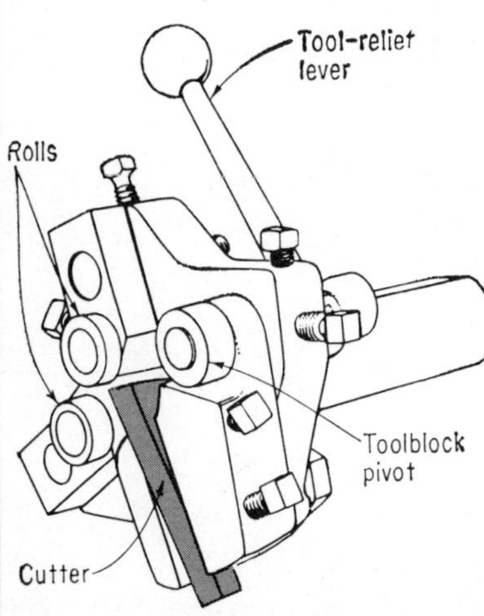

Roller turner has rolls to support the work against the cutting forces

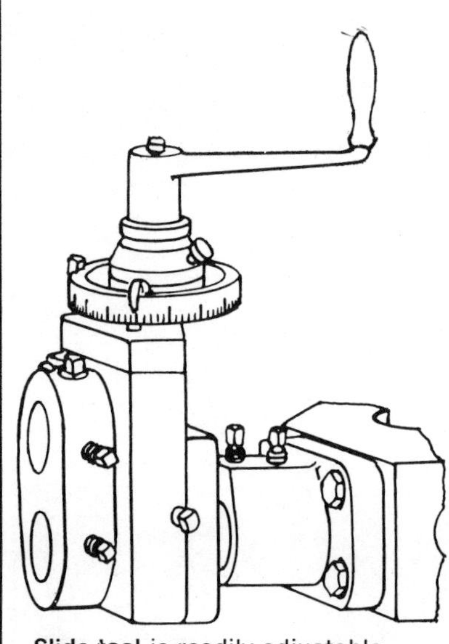

Slide tool is readily adjustable and can also be hand-fed in cuts

shafts rotate at high speed for withdrawing the tools, indexing the spindles, feeding stock, and advancing the tools into cutting position. This time is fixed for any one machine and may be from ½ sec to 8 sec, depending on the design of the machine.

The cutting part of the cycle is adjustable for any operation, and its duration is based on the amount of time (actually the number of spindle revolutions) needed to make the longest cut at its optimum feed rate. But the 'unusual' thing about these machines is that all cutting tools are given this same amount of time because this is the time available during the cutting part of the cycle.

All tools start to feed at the same time and finish at the same time. Therefore, on the shorter cuts, some of the tools will 'cut air' part of the time and others (those that are individually cammed) will cut at slower feed rates. This does not represent lost time, however, because the whole operation is being done simultaneously.

Various types of cams

The cams for these machines may be disk-type or drum-type cams, or both, depending on the design of the machine, but they do not have to be specially made for each job. Instead, standard cams are installed to move each toolslide the approximate distance required.

On some machines, however, there is an adjustable linkage between the cams and the slides that reduces or entirely eliminates the amount of cam-changing required. And some machines use an adjustable gearing device instead.

The basic cutting paths of these machines are very simple, of course, but they can be modified considerably with special tools and attachments for internal recessing, taper turning and boring, and contour tracing. In addition there is a great variety of tooling for such operations as sawing, milling, broaching, and hobbing. And there are drill speeders and accessories for leadscrew threading and single-point threading.

Like the single-spindle machines, these can also stop one or more of the spindles for cross milling and drilling. Pick-off attachments for doing subsequent operations after cutoff are also available but are not used as much as on the single-spindle machines.

The multiple-spindle machines cover a wide range of sizes. Different bar machines may work with bars as small as about ½ inch or as large as 8 inches, and chuckers range from a 6-inch swing to 15 inches. Power ratings range from 7½ to 75 hp, depending on both the capacity of the machine and how many spindles it has.

Turret tools

Most of the lathes in the indexing-tool group make all their longitudinal cuts from a tool mount that is centered on the axis of the spindle. Therefore, in order to perform turning operations, the cutting tool has to be offset from that axis by a distance equal to half the diameter to be turned.

One of the basic toolholders for this purpose is the 'knee turner,' which is used for relatively small diameters and short turning cuts. In addition to mounting a single-point cutter, it also has a center hole in which a drill or boring bar can be mounted so that two cuts can be made simultaneously.

For longer turning cuts, the 'roller turner' is a popular device. This toolholder carries a single-point cutter and two backup rollers that support the work against the cutting forces.

The 'slide tool' has a handwheel that can be adjusted quickly for the diameter of longitudinal cuts. It can also make hand-fed recessing, grooving and facing cuts.

But these are only the basic toolholders for turning from the turret. For larger diameter work and for simultaneous cutting, a number of such toolholders can be mounted in combination on a multiple turning head. ■

Optimum cutter geometry

Here are formulas that let you 'tailor' cutter geometry to suit a variety of workpiece materials. You can calculate optimum angles, tool life, and production gains

Cutting-tool performance, the limiting factor in machining any material, is critical for aerospace alloys. Because the tool life of commercially ground cutting tools is often inadequate, the aerospace industry is expending much effort to improve cutting tool materials and especially cutting-tool geometries.

The most substantial gain is realized by optimizing cutter geometry. For example, machining costs can be reduced as much as 16% for 4340 and 56% for Rene 41 when an optimum cutter geometry is substituted for the 'best standard' cutter geometry.

According to Technical Report AFML-TR-68-350, it would appear, at first glance, that to cover most engineering materials, the cutting tool industry would have to standardize approximately 135 different optimum cutter geometries for face mills alone. Involved with face-mills are fifteen effective rake angles, three inclination angles, and three corner angles, or 135 different geometries. However, 20 to 30 geometries will probably handle 90% of requirements for machining all engineering materials, not only a few aerospace alloys. These requirements can be established by use of empirical formulas presented here and a computer program.

If an optimized system of standard face-mill geometries is set up for use by American industry, it is suggested that cutter densities (number of cutter teeth per inch of cutter diameter) also be established, as well as corner-angle/optimum-inclination-angle relationships. In this connection, cutter density affects metal-removal rate if a constant chip load is maintained. Cutter density may

By F J McGee, P Albrecht, and H N McCalla
Vought Aeronautics Division
LTV Aerospace Corp, Dallas

Extracted from Air Force Technical Report AFML-TR-68-350, which was prepared for the Manufacturing Technology Div, Air Force Materials Laboratory, Air Force Systems Command, Wright-Patterson AFB, Ohio

Optimized cutter geometries for several types of cutters might well be established for all engineering materials, not only the difficult-to-machine aerospace alloys. Values must be established for carbide and HSS cutters

be a function of horsepower, but directly or indirectly it is expected to depend on workpiece-material properties.

Different types of cutter material—HSS, carbides, and ceramics—will significantly affect optimum cutter-geometry values. Inclination angles will probably remain stable, but optimum effective rake angles may vary widely for different types of cutter material.

Optimum cutter geometries for different machining processes may show some similarity, but large errors can occur from programming one machining process with formulas from another process. Because the equations presented in Table II represent face-milling applications, analytical solutions should be developed for lathe turning and plain milling.

The current work on optimum cutter geometries is not the first in the field. Eight years ago Chance Vought engineers reported a technique for finding the optimum cutting tool geometry for any job—in any material (**AM**—Oct 2 '61, p95). They said at the time that in three-dimensional cutting there are four *functional* angles of importance. These are not the *basic* angles used by tool de-

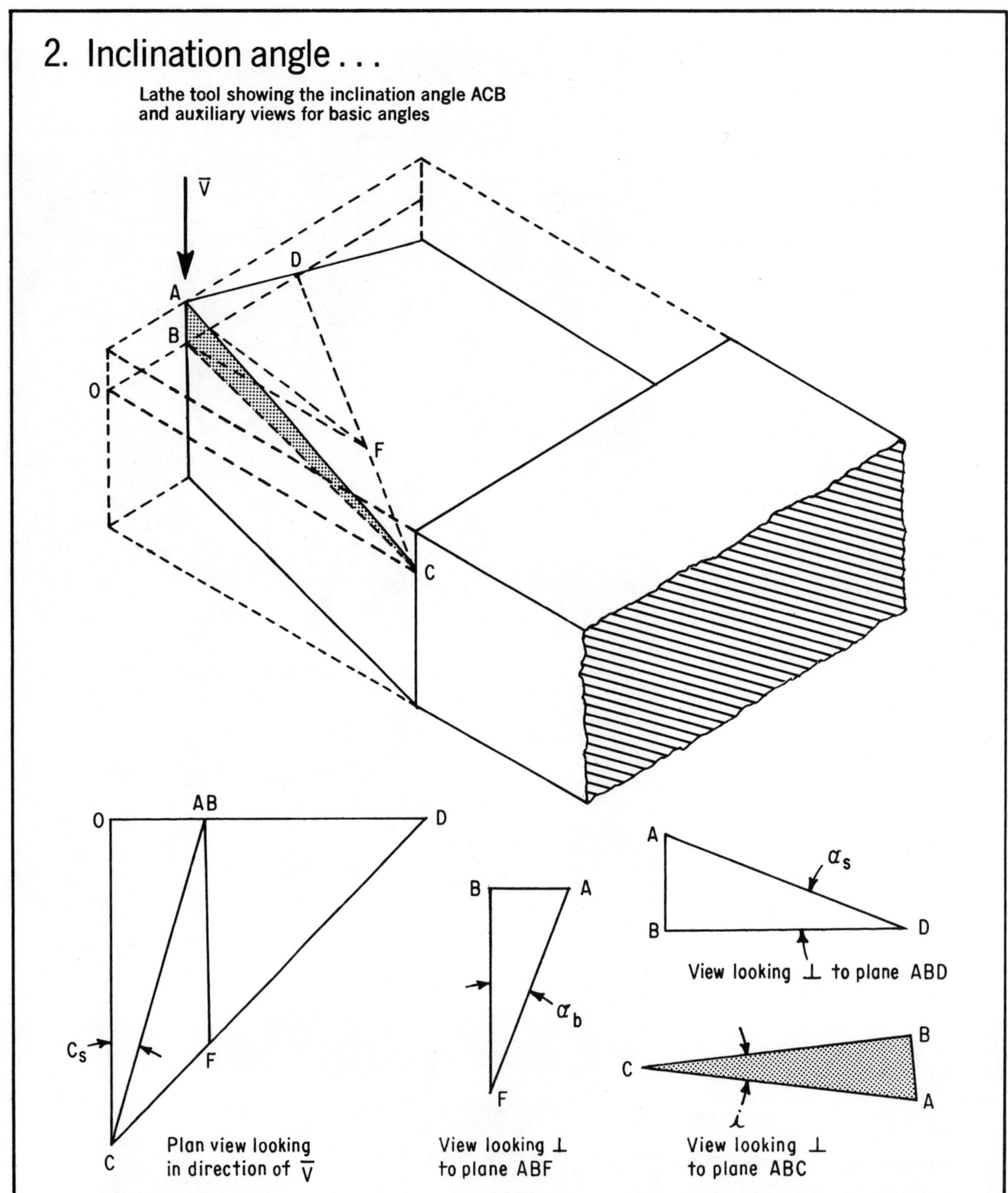

signers, machine operators, tool manufacturers and others. These basic angles—back rake, side rake, end clearance, side clearance, end cutting edge, and side cutting edge—are needed to make and specify tools.

Functional angles are expressed in terms of basic angles. These four angles are:

Inclination angle, i, Fig. 2, is the angle formed by the tool cutting edge and a normal to the cutting velocity vector $\bar{V}$.

$\tan i = \tan \alpha_b \cos C_s - \tan \alpha_s \sin C_s$

Velocity rake angle, α_v, Fig. 3, is the angle formed by the tool face and a normal to the machined surface in a plane containing the cutting velocity vector.

$\tan \alpha_v = \tan \alpha_s \cos C_s + \tan \alpha_b \sin C_s$

Normal rake angle, α_n, Fig. 4, is the angle formed by the tool face and a normal to the machined surface in a plane perpendicular to the cutting tool edge.

$$\tan \alpha_n = \tan \alpha_v \cos i = (\tan \alpha_s \cos C_s + \tan \alpha_b \sin C_s) \times \cos i$$

Effective rake angle, α_e, Fig. 5, is the complement of the angle between

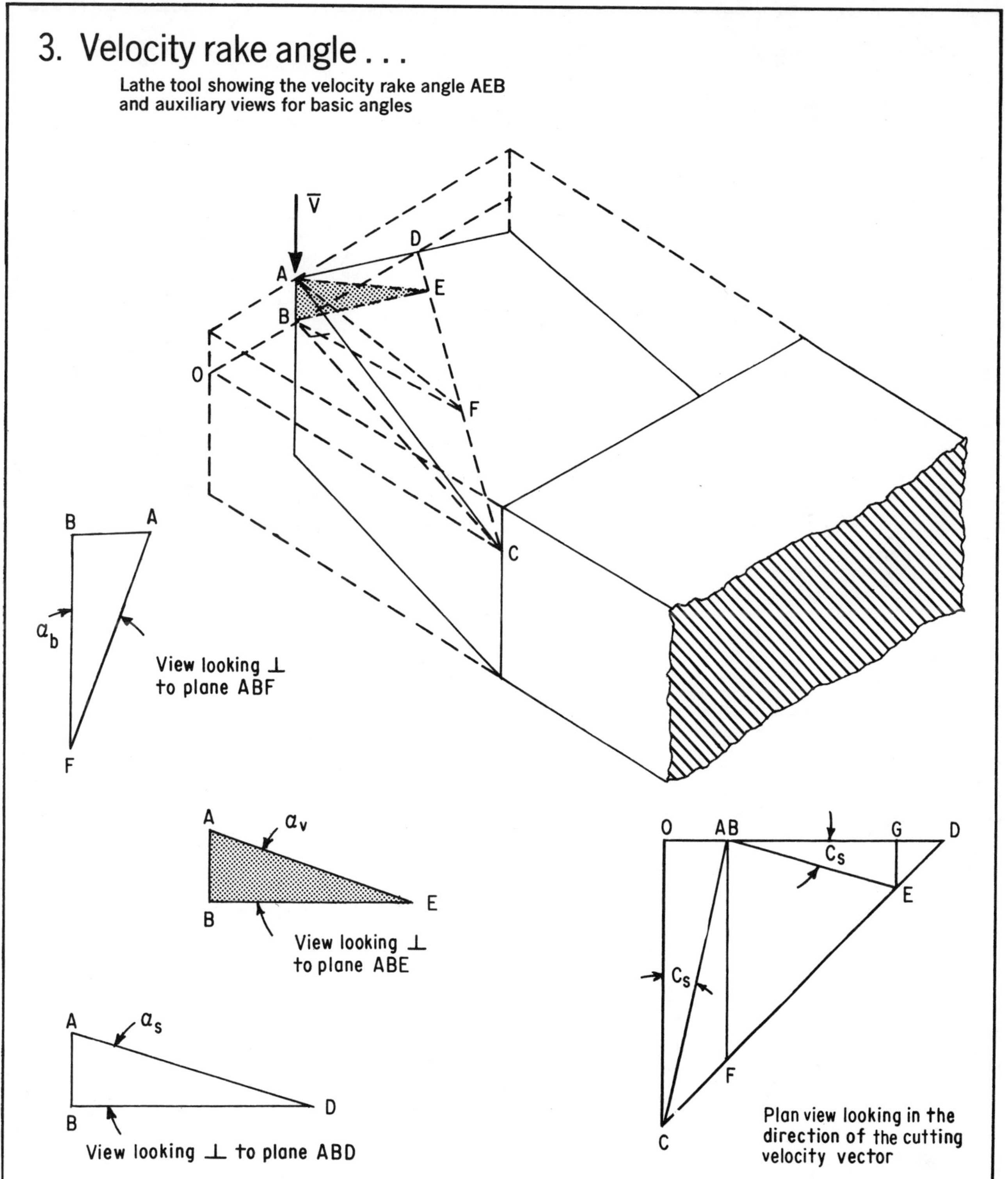

the cutting velocity vector, $\overline{V}$, and the chip velocity vector V_c. The angle $(90° \pm \alpha_e)$ is the angle through which the chip is deflected by the cutting tool.

$$\sin \alpha_e = \sin^2 i + \cos^2 i \sin \alpha_n$$

$$\tan \alpha_e = \tan \alpha_s \cos (C_s + i) + \tan \alpha_b \sin (C_s + i)$$

Except for nomenclature, the tooth geometry of a face milling cutter is essentially the same as that of a lathe tool. As illustrated in Fig. 6, axial rake, radial rake, and corner angles (CA) on a face mill are equivalent to the back rake, side rake, and side cutting edge angles, respectively, for a lathe tool. Functional angles for face mills are computed by substituting angles in the above formulas. The functional-angle formulas for face mills are:

$$\tan i = \tan \alpha_a \cos CA - \tan \alpha_r \sin CA$$

$$\tan \alpha_n = (\tan \alpha_r \cos CA + \tan \alpha_a \sin CA) \cos i$$

$$\tan \alpha_e = \tan \alpha_r \cos (CA + i) + \tan \alpha_a \times \sin (CA + i)$$

Plain milling cutters seem different from lathe tools and face mills, but Fig. 7 shows that the basic angles on a plain milling cutter are functional angles. The radial rake angle is equivalent to the velocity rake angle, and the helix angle

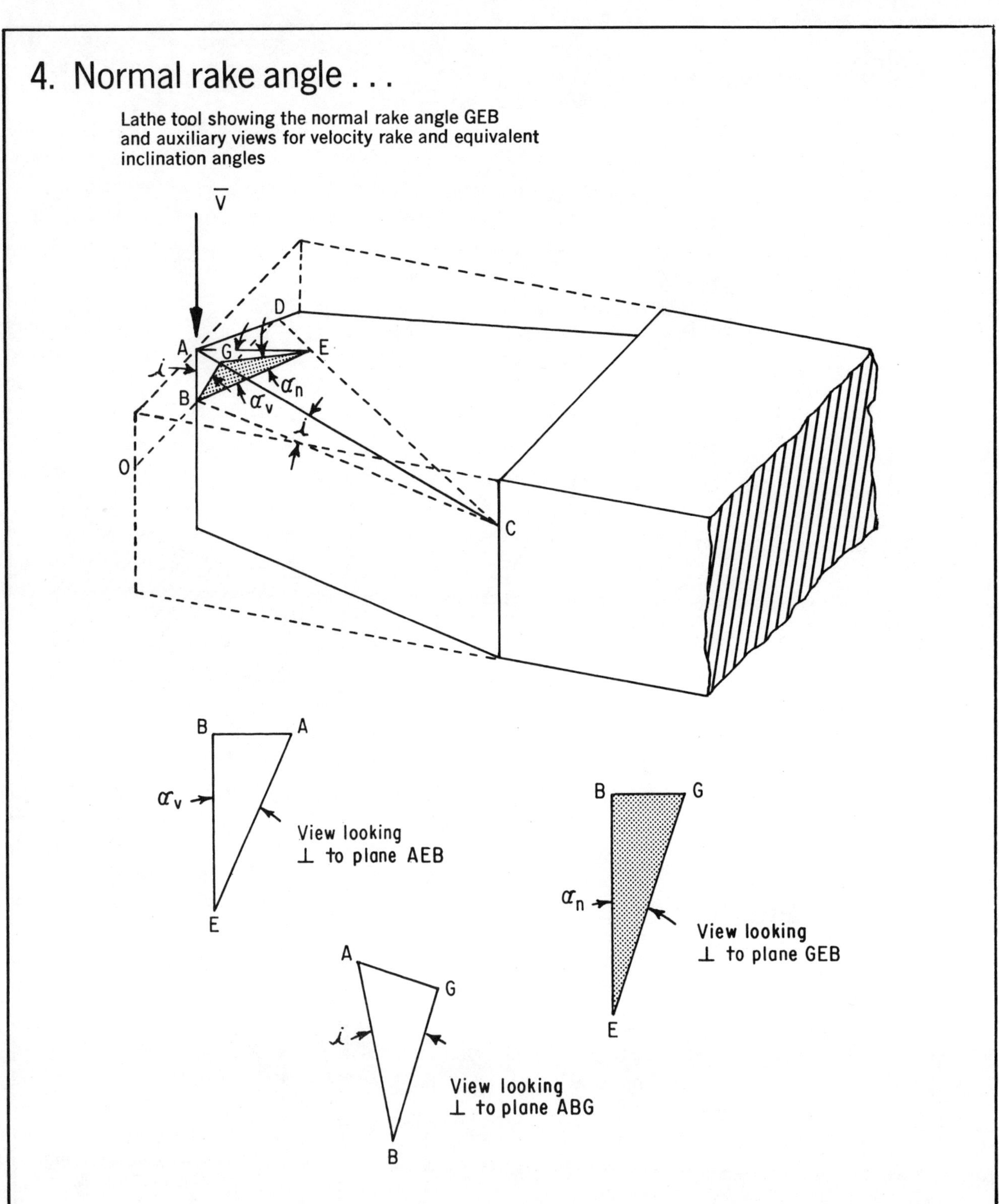

4. Normal rake angle . . .

Lathe tool showing the normal rake angle GEB and auxiliary views for velocity rake and equivalent inclination angles

is equivalent to the inclination angle. Thus:

$$i = h$$

$$\tan \alpha_n = \tan \alpha_r \cos h$$

$$\sin \alpha_e = \sin^2 h + \cos^2 h \sin \alpha_n$$

Empirical formulas

It was realized from the principal cutter-geometry equation ($\sin \alpha_e = \sin^2 i + \cos^2 i \sin \alpha_n$) that optimum cutter geometry can be defined if optimum values for any two of the functional angles are established and the optimum value for the third is determined mathematically.

Because metalcutting is a plastic-flow process, the optimum values for each of the three functional angles are governed by plastic-flow properties. In fact, the stress required to cut a material is a function of the yield strength and the strain-hardening exponent. Thus, from the properties of six experimental materials (Table I), an empirical formula for normal rake angle was derived:

$$\alpha_n = (57.5° - 0.266\,S_{ty})(1 - m)$$

An optimum inclination angle should reduce strain hardening. If optimum inclination angles for the several experimental materials are plotted against their strain hardening exponents, m, a

5. Effective rake angle . . .

Lathe tool showing effective rake angle AHB and auxiliary views for basic and functional angles

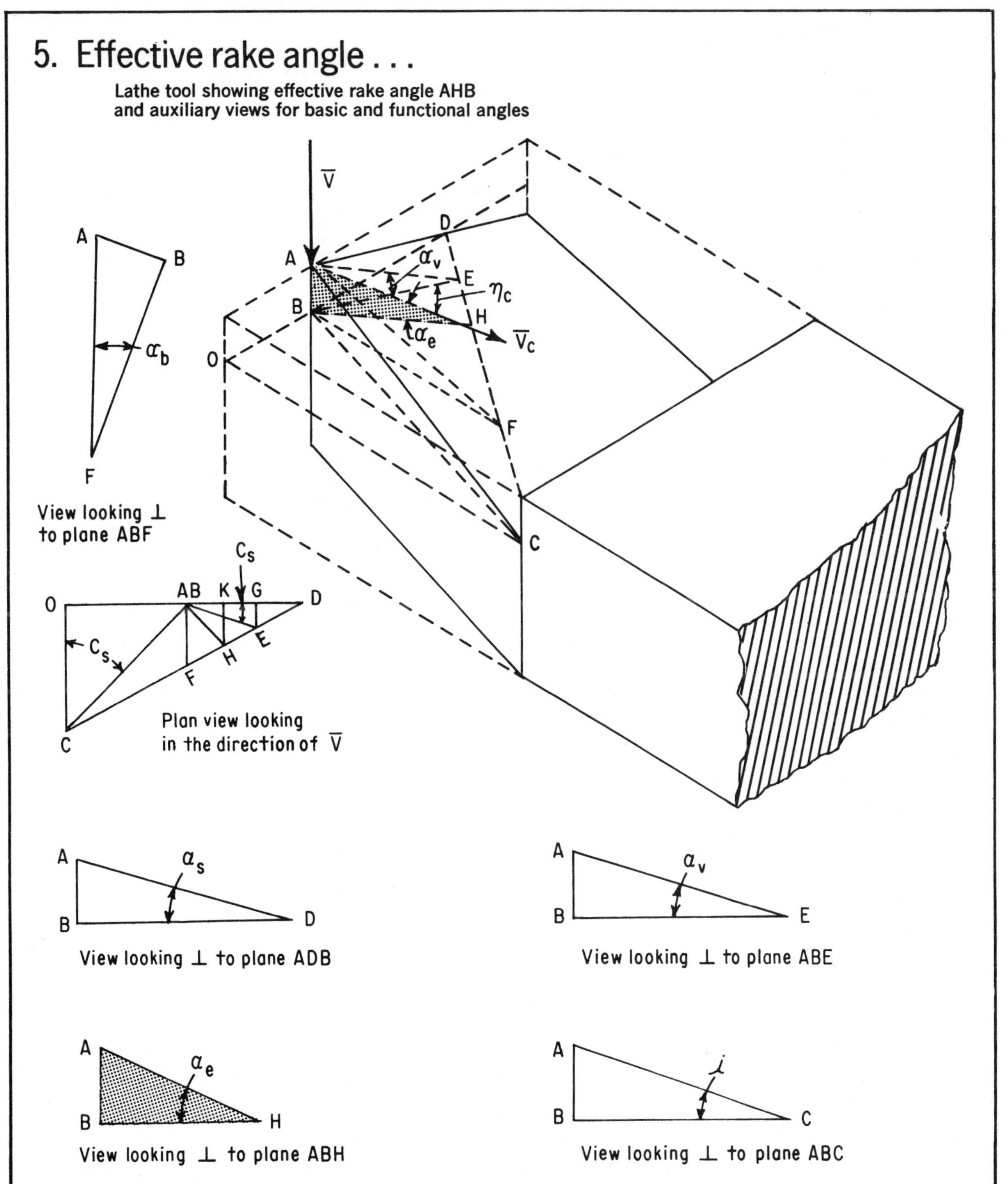

fairly good straightline relationship is shown. But by use of a digital computer and a multiple linear regression technique, one gets a more valid formula:

$$i = 2.9806 - 64.15(m) - 0.1596(S_{ty}) + 0.1071(S_{tu}) - 0.006(R.A.) + 0.2089(e)$$

where: m = strain hardening exponent
S_{ty} = tensile yield strength, ksi
S_{tu} = ultimate tensile strength, ksi
R.A. = reduction of area, %
e = elongation, %
i = optimum inclination angle, degrees

By carrying out further calculations, formulas were developed for effective rake angle, machining shear stress, cutting speed yielding a 60-minute tool life, the strain hardening exponent, and other values. Eventually all of the experimental data were correlated into mathematical relationships for optimizing cutter geometries and overall machining performance of various metals. By computer analysis, the previously

6. Face mill cutter geometry and equivalent lathe tool angles

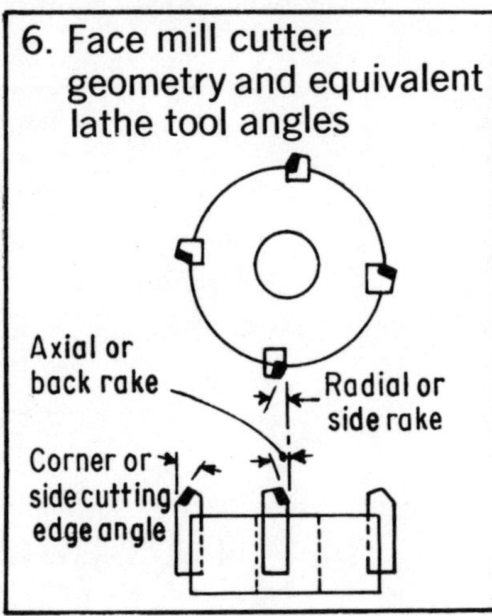

7. Plain milling cutter geometry showing equivalent functional angles

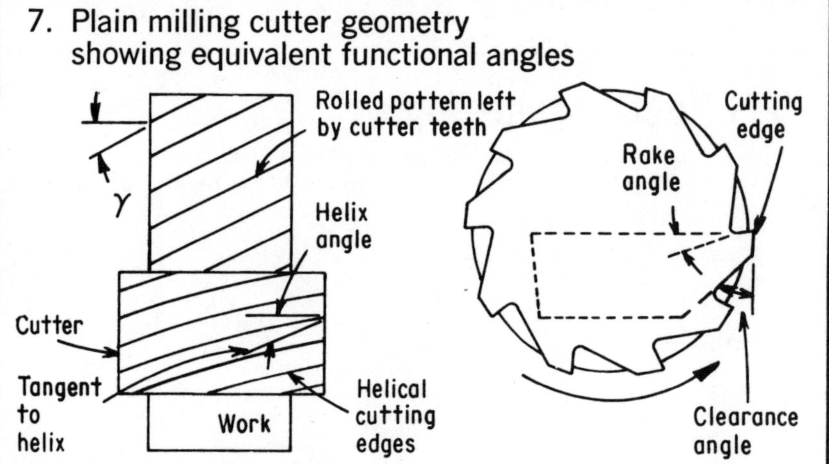

Table I. Properties of materials used in experimental program

Property	4340	L-605	Rene 41	VJ 1000	Inconel-X	321 steel
m	0.117	0.537	0.215	0.06	0.20	0.52
S_{ty}, psi	175,800	64,700	128,400	230,400	112,000	36,900
S_{tu}, psi	241,600	134,300	177,800	270,900	168,100	79,700
S_{su}, psi	147,600	85,000	118,300	167,600	106,000	60,500
τstd, psi	199,900	236,600	225,900	222,400	224,600	142,800
τopt, psi	174,200	180,800	162,500	235,100	145,000	108,600
Reduction in area, %	27.9	34.1	9.9	16.1	26.4	78.1
Elongation, % in 2″	9.2	48.7	10.9	5.7	23.1	63.1
V_{60}, std**	170	69	43	135	59	318
V_{60}, opt	240	165	86	170	98	480
Carbide	C-7	C-2	C-2	C-7	C-2	C-7
$\phi_n$std	30°	28°	31°	32°	30°	23°
$\phi_n$opt	39°	42°	40°	44°	36°	35°
i opt	−5°	−17½°	−10°	−7½°	−5°	−15°
$\alpha_e$opt	10°	22½°	20°	−2½°	17½°	25°
$\alpha_n$opt	9.5°	18.75°	18.75°	−3.5°	17.2°	22.4°
S_{ty}/S_{tu}	0.737	0.481	0.724	0.862	0.667	0.464
E, psi	30.0×10^6	32.6×10^6	27.0×10^6	29.2×10^6	32.8×10^6	28.2×10^6
Hardness	44-54C	19-21C	41C	54C	37C	74B

Definitions for symbols:
m = Strain hardening exponent
S_{ty} = Tensile yield strength, psi
S_{tu} = Ultimate tensile strength, psi
S_{su} = Shear strength, psi
τ std = Machining shear stress, psi, standard cutter
τ opt = Machining shear stress, psi, optimum cutter
V_{60} std = Cutting speed for 60 min. tool life, standard cutter
V_{60} opt = Cutting speed for 60 min. tool life, optimum cutter
ϕ_n std = Shear angle normal to cutting edge, standard cutter
ϕ_n opt = Shear angle normal to cutting edge, optimum cutter
i opt = Inclination angle, optimum cutter
α_e opt = Effective rake angle, optimum cutter
α_n opt = Normal rake angle, optimum cutter

Table II. Machining formulas for carbide face mills

Machining Property	Independent variables relationship
m	$1.990 - 0.00178\,(S_{tu}) + 0.0019\,(S_{ty}) - 1.73147\,(S_{ty}/S_{tu}) - 0.02503\,(E) - 0.00186\,(e) + 0.00353\,(R.A.) + 0.01534\,(k) =$
$V_{60}opt$	$2481.7 - 11.229\,(S_{tu}) + 18.75755\,(S_{ty}) - 2901.29785\,(S_{ty}/S_{tu}) - 24.9359\,(E) =$
$V_{40}opt$	$472.3 + 45.651\,(m) - 0.819\,(S_{tu}) - 23.615\,(E) + 34.153\,(k) =$
$V_{60}stdp$	$1742.8 - 3.83701\,(S_{tu}) + 8.27832\,(S_{ty}) - 1708.32251\,(S_{ty}/S_{tu}) - 31.41408\,(E) =$
$V_{40}stdp$	$901.05 + 0.70688\,(S_{tu}) - 443.91113\,(S_{ty}/S_{tu}) - 31.66241\,(E) + 15.61816\,(k) =$
$V_{40}stdn$	$-786.4 + 1102.78296\,(m) + 7.17159\,(S_{ty}) - 16.93771\,(k) =$
$V_{20}stdn$	$-996.1 + 828.304\,(m) + 1024.682\,(S_{ty}/S_{tu}) - 3.432\,(E) + 19.933\,(k) =$
i opt	$8.24 - 0.05361\,(S_{tu}) + 0.21565\,(S_{ty}) - 31.20369\,(S_{ty}/S_{tu}) - 0.8680\,(E) + 0.13825\,(e) = (-12\tfrac{1}{2}° \text{ to } -17\tfrac{1}{2}°)$
$\alpha_e opt$	$-541.16 + 39.20801\,(m) + 3.13364\,(S_{tu}) - 5.36814\,(S_{ty}) + 790.63267\,(S_{ty}/S_{tu}) + 0.71965\,(E) - 0.85957\,(e) + 0.60341\,(R.A.) + 6.49933\,(k) = (-15° \text{ to } 25°)$
$\alpha_n opt$	$-69.89 + 0.83848\,(S_{tu}) - 1.44835\,(S_{ty}) + 149.9062\,(S_{ty}/S_{tu}) - 1.11727\,(e) + 0.90794\,(R.A.) =$
C opt	$916.5 - 25.25804\,(S_{tu}) + 11.81021\,(S_{ty}) + 113.61975\,(E) - 26.90935\,(e) =$
C stdp	$1852.7 - 1105.365\,(m) - 8.92756\,(S_{tu}) - 21.45583\,(R.A.) + 84.64946\,(k) =$
C stdn	$-133.4 + 1280.1604\,(m) - 10.97843\,(S_{tu}) + 15.82188\,(S_{ty}) =$
n opt	$0.402 - 0.47653\,(m) - 0.00182\,(S_{tu}) + 0.02521\,(E) - 0.01036\,(R.A.) =$
n stdp	$-1.1854 - 0.00207\,(S_{tu}) + 1.21070\,(S_{ty}/S_{tu}) + 0.06016\,(E) - 0.01372\,(R.A.) =$
n stdn	$-0.7076 - 0.0061\,(S_{ty}) + 1.75976\,(S_{ty}/S_{tu}) + 0.017\,(E) =$
$H_{60}opt$	$-0.642 + 2.97128\,(m) - 0.02772\,(S_{tu}) + 0.03114\,(S_{ty}) + 0.15277\,(k) =$
$H_{60}stdp$	$0.267 + 2.86823\,(m) - 0.0115\,(S_{tu}) + 0.01374\,(S_{ty}) - 0.01473\,(e) =$
$H_{20}stdn$	$-7.543 + 5.69114\,(m) + 6.98553\,(S_{ty}/S_{tu}) - 0.00783\,(E) + 0.17809\,(k) =$

NOTE: Similar formulas with different multipliers are available for use with HSS face mills

Definitions for symbols (also see Table I)

- C opt = Constant, Taylor cutter-life equation, optimum cutter
- C std = Constant, Taylor cutter-life equation, standard cutter
- n opt = Slope of Taylor cutter-life curve, optimum cutter
- n stdp = Slope of Taylor cutter-life curve, standard positive rake cutter
- n stdn = Slope Taylor cutter-life curve, standard negative rake cutter
- H_{60} opt = Horsepower per tooth, at cutting speed for 60-min. tool life, optimum cutter
- H_{60} stdp = Horsepower per tooth, standard cutter
- V_{60} stdp = Cutting speed at 60-min. tool life, standard positive rake cutter
- V_{40} stdn = Cutting speed for 40-min. tool life, standard negative rake cutter
- k = Thermal conductivity, btu/ft² —°F — hr/ft
- e = Elongation, % in 2 in.
- C_s = Side cutting edge angle, degrees
- h = Helix angle, degrees
- T = Cutter life, minutes
- $\bar{V}$ = Cutting velocity vector
- $\bar{V}_c$ = Chip velocity vector
- η_c = Chip flow angle
- τ = Machining shear stress, ksi
- ϕ_n = Shear angle normal to cutting edge

given formulas, and others, were given finite multipliers for the various property values. This work led to the summation of important machining formulas for carbide cutters, Table 2. Another table was prepared for HSS cutters.

Practical application

Suppose that there is no previous machining experience for 8Al-1Mo-1V titanium alloy and that we have this certified mill analysis:

Ultimate tensile strength	132.8 ksi
Tensile yield strength	122.9 ksi
Modulus of elasticity	17.6×10^6 psi
Elongation (2 in.)	10.6%
Reduction in area	19.4%

And from the literature we find that the thermal coefficient, k, is 7.2 btu/ft²—°F—hr/ft at 1000 F.

We can now find the strain-hardening exponent:

m = 0.103

Cutting speed for optimum geometry cutter (carbide)

$V_{40}opt = 198.5$ fpm

Cutting speed for standard positive rake cutter:

V_{40} stdp = 139.5 fpm

Cutting speed for standard negative rake cutter:

V_{40} stdn = 86.6 fpm

Obviously the optimum-geometry cutter yields significant economic gains.

Optimum inclination angle:

i opt = −15.05°

Optimum effective rake angle:

α_e opt = 12.7°

Optimum normal rake angle:

α_n opt = 7.9°

Various other values such as axial rake, radial rake, most economical cutting speed, cutting speed for maximum production, and horsepower per tooth can be calculated from the equations.

Cutter-design suggestions

Recommendations for improvement of both HSS and carbide cutters can be made as the result of the research reported. The optimum inclination angle for HSS cutters is virtually constant at −12½° for a wide spectrum of materials. Thus, one functional angle is fixed. Data show that the optimum effective rake angle ranges from 17½° through 32½° for all practical purposes. However, after analysis of the results, it is suggested that a 25° effective rake angle might be chosen as the high limit and a 17½° effective rake angle as the low limit.

Recommendations for carbide cutters are not as simple as for high speed-steel cutters. From data it is evident that optimum inclination angles for carbide cutters will range from −12½° to −17½°. Also the optimum effective rake angles probably range from −10° to +25°. From charts it is evident that the cutting-tool industry has done a good job of producing a nearly adequate range of effective rake angle cutter geometries for the total spectrum of materials. However, the principal shortcoming in existing cutter geometries lies in their inclination angles. The combination can produce excessive thrust forces. For the superalloys like Rene 41, and L-605, large negative inclination angles are mandatory to achieve reasonable cutter life. For these reasons it is recommended that the cutting-tool industry take a new look at negative inclination angles and consider an inclination angle of −12½° and an effective rake angle of +15°. It is expected that this cutter geometry will perform outstandingly when machining the titanium alloys in the annealed condition and creditably when machining stainless steels and superalloys.

Proposed cutters

There is evidence which suggests that optimum inclination angles decrease with corner angles. In one sense this result is unfortunate because at first glance it might seem necessary to establish optimum inclination angles for all major corner angles. On the other hand, thrust forces might be expected to decrease with each change. Additionally, such a trend would make it possible to design a simple system of cutters that will provide near optimum geometries for the total spectrum of materials. This system is presented in Table 3 and Fig. 8.

These six cut-and-try efforts are not a final solution by any means. At best, more trials will have to be made before a workable optimized system of cutter geometries, involving only four types of cutter bodies and five types of carbide inserts, can be established. These points and the tooling scheme are offered only as a guide. ■

[Those wishing a copy of the 514-page report may request No. AFML-TR-*68-350 from DDC, Document Service Center, Cameron Station, Alexandria, Virginia 22314. Distribution of this report is subject to special export controls, and each transmittal to foreign governments or foreign nationals may be made only by prior approval of the Air Force Materials Laboratory, Wright-Patterson Air Force Base, Ohio.]*

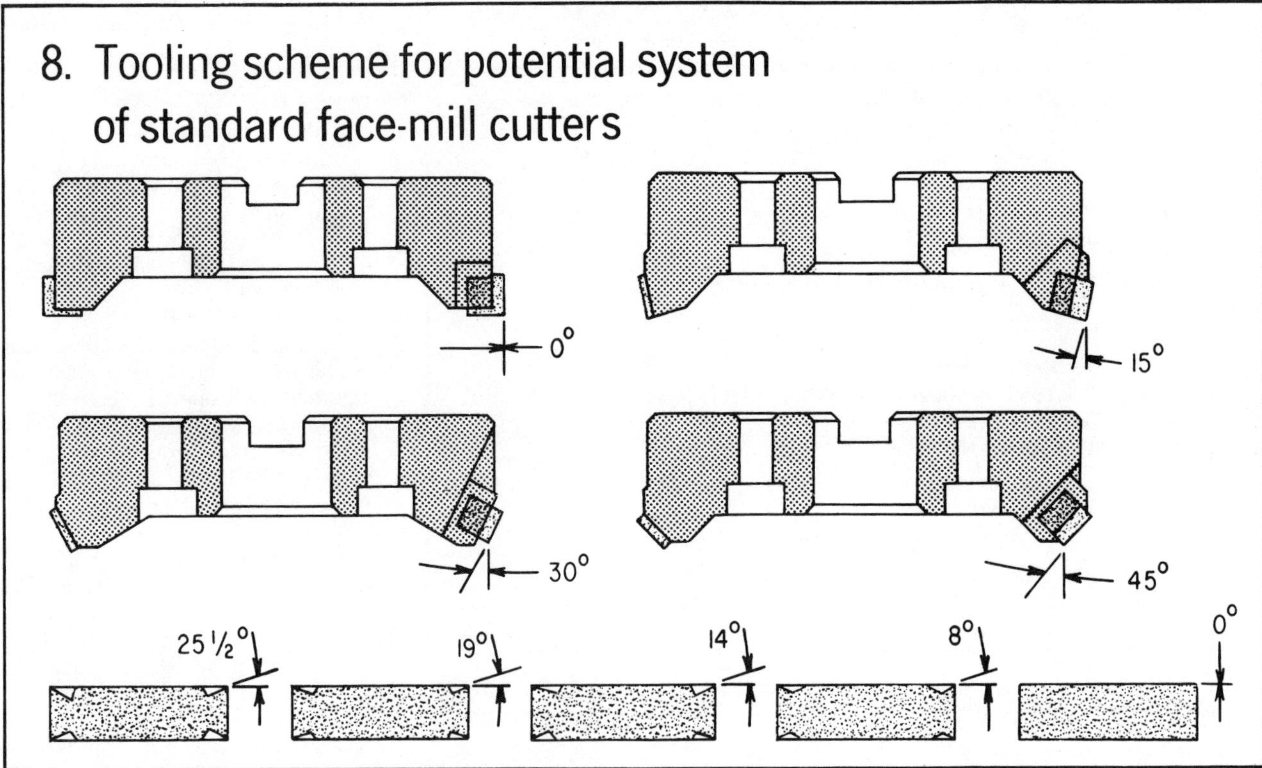

8. Tooling scheme for potential system of standard face-mill cutters

Table III. Potential system of optimized cutter geometries for carbide face-mill cutters

Axial rake	Corner Angle 0°		15°		30°		45°	
Radial rake	i	α_e	i	α_e	i	α_e	i	α_e
−2½°/18½°	−2½°	18.58°	−7.33°	18.05°	−11.58°	16.88°	−14.98°	15°
−5°/12°	−5°	12.37°	−7.95°	11.3°	−10.30°	9.41°	−11.95°	7.43°
−7°/7°	−7°	7.79°	−8.55°	6.18°	−9.5°	4.1°	−9.85°	1.7°
−7½°/1°	−7½°	1.97°	−7.5°	0°	−7°	−2.03°	−6°	−3.97°
−7½°/−6°	−7½°	−5°	−5.7°	−7.13°	−3.5°	−8.7°	−1.08°	−9.5°

Turning with inserts

Economy and productivity have made indexable inserts the primary tooling method for lathes, but continuing advances in materials, coatings, and geometry complicate the questions of application

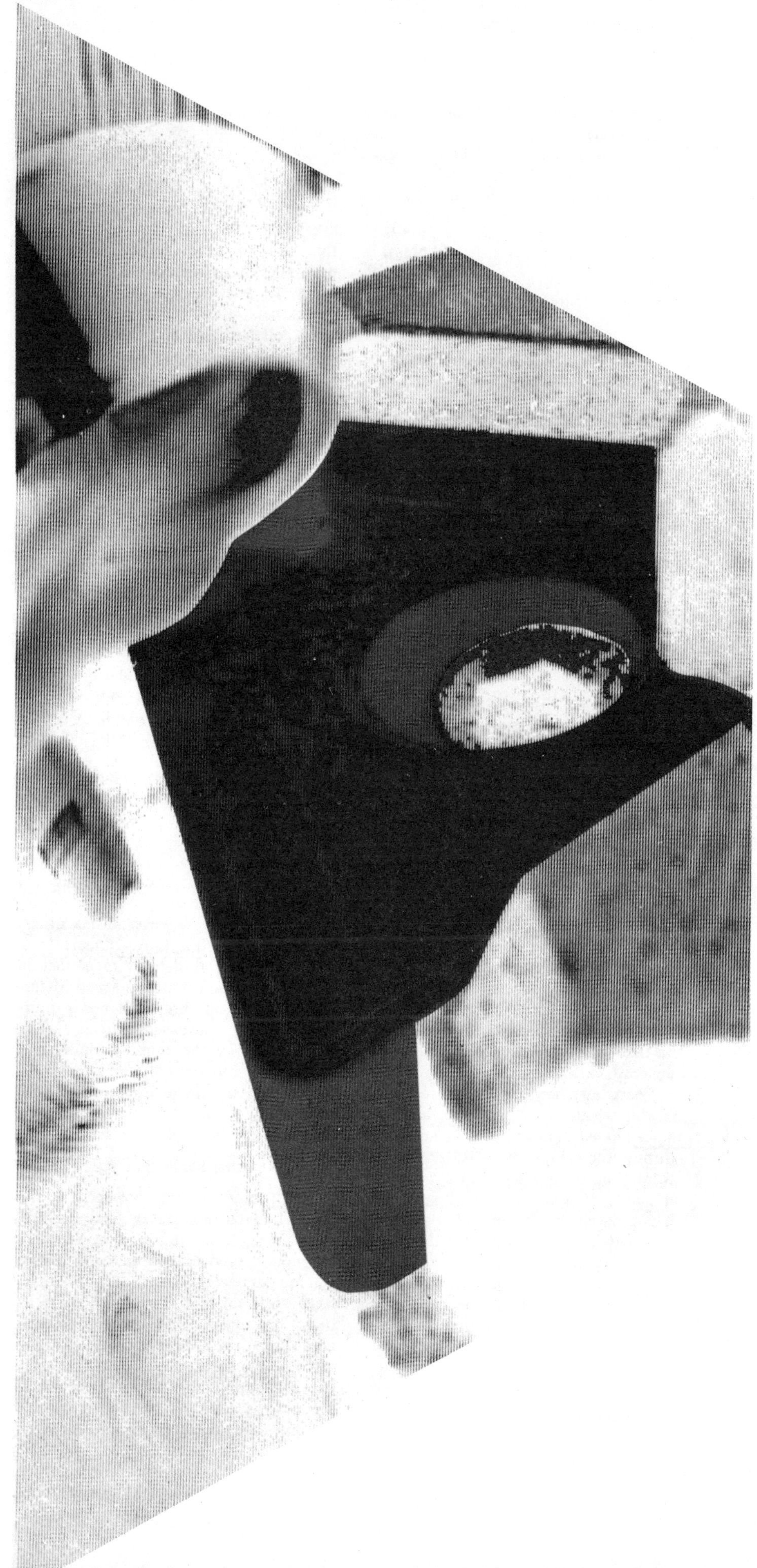

APPROXIMATELY one out of every five metalcutting machine tools is a turning machine. And, like any other metal-removal equipment, the ultimate economic performance of a lathe depends on the cutting tool that actually takes the chips off the workpiece. Regardless of machine cost, be it a few thousand dollars or a few hundred thousand, the profitability of that investment and the ancillary costs of operating it are inextricably tied to the effectiveness of a cutting tool that may cost only a relatively few cents per cutting edge.

It's true, of course, that many other factors enter into the profitability equation in any manufacturing situation. But it is unlikely that any factor offers a "leverage" of equal magnitude. Consequently, the economic impact of what is essentially a technical selection process—the choice of the tooling—is of major importance despite the relatively low cost of the consumable items.

The interdependence of turning machines and turning tools is also evident in their parallel development. It's virtually a chicken-and-egg relationship in which progress in the one field both spurs advances in the other and simultaneously helps to make them possible.

These parallel development paths have

By R.L. Hatschek, senior editor

resulted in remarkable increases in the metal-removal capability of the turning process over the years. This progress is well illustrated by some simple data developed by Sandvik Inc that put numerical values on machining time for turning a straight cylinder of 4-in. diameter and 20-in. length at various points in time. At the turn of the century, with carbon-steel tooling, the job would have taken 100 minutes. High-speed steel reduced this time to 26 minutes early in the 20th Century, and Stellite tooling cut the job to 15 minutes at about the time of World War I.

The introduction of cemented tungsten carbide made it possible to do the job in 6 minutes at about the beginning of the Great Depression, a metal-removal rate that was doubled by the premium carbides available in the 1950s. The time was again halved to 1.5 minutes with the advent of the first coated carbides at about the beginning of this decade, and progress in coatings has more recently brought the machining time down to an even minute with aluminum-oxide-coated inserts. Nor are these inserts the end point in cutting-tool development.

Help from lathe development

Credit for that 100:1 increase in productivity cannot be assigned entirely to the development of better cutting tools; it must be shared with the designers and builders of lathes and lathe accessories. Even the best of today's cutting tools could hardly permit a 19th Century machine tool to remove chips at such a rate. Even if the horsepower were there, the machine's drive system and bearings wouldn't withstand the cutting forces without severe vibration and deflection. Similarly, it's not likely that the chuck could maintain a secure grip on the workpiece at rpm levels needed to generate carbide cutting speeds.

There have been tremendous advances in the capabilities of machine tools and such related equipment as lathe chucks during the years that the toolmakers were progressing from carbon steel to modern carbides and ceramics, and which development came first is a question that is more of historical than practical interest.

It is valid, however, to note that optimum utilization of the capabilities available in today's cutting tools demands modern machine tools.

Savings in noncutting time

Another aspect of progress in the technology of turning tooling has been the reduction in the nonproductive time required for changing dulled tools. It was once the universal practice to grind the desired—and often highly individualistic—tool-point geometry on a length of cutting-tool material, clamp this in a toolholder, and then clamp the toolholder on top of a carefully adjusted rocker in the toolpost. Extreme care was demanded at every tedious step, and then every step had to be repeated every time the tool required resharpening. The necessary changes in machine settings to bring the next workpiece to dimension were a part of the job, of course, but not really much of a problem with a manually controlled machine. Repeatability in the sense that it is required by a numerically controlled or otherwise automated machine tool wasn't much of a consideration. What was important was the skill of the machinist.

There is still a place in industry for the individually ground toolbit, as there is for the skilled machinist, but less and less is that place on the production floor in manufacturing. More and more, the lathe tool of preference for production turning operations is the indexable insert—a relatively small wafer of hard material that can be quickly shifted from a dulled edge to a fresh one, presenting several of these uniform edges inexpensively enough to eliminate any economic need for resharpening in all but the most exceptional cases.

As noted above, it takes a good machine to fully utilize the capabilities of today's inserts. But it is obviously not economically feasible to replace all of the lathes in any particular plant just to match the latest advance in cutting-tool capability.

In recent years, however, it has almost seemed that cutting-tool suppliers were engaged in a kind of horse race—one in which maximum cutting speed determined the victor. For each newly proffered insert, increased cutting speed appeared to be the primary claim, and extended tool life was offered almost as a consolation for those users whose machine tools weren't quite up to the capabilities of the tooling.

There's more than one way . . .

Metal-removal rate, however, is the product of cutting speed, feedrate, and depth of cut; an increase in any one produces a proportional increase in chip volume per second. There are, of course, constraints on feed and depth—just as there are on cutting speed—but the constraints are different and, in many cases, allow users of less-than-the-latest turning machines to increase their productivity.

Many of the more recent developments in insert technology appear to be beneficial in this regard. Where earlier developments, such as coatings, provided increased speed capability, more-recent work on substrates and geometry is facilitating the use of increased feeds.

1. Disk turret of Pratt & Whitney Star-turn 1800 lathe bristles with insert-type tooling at 17 of 18 stations. Most are in left-hand holders to suit 'reverse' rotation of slant-bed machine

There was no intent in the paragraphs above to disparage the importance of tool life, and it is a fact that cutting-tool suppliers have made many advances in this area. Indeed, most of the advances in materials, coatings, and such geometry considerations as cutting-edge treatments present the user with a choice: increased metal-removal rate or increased tool life. And it's usually quite possible to establish turning conditions in some middle ground that permits some gain in both areas.

The point to be made is that extended tool life provides savings in tool costs, an obviously desirable end in itself, but increased metal-removal rates can enhance the profitability on the machine-tool investment and operating costs. This is the central theme of the High Efficiency Machining (Hi-E) concept so strongly espoused by General Electric's Carboloy Systems Dept. Limits exist at both ends of the tool-life/metal-removal tradeoff, but it is possible to approach an area of optimization in the middle range.

Problem has too many solutions

Cascading progress in the technological development of cutting tools has generated still another problem for users: virtually an embarrassment of riches. Cutting-tool producers have eased, if not solved, so many specific problems that the sheer volume of these solutions has become a problem itself—both in the selection of an optimum cutting tool and in stocking the toolcrib. This is also a double-edged problem for insert producers in their own manufacturing operations and in stocking and distribution.

Although it's unlikely that anyone will come up with a universal "job-shop" insert that will suit the multiplicity of metalcutting requirements of manufacturing industry, cutting-tool producers are certainly striving in that direction. Witness to this is the increasing concentration of emphasis on the broad utility of a few grades of insert material and the development of special chipbreaker configurations designed for effective use at a variety of feedrates, both of which were evident in several tooling exhibits at the 1978 International Machine Tool Show.

The principal decisions to be made in the choice of lathe tooling are the insert material or grade; its size, shape, and other geometric features; the toolholding system to be used; and the operating conditions of speed, feed, and depth of cut. The selection for any particular turning operation requires a complex assortment of considerations primarily involving the interactions of workpiece, machine tool, and tooling as a system. Many of the factors will affect more than one feature of the cutting tool under consideration, and many of the decisions will have a feedback effect on preliminary decisions.

A discussion of insert materials seems as logical a place to start as any.

Insert materials

Most lathe chips today are generated by cutting tools made of cemented carbide, a hard material that made its first commercial appearance in Germany in 1926 and in the US in 1928. Two grades were initially available in the US; both were straight mixtures of tungsten carbide and cobalt, one containing 94% WC and 6% Co, the other 87% WC and 13% Co.

A half century later, a tremendous number of grades are available, not standardized as to content but rather the proprietary products of their manufacturers, each of which has adopted special compositions, various additives, and process variations in its attempts to market a superior line of products for the metalworking industry.

Since there are no compositional standards by which users can specify cemented carbides—unlike high-speed steels, for example, for which alloy standards are precisely defined—a practice of grading carbides by application class has come into being. The system originally devised at Buick Motor Div of General Motors in the early 1940s forms the basis for the industry code currently used in the US (generally called the "C-classification" system).

This system, insofar as it applies to cemented carbides for metalcutting uses, establishes eight classes or categories of application:

C1 to C4 are essentially abrasion-resistant compositions for cutting cast iron, nonferrous metals, and nonmetallics. The numerals progress upwards in terms of increasing hardness and wear-resistance and downwards in terms of increasing strength and binder content. Thus a grade-C1 carbide would provide, say, a typical hardness of Ra 91.0 and a typical transverse rupture strenth (TRS) of 290,000 psi and would be suited for roughing cuts in materials of the types indicated above. C2, then, would be harder but not as strong (Ra 92.5 and 250,000 psi), thus giving better wear-resistance but somewhat less toughness for what are usually termed "general-purpose" applications. Grade C3, at typical values of Ra 92.8 and 200,000 psi, would be more suited to light finishing operations, and C4 (Ra 93.0 and 175,000 psi) would be better adapted to precision-finishing jobs in the range of materials indicated, characteristically producing short, easily broken chips.

Additives for steel cutting

The so-called "steel-cutting" grades are compounded to provide increased resistance to deformation and cratering, which is generally done by adding either titanium carbide or tantalum carbide or both. Progressing from C5 to C8, hardness and wear-resistance are increased, while a reduction in toughness is accepted as a necessity. Some typical values would be Ra 90.5 and 280,000 psi for a C5 roughing grade, Ra 91.0 and 250,000 psi for a C6 general-purpose grade, Ra 92.2 and 210,000 psi for a C7 finishing grade, and Ra 92.5 and 170,000 psi for a C8 precision-finishing grade.

Somewhat similar is the ISO classification system, which establishes three broad groups: K, P, and M. K includes straight WC-Co compositions for use on ferrous metals producing small chips, nonferrous metals, and nonmetallics. P designates highly alloyed carbides for steels and other ferrous metals that produce long chips. And M is an intermediate or multipurpose class of more-moderately alloyed carbides that may be used on steels, cast irons, high-temperature alloys, and the like. Numerals in the ISO system run from 01 to 50, increasing in strength and decreasing in hardness.

One of the difficulties in evaluating cemented carbides for cutting-tool applications is that the "formal" properties of hardness and transverse rupture strength rarely show precise correlation with the properties that actually are desired: resistance to wear and resistance to chipping, cracking, breakage, and deformation. Furthermore, hardness and TRS are measured at room temperature and under static conditions, whereas a cutting tool must perform at elevated temperatures and in a dynamic environment of constantly changing forces.

An alternate laboratory test for edge strength, used by Kennametal, puts the tool into a very severe, interrupted cut in a carefully controlled manner. The "workpiece" is a 41L50 bar with four slots cut axially along the bar slightly off center so that the impacts occur at the centerline. A constant depth of cut is used, usually 0.100 in., and cutting speed is held as constant as possible at 350-450 sfm, depending on lathe settings. An initial feedrate of 0.015 ipr is used, and 10 sq in. of the bar's surface area is machined at these settings. The feed is increased by 0.005 ipr, then, for successive runs of 10 sq in. each until the cutting edge fails. When an insert breaks or chips, the total machined length is recorded, and the total machined surface area is calculated in sq in.

This technique produces an accelerated test and provides data that are not linear; if one edge machines twice the area as another, it is more than twice as strong. To avoid test-to-test discrepancies, control inserts of known performance are also run. And, to ensure consistency, five inserts are tested from each of three production lots of powder.

In evaluating whether an experimental composition merits further investigation, Kennametal requires that five inserts must average 60 sq in. in the slotted-bar test with good consistency—and note that, at 60 sq in., the feedrate reaches 0.045 ipr. Resistance to edge wear is similarly tested by Kennametal in accelerated and carefully controlled cutting conditions.

Coatings for inserts

Unquestionably the most important advance in carbide technology during the past decade has been the development of coated inserts, and this is especially so because the benefits of increased metal-removal rate and/or greater tool life are realized in the most important application areas, the turning of ordinary cast irons and steels.

The initial coated inserts—introduced by Sandvik in 1969—consisted of metal-cutting grades of tungsten carbide with thin, vapor-deposited surface layers of titanium carbide (TiC). The list of coating materials has since been expanded to include titanium nitride (TiN), titanium carbonitride (TiC-N), hafnium carbide (HfC), hafnium nitride (HfN), and aluminum oxide (Al_2O_3). In many cases, the coatings are multilayered, such as the Kennametal KC850 insert shown on the front cover of this issue, which incorporates a triphase coating of titanium carbide, titanium carbonitride, and titanium nitride.

The improved cutting performance of coated inserts results from several factors: greater surface hardness, reduced friction, and chemical inertness provided by the microscopically thin layer. The coating provides a cobalt-free diffusion barrier between the chip and the tool, which virtually eliminates cratering as a tool-life factor and reduces the apparent coefficient of friction.

Even though enhanced wear-resistance is one of their attributes, coated inserts are not generally recommended for highly abrasive applications. They are rarely, if ever, recommended for machining nonferrous metals or nonmetals. And heavy roughing or severely interrupted cuts have not generally been within the repertoire of the coated grades.

2. Insert at top shows normal wear of steel-cutting grade used to turn steel. Bottom insert shows excessive cratering that results from turning steel with a cast-iron grade. Photo by Kennametal

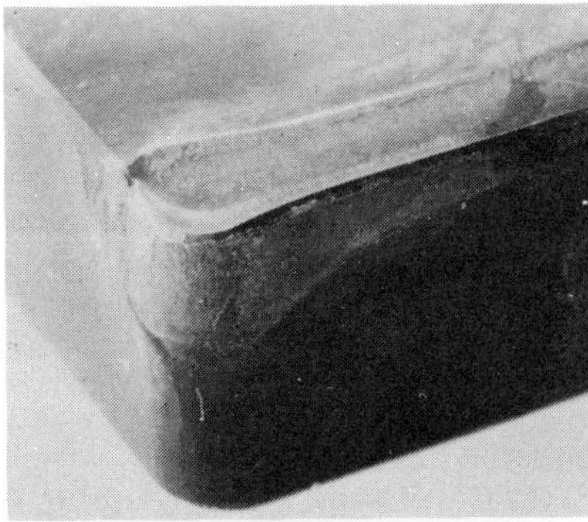

3. Thermal deformation shown by this insert suggests switch to a more heat-resistant grade. Kennametal photo

The relative merits of the individual coating materials are subject to some disagreement among producers of inserts, but differences in the quality of the coating process, of substrate materials, and of other variables are likely to cause greater performance differences than the coating material.

The early TiC coatings had several limitations. Thicknesses varied from one surface to another of an insert and from insert to insert—even within a single furnace load. Cutting performance of too-thin coatings was erratic. Too great a thickness was likely to result in checking and spalling because of thermal-expansion differences between coating and substrate. And compositional variations caused problems of edge strength, again resulting in unpredictable tool life.

Once control of these coating problems became better understood, work turned to improving the edge strength of coated inserts. New substrate materials were developed for coated grades—some totally unsuited to metalcutting without the special surface layer. Because coatings permit greater turning speeds, the temperatures within the cut were rising, and greater importance was placed on high-temperature properties of toughness, resistance to deformation, thermal-shock-resistance, and hot-hardness.

Another phenomenon relating to edge strength is an embrittlement resulting from a depletion of carbon content near the surface of a coated insert. At least one insert producer (Kennametal) has developed special processing techniques that result in an intentional variation of composition with respect to depth below the surface to prevent the carbon deficiency at the cutting edge.

Progress in all of these areas is extending the application range of coated inserts ever farther into the realm of heavy roughing cuts. There is still no single "general-purpose" insert, but the number of grades necessary to cover the full range of turning operations is growing smaller.

Coated carbides have been estimated by various sources to account for 30-40% of present insert sales—perhaps approaching 50% in the case of turning, according to Sandvik, which anticipates a "saturation level" of 80-85%.

Solid titanium carbide

If these coating materials work so effectively, goes one line of reasoning, why not make inserts out of solid "coating" and avoid the extra processing steps. These are available, of course, in the form of solid-titanium-carbide (actually highly alloyed) inserts and ceramic (Al_2O_3) inserts.

TiC is, in fact, the hardest of the pure metallic carbides and, if nickel instead of cobalt is used for a binder, provides inserts with excellent wear-resistance in high-speed machining of steel and cast iron. The possibilities of solid TiC inserts can be exemplified by two new grades just introduced by Adamas Carbide, Titan 50 and Titan 100, the development of which was principally in the area of binder composition and special additives. Titan 50 is claimed to be the highest-toughness titanium carbide available (TRS of 275,000 psi with a hardness of Ra 90.5). Titan 100 (170,000 psi, Ra 93.3) is touted for its wear-resistance and is claimed to be competitive with ceramics and aluminum-oxide-coated inserts in machining both cast iron and steel at surface speeds in the 1200- to 1700-sfm range. Performance superior to that of TiC-coated and TiN-coated inserts at all competitive speed ranges is also claimed. And an additional advantage is that the solid-TiC inserts can be used with sharp edges; they do not require the edge-hone prior to coating of all coated inserts.

Ceramics and cermets

Another coating material being used in solid form for inserts is aluminum oxide, although these have been made since the 1950s, which considerably antedates the introduction of aluminum-oxide coatings. A second type of ceramic material, consisting typically of about 70% aluminum oxide and 30% titanium carbide, has more recently come into use for metalcutting inserts. The former type is generally referred to as ceramic, cold-press ceramic (the powder is pressed at room temperature, and the compact is later sintered in much the same way as cemented carbide inserts), or white ceramic. The latter type is more often called cermet, composite ceramic, hot-press (its production process combines the pressing and sintering into a single operation), or black ceramic. Hot-pressed pure ceramic (99.9% Al_2O_3) inserts, which are an intermediate gray in color, are also available.

High hardness (in the range of Ra 93-94 for the white and as high as Ra 94.5-95 for the black), low chemical reactivity, and excellent high-temperature characteristics are the benefits to be gained from these materials. In payment for these, the user gives up some toughness. Typical values of transverse rupture strength for recently introduced

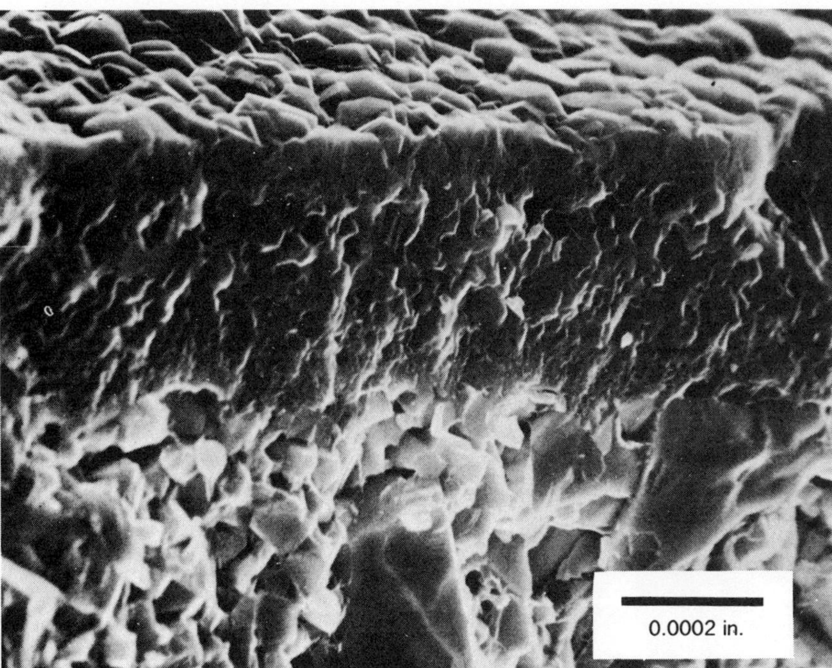

4. Scanning electron micrograph reproduced here at magnification of about 3800X shows thin surface coating of aluminum oxide deposited on intermediate coating of titanium carbide on tungsten-carbide substrate. Grade is Sandvik's GC 015

cold-pressed grades are of the order of 100,000-110,000 psi and, for hot-pressed composites, in the range of 130,000-135,000 psi.

Low in comparison with some of the other insert materials, these strength levels still represent considerable progress from the levels of earlier days and result largely from advances in grain refinement and elimination of microporosity. Surface treatments, such as chemical etching, also are beneficial, apparently by reducing the notch effects of microscopic discontinuities on the surfaces.

High-speed machining is the forte of ceramics and cermets. Recommendations for one white ceramic, for example, include speeds of up to 2000 sfm for alloy steels of Rc 23-24 hardness and up to 2500 sfm for cast irons in the Bhn 170-235 hardness range. The same producer recommends using a black cermet for cast irons over Bhn 235 and for alloy steels with hardnesses as great as Rc 66.

Various estimates put present use of ceramic inserts at about 1%, more or less, in the US and as high as 5-7% in Europe and Japan. The more brittle nature of these inserts demands a much more highly engineered application—ceramic inserts are just not as forgiving of overloads as other inserts—and their use tends to be in high production industries, such as automotive. Continuing improvements in the properties of ceramic inserts, however, seem to ensure further increases in use.

That this is a very strong possibility is illustrated by the fact that virtually all of the major suppliers of carbide inserts in the US also produce ceramic and cermet inserts, though they are not all sold under the producer's principal brand name. There are, of course, also other producers specializing in ceramics, notably American Feldmuehle and Babcock & Wilcox (Nucermet).

High-speed steels

For applications challenging the toughness of even cemented carbides, inserts are also made of chill-cast cobalt-based alloys and even of high-speed steel. One example of the former is Tantung G (Rc 60-63 and 325,000 psi TRS), offered in positive-rake triangles and squares by VR/Wesson Div of Fansteel (which also offers carbide and ceramic inserts). An example of hss inserts is provided by Teledyne Vasco, which offers a variety of positive-rake triangles and squares in both T15 alloy (Rc 67) and M42 alloy (Rc 68).

Latest development in the field of high-speed-steel inserts is the introduction of Special T inserts by Union Twist Drill Div (Litton) this month. These are produced, again in positive-rake squares and triangles, by a powder-metallurgy technique quite similar to the production of carbide inserts. T15 hss powder (with no binders or lubricants) is impact-pulverized to reduce particle size, loaded into a die cavity, and mechanically compacted. It is subsequently sintered to 100% density in a vacuum furnace, finish-ground, and given a ferrous-oxide (blue) treatment to reduce any chip-welding tendencies. Said to be approximately 30% stronger than carbides and priced approximately 25% lower, these inserts are recommended by Union for applications in which carbides may fail because of interrupted cuts, underpowered spindles, and repetitive starting and stopping.

Polycrystalline tools

At the exotic end of the cutting-tool spectrum are inserts of compacted polycrystalline diamond and cubic boron nitride (CBN). Both are the products of ultrahigh-pressure and -temperature techniques, and both are available in insert form, but they have distinctly different and, at present anyway, rather narrow areas of suitable application.

The diamond product offers exceptional properties in highly abrasive operations, such as turning of glass-fiber-reinforced plastics, ceramics, and high-silicon aluminum alloys. Especially in view of increasing use of high-silicon aluminum diecasting alloys, the automotive industry is currently making increased production use of polycrystalline-diamond tooling. However, diamond is carbon, and it is totally unsuited for machining materials with which carbon reacts at cutting temperatures—such as iron or steel.

This problem does not affect polycrystalline CBN, which is best known under the Borazon tradename used by its only present producer, Specialty Materials Dept of General Electric, which does not make inserts itself. Although it is not as hard as diamond, Borazon is inert to superalloy constituents at temperatures beyond 1800F and is suitable for turning hardened steels, chilled cast irons, and nickel- or cobalt-based superalloys. Insert prices are extremely high but are economically justifiable in many special cases.

Insert geometry

In the narrowest sense, there's really no such thing as a "single-point" tool. The tool, insert-type or any other, is a three-dimensional solid, the shape and attitude of which permit it to shear chips off the workpiece. The geometry of the tool shapes both the workpiece and the chips; it is simple yet complex; it is highly standardized in many ways and simultaneously is the subject of considerable recent development.

The *American National Standard Identification System for Indexable Inserts for*

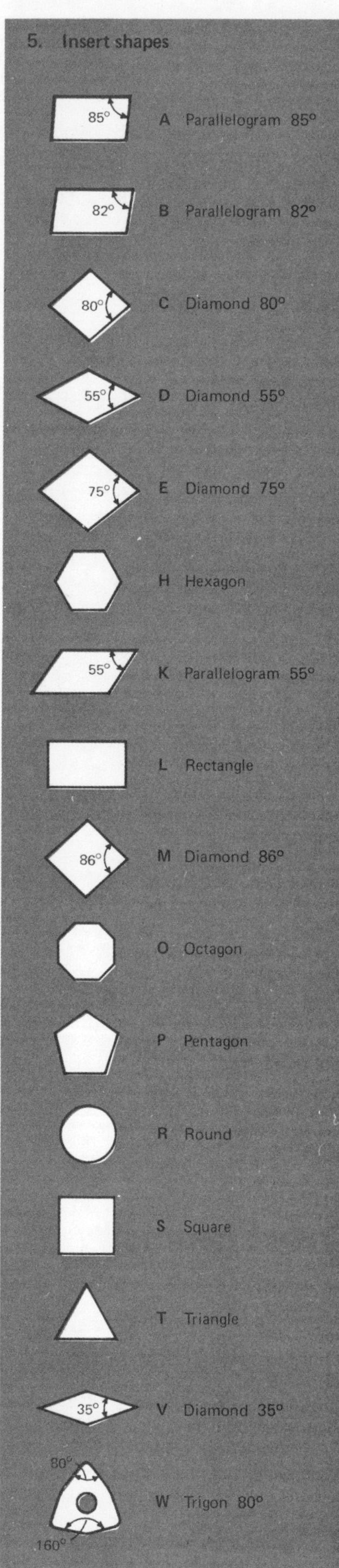

Cutting Tools (ANSI B94.4-1976) provides a ten-position code (though not all positions are always filled) that defines most of the features of insert geometry. Fig 5 illustrates the 16 basic shapes, of which six are fairly common to turning operations: triangle (T), square (S), round (R), and three of the rhombic diamonds (C, D, and V).

The next three letters in the code define, respectively, the clearance angle of the insert (N for 0° and P for 11° are most common in turning tools), the tolerance class of the insert, and the type (whether or not the insert has a hole for a pin lock, chip grooves in the rake face, etc). Three further positions, generally all numeric for turning inserts, indicate size, thickness, and point radius. An eighth position, rarely used for turning inserts, describes special (faceted) cutting points; a ninth shows edge-hone radius and rake-face polish, if applicable; and the tenth indicates whether the tool is for right-hand or left-hand cutting, again only if applicable.

Details of the coding system are reproduced in most insert manufacturers' catalogs and need not be reproduced here. An example, however, will help to clarify the system. The insert shown on the front cover of this issue of AM is a TNMG-543—T for triangle, N for 0° clearance (or negative rake, mounted at −5° in the holder to provide 5° actual cutting clearance), M for a tolerance of ±0.005 in. over the cutting point (which varies with size and shape) and ±.005 in. on thickness, G to indicate chip grooves and lock-pin hole, 5 to indicate the ⅝-in. diameter of the inscribed circle, 4 to indicate 4/16-in. (or ¼-in.) thickness, and 3 to indicate a point radius of 3/64 in.

What's in a shape?

A single-point tool—now in the broadly accepted meaning of the term—generates the part shape. Although it's analogous to a pencil, which can draw any outline on a piece of paper, the three-dimensionality of the lathe tool does handicap its ability to generate some workpiece configurations. That's why lathes have turrets. A rudimentary understanding of how a lathe shapes a workpiece, however, is about all that's needed to determine whether any particular insert shape is capable of performing a particular operation. Short of noting that 35° diamonds (V) and 55° diamonds (D) are often used in tracer operations because they minimize cutting interferences, this article will not delve deeply into the shape limitations posed by workpiece configuration. It should be noted, however, that a variety of special-purpose inserts and holders also exist for such operations as threading, O-ring and retaining-ring grooving, pulley grooving, and cutoff.

For the broad range of turning operations, the important factors largely determined by an insert's basic shape are the geometric strength of the cutting edge, the number of indexes possible, and cost (a function of the number of usable edges and the volume of material required to make the shape).

The number of indexes, rather obviously, depends on the number of usable corners and whether the insert can be inverted to double that number. With suitable holders, all four (or eight) corners of an 80° diamond are usable. And the number of indexes possible with a round insert depends on depth of cut, four to ten per side being the generally accepted range.

Also obvious is that the maximum strength is provided by the largest included point angle, and that the strength of any particular angle is enhanced by the largest possible nose radius. Thus, the round is the strongest geometric shape. And this is also the reason why pentagons are often recommended when the insert is made of ceramic material.

These facts combine to suggest a general guideline: select the strongest geometric shape that will generate the desired workpiece configuration. This will also provide the largest number of indexable edges, and each insert will contain the minimum volume of carbide for reduced cost. This line of argument also favors negative-rake inserts for both the increased strength of the 90° cutting-edge angle and the fact that such inserts can generally be inverted. This problem, however, is now complicated by some of the rake-face geometries currently available.

Control for the chips

The long and stringy chips produced by some workpiece materials can be troublesome, especially in automated turning machines, and it was fairly early in the history of inserts that the mechanical chipbreaker was introduced. This is simply a wedge-shaped piece clamped on top of the insert to simulate the action of a chipbreaking notch in the rake-face of a hand-ground lathe tool. One advantage of these mechanical chipbreakers is that they are typically adjustable in position so that the chip-curling effect can be tailored to the specific operation.

Next came the molded-groove type of chipbreaker insert, which, with a lock-pin hole for securing the insert in the toolholder pocket, provided a chip-curling action (though not adjustable) while leaving the top surface of the tool unobstructed. This idea led to the positive/ne-

6. Currently offered line of carbide inserts by Sandvik includes several with geometry varying from point back

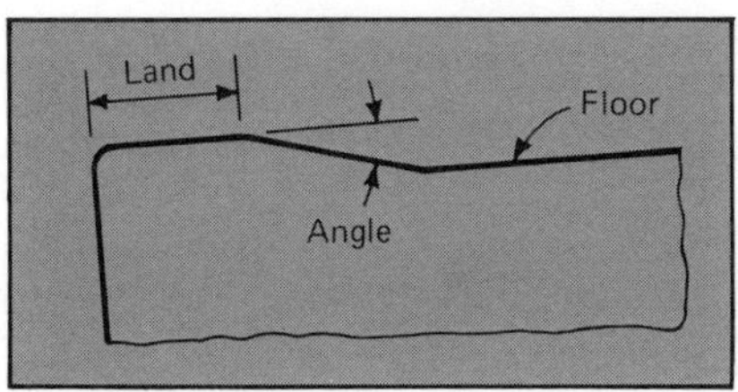

7. Land-angle insert geometry (Kentrol) controls chips over range of feeds, reduces horsepower requirement and lowers temperature of chip and tool

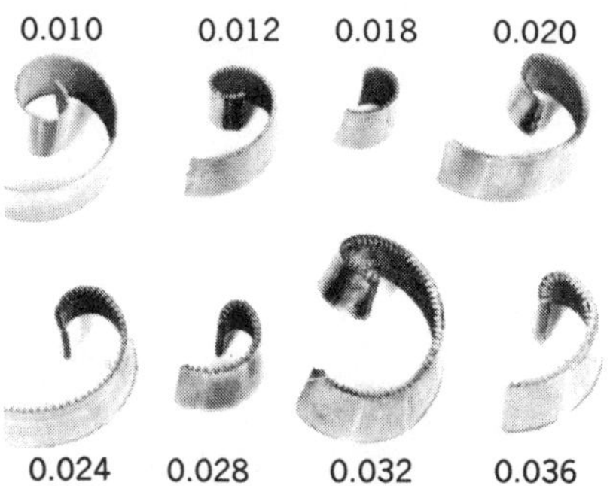

8. Land-angle insert (TNMM-433) cut these 1045-steel chips at 370 sfm, 0.350-in. depth, and feeds indicated

gative insert, in which the chipbreaker groove is closer to the cutting edge and is configured to give the insert an acute cutting-edge angle without adding any clearance to the insert's flank surfaces. The concept provides the advantages of positive-rake cutting (primarily reduced cutting forces and spindle-power requirements) while the insert is held in a standard negative-rake toolholder and of permitting a two-sided insert to double the potential number of indexes.

Variations in both depth of cut and feed affect the performance of molded chipbreakers. As a result, new designs have been conceived in attempts both to broaden the application range of inserts and to produce inserts tailored for greater efficiency in specific applications—such as a roughing geometry and one for finishing cuts. Interestingly, some of these present variations in geometry from the point back along the cutting edge (Fig 6).

Chipbreaking at high feeds

Early in 1975, Kennametal introduced "land-angle" geometry (tradenamed Kentrol), which departed considerably from earlier chip-control concepts. The geometry features a raised cutting edge and land with an angled ramp behind it descending into a flat central area (the floor) covering the remainder of the insert's rake face (Fig 7). Both tangential cutting force and frictional force of the chip flowing over the rake face are reduced by this design, in turn reducing horsepower requirements and temperatures of both chip and tool. Heavier cuts are possible, and well-formed chips are generated over a wider range of operating conditions (Fig 8).

Generally similar in appearance but differing in detail is the high-feed chip "groove" recently introduced by Carboloy on ProMax 515 TiC-coated and ProMax 570 Al_2O_3-coated inserts, both of which are also offered (along with other grades) with "standard" grooves. Carboloy geometries are based on the following considerations:

- Increasing the land width increases the required cutting forces, increases the strength and resistance to breakage, and shifts the starting feed for chip-breaking to larger values.
- Increasing the groove width reduces cutting forces and shifts the range of chip control to higher feed levels.
- Increasing depth of the groove reduces forces, increases the natural tendency for the chip to curl, and reduces the strength of the cutting edge.
- Increasing the ramp angle decreases cutting forces and reduces the resistance to breakage.
- Chip grooves with larger lands may be lightly honed.
- Chamfers are effective in strengthening chip grooves in interrupted turning at high chip loads.
- Chamfers perform best when honed to remove sharp edges.
- Coated inserts are usually used at lighter chip loads than uncoated inserts; hence, chip grooves in coated inserts can have narrower lands.
- Use of wide grooves can extend the capability of coated inserts to higher feedrate values.

These lead to the general conclusion that optimum chip-groove geometry depends on land width, groove width, groove depth, groove-entrance angle, edge preparation, and coating system.

Carboloy's application literature, incidentally, states that standard chipbreaker grooves offer excellent chip control on nearly all materials at feeds of 0.008 ipr to 0.035 ipr and that, "at feedrates in the 0.035- to 0.075-ipr range, high-feed chip grooves reduce cutting forces up to 25% and temperature to 35% compared with standard chip grooves."

What edge preparation can do

Geometric modification of the cutting edge itself can have profound effects on an insert's performance and durability. As noted earlier, the sharper the cutting-edge angle, the geometrically weaker it is. And microscopic inspection of a sharp edge will show minute discontinuities, which are actually stress-raisers at which cracking and chipping can begin.

Three basic modifications are available, sometimes in combination, to strengthen the cutting edge: honing, chamfering, and creation of a negative land.

Honing, which is standard practice for coated inserts and is done prior to the coating process to improve coating performance at the cutting edge, consists of rounding off the edge, generally by tumbling the inserts in an abrasive slurry. Light (0.001- to 0.003-in.), medium (0.003- to 0.005-in.), and heavy (0.005- to 0.007-in.) hones are indicated in the ANSI insert-identification system. Honing produces a smoother edge; in general, it's recommended that the hone on a carbide insert should not exceed about 30% of the feed per revolution.

Chamfered edges, in which the insert is beveled at an angle of from 25° to 45° (measured down from the rake face), are typically used for heavy-shock applications, interrupted cuts, and scale. Again, the chamfer width should not exceed more than about 30% of the feed.

The negative land, sometimes referred to as a K-land, is similar to the chamfer but is at a considerably shallower angle—generally 3° to 15°—and acts more as a modification of the rake angle.

The importance and effectiveness of these geometric edge-strengthening modifications increase as the brittleness of the cutting-tool material increases. Thus, they are of special value for ceramic and cermet inserts. In a technical paper presented earlier this year, Donald O. Wood (American Feldmuehle Corp) said, "In the application of ceramic cutting tools, the edge chamfer selected for a particular usage is very critical. The primary function of the edge chamfer is to protect the cutting edge. The basic rule of edge-chamfer width states that feedrate should be generally 1.5 times the edge-chamfer width. The rule is followed in all normal turning, boring, and facing operations for maximum tool life. The exception exists in interrupted-cut applications, in which the basic rule is that the chip should form on the center of the edge-chamfer width."

In line with this, American Feldmuehle offers a wide range of edge-preparation specifications: chamfers at 10°, 20°, and 30°; chamfer widths ranging from 0.002 in. up to 0.095 in.; and a number of hone variations in combination with the different chamfers. Some of the company's typical recommendations are as follows:

- 0.008-in. x 20° chamfer with a hone of 0.001-0.002 in. or 0.002-0.004 in. for standard roughing and finishing
- 0.004-in. x 20° chamfer with a hone of 0.001-0.002 in. for light roughing and finishing.
- 0.002-in. x 20° chamfer with a hone of 0.0005-0.001 in. for light finish-turning of hard materials.
- 0.012-in. x 30° chamfer with a hone of 0.001-0.002 in. or 0.002-0.004 in. for high-speed roughing.

Into the cut

As noted earlier, the great variety of rake-face configurations currently available from the different insert suppliers—and there's no set of standards applying to the top of the insert—complicates the choice of positive or negative rake. For example, improvements in materials, coatings, edge preparation, and groove design tend to offset the strength disadvantages of an acute cutting-edge angle. Yet the objective of heavier feeds increases the loads on inserts, requiring flat seating surfaces for maximum support, which precludes the possibility of inverting even some negative-rake inserts to get the cost benefit of three or four more indexes.

The letter N in the second position of an insert's identification code, once freely interpreted to mean negative-rake, now really serves primarily to designate the angle at which the insert is secured in the toolholder, and those designed for N-type inserts are by far the most widely used.

But rake is only one of the angles determined by the toolholder; another of significance is the side-cutting-edge angle (SCEA). This angle (see Glossary for definition) is also sometimes referred to as the lead angle, and there are proponents (such as TRW/Wendt-Sonis) of a still newer term: attack angle. By what-

Glossary

Abrasion resistance—The ability of a material to withstand a change in dimensions due to a rubbing action with another material. As tested for carbide, a steel wheel with an aluminum-oxide slurry is rubbed over the test piece. (Cemented Carbide Producers Assn Physical Test P-112, 'Abrasive Wear Resistance.')

Abrasive wear—The wear that occurs on a tool in use due to rubbing action in machining. (See Flank wear)

Anvil (Seat, Shim)—A removable part of a toolholder designed to provide support for the cutting insert. (See Insert package)

Attack angle (Lead angle, Side-cutting-edge angle)—Term preferred by some to avoid confusion over the meaning of Side-cutting-edge angle. (See Lead angle)

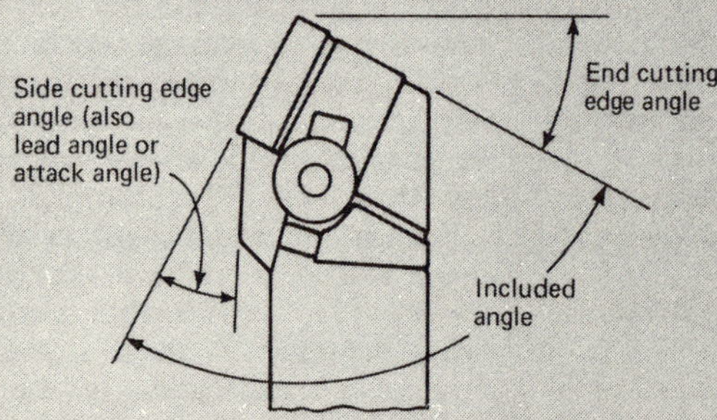

Back rake (Top rake)—The angle of inclination of the face of the tool away from the end cutting edge. It is measured in a plane perpendicular to the base of the tool and parallel to the side cutting edge. (See Rake)

Bar (Boring bar)—A toolholder specifically designed to support cutting tools in a boring operation. Deep-hole boring often requires a long, round (or 'bar-shaped') toolholder that is supported on one end.

Buildup—The welding of chips to the cutting tool. It is a major cause of surface roughness. (See built-up edge)

Built-up edge—An adhering deposit of work material on the tool face (rake face) adjacent to the cutting edge. (See Buildup)

Chamfer—A bevel on the cutting edge of a cutting tool for the purpose of increasing its strength. The angle is measured from the cutting face downward and will generally vary from 25° to 45°.

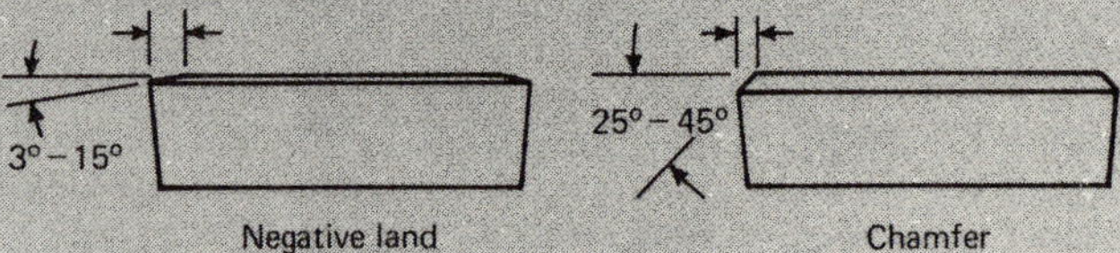

Chipbreaker—A groove or irregularity in the face of a tool, or a separate piece fastened to the tool or toolholder, to cause the chip to break into short sections, or curl.

Chipbreaker insert—An insert with a built-in chipbreaker. This generally consists of a groove around the top face of the insert near the periphery.

Chipping—The breakdown of cutting edges by loss of fragments broken away during the cutting action.

Clearance—The angle below or behind the cutting edge, which allows the cutting-edge to be forced into the work. Without clearance, the tool will not cut. It is also the term used for secondary relief in some cases. (See Relief)

Crater—A grooved area or depression caused by chips rubbing-away or eroding a groove or well in the top (the rake face) of the insert behind the cutting edge of a tool.

Cratering—The action whereby an insert face gets a depression in it from being eroded by chip contact. (See Crater)

Cutting edge—That part of the face edge along which the chip is separated from the work. The cutting edge consists of the side cutting edge, the nose, and the end cutting edge.

Cutting speed—In turning, the peripheral velocity of the

ever name you call it, standard toolholders (and boring bars as well) are available in styles that provide a broad range of values in SCEA—typically from $-5°$ to $+45°$ with $-3°$, $0°$, $+15°$, and $+30°$ as intermediate values for the most common insert shapes at least.

Larger angles preferred

Although workpiece shape and the specific cutting operation will often determine what SCEA cannot be used, most experts recommend the highest value possible for a number of reasons. Larger angles produce thinner and wider chips for any given feed. This spreads the load and reduces pressures on the cutting edge, as well as producing a chip that's easier to curl. Heating effect of the cut is also spread, and heat dissipation is improved. If a straight turning cut is started with a generous positive SCEA, initial contact will be well up on cutting edge, which protects the more vulnerable nose radius of the insert and also reduces the shock of the insert's entry into the workpiece.

One negative aspect to large side-cutting-edge angles is that they result in higher radial forces between the work and the tool. With slender or thin-walled workpieces, this increase in radial force can result in deflection or chatter.

Tooling practice with inserts follows the same basic principles as any other type of cutting tool. The only differences are in the basic hardware and the fact that some problems can be made more severe by such factors as higher machining speed. Always desirable for any machining operation, tool rigidity becomes even more important with carbide inserts and still more so with cermets and ceramics.

NC lathes are basically designed to use insert-type tooling, and the machine builder's engineers will have put considerable thought and calculation into the size of toolholders that should be used. But retooling some older machines may require some calculations, even though suppliers of toolholders have also done their homework in matching capabilities of holders and inserts. Shank height should always put the cutting point of the insert on center.

Keep tooling short

A toolholder, of course, is a cantilevered beam, and its deflection under load is proportional to the cube of its length (L^3). This underlines the importance of minimizing overhang. One rule of thumb states that the toolholder should never project any distance greater than its height, but less is better. For boring bars, the general rule is to select the largest possible diameter that fits the ID of the workpiece and provides some required

workpiece at the cutting radius. It is measured in surface feet per minute, sfm.

Deformation—The permanent change in the shape of a cutting tool due to cutting forces and temperature. This generally occurs in high-speed or heavy machining.

Depth of cut—The distance between the bottom of the cut and the uncut surface of the work, measured in a direction at right angles to the machined surface of the work. This is the difference in height between the machined and unmachined surfaces.

Edge preparation—A conditioning of the cutting edge, such as honing or chamfering. (See Chamfer, Honing)

End-cutting-edge angle—The angle between the cutting edge on the end of the tool and a line perpendicular to the side edge of the straight portion of the tool shank.

End relief angle—Angle between the portion of the end flank immediately below the cutting edge and a line drawn through that cutting edge at right angles to the base of the toolholder. (See Clearance, Relief)

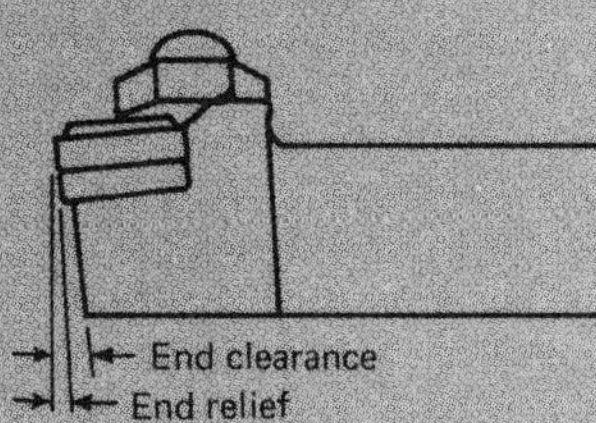

Entrance angle—The angle that the side-cutting edge of a tool makes with the machined surface of the work, measured on the cutting-edge side of the tool point.

Face (Rake face)—That surface of the cutting tool on which the chip impinges as it is separated from the work.

Feed—In turning, the distance moved by the tool into the work for each revolution of the work. It is measured in inches per revolution, ipr.

Flank—That surface adjacent to the cutting edge and below it when the tool is in a horizontal position for turning.

Flank wear (Abrasive wear)—The wear that occurs along the flank of a tool, below and immediately adjacent to the cutting edge, while cutting. This wear reduces the clearance angle of the tool until failure finally occurs. (See Wear land)

Heat check—Many small cracks in the surface caused by grinding too rapidly, without proper coolant, or with too much pressure during grinding.

Honed—Abrasively finished, generally by rubbing using a wet compound and abrasives.

Honing—The process of rounding or blunting the cutting edge with abrasives for increased edge strength. It may be done by hand or by machine. Standard hone specifications: A, light hone, 0.001- to less-than-0.003-in. radius; B, medium hone, 0.003 to 0.005 in.; and C (or H), heavy, 0.005 to 0.007 in. (See Edge preparation)

Included angle—A measurement of the total angle within the interior of a piece. The angle between any two intersecting lines or surfaces. The corner angle of an insert.

Inscribed circle (IC)—The circle that can be constructed internal to any closed figure or shape so that all sides of the figure are tangent to the circle. The inscribed circle is most often used to describe the dimensions of a triangle, pentagon, hexagon, octagon, or trigon. The IC of a square or parallelogram is equal to the perpendicular distance between opposite sides; IC of a round insert is the same as the insert diameter.

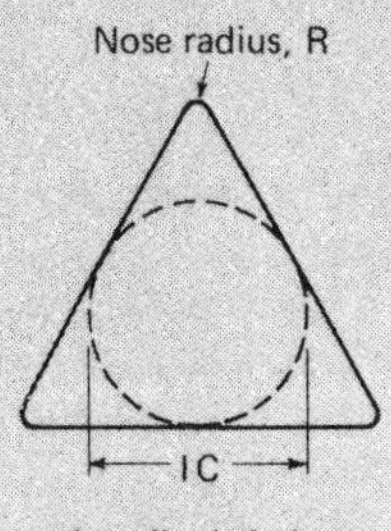

Inscribed circle

Insert—The cutting tip, made of hard material, that is mechanically affixed to the toolholder for use. It can generally be indexed to present more than one cutting edge.

chip clearance. When length does not exceed four times bar diameter, ordinary steel-shank tooling should prove adequate. When length exceeds this ratio and trouble is encountered, special bars with tungsten-carbide shanks and/or special antivibration mechanisms may be required.

Widely used on numerically controlled lathes are qualified toolholders, which facilitate the close positioning of the insert point to a known set of dimensions within the machine's X-Z coordinates. Such holders are typically ground on two surfaces—the back end to determine length and either the side opposite or the side adjacent to the insert pocket for tool width. Qualification is generally to ±0.003 in. with respect to a gage insert and may be measured either over a nose radius or to a sharp corner, depending on the geometry of the measurement. Inserts also have tolerances, of course, and these (which vary according to insert shape and size in the more commonly used categories of insert precision) will add to those of the toolholder. Control manufacturers, therefore, provide NC systems with adjustment capability that is independent of the part program. These are tool-offset pairs (X and Z), and they are generally provided in sufficient numbers to permit assignment of at least one pair for each location in the machine's turret or turrets.

A difference on coolants

Expert opinion varies on the value of coolants for turning with carbide inserts. Most recommend the use of cutting fluids; a few believe that dry machining is preferable. Both camps are as one, however, in stating that a lack of coolant is better than improperly applied coolant. Proper application should start before the cut begins and should not be turned off until the tool is withdrawn from the work, and it should consist of a steady, heavy flow that floods workpiece and toolholder as well as the insert. An intermittent flow, even one that may be interrupted by individual chips coming off the workpiece, is worse than none at all because it allows thermal cycling of the insert, which is likely to lead to chipping, cracking, and premature fracture of the tool.

The argument in favor of using coolant, of course, is that tool wear is accelerated by elevated temperatures, and anything you can do to hold down the temperature of the insert will extend its life. In addition, of course, the lubricity of the cutting fluid reduces friction and thus the horsepower necessary for the operation.

The opposing argument, stated at a recent Sandvik metalcutting seminar by Dr Evert Lundgren, the company's

Glossary

Insert package—The complete assembly that fits into the insert pocket of a toolholder. This includes such components as a seat, setscrew, insert, chipbreaker, clamp, and clamp screw.

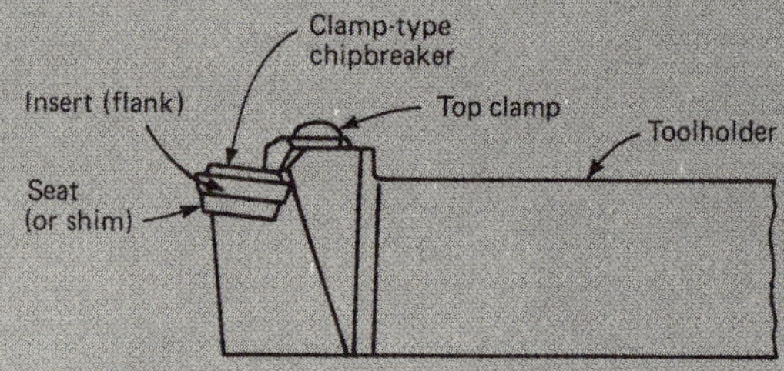

Insert pocket—The space that has been machined out of a toolholder to receive the insert package (seat, insert, chipbreaker, etc)

Lapping—An abrasive finishing process involving rubbing with a wet compound. Commonly used to achieve a high degree of flatness on inserts.

Lead angle (Side-cutting-edge angle, Attack angle)—The angle between the side cutting edge and the projected side of the tool shank or holder, which leads the tool into the work. (See Attack angle)

Machinability rating—A numerical value expressing the ease or difficulty of machining a particular workpiece material in comparison with AISI B1112 cold-rolled steel being turned at 180 sfm, which is rated at 100%.

Negative land (K-land)—A bevel along the cutting edge causing the rake to become more negative. It may range from 3° to 15° and is measured off the top face of the insert. (See Chamfer)

Negative rake—An rake angle that is less keen or more blunt than zero rake. (See Rake)

Neutral rake— A rake angle of zero degrees. This angle is perpendicular to the surface of the work and neither positive nor negative.

Nose—The corner angle formed by joining the side and end cutting edges of a tool.

Nose radius—The radius on the tool between the end and side cutting edges.

Pin-type insert—An indexable cemented-carbide cutting-tool insert with a hole in the center used to locate the insert and clamp it in place by means of a pin in the toolholder. Need for a top clamp is thus avoided.

Pocket—See Insert pocket.

Positive rake—A rake angle that is keener or more acute than zero rake. (See Rake)

Precision insert—An insert (Code G in third position) that is ground on all surfaces to specified dimensions ± 0.001 in. on cutting point and ±0.005 in. on thickness. (See Super-precision insert)

Prehoned—Machine-honed to a uniform size by the insert manufacturer. (See Honing)

Rake—The angle of inclination between the face of the cutting tool and the work. If the face of the tool lies in a plane

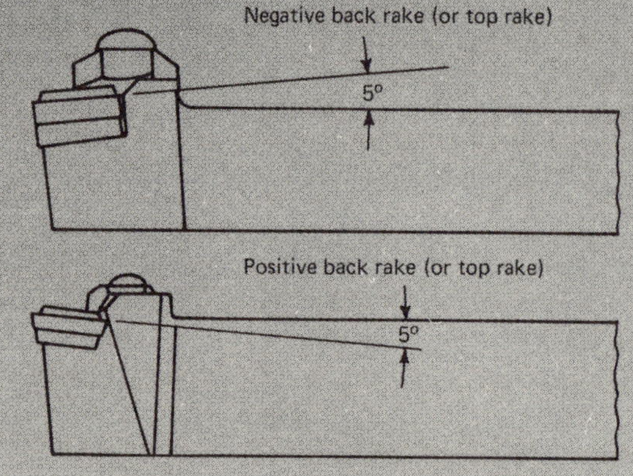

through the axis of the work (on a round workpiece), the

authority on insert geometry, is that high temperatures improve the machinability of the workpiece material. He cited hot machining, in which high-strength materials are preheated just ahead of the tool to improve machinability, as an analogy, and he pointed out that high temperatures reduce the forces necessary to deform the chip. Although he admitted that knowledge of what happens at the cutting edge is far from complete, Lundgren described experimental results showing the formation of an oxide layer on some inserts that acts as a lubricant and a chemical barrier, lowering cutting forces and reducing insert wear.

When ceramic inserts are used, the general rule is to turn off the coolant—both because of the excellent high-temperature properties of ceramic inserts and because they are less resistant to the effects of thermal shock. There are cases, however, cited by TRW/Wendt-Sonis, in which air-blast has been used, in which water-soluble mists for copper, brass, bronze, and aluminum have been successful, and in which sulfurized oil has been used for machining 300 and 400 stainless steels.

Metal-removal rate vs tool life

In turning, as in most endeavors of the real world, there's no such thing as a free lunch. Whenever you increase the feed or speed or depth of cut, you pay for the added productivity by giving up some tool life. The widespread use of throwaway inserts has greatly reduced the relative cost of each cutting edge, and the minimal machine downtime required for indexing an insert to each new cutting edge has also reduced the cost of tool-changing. The net result of this is that inserts justify a faster rate of tool consumption than traditional forms of cutting tools. Where 60 minutes might be desirable with a brazed tool, 15 minutes would be economical with inserts. But tool costs are still real.

Increased metal-removal rate, however, can be achieved by increasing any one of the machining parameters that affect it—speed, feed, or depth of cut—and these do not all cost the same in terms of shortened tool life. In its literature on the selection and application of cutting tools, Carboloy provides examples to illustrate this.

Cutting speed, the publication states, has by far the most pronounced effect on tool life. A 50% increase in cutting speed will commonly reduce tool life by approximately 90% and, therefore, is the most costly route to higher metal-removal rate in terms of tool life.

Boosting feedrate by 50%, however, will decrease tool life by only about 60% in the Carboloy example and is considered much more cost-effective, therefore,

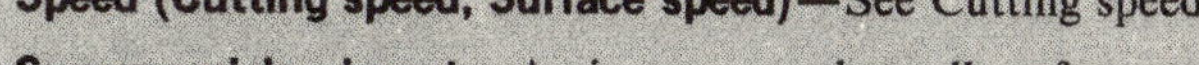

tool is said to have zero, or neutral, rake. If the inclination of the tool face makes the cutting edge keener or more acute than when the rake angle is zero, the rake is defined as positive. If the inclindation of the tool face makes the cutting edge less keen or more blunt than when the rake angle is zero, the rake is defined as negative. (See Back rake, Side rake)

Rake face—See Face

Relief—The clearance angle behind or below the cutting edge, allowing the cutting edge to be forced into the work. It is sometimes divided into primary relief (adjacent to the cutting edge) and secondary relief (beyond the primary relief). Depends on both the clearance angle of the insert and the insert attitude in the toolholder.

Seat (Anvil, Shim)—A removable part of a toolholder, designed to support the cutting insert. (See Anvil, Insert package)

Sfm (Surface feet per minute)—Peripheral velocity of the workpiece at cutting radius. (See Cutting speed)

Side-cutting-edge angle (Lead angle, Attack angle)—The angle between the side cutting edge and the projected side of the shank or holder. (See Lead angle)

Side rake—The inclination of the face of the tool away from the side cutting edge. It is measured in a plane perpendicular to the top plane of the tool and the side cutting edge. (See Rake)

Side rake (positive)

Side clearance

Side relief

Side relief angle—The angle between the side flank immediately below the side cutting edge and a line drawn through the side cutting edge perpendicular to the base.

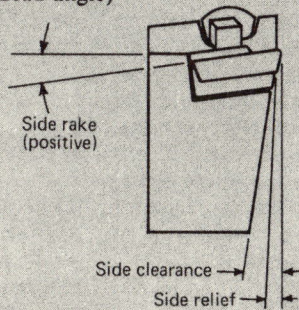

Speed (Cutting speed, Surface speed)—See Cutting speed

Superprecision insert—An insert ground on all surfaces to a closer tolerance than a precision insert. There are several classes of superprecision inserts, with tolerances on the nominal point dimension ranging from ±0.0002 in. to ±0.001 in. and on thickness dimension from ±0.001 in. to ±0.005 in.

Toolholder—A tool component that mechanically holds the insert and that, in turn, is mechanically affixed to a tool-carrying component of the machine tool, such as a turret. Loosely, a tool-shank for insert-type lathe tooling.

Top rake (Back rake)—The angle of inclination of the tool face away from the end cutting edge. It is measured in a plane perpendicular to the base of the tool and parallel to the side cutting edge. (See Rake)

Transverse rupture strength—Breaking strength of a material in a standard bending test (ASTM B406-70 for carbides). Measured in pounds per square inch (psi) in the US. Generally used as an indication of the toughness of a cutting-tool material although certain limitations exist. For example, the test is static and may not provide a good prediction of performance in a dynamic cutting mode.

Utility insert—An insert ground on the top and bottom surface only, with a common tolerance of ±0.005 in.

Wear land—A flat area worn on the relieved flank face of the insert, below or behind the cutting edge. The depth of wear affects size and finish, and the width of the wear land is a good indicator for comparing insert performance or for determining the proper time to change or index the tool. (See Abrasive wear, Flank wear)

Working angle—Those angles between tool and work that depend not only on the shape of the insert but also on its position with respect to the work as determined by the toolholder.

Zero rake (Neutral rake)—See Rake.

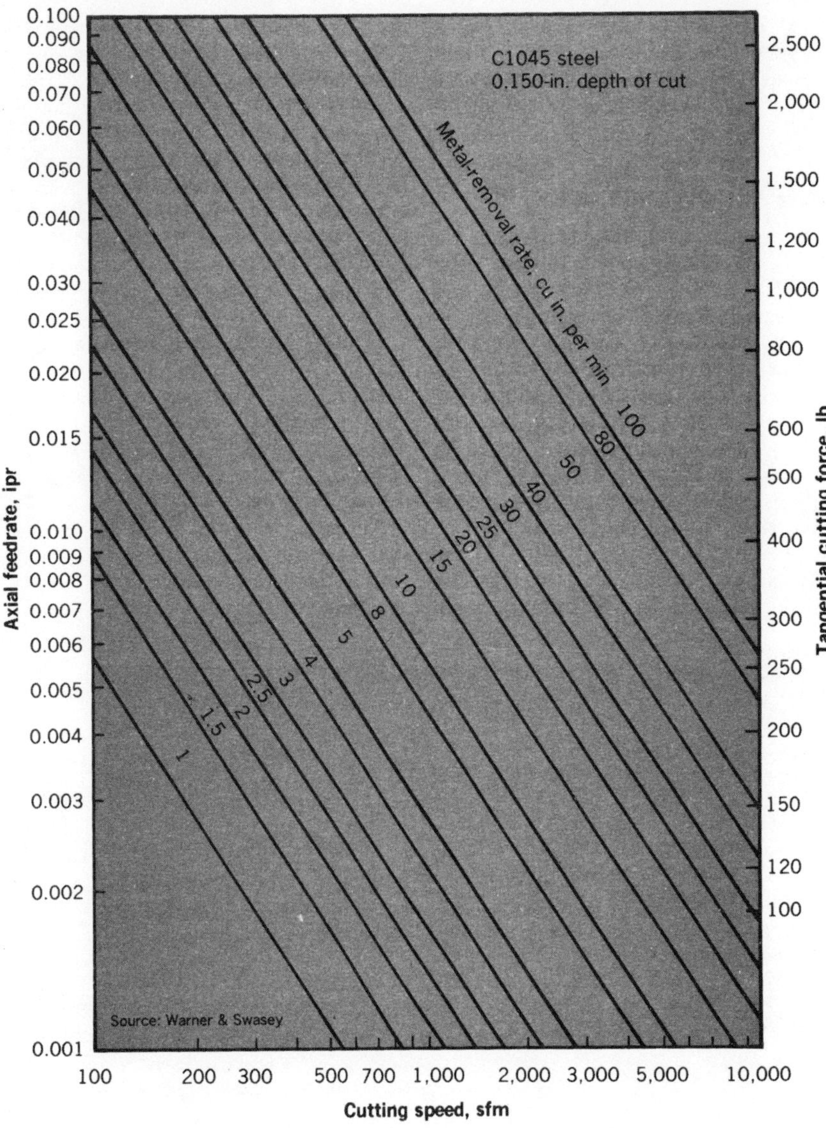

9. Relationships between speed, feed, metal-removal rate, and tangential cutting force are plotted on log-log grid for constant cutting depth of 0.150 in. in C-1045 steel of 204 Bhn. At low values of speed, an increase reduces cutting force, and, at higher values, the force is virtually independent of speed. Correlation between cutting depth and force is nearly linear—doubling depth very nearly doubles force. Rule of thumb is that it takes about 1 hp at the spindle for each 1 cu. in. per min of metal removal for plain carbon steels with sharp cutting tools and about 25% more for dull ones

than turning up the lathe's spindle rpm.

The effect of varying depth of cut is rather more complex because it is a simultaneous function of feed. However, the Carboloy example shows that at a cutting depth of ten times feedrate, a full 100% increase in depth reduces tool life by only about 25%. Below this 10:1 depth-to-feed ratio, an increase in depth of cut has a greater impact on tool life because more of the cutting heat is concentrated on the inserts' nose radius.

These facts lead to a rule of thumb for setting turning conditions. Select the heaviest depth of cut possible. Select the highest feedrate possible. Then select a machining speed that yields an economically acceptable tool life.

What's possible in each case depends on many factors other than the insert itself. These include machine horsepower, general condition, and speed and feed settings available; workpiece rigidity, material, fixturing, required finish, and tolerances; and the nature of the operation itself (you can't increase depth of cut very much if there's only 1/16 in. of stock to remove, for example).

Getting up to speed

There are a number of sources of data correlating work material, tool material, and cutting speeds. All are useful. All have some limitations. The basic problem, of course, is that there are too many variables involved.

One of the better handbooks is the *Machining Data Handbook,* published by Metcut Research Associates (Cincinnati), which provides tabular information on feeds, speeds, and depths of cut for a wide range of machining processes in addition to turning and does so for a large number of workpiece materials. More than 700 pages of tables are included, plus some 300 pages of additional useful material. The book is necessarily limited, however, in the number of tool materials and tool geometries that it covers. Another limitation, shared by all such handbooks, is that developments in the field have been coming at a pace that is difficult to match in the publication of books.

Many material suppliers also offer literature on the machinability and machining characteristics of their products. These are far more limited in scope, of course, and they also tend to lag behind the advancing technology of cutting tools. Nevertheless, if you are using that particular material or a similar one, this source can often be useful.

Machine-tool builders provide still another source of knowledge in this area and, in many cases, provide manufacturing-engineering services and full tooling packages for specific production applications. Their wide experience and recommendations can be extremely valuable.

And all producers of inserts provide recommendations on the application of their own grades to a broad range of workpiece-material classifications, including suggestions on cutting data. Insert producers' literature is as up-to-date as the latest product introduction, and much of it is genuinely helpful in the selection of an optimum insert from each particular product line. The difficulty, of course, lies in crossing the lines of different producers.

Any of these sources can provide a reasonably good set of "starting" values for cutting speed, tempered, of course, by the limitations of machine and workpiece. Getting to the optimum speed is a matter of economics.

Insert economics

Three principal cost elements enter into any turning operation: setup cost, (nonproductive time), machining cost, and toolchanging cost. The per-part cost of machining the workpiece is the sum of these three (which include both direct and overhead components). A graph of the total plotted against machining speed will pass through a minimum value, because setup cost is constant for any speed, machining cost decreases with increasing speed, and toolchanging cost rises with increasing speed. (See upper curve in Fig 10.)

In other words, if tooling is babied along with low cutting speeds, the unit production cost goes up, because the hourly burden of labor and overhead is spread over too few workpieces. And, if cutting speed is revved too high, the unit production cost is forced up by the higher cost of cutting tools consumed and—more important—the cost of increased downtime for changing the worn-out cutting edges.

For any given job, with all other factors held constant, the relationship between tool life (T) and cutting speed (V) is expressed by the Taylor law:

$$VT^n = \text{constant.}$$

Using this relationship, it is possible to mathematically derive an equation to find the tool life resulting in minimum unit cost (T_c, expressed in minutes) in terms of toolchange-time (TCT, in minutes), cost per cutting edge (e, in minutes), cost of labor and overhead (L, in dollars per minute), and the Taylor tool-life exponent (n). This equation is:

$$T_c = \left(TCT + \frac{e}{L}\right)\left(\frac{1}{n} - 1\right)$$

In a similar manner, a plot of part production against cutting speed will pass through a maximum value (lower curve, Fig 10). In the low-speed range, it's obvious that a boost in speed will also boost output; at the high-speed end of the curve, production rate falls off because of excessive toolchanging time. Again, the tool life for maximum production (T_p) can be expressed mathematically as a function of toolchange time and the Taylor exponent:

$$T_p = TCT\left(\frac{1}{n} - 1\right)$$

Cost of inserts per cutting edge and the cost of labor and overhead are easily available. Toolchange time can quickly be determined with a stopwatch; it's the average total downtime resulting from the need to index an insert, and it must include any time required for part measurement and adjustment of the tool so that its cut is within tolerances. The Taylor exponent can be determined experimentally and will typically be in the range of about 0.22 to 0.30 for tungsten carbide, 0.38 and up for ceramics, etc (also see Fig 13).

This somewhat academic discussion leads up to the high-efficiency-machining concept put forth by Carboloy some years ago. The concept states that there is a range of economically justifiable metal-removal rates (the Hi-E range) that lies between the minimum-cost point and the maximum-production point as indicated on Fig 10.

Carboloy has published a series of tables of precalculated solutions for these equations in its booklet discussing tooling selection (Publication GT6-262). Each table applies to a specific tool-changing time (1, 2, 3, 4, 5, and 10 minutes) and is arranged in rows of tool-cost figures (from 10¢ per edge to $2 per edge) and columns of burden rate ($10 per hour to $90 per hour). The tables are apparently based on a Taylor exponent value of approximately 0.22.

Let's say your burden rate is $30 per hour and you're using inserts that cost 70¢ per edge. If toolchange time is 2 minutes, tool life is 7 minutes for maximum production and 12.0 minutes for minimum cost. If TCT is 5 minutes, T_p is 17.5 minutes and T_c is 22.6 minutes. If TCT is 10 minutes, T_p is 35 and T_c is 40.4.

To select a machining speed within the high-efficiency range, therefore, it's just a matter of adjusting spindle rpm so that average insert life per corner falls between the paired values.

Try the hardest first

At this point, it is time to look to the recommendation charts of the insert producers to match the specific grade with the workpiece material and the machining parameters selected. Standard practice is to pick the hardest and longest-wearing grade that will stand up to the rigors of the particular operation, and back off to a tougher but softer grade only if forced to.

More-sophisticated approaches are also possible, and Carboloy, among others, offers computerized solutions to assist customers in the selection of optimum tooling and turning conditions for specific operations.

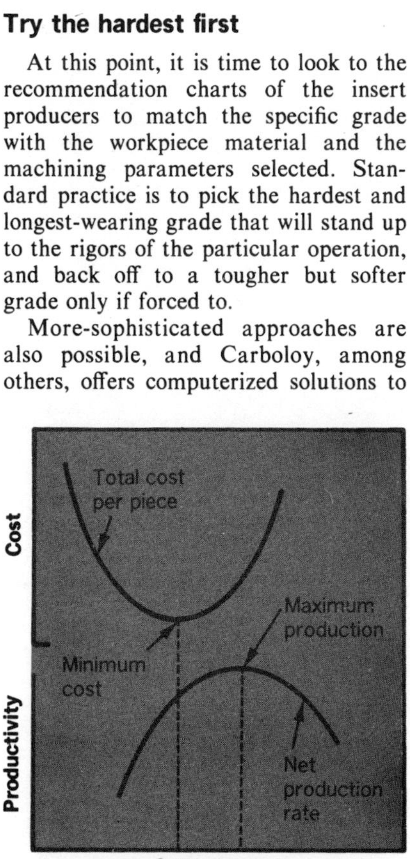

10. Total part cost and production rate vary with cutting speed as shown here, with maximum production at a higher speed than minimum cost. The speed range between these two values is called the high-efficiency range

Few workpieces, however, can be machined with a single insert. Part configuration in all but relatively simple cases generally requires inserts of different shape or holders that present the inserts at different attitudes. Turrets fill this need and, at the same time, complicate the question of optimum tool life. Because each tool in the turret performs a different operation, or series of operations, it's a rare occasion when all will wear out simultaneously.

It's obviously poor operating efficiency to shut down the lathe as each insert comes to the end of its useful life. A far better economy will be achieved by indexing all inserts at the same time or at least in groups. The problem is parallel to the toolchanging problems and practices associated with multispindle machines and transfer lines, and some of the techniques used there for balancing tool life may prove practical if adapted to the tooling of lathe turrets. Also parallel to the transfer-line case is the need for consistency of tool life.

The insert with the shortest average working life—generally the one that's in the cut for the longest total time in the particular machining cycle—is the key tool. It may be justified to reconsider this operation to find out whether a premium grade of carbide, or perhaps a ceramic tool, is justified. And, if this one is a premium grade, it may not be necessary to use a similar premium grade for some of the other less demanding operations. Speeds and feeds may also be juggled in order to get balanced life from all inserts.

The tool-life end-point criterion may also be re-evaluated. For roughing cuts especially, it may well be worth while to go slightly beyond the normally accepted limit. And there may well be situations in which it's more economical to index an insert before it reaches its full measure of flank wear.

It isn't necessary to achieve a perfect match of insert life; convenient multiples may do the job just as effectively—for example, index this insert for each new workpiece and that one for every third workpiece. Or the numbers might be ten workpieces and 30.

Why not double tool?

Another approach that can be adapted from transfer-machining practice is to provide multiple tooling for the longest operation. On transfer machines, it's done to shorten the machining cycle time; in NC turning, it could be done to balance tool life. The idea is to use identical tooling in two turret stations—

provided there's an extra station available, of course—and, for example, break down the roughing operation into two sets of passes, one with each tool. This would cost some extra cycle time, but a turret index is faster than an insert index—even if it's a roughing operation that doesn't require readjustment of the insert's cutting diameter.

What does an insert cost?

One point well worth remembering is that the cost of insert cutting edges is really very small in comparison with other machining costs. If labor and overhead come to $30 an hour, for example, a one-minute tool index costs 50¢, and that's considerably more than the per-edge cost of many inserts. It's true that insert prices are going up with worldwide inflation. But it's also true that the other costs are rising even faster. And continuing progress in the materials, design, and manufacture of inserts constantly adds to their value and their economic effectiveness. Finally, it is the performance of the cutting tool that determines the ultimate efficiency of the machine tool—its profitability is balanced on the point of the insert.

Troubleshooting

Every job is different. Every workpiece has its own idiosyncrasies of material and shape. Every machine has its own characteristics of design and condition. Every toolcrib contains its specific complement of toolholders, inserts, and accessories. Every process planner, every supervisor, and every operator is human. These, and still other variables, combine—often unpredictably—to cause machining problems in the plant.

Isolating the cause or combination of causes is the first step in solving such problems. The first place to seek answers is from the operator and the supervisor in charge: what are their opinions about the problem and its potential causes? Then examine the part, the cutting tool, the machine, and the setup.

If the solution isn't immediately obvious and if it still seems related to the cutting tool, the checklist below, gleaned from a variety of sources, should help to correct most problems. In applying the suggestions, it's preferable to change a minimum of variables at any one time, ideally just one at a time. That way, you will be able to pin down the solution to the specific action that was effective, and you may avoid the hazard of having one change offset another.

The single most important factor in improving tool life, according to the *Carbide Technical Manual* published by Adamas Carbide Corp, is proper determination of why the tool failed prematurely. Here is an example:

Conventional procedure in machining tests, where time is plotted, involves running a test tool for a specified duration, after which the tool is examined and conditions are noted. The same tool is then put back into the work for the next specified interval and examined again. This is repeated with the same tool until the end of the test. Flank wear is the usual measure of cutting-tool wear. The longer a given tool lasts in terms of time or pieces produced up to a specified flank wear, the better the tool is.

Why did the tool fail?

But was the tool's failure caused by flank wear? Edge chipping and rake-face pullout can often be mistaken for flank wear. Especially on tough jobs, tools very often do not get the chance to fail by

Troubleshooting checklist

Edge chipping
Increase speed
Decrease feed
Use prehoned insert
Increase nose radius
Decrease relief and clearance
Increase side-cutting-edge angle
Decrease end-cutting-edge angle
Use negative rake
Increase chipbreaker width
Switch to tougher (softer) grade
Try uncoated grade
Reduce overhang or increase toolholder cross-section
Check machine looseness
Check center height

Insert cracking, breakage
Decrease feed
Reduce depth of cut
Use prehoned insert
Increase nose radius
Use thicker insert
Increase included angle; use square instead of triangle, etc
Check insert pocket for wear, chips, or dirt
Reduce overhang or increase toolholder cross-section
Increase side-cutting-edge angle
Decrease relief and clearance
Use negative rake
Chipbreaker may be too close; increase width
Change to tougher (softer) grade
Check machine for looseness
Check center height

Cratering
Decrease speed
Reduce feed
Increase chip-control width
Use coated carbide
Switch to harder grade
Switch to grade with TiC and/or TiC additions (more crater-resistant)
Try ceramic or cermet
Use (or increase) positive rake

Excessive flank wear
Decrease speed
Increase feed
Increase depth of cut
Increase relief angle
Use coated inserts
Try higher hardness grade
Try solid-TiC grade
Try ceramic or cermet
Inspect under magnification; problem may be micro edge-chipping
Decrease nose radius
Use lighter hone
Check center height
Eliminate vibration

Built-up edge
Increase speed
Increase feed
Use (or increase) positive rake
Use insert with polished rake face
Use coated carbide
Use grade with higher Ti and/or Ta content
Use ceramic or cermet
Increase lubricity of coolant

Cutting-edge burning
Reduce speed
Increase/decrease feedrate
Increase/decrease depth of cut
Change to grade with higher Ti and/or Ta content
Change to ceramic or cermet insert
Increase coolant flow
Check cooling properties of cutting fluid

Chip control
Chips too long
- Increase feed
- Increase speed (for hotter chip)
- Increase depth of cut
- Use narrower chipbreaker

flank wear. Instead they fail from chipping, breaking, or rake-face pullout.

During a cut, there is a normal tendency for metal transfer to take place. Workpiece material welds to the tool's top surface forming a built-up edge. Failure then occurs, or is initiated, when this welded chip is torn from the tool and carries with it a fragment of either the cutting edge or the rake face. Since these may be minute particles, the visual evidence resembles flank wear.

Another problem is that interruptions or inclusions in the workpiece material may cause a notch to develop in the cutting edge, which is weakened by the notch and may then fail by chipping or breaking.

A third factor to consider is the rapid heating and cooling of an insert that may occur when coolant is used or in the case of an interrupted cut. Such thermal cycling may cause fine thermal cracks to develop in the cutting edge, which again may ultimately cause the tool to fail by breakage.

In a conventional test in which the same cutting edge is used continuously until a given flank-wear is reached, the Adamas publication cautions, it is often virtually impossible to determine the real cause of tool failure since the edge may actually fail by breakage before the true cause of the problem can be determined.

Here's a test procedure

Adamas has developed the following procedure in order to avoid these testing difficulties and uses it in all research tests concerning the effects of such considerations as grade selection and edge preparation on tool life:

- Establish the expected life, in minutes, for a tool on a given job.
- Select a test time of something less than the expected life—say, half, a third, a fourth, etc. The duration of this test time must be short enough so that the test does not destroy the cutting edge.
- Run the series of test inserts, which may include various grades, various edge preparations, or whatever other variables may be under investigation, using only one cutting edge on each insert for the specified test time.
- Compare the test inserts to determine which one is performing best.
- Determine whether additional edge preparation or other factors may then be used to improve performance on the hardest grade or whether a tougher grade is necessary.
- Finally, run several inserts (or corners of the same insert) of the same grade and edge preparation to destruction to determine tool life.

In this way, Adamas finds, it is often possible to select harder grades and achieve insert life considerably in excess of that originally obtained.

Test various chip-groove shapes
Chips too tight
Reduce feed
Reduce depth of cut
Increase chip-control width
Test various chip-groove shapes

Chatter
Increase/decrease speed
Increase/decrease feed
Increase/decrease depth of cut
Reduce nose radius
Vary side cutting-edge angle
Reduce overhang, use larger toolholder
Check tool height
Decrease rake angle
Increase relief and/or clearance
Check machine for looseness

Poor finish
Check insert condition and index
Increase speed
Reduce feed
Use larger nose radius
Increase lead angle
Check for tool rubbing and correct
Make sure chips are not hitting work surface
Check cutting-fluid effects
Check machine for looseness

Turning with a computer

No computer ever cut metal, but electronic data processing does provide the means for simulating the machining process, and this, in turn, permits specific machining parameters to be selected on a more scientific basis before the machine tool itself ever cuts a chip.

The merit of this possibility is increasing in today's world of numerically controlled machines, not only because the feeds and speeds have to be punched into a tape or stored in semiconductor memory, but also because of the increasingly high costs of manufacturing equipment, its tooling, and the labor and overhead required for its operation. Seat-of-the-pants selection of machining parameters—whether by the machinist, the supervisor, or the manufacturing engineer—is becoming too costly to gamble on. And the number of workers possessing the necessary skills, judgment, and experience is constantly dwindling.

The computer's apprenticeship

Before the task can be turned over to the computer, however, somebody has to tell the computer what machining is all about; this demands what is called "a mathematical model." Then reliable values have to be filled in to accommodate the many variables involved.

One active researcher in the field of mathematical modeling of the turning operation is B.K. Srinivas, of Warner & Swasey's research department, who

11. Insert-testing program was carried out on instrumented W&S SC-15 four-axis NC turret lathe. Cutting force and feed force were recorded automatically

12. Strain gages were bonded to back and bottom of 1-in.-sq-shank holder to sense tangential and axial forces of cut on SNG-432 inserts. Tool-life criterion was 0.015-in. flank wear

13. Materials compared

Insert material	Exponents for Tool life (n)	Feed-rate (a)	Depth of cut (b)
Uncoated WC	−0.30	−0.31	−0.13
TiC-coated WC	−0.27	−0.43	−0.15
Solid TiC	−0.31	−0.41	−0.23
Niobium nitride	−0.38	−0.40	−0.17
Aluminum oxide	−0.38	−0.48	−0.12

Source: Warner & Swasey

points out "some of the factors" affecting tool life:

- Cutting speed
- Feedrate
- Depth of cut
- Machinability rating of the workpiece material
- Tool geometry
- Tool material
- Type of cut, whether continuous or interrupted
- Rigidity of the overall tool/workpiece system
- Total machining tolerance
- Coolant (or lack of it)

In a technical paper describing a series of actual turning tests to evaluate the performance of several cutting-tool materials, Srinivas cites earlier work done by Carboloy that showed these relationships: a 50% increase in cutting speed cuts tool life approximately 80%; a 50% increase in feedrate results in about a 60% loss of tool life; and depth of cut has very little effect unless it is less than 10 times the feedrate. Srinivas then cites the classic tool-life equation developed by Frederick W. Taylor early in the century (VT^n = Constant) and presents an extended version of greater utility:

$$V = KT^n f^a d^b$$

in which V is cutting speed, T is tool life, f is feedrate, d is depth of cut, K is a constant, and n, a, and b are experimentally determined exponents.

Srinivas then described the testing procedures by which the three exponents were determined for five different tool materials in turning 1045 steel (197 Bhn) on a W&S SC-15 four-axis NC turret lathe (Fig 11). All tools were ½-in.-square SNG-432, negative-rake, flat-topped inserts held in a specially instrumented toolholder (Fig 12). Low, medium, and high values were selected for speed, feed, and depth of cut, and cuts were made (without coolant) at all combinations of the three (27 in all). Three different chipbreakers were used for the three different feeds, but all other variables were held constant in three tests at each combination for each of the five insert materials, which were an uncoated tungsten carbide, titanium-carbide-coated tungsten carbide, solid titanium carbide, niobium nitride, and aluminum oxide. Results are shown in Fig 13.

These values were used to plot tool-life curves (on a log-log grid), as shown in Fig 14, for tungsten carbide, titanium carbide, and aluminum oxide. The plots in this graph are for 0.025-ipr feed, 0.200-in. depth of cut, and the tool-life end-point criterion is a wear-land width of 0.015 in. Tool geometry, machine tool, and workpiece material are as described.

The extended tool-life equation, however, requires a considerable amount of further extension before it can be used in a computer to establish the most desirable set of machining values for a production operation. Factors that were held constant in the W&S tests—workpiece material, tool geometry, and setup rigidity to mention but three—must be incorporated in the mathematical model. It is necessary both to establish how much effect each factor exerts on the machining operation in combination with the others and to place values on these variables. The task is a complex one, but it is being done. ■

Acknowledgements

AM would like to thank the many companies that contributed material for this report, especially the following: Adamas Carbide Corp, American Feldmuehle Corp, Babcock & Wilcox Automated Machine Div, Carboloy Systems Dept (General Electric), Ex-Cell-O Tool & Abrasive Products Div, General Electric Specialty Materials Dept, Greenleaf Corp, Iscar Metals Inc, Kennametal Inc, Lodge & Shipley Co, Sandvik Inc, Teledyne Firth Sterling, Teledyne Vasco, TRW/Wendt-Sonis, Union Twist Drill Div (Litton), Valenite Div, VR/Wesson Div (Fansteel), and Warner & Swasey Co.

14. Tool-life curves

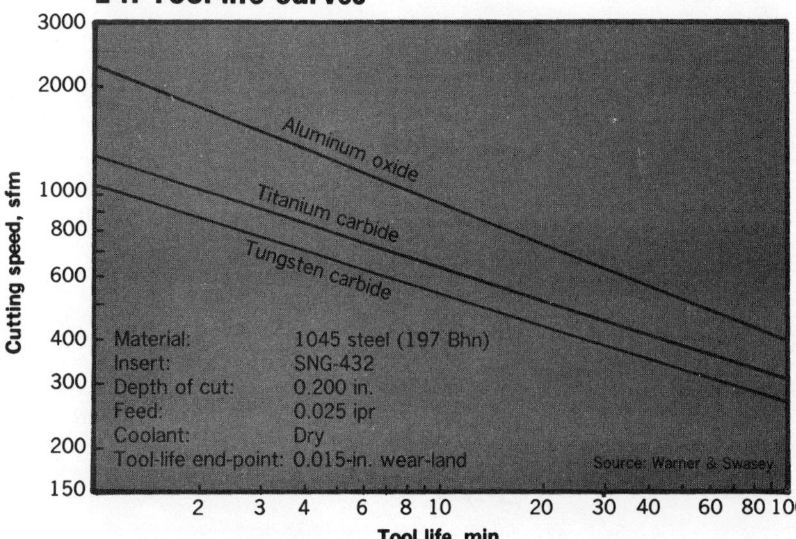

American Machinist

Modern machine tools, current manufacturing trends, and today's economic realities are having significant effects on basic metalworking operations. Take a new look at . . .

WORKHOLDING

HOW TO GRIP THE WORKPIECE is one of the most fundamental questions that must be answered in planning each of the steps in any manufacturing process. For most components, the question must be answered many times as each piece progresses from step to step in the process of being modified from its input configuration until at last it has been completed.

Further, the question applies to all steps, including such operations as assembly and inspection as well as metal-removal operations. In this sense, workholding is a universal consideration in the metalworking industry—and it can often be a problem.

Because the solution of workholding problems for special machines is inherent in their design and is typically handled by the machine-tool builder's specialists, the emphasis of this report will be on workholding for general-purpose machine tools.

For the majority of cases, there are probably more than enough "standard" solutions so that the manufacturing engineer, process planner, or tool designer may not consider workholding a serious problem in the organization of the machining sequence for the particular workpiece at hand.

By R.L. Hatschek, senior editor

Consider the costs

Nevertheless, in answering even the simplest workholding question, there are a variety of choices and considerations that will ultimately affect the economics of the machining operation. This is true whether a setup is built for a single prototype workpiece or a special fixture is made for a long production run. In the first case, a few extra minutes may not be very significant in terms of cost; in the latter case, even seconds may make the difference between red ink and black.

Key to the profitability of any machining operation is chip-cutting time, and, while inflation continues to squeeze more dollars into each minute, there is no way to get more than 60 seconds out of each minute. A major consideration in workholding design, therefore, is minimization or elimination of any part-setup time that interferes with spindle-cutting time. And this point is valid whether the lot size is large or small and whether the machining cycle time is long or short. What differs in these contrasting situations is the means of achieving a higher ratio of cutting time to fixturing time.

But rapid setup and quick workpiece interchange are secondary considerations, no matter how economically important they may be. Primary, of course, are the requirements for positioning the workpiece with sufficient accuracy to meet whatever tolerances are specified for the particular operation, for doing this repeatably for part after part, and for adequately securing the component being machined.

Multifixture on a machining center

Documation Inc (located in Melbourne, Fla), a heads-up manufacturer of computer peripheral equipment, has plunged heavily into the use of numerically controlled machine tools. It also pays considerable attention to the tooling of its NC equipment in order to maximize the return on its capital investment in these machines.

The photo shows a three-position fixture for machining a cast-aluminum drive-roller support on a Cim-Xchanger 720 horizontal machining center in a cycle time of 19 min. Previous cycle time—on a vertical-spindle NC machining center with an indexer—was 42 min. Typical lot sizes are 300 parts.

Operations include drilling, milling, boring, counterboring, reaming, and tapping. There are 20 drilled holes in the part (ranging in ID from 0.089 in. to 3/8 in.), 10 of which are tapped (4-40 and 8-32), and four milled slots. Tightest tolerance on the part, which measures approximately 3 x 3 x 4 in., is +0.0002 in., −0.0003 in.

The special fixture is a fabricated box that clamps three of the castings in three different attitudes for machining the various surfaces. T-bolts and strap clamps secure the fixture, and equally simple, manual clamping devices hold the three workpieces to the fixture.

Despite its simplicity and reliance on traditional clamping, the fixture works in conjunction with the sophisticated machine tool to demonstrate a number of modern trends in workholding:

- Multiple operations are performed in each clamping to avoid any buildup of positioning errors.
- Use of an indexer—built into the machine, in this case—to give access to several sides of the workpiece.
- Simultaneous setup of several workpieces to extend the machine's cutting cycle, which both reduces the spindle's non-cutting time for workpiece interchange and provides the operator with a longer time span for other tasks while the machine does not require attention.
- Dual workstations—the machine features an automatic pallet-interchange system—to allow the operator ample time for unloading machined components and reload the fixtures while the machine continues to cut metal at the other station. (Note that, in this case, an entirely different workpiece is fixtured on the pallet in the background.)

It is necessary, however, for the tool designer to keep economics in mind during the planning of any workholding scheme for the simple reason that, for example, an anticipated long production run will more easily justify added expense in the tooling phase because time savings per part will ultimately add up to a bigger return on investment—including the investment in the machine tool as well as the fixturing—and reduced labor.

Individual workpieces bring their own idiosyncrasies to help complicate the basic problems. Some workpiece configurations seem almost to have been diabolically conceived in the product-design department with the purpose of thwarting all conventional efforts to hold them for machining. Some parts are thin-walled or fragile, requiring special support so that they can be machined without distortion. Others may have tender surfaces that must not be marred by the clamps or jaws that are needed to grip them for still further machining operations. And still other workholding problems can be posed by limitations of the particular machine tool that must be used, interferences between cutting tools and clamping devices, or the need for automated work-handling between machining stations.

What are the trends?

General trends affecting workholding and its economics can be broken down into three broad categories: Changes in the work itself, in the way products are manufactured today, and in the way modern plants are being run; developments in the design and application of machine tools; and, of course the continuing evolution of tooling components and practices.

Although the almost infinite variety of workpieces being machined makes it impossible to generalize with any statistical certainty, some trends can nevertheless be perceived. The market life of products tends to be shorter today than previously, which means that tooling must be amortized over shorter periods and fewer workpieces. Proliferation of models and buyer options, increasing emphasis on modular design of products, and new attention to the family-of-parts concept all contribute to a demand for greater versatility in fixtures and for multipurpose fixtures.

Management is paying increasing attention to the cost of carrying inventories of components and of finished products. This, in connection with several of the trends noted above, tends to reduce the lot sizes in which parts are machined and to increase the requirement for fixturing versatility. Neither is

The case of the cockeyed casting

Unlike such special machines as transfer lines for boring V-8 engine blocks, general-purpose machine tools are basically designed for orthogonal work. And, when the workpiece departs significantly from rectangular geometry, it's generally the fixturing that provides the necessary angularity for machining.

A number of other problems are introduced when that cockeyed workpiece is the bed casting for a slant-bed lathe. The casting is a big one, and heavy. Precision requirements are typically tighter than ordinary. Fixturing must accommodate a variety of different models—five in the case illustrated, with bed lengths ranging from 100 in. to 15 ft. And lot sizes just don't approach Detroit-style numbers.

Using NC machines to build NC machines, Cincinnati Milacron does the milling, drilling, tapping, boring, and backfacing of CinTurn NC-turning-center bed castings on a horizontal-spindle traveling-column 72-in. x 240-in. 25HC Cim-Xchanger machining center. Two work stations are used, permitting unloading and reloading of new workpieces to be done at one station while machining progresses at the other.

At the first station, the casting is set up on an indexing base with normal—if heavy-duty—bolt-down clamping components. One end of the casting is milled, drilled, and tapped, and then the part is indexed for similar operations on the other end. Twelve different cutting tools are used.

To accommodate the various bed lengths, the 48-in. index table is mounted on precision ways that provide 60 in. of travel. Movement is facilitated by air bearings and is powered by a rack-and-pinion drive.

The second work station is a stationary table upon which is mounted a power-indexing headstock and tailstock, which hold the casting horizontally between centers. Machining is done in this setup at four different angular positions, and, at each of these angles, support wedges are placed under the casting to ensure stability for the ensuing operations. Spindle power of the machining center is 25 hp. Typical operations in this setup include milling of all faces, pockets, straight slots, and T-slots; drilling and tapping all holes; and backfacing for leveling screws and way-mounting holes. Stock removal in rough-milling is approximately 1/4 in., and 0.030 in. is the allowance for finishing passes. Twenty-eight different cutters are used for operations performed at this station.

Prior to adopting this technique for machining the lathe beds, the company used five different machine tools. Now, the company notes, setup time has been reduced by about a third, scheduling leadtime has been halved, and it's no longer necessary to use an overhead crane for jockeying the castings into each of the various angles required for machining.

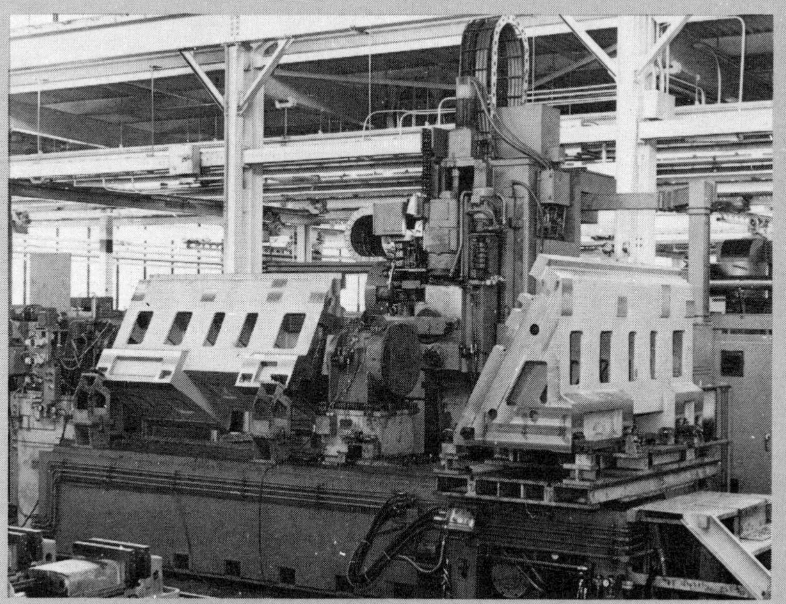

tooling inventory immune to this inventory-reduction drive, which again places a premium on more versatile or multi-use fixtures and tooling components.

At the same time that these pressures are pushing fixture design toward greater simplicity and lower cost, trends in modern machining practice are also affecting workholding. Most significant of these are the parallel trends to higher machining speeds and higher machine-tool horsepower.

Cutting-tool materials are constantly being improved, allowing more and more power to be applied in the cut for faster metal-removal rates, and machine-tool builders are responding with machines that have the power and speed to use these new tools effectively. The result is that cutting-force levels are rising for all types of metal-removal operations, and these higher forces must be resisted by whatever devices are used for holding the work.

Increasing application of numerically controlled machine tools also is affecting workholder design, and these effects go somewhat beyond the fixturing of work on NC machines. Over the past two decades, NC has brought with it an increased attention on the part of management to fixturing costs—simply because a major selling point for tape-controlled machines is the reduction they make possible in the cost of building fixtures and maintaining them in a tooling inventory.

The precision-positioning capability of NC machines also eliminates at least some of the need to guide cutting tools. And the more recent capability of computerized NC systems to compensate for errors of workpiece location or alignment—fixture compensation—eases the pressures for precision in workpiece setup. In short, the NC machine has taken over some of the functions of the jig or fixture. "We've taken the accuracy out of the tooling and put it in the NC" is the way builders of NC machines put it, "and this allows users of NC machines to get better use of their tooling dollars."

Special two-faced vertical vise—back and front are identical—adapts to a variety of workpiece sizes and is especially useful on horizontal machining centers with rotary or index tables. Central column of vise, which was made by Cincinnati Milacron for use in its own operations, is an iron casting

Fixturing for precision in tender parts

Kerns Mfg Corp is an aerospace subcontractor specializing in the production of intricate, precision components and subassemblies for a variety of jet engines. Many of the parts are made of high-temperature-resistant superalloys, which have never been noted for their high machinability. Airfoil contours abound in these components, often helically twisted, making them troublesome for both location and clamping. And, since weight is the enemy of anything that must fly, wall sections are often extremely thin.

Furthermore, there's a high penalty in any part scrappage, both because of the expensive materials generally used and because of the considerable value previously added in forgings or castings supplied to Kerns or in the complex assemblies brazed or welded together by Kerns itself prior to performing many of the final machining operations.

Doing this kind of work for the past 25 years has given Kerns—no stranger to 0.0001-in. tolerances—a special expertise in fixturing "delicate" workpieces that could easily be deformed by inappropriate workholding techniques. The company, a member of the National Tool, Die & Precision Machining Assn, designs and constructs all of the special jigs and fixtures required for assembly, machining, and inspection.

Exhaust fairing is welded assembly of chrome-nickel steel with 0.075-in. wall thickness, requiring relatively complex setups on these two milling machines

A walk around the Kerns facility in Long Island City, NY, quickly demonstrates these special tooling skills.

An exhaust fairing is one typical Kerns part. The component is built up of heat-resisting chrome-nickel steel. The exterior is a streamlined shape formed of 0.075-in.-thick sheet, assembled to the interior structure by heliarc and resistance welding (Kerns uses a wide range of both conventional and highly specialized welding and brazing techniques, including electron-beam welding).

One machining operation on this assembly consists of simultaneously milling the ends of two projecting members of the internal structure (see photo) in a horizontal mill. The part rests against

Relatively light fixturing proves more than adequate for milling, drilling, boring, and tapping operations on aluminum casting while providing access to four sides. Part is photogrammetric-equipment housing for Chicago Aerial Survey; machine is Kearney & Trecker Milwaukee-Matic 200 machining center

Tapered diamond 'cones' are hydraulically advanced into cored holes on both sides of this iron casting for location and clamping. Hydraulic rests support work under the three pads that are to be abrasive-belt-ground for 'feet' for machining. Fixture holds four parts in AEM Abrasagrinder at Gresen Mfg

Putting the positioning capability in the control system, however, does result in changes of workholding methods. Consider a casting, for instance. As one fixturing expert put it, "I see no trend to better castings. They're no worse than they ever were, but practical foundry tolerances on dimensions and core-shift can be more troublesome because of the simplicity of the workholding devices we use on NC machines.

"Where we previously used a drill jig located in a cored hole in the side of a box-shaped casting to drill a bolt circle around that core, we now have to locate the drilled holes by NC coordinates from the base. It creates a different problem. We could go into the cored hole, check its coordinates, and dial in zero shifts for that part of the program, but that kind of procedure would make the overall operation less automatic. We are currently looking into ways to automate the solution of this problem with an automatic offset adjustment."

The advantages of more-adaptable

contoured blocks in the fixture; its leading edge is snugged against a locating pad by a knurled thumbscrew; and two strap clamps with contoured pads hold the work against the fixture with hand-knobs. Finally, a support block clamps the vertical projections firmly with a hex-nut to prevent chatter and vibration while the machining operation is being performed.

Another operation on this same part consists of counterboring an internal tube in the opposite end in a vertical mill. Fixturing is similar but includes adjustable rest pads, beyond the table edge, that bear against a steel plate on the machine's saddle ways to support the component against the vertical thrust of the operation.

In contrast to the relative complexity of the fixtures described above, very simple workholding devices are used for some critical components. The use of aluminum plate disks affixed to the spindles of both vertical and horizontal lathes in the turning department is an example. The face of such a disk is then machined to provide a nest in which the workpiece can be clamped axially with either several strap-type clamps screwed down in tapped holes in the plate or a single aluminum disk (as shown in the photo) held with a central, manually tightened capscrew.

The component in the photos of this technique is a stationary turbine seal

Aluminum-plate disk on lathe spindle is simple and adaptable and provides axial clamping to prevent distortion of thin-walled turbine-seal assembly

workholding—such as reduced cost and lead-time to build tooling, reduced tooling inventory, and quicker changes of setups—are also proving attractive to users of non-tape-controlled machines as well. And the enhanced positioning capability offered by such "near-NC" developments as digital readouts has permitted workholding practices for non-NC machines to parallel those used in NC machining in many cases.

And a reverse trend is taking place where NC machines are being incorporated in production lines. The table tooling of these machines has tended to become more specialized to the specific workpieces being produced. And, in many of these cases, hydraulic workholding devices are being applied to gain in NC the advantages of speedy interchange and consistently uniform holding forces that hydraulic clamping systems are providing in conventionally controlled production lines.

It's good machining practice, both for increasing spindle-utilization efficiency and for maintaining tolerances, to minimize the number of setups needed to complete any given part. "Grab it once and machine it all over" is the ideal, and it is the basic concept of the machining center—especially one equipped with a work-indexing capability so that several sides of the part can be presented to the machine's spindle.

This tests the tool designer's ingenuity in devising fixtures that allow greater access to the part while clamping it against machining forces from several different directions. Best tooling practice, of course, is to provide fixed banking points in the fixture and to avoid arranging clamps so that they must resist cutting forces. But the advantages of machining from all sides generally outweigh this consideration.

Some cutter tricks

Over-design is the usual solution to this problem, but it is sometimes possible to "adjust" cutting forces with cutting-tool geometry. Positive-rake tools tend to pull on the work; negative-rake cutters tend to push. And cutting forces can be "aimed" by judicious selection of conventional or climb milling.

For any fixture, the question of how strong the grip must be is extremely complex. Textbooks on jig and fixture design go into considerable detail on the calculation of clamping forces and the permissible loads for T-bolts, strap clamps, and similar devices. Catalogs of tooling component suppliers—and many of these are quite useful books—provide considerable data on the force capabilities of the seemingly infinite variety of toggle clamps, hydraulic clamps, and other tooling components.

But none of these sources goes into the question of how much clamping force should be applied, Nor are tool designers themselves particularly articulate on this subject.

As one tooling manager put it, "I've never seen any good information on the subject. Our fixture designers do it by the seat of their pants, by their long

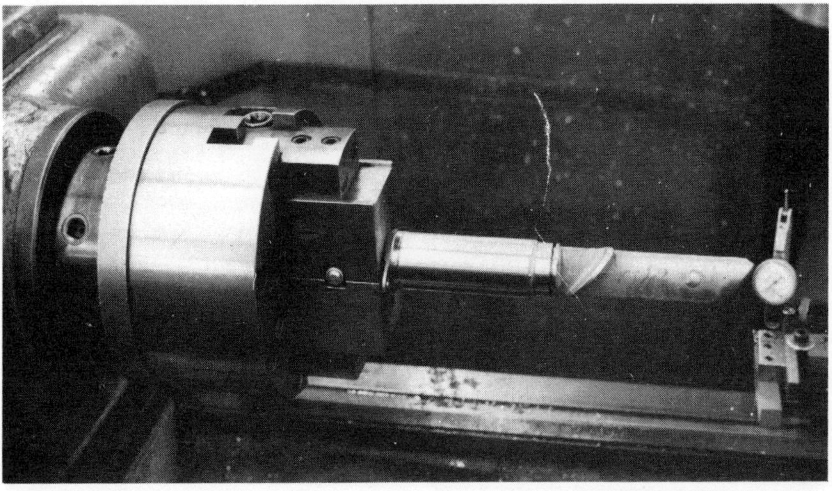

Special top jaws on standard two-jaw chuck are internally machined to clear projections (only one showing) at chucked end of stainless-steel jet-engine pressure probe

Indexing fixture holds delicate part for straddle milling a square in two passes

made of 300-Series stainless steel. It consists of an exterior ring-forging and two internal formed-sheetmetal baffles spotwelded in place. The operation in this setup is the final finish-turning of the OD.

Also noted during AM's visit to the Kerns turning department was a 24-in. Bullard vertical turret lathe with a new 4-in.-thick aluminum disk attached to its table. By starting with such heavy sections, it's possible to remachine the nesting surface many times to accommodate new parts before the aluminum disk must be replaced with a new one.

A standard two-jaw chuck with special top jaws locates and clamps one end of a jet-engine pressure probe that is made complicated by several projections jutting out at various angles from the chucked end. This part is a furnace-brazed assembly of two precision castings and seamless tubes. The material is 347 stainless steel.

Kerns uses an indexing workholder for straddle-milling a square on the end of a variable-vane assembly in two successive passes on a horizontal mill. The vane is an assembled component in which the hollow blade, formed of 0.022-in. 410 stainless-steel sheet to a helically twisted airfoil contour, is furnace-brazed to the internal stainless-steel casting that forms its spindle. The part is inserted in the vertical, cylindrical fixture, the airfoil trailing edge is pushed against a locator, and the spindle is clamped by tightening a setscrew. A peripheral shot-pin provides rotary location for the fixture in two positions at 90°, and two cam-clamps lock the rotating fixture element. A snug-fitting aluminum collar around the base prevents entry of chips, and a simple, sheet-plastic "umbrella" placed over the protruding spindle before machining (not shown in photo) provides additional control for the flow of chips and coolant.

Several 'plastic' fixtures are used in machining this part from a solid Inconel forging at Kerns Mfg. Raw casting was coated with release agent, positioned by capscrews up through baseplate, held by strap clamps, and dammed with sheet metal. Fixture then inverted, and Devcon Plastic Steel poured in through holes in bottom. Fixture shown is for first operation, straddle-milling edges. Note cutter targets

Forged steel ells are clamped by special jaws in 16 Cushman two-jaw chucks for machining at Henry Vogt Machine Co. Multistation machine by W.F. & John Barnes

Multiple setups help to extend the cycles of NC machining centers for improved spindle-utilization efficiency. Setup at right allows work-changing at one fixture while machining is done at the other

Multiple setups are also utilized by Kerns, especially on the company's two Hillyer numerically controlled machining centers, to increase spindle-utilization efficiency. In one case, a relatively simple inlet cover made of 410 stainless requires four holes to be drilled—hardly a likely candidate for NC machining. However, by ganging ten of these in a single fixturing, a more justifiable workload is achieved. The ten parts are secured to the simple fixture, which is raised above the machine table for chip-escape, by capscrews and C-washers; and the fixture is held by standard strap-clamps, T-slot bolts, and serrated step blocks right out of the tooling catalog.

A dual-multiple setup occupied the table of Kern's second Hillyer at the time of AM's visit. Here the workpiece is a turned aluminum component that requires a number of milling, drilling, and tapping operations. Twin fixture plates are fixed to a common aluminum subplate on the Hillyer's table. Each of the fixture plates holds three of the bell-shaped workpieces by means of tapped-end strap clamps using inverted capscrews as heels. Not only does this arrangement increase the duration of the machining center's uninterrupted cycling time, but it also permits new work to be set up on one plate while machining takes place at the other.

Other tooling tricks in the Kerns arsenal include the use of low-melting-point materials, which are poured molten into a workiece or fixture to add rigidity for machining thin-walled elements of complex geometry and are then melted out for reuse after machining. These materials include bismuth-based metal alloys, such as Cerrobend (produced by the Cerro Metal Products Div, Bellefonte, Pa), special wax-like tooling compounds, such as Rigidax (produced by M. Argueso & Co, Mamaroneck, NY), and beeswax on some special jobs.

experience. They probably have a tendency to over-design a bit—for commonsense, or safety, or their own personal backgrounds. They tend to have more shop experience than theoretical engineering knowledge. They've come up through the toolroom, and they've seen enough jobs out on the floor that they've got a pretty good feel for what it takes. That's one of the reasons that it takes a long time to make a good tool designer, and that's why they're so valuable to their employers."

The "style" of fixturing has also continued to evolve. The use of castings as components of jigs or fixtures is unusual today. These have been replaced by welded assemblies, which often include components that are bolted in place. Aluminum tooling plate is more widely used, especially for fixture bases and for subplates that are left permanently in place on a machine table. The ease of machining aluminum facilitates the creation of part-holding nests simply by milling out a suitable cavity. Although this technique isn't widely used, it can sometimes yield advantages when it's done right on the machine for which the fixture is being built.

Another nest-building technique makes use of castable epoxy resins (such as those produced by Ren Plastics). This method is especially useful for holding complex castings for machining, and it has the advantage of damping out some of the noise that can be generated when castings start to "ring." With different resin mixtures, it's possible to produce a resilient surface that will adjust to casting variations while the rest of the plastic is rigid.

Holding work magnetically

"I just don't know how we'd hold the work on a surface grinder if it weren't for magnetic chucks," says one tooling expert at a major machine-tool plant, "but that's about all we use them for."

Suppliers of magnetic chucks, naturally enough, disagree with the limitations expressed—and they're quick to point up successful applications of magnetic workholding for milling, planing, and turning.

The ability to simply place the workpiece on a table and flip a switch or a lever to hold it with no interfering clamps is certainly appealing. The invisibility of the magnetic holding flux, however, and the complexities of magnetic theory make this far from a simple technique. And many variables affect the attractive force between chuck surface and work.

Among these variables, as pointed out by James Neill Ltd, the British manufacturer of Eclipse magnetic chucks, are such workpiece factors as the contact area and surface quality, cross-sectional area of the part, the particular material of which it's made, and its specific heat-treat condition. The work, of course, must be ferromagnetic (excluding nonmagnetic austenitic stainless steels as well as such commonly used nonferrous metals as aluminum and copper alloys). Using mild steel as a comparison base of 100%, Neill points out that cobalt-iron would have a relative potential magnetic attraction of 54% greater, 0.9%-carbon steel 29% less, cast iron 50% less, and nickel 77% less. Still against the mild-steel standard of 100%, an oil-hardening tool steel with 1% C, 1% Cr, and 1% Mn would provide 84% in the annealed state, 44% hardened, and 49% hardened and tempered. Further, the best grip results when a smooth workpiece intimately contacts the chuck surface with minimum air gaps and the cross-sectional area of the chuck is large enough to carry all the magnetic flux the chuck provides.

With proper application—and the aid of such accessories as specialized chuck blocks and top plates that adapt the setup for particular workpiece geometry—magnetic chucks can be used effectively for chipmaking operations. Securely fixed end- and side-stops are mandatory to prevent sliding of the workpiece in such operations as milling, and the process should be planned so that cutting forces are directed down into the chuck and into the stops. Climb-milling, for example, may be preferable to conventional milling in heavy stock removal for this reason—remembering that the end-stop should be at that end of the work where a climb-cut begins. Similar care is required in planning face-milling and end-milling cuts.

O.S. Walker Co (Worcester, Mass) points out the special ability of swivel chucks for angular grinding or milling and has supplied application photos of electromagnetic, permanent-magnet, and electro/permanent-magnet chucks. This last type uses electrical pulses to magnetize and demagnetize permanent magnets, providing improved characteristics and eliminating the need (in electromagnetic chucks) to leave power on as long as a setup is in place.

Multiple wedge-shaped parts are face-milled in swivel magnetic chuck

Electromagnetic chuck holds two parts for gang-milling

Permanent-magnet chuck holds workpiece for horizontal milling

Electric permanent-magnet chuck holds long work for planing

Photos courtesy O.S. Walker Co

Holding rotating work

TURNING MACHINES represent one of the largest classes of metalworking machine tools, and, since it's the workpiece that rotates instead of the cutting tool, the workholding picture is radically different from that seen in other classes of machine tools (with the exception of such machines as cylindrical grinders, of course).

Even though workholding devices for lathes have a relatively small overlap with those for rotating-cutter machines, many of the trends are parallel. Modern turning machines are especially involved in the machining-horsepower race with its emphasis on higher spindle speeds and heavier metal-removal rates.

The manufacturers of lathe chucks have been in the middle of the beneficial shoving match in which cutting-tool producers have pushed machine-tool builders to provide more capability, and vice versa. The plain fact is that traditional chuck designs were approaching the danger point in their capability to hold workpieces securely at the higher speeds possible with modern turning equipment.

The problem is simply the centrifugal force developed at high rpm by various components of the chuck—primarily the master jaws and top jaws—which opposes the direction of the normal OD-gripping forces.

How much grip is lost?

For standard wedge-type power chucks, Cushman Industries calculates that a 6-in. chuck loses 50% of its initial jaw force (at maximum recommended drawbar pull) at 2650 rpm and loses 75% at 3250 rpm. A 12-in. chuck loses 50% at 1850 rpm and 75% at 2250 rpm. Similar speeds for an 18-in. chuck are 1530 and 1880 rpm; and for a 24-in. chuck, 1120 and 1370 rpm. These calculated figures, the company notes, have been confirmed in the laboratory.

Yet a recent survey of lathe builders and users by S-P Mfg Corp reveals that anticipated speed requirements in the next few years will go to 5000 rpm for a 10-in. chuck, 3500 rpm for a 12-in. chuck, 3000 rpm for a 15-in. chuck, and 2500 for the 18-in. size.

Counter-centrifugal chucks, which are now available from a wide variety of chuck manufacturers, provide the solution (AM—Jun'77,p87). These power chucks, in both wedge- and lever-actuated configurations, simply fight fire with fire; they incorporate counterweights that are pivoted in such a way that centrifugal force acting on their mass tends to increase the grip, thus offsetting the outward forces developed by jaw mass. Various configurations are offered—even including chucks in which the counterweights are removable, thus producing a chuck that's convertible—but they all operate on this simple, basic principle.

One characteristic of counterweighted chucks is that a plot of gripping force against rpm as the chuck is slowing down does not match the plot of gripping force as the chuck is accelerated. Instead, as the spindle slows down (with drawbar pull held constant), the counterweighted chuck's gripping force per jaw will actually increase above the jaw-force levels measured at any particular rpm as the spindle was speeding up.

This "overtightening" upon slowdown obviously does not present any safety hazards, but it does present the possibility of distorting or marring a thin-walled or fragile workpiece and should be taken into account when the particular situation warrants it.

How tight the bite?

The question of just how much jaw force is required for safe chucking of a workpiece is an extremely complex one, and one that cannot be answered categorically for all workpieces. Among the pertinent variables are such considerations as part geometry and overhang, strength and rigidity of the workpiece, any need to avoid marring the surface, the material itself and any variations in its properties (such as hard spots in castings), cutting tool geometry, and condition (sharpness) of the tool.

Nevertheless, the basic laws of physics can be applied if a relatively simple workpiece is considered with the machining operation fairly close to the face of the chuck or with workpiece support from the tailstock or steadyrest. Formulas are shown on the next page for calculating the approximate tangential cutting force and horsepower required for turning. A final formula then uses these results to determine required gripping force per jaw in a chuck. Cushman Industries cautions that a safety factor of at least 2 should be used in these calculations and suggests using a friction coefficient (U) of 1.0 for diamond-serrated jaws and 0.2 for smooth jaws. Material constants (K) for a variety of materials are also listed on the next page. These have been adapted for use in the given formulas from data in the *Machining Data Handbook* (Metcut Research Associates, Cincinnati) on the power requirements for turning with sharp tools and reflect horsepower available at the spindle uncorrected for spindle-drive efficiency. For dull tools, the values should be increased 25%.

A very simple workpiece is illustrated in two different setups that could represent a first operation and a second operation, and calculated results for both

Quick-change of jaws is emphasized by Universal Engineering for wedge-bar-type chucks, both manual (shown) and power-actuated. Laterally moving wedge bars disengage at end of travel to allow jaw repositioning in 10 sec, reversal in 20 sec, or complete change in 30 sec. Repeatability is 0.002 in. with prebored jaws.

Facing operation

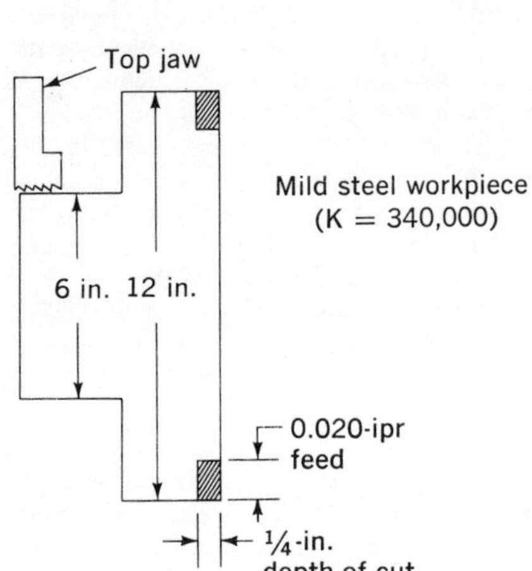

Turning operation

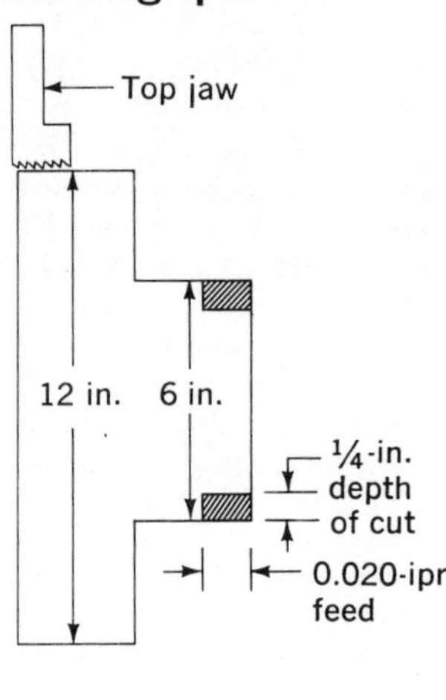

Calculated jaw-force requirements

Operation	Tool	Speed Rpm	Speed Sfm	Tangential cutting force (lb)	Horsepower	Required force per jaw (lb) Serrated jaw (U = 1.0)	Required force per jaw (lb) Smooth jaw (U = 0.2)
Facing	Carbide	150	471	1700	24.3	2268	11,340
	Ceramic	325	1021	1700	52.6	2266	11,330
Turning	Carbide	300	471	1700	24.3	567	2,835
	Ceramic	625	981	1700	50.6	567	2,835

Some useful formulas

Tangential cutting force:

$$F = d \times f \times K$$

Horsepower of cut:

$$P = \frac{F \times C \times N}{63{,}000}$$

Required jaw force (per jaw):

$$J = \frac{63{,}000 \times P \times S}{U \times N \times G \times n}$$

$$\text{or: } J = \frac{F \times C \times S}{U \times G \times n}$$

$$\text{or: } J = \frac{d \times f \times K \times C \times S}{U \times G \times n}$$

Where:

- C = radius of cut (in.)
- d = depth of cut (in.)
- f = feed (ipr)
- F = tangential cutting force (lb)
- G = gripping radius (in.)
- J = required force per jaw (lb)
- K = material constant
- n = number of jaws
- N = spindle speed (rpm)
- P = spindle power (hp)
- S = safety factor (minimum 2)
- T = torque (lb-in.)
- U = coefficient of friction (1.0 for serrated jaw, 0.2 for smooth jaw)

Machining force constants for various materials

Material	Hardness	K-constant
Plain carbon, alloy and tool steels	85-200 Bhn	340,000
	35-40 Rc	430,000
	40-50 Rc	465,000
	50-55 Rc	620,000
	55-58 Rc	1,050,000
Cast iron	110-190 Bhn	400,000
	190-320 Bhn	430,000
Stainless steel	135-275 Bhn	215,000
	30-45 Rc	430,000
Titanium	250-375 Bhn	370,000
High-temperature alloys, Ni Co base	200-360 Bhn	770,000
Aluminum	30-150 Bhn (500 kg)	77,000

Above data calculated from machinability data assuming sharp cutting tools (for dull tools, add 25%). Reflects horsepower available at spindle.

setups with both carbide and ceramic tools are tabulated. A safety factor of 2 has been used. In these examples, the spindle speeds are well below those at which a 15-in. chuck (the most probable size for this hypothetical workpiece) would lose any significant amount of gripping force as a result of centrifugal force.

Note that required jaw force is actually independent of rpm and horsepower and depends entirely on tangential cutting force, the ratio of cutting radius to chucking radius, and whether the top jaws are smooth or serrated. Radii and material are design characteristics. The highest possible metal-removal rate within the capability of the machine and its cutting tools is the objective—so long as the workpiece is not distorted or marred to an unacceptable degree and the safety of the turning operation is not jeopardized.

Chuck manufacturers are unanimous in pointing out that proper maintenance and lubrication are necessary for consistent performance of their products. With few exceptions (such as the highly precise diaphragm-type chuck), lathe-chuck designs incorporate sliding surfaces loaded at heavy pressures, and any increase in frictional forces due to dirt or inadequate lubrication will reduce the gripping forces produced by any given actuating pressure or torque. S-P Mfg indicates that this loss can be 50%.

Regularly scheduled inspection of chucks, therefore, is recommended and can be done with a jaw-force gage or jaw-force analyzer, and, whenever jaw force for a given cylinder pressure or wrench torque drops below the normal level, the chuck should be cleaned and lubricated.

Cushman Industries points out that the frequency of cleaning and relubrication is entirely dependent on operating conditions—under clean conditions, some Cushman chucks have been operated up to 35,000 cycles before relubrication—and suggests weekly inspection

Electronic jaw-force analyzer demonstrates effect of centrifugal force on standard 18-in. wedge-operated power chuck with Acme serrated master jaws and soft blank top jaws. Static reading of 19,100 lb per jaw drops to 10,000 lb at approximately 1500 rpm. Bar arrangement above case is to hold antirotation rod of slip-ring assembly. The Cushman analyzer, which handles two-, three-, or four-jaw setups, is priced at $1962 for static applications or $2834 with the slip-ring assembly for dynamic use

Special indexing chucks—this one by Cushman on a Warner & Swasey SC-28 NC chucker—prove valuable for machining components with intersecting axes. Indexing can be automatic

Face-drivers allow turning the complete length of work held between centers. Gripping the driver in a standard chuck allows maximum versatility in changeover of operations

of a new chuck to determine what the frequency should be.

When workpieces are thin-walled or fragile, standard top jaws on a standard chuck may well not be the best way to hold the part. One frequently used solution is simply to make wide, pie-shaped top jaws that will contact virtually the entire OD or ID circumference of the part. Because of the large volume of metal in such jaws—and especially if they are to be used at high spindle speeds—aluminum jaws should be considered for reduced weight.

For turning shaft-type parts between centers, the face-driver introduced in this country a few years ago by Madison Industries merits consideration. Conceptually similar to the spur-center commonly used in wood-turning, the Madison-Kosta face-driver avoids the necessity to grip the OD of a part with chuck jaws or a drive dog. Thus, it becomes possible to turn the complete length of the work in one setup—even chamfering the driven end.

Although the concept is similar, the metalworking face-driver is considerably more sophisticated than the wood lathe's spur-center. The workpiece (center-drilled on both ends) is first positioned by its initial contact with a spring-loaded center in the face-driver. As the tailstock is further advanced, compressing the center's spring, the end of the workpiece comes into contact with the sharp-edged drive pins (five or six, depending on model), which bite into the face of the work. Because the pins are backed up by hydraulic pressure (either oil or an elastomer in different types), the bite is uniform even if the face of the workpiece isn't.

"Bite" diameters of individual face-drivers are adjustable by means of interchangeable drive pins, and the range of the entire Madison line runs from a minimum of 0.28 in. to a maximum of 13.57 in. Various taper-shank and flange mounts are available, but perhaps the most versatility is provided simply by chucking the face-driver on the OD of its flange.

For small-diameter work, such as bar-

Special mandrels can pay dividends

"The extra investment in a special expanding mandrel is normally far outweighed by substantial reductions in the payoff period," states John Raymond, of Warner & Swasey's Balas Div, Cleveland.

Balas designs and builds a wide range of special expanding mandrels (the division does not offer a standard line). The Balas specials are constructed so that the mandrel body is mounted directly on the spindle nose or is adapted to it with a plate. The sleeve is keyed to the body to prevent it from rotating on the body under heavy machining cuts. Each end of the sleeve has a closing angle that contacts the body on each end by pulling a drawbar to expand the bushing along the entire sleeve. Uniform gripping is provided along the entire length of the sleeve by the opposed-split design of the Flexi-Grip collet. Drawbar-actuating methods include hand, lever, wrench, air, and hydraulic mechanisms—each having a proper application dependent on the type of machine tool, workpiece, and job-lot size.

As a general rule, Raymond states, special expanding mandrels are used for second-operation work to maintain close concentricity between ID- and OD-machining operations that must be done in separate chuckings. However, he adds, they are applicable to a wide variety of work—from gear cutting to assembly fixturing—especially where tight overall-length tolerances can benefit from the elimination of longitudinal movement by dead-stop location.

A recent example of this advantage is the use of special expanding mandrels on the six-spindle Conomatic Chucker shown in the photograph. The machine is used on automotive starter-motor frames that are being finished to length, and, hence, the workholding system must provide rapid workpiece interchange and repeatable control of the axial position.

Most expanding-type mandrels must be designed and built as specials, Raymond points out. The special design is based on the relationship of the inside diameter, length of the bore-gripping surface, type of machine (as it pertains to spindle mounting), and the amount of expansion required. Typical applications range from part diameters of 1½ in. to 8 in. and lengths that range from ½ in. to 30 in.

Raymond also notes the cost-saving possibility, when similar parts or a family of parts are to be produced, of utilizing a number of expanding sleeves of different OD that will all mount on the same body.

Probably the biggest single user of specially designed expanding mandrels, according to Raymond, is the automotive industry, which uses them in such diverse applications as gear cutting for transmissions and machining water-pump and distributor parts. "But," he adds, "other major industries are also beginning to shorten their payoff periods with specially designed expanding mandrels."

---

turning operations—and especially in automatics and multispindle machines—collets are most likely to be the workholding devices. They may be either push type or pull type, they may be manually or power-actuated (often by air in the latter case), they may be of single-piece construction or they may consist of master collets with separate jaw pads (in which case the master collet can be adapted for a range of diameters), and they may be highly standardized or highly specialized. In any of these cases, collets provide a simple and effective OD-gripping system that is capable of extremely high precision. Expanding mandrels, virtually inside-out collets, offer similar attributes for internal gripping of IDs.

Hydraulically expanded mandrels, some of which are actuated manually by tightening a screw and some by a power cylinder, are claimed capable of repeating to 0.0001 in. but are restricted to a clamping travel of only about 0.003 in. per in. of diameter. An inverse design is also possible, incorporating a sleeve that is compressed hydraulically for OD gripping, and it provides similar accuracy capabilities. ■

Fundamentals of sawing

Proper selection of the first metalcutting operation has strong impact on all subsequent work. Here is what you should know about this primary technique

Kerf loss differs dramatically for circular, hack, and band cuts. But accuracy and the possibility of the need for a secondary squareness operation should also be considered

There would be no need for sawing machines in a metalworking shop if all raw stock was delivered in ready-to-machine shapes and sizes. The lathe, milling-machine, and press-brake operators could merely go over to the stockpile, pick out a suitable workpiece, and perform the necessary finishing operations on it.

Such a situation seldom exists, of course. There's almost no stock that can be started through a machining schedule without first being cut in some way. Bar, angle, flat, strip, sheet, tube, billet, wire, and even cast stock must be cut to elementary shape to start with.

A shop, on occasion, can avoid this. Stock can be bought (in some circumstances) in prepared lengths and shapes. But the shop has to pay for this service in one way or another, and in almost all cases it is simpler and more economical to do the basic cut-to-size operation in-house. (Even if it is done in the plant, cutoff still does not have to be done on owned equipment. Some distributors offer leasing services specifically designed to reduce sawing-department bottlenecks.)

But however–or wherever–the sawing job is done, it still must be accomplished. Thus, it is the primary, if not the most basic, metalworking operation.

Something so basic is likely to be taken for granted, especially at a time when more and more emphasis is being placed on exotic techniques and intricate workpiece designs. But it is important to realize that, since sawing *is* the first operation done on a piece of stock, it governs the rest of the machines in the shop. The way in which the sawing is done–its accuracy or lack of it, its speed or slowness, whether it is consistent or haphazard–has a profound effect on the flow of the workpiece from the receiving bay to the shipping department.

Where the saws are

The *11th American Machinist Inventory of Metalworking Equipment* indicates that there are close to 125,000 cutoff and sawing machines in use in this country. Most of them are used by the fabricated structural metal products industry, which accounts for 17% of the total. Next largest industrial users of cutoff and sawing equipment are the miscellaneous machinery manufacturers (12% of total) and the tool, die, and accessories industry (8%).

The rest of the total is used by many different, and often nonrelated, industries. Everyone uses saws, and everyone develops specific sawing techniques.

Over the years, sawing processes have remained fairly constant. But in maintaining the same theory and practice, there is much opportunity for improvements. Many processes have been adapted, changed, simplified, or modified, for better, faster, more productive results.

Here, then, are the fundamentals–and some ideas for saving on a universal and much-repeated operation.

Cutting off and cutting out

One major advantage of sawing over all other kinds of machining is narrowness of cut. This advantage is put to good use in two basic metalworking functions: cutting off and cutting out.

Most sawing machines perform the cutoff function. They take a piece of stock and cut it to a workable length. Machines that accomplish this job include bandsaws, hacksaws, and circular saws. There are other methods that can handle the cutoff function, but, for a variety of reasons, the saw is usually the most efficient.

The second major function of saws is cutting out. By removing large chunks of metal in solid pieces rather than in tiny chips, considerable savings in horsepower consumed, tool life expended, and machining time are realized. Vertical contouring bandsaws and hole saws (a type of bandsaw) use this cutting-out principle.

But regardless of the basic operation performed, there are three economic factors to be considered: tool life, cutting rate, and accuracy. These interdependent and inseparable factors can be seen as three weights spinning about an axis in dynamic equilibrium. When the weight assigned to one is changed, there must be a counterbalancing change in the other two, just to maintain equilibrium.

If cutting rates, for example, are increased, tool life and accuracy diminish. High accuracy, on the other hand, means scrapping a dull tool and taking more time for machine setup and operation.

This equilibrium holds true for any one machine and for any one type of workpiece material. To a certain extent, it is also the basis for comparing the cost-effectiveness of a particular sawing method vs another.

But there are other factors that can affect the choice of cutoff methods. By their very nature, different cutting methods are more individually responsive to technological improvements. Such outside improvements can materially affect the tool-life/rate/accuracy equilibrium.

Some of these improvements, for example, have been the result of research into blade materials. The chart, facing page, graphically illustrates the effect of blade-material change on cutting rates and blade life, with no sacrifice in accuracy. It can be seen from the chart that the newer, welded-edge blades (hss welded to alloy backstrip)–although they cost more in initial investment–are superior in the long run on a cost-effectiveness basis.

Whether advances such as new blade compositions are adopted by machine users is another story; part of the problem may be lack of awareness. Nevertheless, misuses hang on, and being able to realize the true cost may be half the battle. Or, as one machine builder puts it: What other machine tool is still using carbon-steel tooling?

Like the selection of the kind of blade to use in a particular machine, the selection of the basic type of machine itself may not be readily apparent.

By Joseph Jablonowski, assistant editor

Sawing is essentially a milling operation, but one application of that process can be more suitable for your type of work than another. Different machine types cut with different rates, kerf losses, squareness, surface finishes, safety, handling ease, horsepower consumption, etc.

So the choice of a means of cutoff (or cut-out) can be a complex one. And, to complicate the choice, there are non-sawing techniques also.

Other methods

Sawing is only one method of cutting off materials prior to further machining. Whereas all sawing involves the cutting action of a series of small teeth, other basic machining methods can be adapted so that essentially the same job can be accomplished.

Some of the techniques that cannot be classified as sawing, but nevertheless are used to cut off metals and other materials, include:

- Single-point cutoff on a lathe;
- Shearing for certain shapes and sizes;
- Abrasive cutoff;
- Friction "sawing"; and
- Wire "sawing."

While these operations will not be fully discussed in this re-

Progress in blade performance

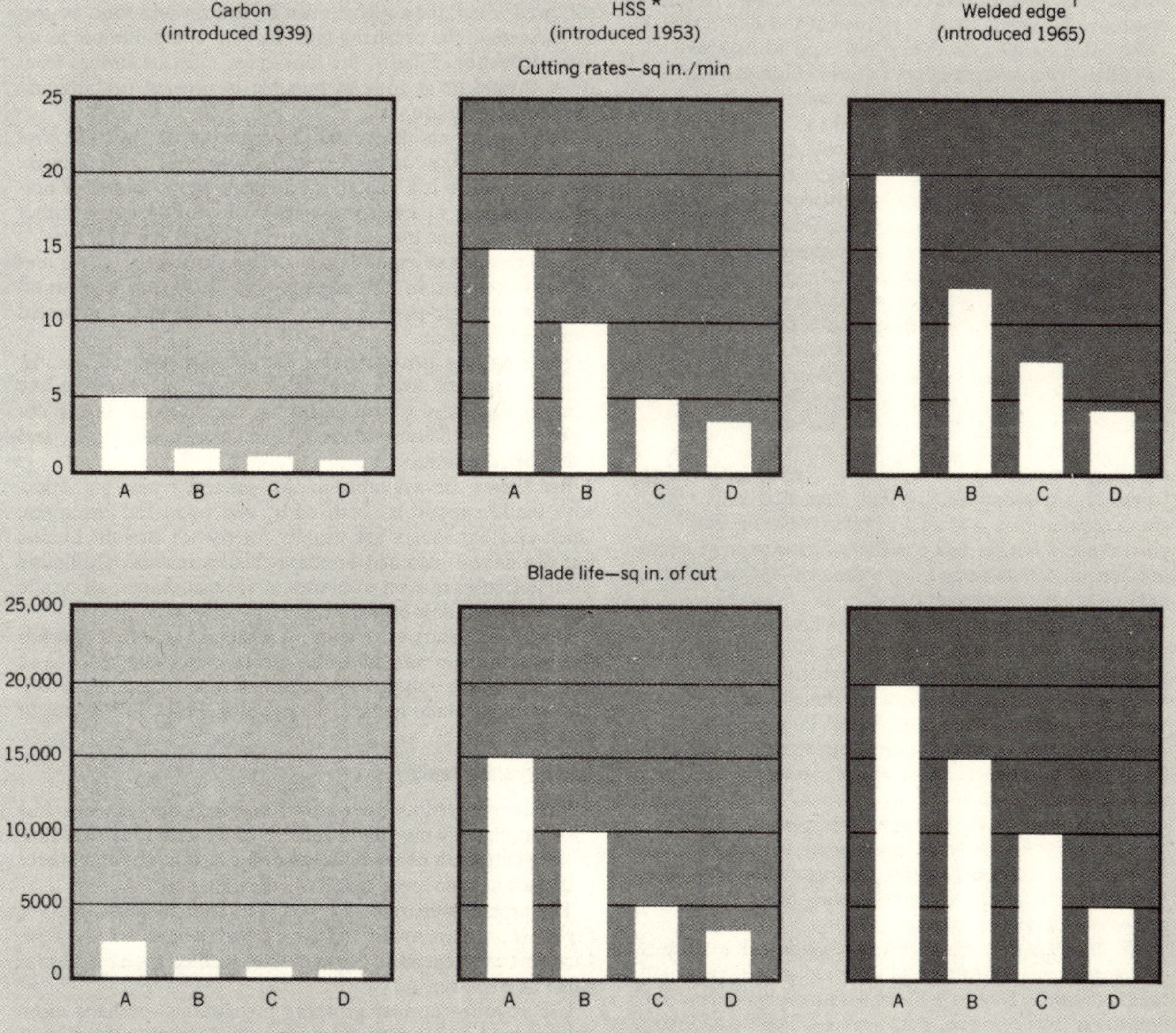

A—cold roll carbon steel; B—alloy steel; C—tool steel; D—stainless steel

* The HSS is a solid, one piece hss-steel blade

† The welded edge has an hss cutting edge and alloy backing

data courtesy DoAll

The next big advance

In its 75th Anniversary issue, in 1952, *American Machinist* took a look at the future and made a few predictions. At that time, the editors identified greater mechanization, wider application of contour sawing, automatic feed and gaging, and added blade types for contouring as items to look for in "What's ahead in cutting off."

In the years since that crystal-ball gazing, most of those predictions have become reality. The outlook also foresaw a technique called jet-blast slicing, originally developed for drilling human teeth, as a technology that was going to take off. Well, one can't always be 100% correct.

Unable to resist the temptation to publish more predictions, early this year we asked over 100 builders, suppliers, and importers of sawing machines and accessories what they foresee as the next big advance in sawing. Here are some of the more-interesting replies:

A builder of both band- and circular-sawing machines, and another builder, primarily known for its bandsawing models, both replied that the blade technology that's now available is not used to its greatest potential. "Machine designs to utilize blade designs" is what they see as the next big step forward.

"Cutting-tool technology," replied one bandsaw maker. A West Coast importer of cold saws was more blunt—"carbide" was the single-word answer. Application of carbides in sawing, particularly in circular sawing, evoked similar responses. One circular-saw builder predicts that the use of tungsten carbide in blades, instead of hss, will mean better finishes.

An importer is a bit more specific: The use of silicon carbide in circular steel-cutting blades, combined with greatly increased cutting speeds—260 sfm instead of the normal 80 sfm—will mean far better workpiece finishes.

Researchers in Germany, according to another source, are working on indexable silicon-carbide inserts for cold saws. Major problem to be overcome is safety. Carbide inserts, should they ever become unlocked from a fast-spinning blade, would have all the dynamic characteristics of a .45-caliber bullet. Nevertheless, engineers are working on a specific solution to the insert problem. Carbide-tipped (not indexable) sawblades have long been in successful use.

Tied closely to cutting-tool material, coolants are expected to improve in the near future. "Better coolants, enabling hss to cut exotics," is the prediction of a major distributor. An importer concurs: "Better coolants, and more-heat-resistant teeth that are less fragile."

New machine-control options were seen by many respondents as the next big step in sawing. A Dakota manufacturer looks for "more precise cutting, with automatic feeds and better electrical controls." "Controlled peripheral speed of the blade," according to a West-coast importer. "Adaptive controls" will be the next big advancement in cold sawing, says the U. S. agent of a European builder.

Some replies, predictably, touted specialized methods peculiar to individual builders or to small groups of manufacturers. Thus, a builder who deals in hacksaws looks to a "resurgence of hacksawing." Similarly, "friction cutting hasn't been exploited to its full potential," and "I foresee greater use of bandsawing."

port, the following overview will help you to recognize their potential use as alternatives to the more widespread methods of circular, band, and hacksawing.

Single-point cutoff and shearing

Just as the lathe is the primary metalworking tool, single-point cutoff is one of the basic methods of parting stock.

Traditionally, lathe cutoff tools are thin and flat, and take a necking cut. When the cut starts on the OD, cutting can be just as effective as the best turning jobs, with high speed, good feedrate, good chip characteristics, and good finishes. As the tool tip approaches the center of the work though, conditions get less favorable.

The first thing that happens is loss of surface speed—the speed of the work relative to the tool tip. If good cutting characteristics depend on high surface speed, the tool is likely to be affected by forces and pressures beyond its design limits. Tool-design problems include the tendency to jam in a tight neck (because of design or overheating), to wander or flex, or to cut under or over center. These problems intensify as the tool feed nears center.

The solutions to these problems might include variable-speed drive. But short of that, three important rules must be followed. First, the support must be as firm and solid as possible. Second, the overhang must be kept to a minimum to reduce deflection. Finally, the tool shape must be strong; front relief should be as little as possible to prevent rubbing and friction-heat buildup.

Single-point cutoff embodies another problem: The tool actually parts the material continuously until breakthrough, but unless there is a pickoff mechanism or some kind of outboard support to keep the workpiece from moving (other than rotating), the piece will flex and damage the tool.

Special-purpose cutoff lathes are also available. These machines typically spin tube stock at up to 1000 rpm, and cut off either with single-point tools or with circular blades mounted on the front slide.

Shearing is a general name for most sheetmetal cutting, but, in a specific sense, it designates a cut in a straight line completely across a strip, sheet, or bar. Good shearing depends on four factors: blade rake, clearance, sharpness, and machine adjustment.

Bar shears are available in two general types: guillotine, with blade support at both ends; and open-end machines. Open-end bar shears are usually fitted with straight blades, but can have V-notched or cutout blades instead. Guillotine shears often have a set of blades of various shapes, all on the same lever to allow easy feeding.

Most tube shears work with a V-shaped blade that punches through the top tubing wall, shears down the sides, and punches out the bottom of the tube, all in a continuous stroke. This pointed blade is likely to produce a nick in the end of the tube.

Sawing abrasively

Fundamentally, abrasive cutoff machines are grinders. Just as a circular-sawing blade can be compared to a thin-slot-milling cutter, an abrasive cutoff disk is a thin grinding wheel with thickness no more than 1/48 the diameter.

The process is extremely fast. Typical cutoff times are 5 sec for a 1-in.-dia hss round and 2 sec for a 1½-in.-dia brass pipe. Cutting time depends on wheel size, surface speed, and machine horsepower.

Use requires normal grinding precautions—perhaps more than normal, because the wheel speed of abrasive cutoff units is very high. Most wheels are marked by the manufacturer to indicate maximum operating rpm, and this is a statistic that should be taken seriously. At recommended speed, a cutoff wheel has high radial strength, and there is a reasonable margin left to keep down the danger from shock- and vibration-induced breakup. If a wheel is run at a higher speed, most or all of its strength might be taken up by centrifugal force, and even the least amount of shock incurred when the blade enters the work could spell disaster.

Modern abrasive cutoff machines are well guarded, however. Wheels, in addition, do have reserve strength, so the method is not particularly hazardous. Of course, if an operator wants to live dangerously, he can engineer situations that will provide an unusual thrill.

The speed with which material can be cut off with abrasive methods has a price. In a mechanical sense, one of these costs is horsepower. Machines with 50-, 100-, and 150-hp driving motors are not uncommon. Replacement wheels also are a factor. While the initial capital cost of an abrasive machine is often far less than that of a sawing unit with similar capacity, annual abrasive-wheel costs can occasionally exceed the cost of the machine itself.

Considerable amounts of cutoff work are done with small, portable power machines, which usually are equipped with reinforced, resinoid-bonded wheels for dry cutting. Cutoff wheels can also be mounted on ordinary floor-stand grinders.

For high-volume production, cutoff is done on large machines with special coolant systems and complex materials-handling systems. For these big machines, water or water-base emulsions are normal coolants, and they are applied to both sides of the wheel through piping systems. Larger machines also incorporate oscillators to help the wheel eat through the workpiece.

Sawing with a wire

Wire "sawing," like friction sawing, is not a sawing operation at all. But by its nature and use it has become a process that often gives answers to difficult cutting problems.

The process, used millennia ago to cut 22-million-lb stone blocks for the Egyptian pyramids, cuts workpieces by abrasion. A wire or similarly flexible tool is drawn back and forth (or in a continuous motion) across the workpiece, and an abrasive agent is sprinkled into the cut or bonded to the wire.

The technique has come a long way; wire-sawing machines were used recently to cut precious moon rocks, with an absolute minimum of lost kerf. Today's sawing wires are diamond-encrusted, enabling them to cut any material, regardless of hardness. And since the diamond abrasive material is uniformly bound around the entire circumference of the wire, very precise contouring can be achieved.

It is this combination of cutting and contouring capability that makes diamond-encrusted wire saws particularly useful in cutting carbides. In the unsintered (green) state, carbides can be contoured at a rate of up to 4 sq in./min. When cutting sintered carbides, however, some provision must be made for removing the carbide dust from behind the cutting wire, or the wire would have to recut its way out of the workpiece.

But even when the technique is used on sintered carbides, the resulting finish is often so good that additional grinding is unnecessary. Surface finish is a factor of wire tension, since the higher the tension on a wire, the straighter it becomes (with improved cut and finish). Latest sawing wires, made of heat-treatable alloys which are coated with copper to improve diamond bonding, have tensile strengths of over 500,000 psi. This results in a 0.008-in. wire that can be operated at a tension of 3500 grams.

Friction sawing burns

Another case of sawing that is not really sawing is found in friction "sawing," which burns the workpiece apart.

This technique uses either a solid, large-diameter circular blade rotating at high speed or a bandsaw moving at anywhere between 4000 and 15,000 sfm, to create friction instead of chips. Friction heats the workpiece to a point between red heat and its melting point. With the material in this condition, the weakened surface can no longer resist the sliding action of the blade, and the material is literally wiped away.

Oxygen carried in the gullets between the blade teeth causes the blade to burn its way through the work. The resulting action is similar to that of a cutting torch, although many materials that cannot be flame-cut may be friction-cut.

But not all materials can be cut with friction. The process works well on steel and iron, but not so well on such nonferrous metals as aluminum, brass, and magnesium. Some plastics—the thermosetting types filled with hard materials such as glass and mica—can be successfully friction-cut; thermoplastics, of course, cannot.

The method is very fast. In plain carbon-steel workpieces, a friction-cutting bandsaw can cut 1/16-in. stock at 1400 linear ipm and 1-in. stock at 6 linear ipm. Thicker stock takes too much feed pressure for hand feeding and is usually moved by a hydraulically powered table.

Surface accuracy is generally poor, because there is no rake angle to produce chips. By the very nature of the process, excessive amounts of heat are generated and must be dissipated. The workpiece shows discoloration, burn marks, heavy burr, and flashing. The burr must be ground away; this secondary grinding operation often takes longer than the cutting itself. Another problem is the noise level—the workpiece fairly screams when it is cut. But straight lines can be cut, and material can be parted almost as straight as with methods that use cutting teeth.

Blade life in friction cutting is far greater than in conventional sawing. Dullness is no handicap; in fact, a dull blade is often better than a sharp one. A low-carbon, flexible friction-cutting blade is better than a very hard one in a bandsawing machine, since it is less likely to crack from flexing over the drive wheels. And because the blade is moving fast, coarser pitches can be used on thin materials than in conventional sawing, with little danger of tearing out teeth.

Blade life is determined more by set, or blade waviness, than by tooth sharpness. When the set has been worn off the teeth, the blade will still friction-cut, but it is difficult to guide it accurately. Special friction-cutting bandsaw blades are available with increased set to minimize this problem.

Friction cutting can also be done on circular-sawing machines as long as surface speed is fast enough. Most circular friction machines, however, are specialized pieces of equipment and have very high horsepower, accurate speed control, and a high-pressure water pump for blade cooling.

Although convection air currents set up by blade rotation do help in cooling the blade, this often is not enough. Compared to a friction-cutting bandsaw blade, a circular blade has much more metal, and therefore retains heat better. If the blade should heat enough to lose its temper, it would probably be ruined. Water cooling on specially built units prevent this loss.

While heat is the cutting force in circular friction sawing, it is also a limiting factor, especially in cutting heavy stock. Heat increases directly as the length of contact between the workpiece and the blade increases.

This contact arc can be controlled somewhat by the setup. A downstroke machine, most often used for cutting bar, rod, angle, and similar stock, is set up for the most efficient friction cutting, with the contact arc between blade and work nearly horizontal. On square or rectangular stock, the arc length can thus be only as long as the workpiece is wide. Pivot-stroke machines and horizontal-feed units lengthen the arc as the circular blade cuts diagonally through the stock.

Bandsawing

photo courtesy Armstrong-Blum

Start up a conversation about sawing with many production people and the first type of machine that will probably come into their minds is the bandsaw. In the 165 years since the first one was patented, it has developed into one of the most productive machine tools in the shop. And if there does exist a bandsawing bias in America, as many people claim, there is also a wide variety of jobs the machine can handle.

Bandsawing basically serves two purposes: cutoff sawing and contour sawing. As a cutoff machine, the bandsaw vies with several other sawing methods, including circular and hack, and to a lesser extent with other cutoff methods, such as abrasive, shear, and single-point cutoff. In its other main use, that of a contouring machine, the bandsaw competes with a wide range of other machining methods, including milling, shaping, flame cutting, hobbing, broaching, shearing, nibbling, and even lathe turning.

Before examining relative virtues and drawbacks of the bandsaw in either its straight-cutoff mode or in its contouring mode, however, it is important to understand the basic concepts and motions that are common to both.

Londoner William Newberry patented the first endless-band saw near the beginning of the 19th Century. But it wasn't until 1846 that a Frenchwoman, Mlle. Crepin, perfected a workable blade joint to make the design practical. Bandsaws rapidly became popular in the lumber and logging industry, where their thin cuts meant smaller kerf and more boards per log than any other saw of the time.

Cutting bands for metal have been in use since early this century and employ a weld to loop the blade. Early metal-sawing bands were wide—over 1 in.—and were used strictly for cutoff until the mid-1930s, when narrow blades brought contouring capabilities.

Bandsawing, like circular sawing, is a continuous cutting operation. An endless blade (properly called a band) is tensioned between two shrouded, rotating wheels, and a portion of it is exposed to do the cutting. A typical blade (for a cutoff operation) may be 1¼ in. wide by 0.050 in. thick; the loop will be as long as the machine design requires. Teeth are on only one edge of the band; as it travels in a continuous motion, they feed against the workpiece, and chips are milled away.

Early bandsaws were limited by the basic design of the moving blade. The workpiece could only be as long as the machine's throat—the clearance between the exposed portion of the blade and the unit's frame. On most saws, the blade guides, which are stationary and close to the workpiece, could be angled, so throat clearance was not a problem. This practice, however, tended to place undue stresses on early stiff blades.

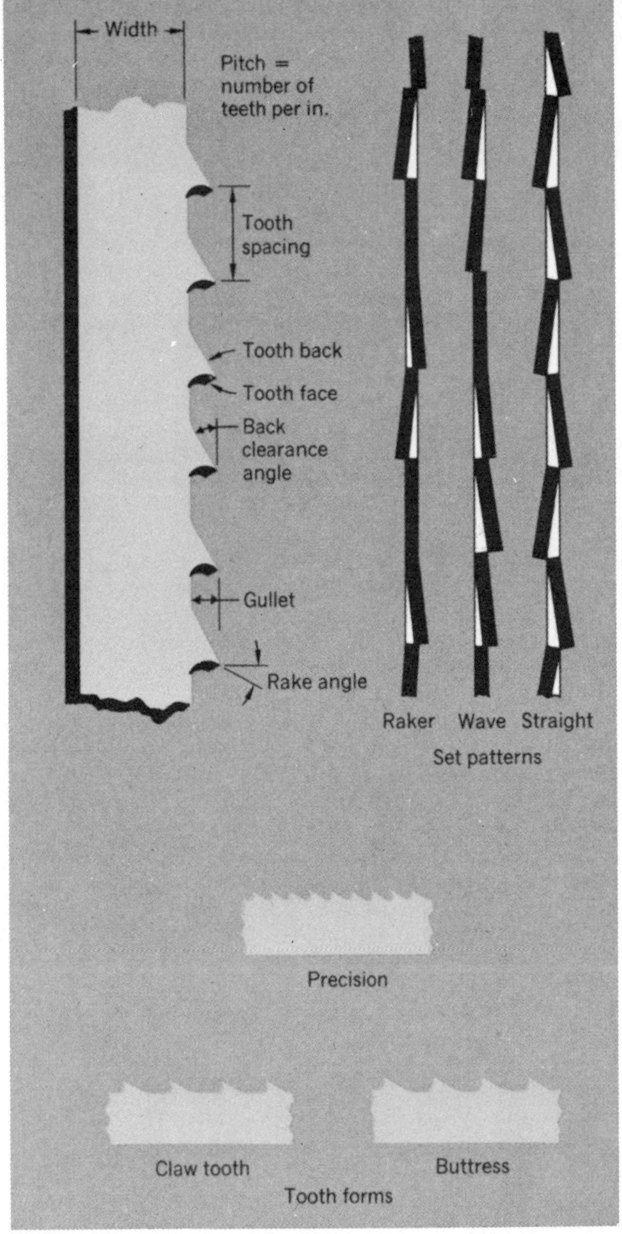

Another saw design has the blade intentionally twisted so the toothed face is in line with the machine's throat. The blade moves from the rubber-faced wheel and points away from the machine, but it passes over the wheels (tires) and through the rear of the unit in the usual alignment.

Many modern production machines are designed to cut at about a 45-deg angle to the blade-travel axis. This usually provides adequate throat clearance. Of course, the diameter of the driving wheels actually determines the limiting size of the machine throat; blade flexibility limits twist.

Blade design

Bandsaw blades have five basic design characteristics: width, gage, tooth form, pitch, and set. Stability in the cut is directly affected by the width of the blade; the general rule for both cutoff bandsawing and contour bandsawing is to use the widest blade possible. The blade's gage is the thickness of its body, and bandsaw gage is generally smaller than that of any other type of sawing blade.

Selection of tooth form and pitch should always take into account the type of material being cut. Using the dictum of three teeth (minimum) in the cut as a guide, the number of teeth to the inch depends on the workpiece. Thin materials need fine-pitched blades. The tooth shape should have enough stability to stand up to the type of material that is being cut.

Set of the blade determines the width of the cut. It is built into the blade geometry to permit a kerf that is wider than the blade gage, so the blade does not jam in the cut. Of the three major set patterns, the raker type of set is most prevalent in bandsawing.

Saw blades are available in many different materials, including carbon steel, intermediate-alloy steel, full high-speed steel, and welded-edge bands (hss teeth electrically welded to an alloy-steel backstrip).

Besides these four basic blades, other kinds of band materials are used in these machines—but generally for nonsawing functions. Because of their wide range of selectable speeds, modern bandsaws are often used to perform other functions (this is particularly true of vertical units). Among them: filing with file segments mounted on a continuous band, shaping, slicing through hardened steels with diamond-impregnated flexible bands, electrical-discharge machining of such items as metal honeycomb structures, and finishing and polishing with abrasive-coated belts. Friction sawing with bands constructed with negative tooth-rake angles, to "burn" through the workpiece, is also possible with very-high-speed bandsaws.

For strictly sawing jobs, however, blade selection is usually limited to the four steel materials listed above. Whether the blade selected is made of carbon steel, intermediate, hss, or a welded-edge composite of hss plus intermediate alloy, several factors affect the life, strength, and cutting qualities of each.

Saw-blade materials

Carbon-steel blades can be constructed in two ways: with a flexible back or a hard back. Both have hardened (Rc 60-64) teeth. In flexible-back versions, the backstrip is heat-treated to give a hardness of Rc 24-25. These are widely used in vertical machines for contouring, because they are available in narrow widths. Resharpening is not considered economical. Hard-back carbon-steel blades are made from high-carbon alloy steel. While the teeth generally have the same Rockwell hardness as flexible-back versions, the backstrip is spring-tempered to Rc 44-45.

These hard backstrips have almost twice the tensile strength of the flexible-back versions, permitting more accurate cuts under heavier feed pressures. High backstrip hardness also gives improved resistance to mushrooming (the tendency to bulge at the point where the band enters the squeezing guides near the workpiece).

Intermediate-alloy blades have more resistance to shock and vibration than most hss types, and they can cut steel faster than carbon-steel blades. They can withstand tooth-tip temperatures of up to about 800 F (compared with 400 F for carbon-steel and 1000 F for hss), and they are used often, because they cost less than hss.

Hss bandsaw blades were developed some 25 years ago, and have considerably improved in performance since then. This type of blade material can saw up to six times faster than standard flexible-back carbon-steel blades, and will last considerably longer. High tensile strength, fatigue-resistant backstrips, and tooth sharpness with high red hardness permit higher tensioning loads for greater feed forces and higher cutting speeds—up to 5000 sfm in some cases.

Bandsaw terminology

Some single-metal hss blades are specially processed, or made from special materials, to increase shock resistance. A fairly typical one is made from an alloy that should be classified midway between the intermediate alloys and single-metal hss blades. It cuts faster than the intermediates and costs less than other hss blades, and it is recommended for structurals, stacked shapes, and workpieces with varying cross-sections.

Welded-edge blades outperform all others. These are composite blades made of fatigue-resistant alloy backstrips, to which hss teeth are attached by careful welding. By their very construction, they eliminate the need for balance between long-wearing teeth and fatigue-resistant backstrips, for which single-metal hss blades must compromise. On single-metal hss blades, only the very tip of the tooth, only a few thousandths of an inch, is as hard as Rc 63; welded-edge blades extend this hardness the entire length of the tooth face—without the hardness extending into the rupture-vulnerable gullets.

While these four different types of blade materials exhibit different cutting characteristics and overall life spans, they share a common tendency at the end of their useful lives. The cutting rate of a bandsaw blade, measured in square inches per minute, remains virtually constant during its life span (except for initial break-in). But toward the end of the blade's useful life—whether that life is 2000 total square inches cut (for a carbon-steel band) or 3700 (for a welded-edge band)—tooth efficiency is reduced, and the cutting rate drops off sharply. For a given operation, about 90% of the wear generally occurs during the last 10% of a band's useful life. When this rapid wear starts showing up, it's time to change the band, to prevent out-of-tolerance sawing.

The theoretical economy of the bandsaw lies in the fact that all teeth cut equally and continuously. Blades, in addition to being supplied in different widths and tooth geometries, can be purchased either in closed circles or in long strips coiled in a box. The latter, of course, must be joined to form a loop. Mlle. Crepin's invention, the joining blade fastener, has long been replaced by a butt-welding operation. The two ends are squared with snips or a grinder and inserted in a welding fixture at the side of the machine. The connection is then trued in a flash grinder. (The close-up photo on the bottom of this page shows the stress pattern of a proper weld.) The now-continuous loop is then put into the machine and is ready for operation.

For special die-cutting jobs on a vertical contouring bandsaw, the unjoined blade is often slipped into a starting hole drilled into the workpiece, and then welded. This procedure, teamed with a slight tilting of the machine table, produces a punch and die simultaneously.

As with many operations in the shop, the specified methods for bandsaw use can often be improved, and are expected to be used only as a starting point. One AM reader, for instance, reported that he observed a tendency for the blade on his machine to break at the weld joint. By constructing a simple punch and die (a job that he did on his contour bandsaw), he trimmed the band ends to V shapes, and got long-lasting weld joints as a result.

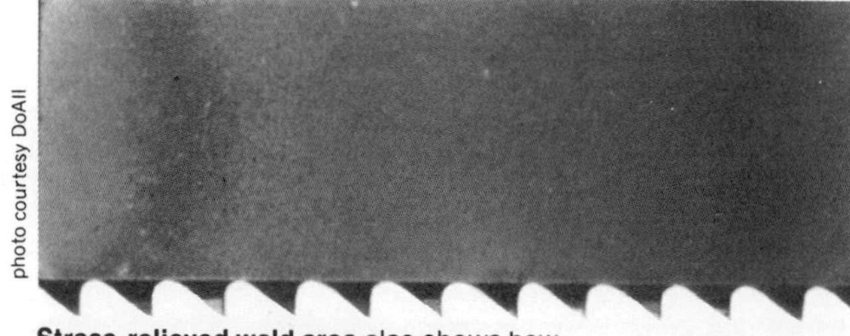

photo courtesy DoAll

Stress-relieved weld area also shows how hss teeth are welded to alloy backstrip

The same type of approach should be taken with blade selection. Data presented by manufacturers are, for the most part, conservative estimates, based on the so-called best-blade principle. This approach assumes that, for each material, regardless of blade size, shape, or condition, there is one blade that is best for a specific situation. Such estimates can often be changed, based on judgment and a reasonable amount of bandsawing skill. Often the best-blade conditions can be exceeded; at other times, they must be adjusted downward to suit existing conditions.

Cutoff bandsawing

Straight cutoff bandsaws are manufactured in both horizontal and vertical designs. Some of the smaller ones are convertible—changing from vertical to horizontal configuration in just a matter of seconds.

Vertical machines use an upright column to carry the bandwheels and the blade. The column is mounted in a traveling carriage, which moves parallel to the long dimension of the table or bed. The column and carriage are attached as an integral unit, to provide rigidity for sawing accuracy. Feed can be automatically powered, or it can be hand-driven for delicate work, such as notching.

The largest nonlumber bandsaw in the world happens to be a vertical machine. This 20-ft-high monster was shipped three years ago to an earthmover manufacturer and is used to slice into tires to inspect for flaws, capitalizing on the ability of bandsawing to cut through nonmetallic materials efficiently.

In a vertical machine, mitering is usually done by tilting the column as much as 45 degrees left or right from upright.

Horizontal machines have the bandwheel column moving downward to the workpiece, in either a straight-down motion or a chop stroke. Feed is often gravity-assisted, and one builder uses a patented compensator that eliminates feed variations caused by the changing force of gravity in chop-stroke models as the sawhead feeds downward.

Cutoff bandsaws are classified into three basic groups: light-, medium-, and heavy-duty models. Light-duty machines are usually horizontal-type saws equipped with fractional-horsepower motors; they use carbon-steel blades up to ⅝ in. wide. Medium-duty machines can be either horizontal or vertical; they are equipped with 1-hp to 3-hp motors and use ¾- to 1-in.-wide blades. Heavy-duty machines are usually all-hydraulic designs; they also can be either horizontal or vertical. These use blades 1 to 1½ in. wide, driven by motors of 5 hp and over.

Straight cutoff

Two major factors govern straightness of cut for bandsaws of all sizes: guide spacing and blade tension.

The rigidity of the blade is a function of guide spacing, with rigidity decreasing as the cube of the increase in the distance between the guides. Thus, the arms on which the blade guides are mounted should be positioned as close to the workpiece edges as the shape will permit. The design of the guides themselves also plays a part: roller guides should be considered as pivotal contacts; carbide-faced guides are generally thought to act more as anchor supports.

The greater the distance between the guides, the greater the probability of a crooked cut. The solution is to reduce the cutting pressure—but if the workpiece is hard or tough, cutting may stop altogether. A compromise between too much and too little cutting pressure must be found.

Tension on the band is equally important in assuring cut straightness. Adequate tension prevents the center of the blade from being deflected to the side and causing a crooked cut. It also prevents reduced tooth penetration in the center of the cut. The higher the tension, the better the cut will generally be; blade fatigue is the limiting factor.

Tension setting is a simple job on most modern bandsaws. Some provide a spring-loaded handwheel and stop to stretch the distance between the two driving wheels. In other, more sophisticated, machines, blade tension is automatically maintained by a hydraulic cylinder operating under preset pressure; when the machine is stopped, blade tension is automatically reduced to relieve fatigue.

For high-volume production work, evacuating the chips from in front of the following teeth is important, especially with materials that are prone to work hardening. Coolants with antiweld characteristics are especially important in high-volume applications, but are not recommended for most cutting in high-speed steels. Aluminum is another metal that often runs better without coolant, in part because of its tendency to contaminate. Here, circular sawing is often advisable.

In production situations, one of the most important variables in selecting a bandsaw is its adaptability to materials-handling systems. When labor rates were low (remember?) and sawing costs relatively inexpensive, no one really noticed if it took 20 minutes to get a bar of material into a machine and another 20 minutes to get the cutoff pieces out of the way.

Today, production-rate bandsaws are incorporated into efficient transfer-table systems that load, measure, saw to length, discharge, and move workpieces onto loading or holding racks. The result is less machine downtime while waiting for a crane or special help, and increased productivity, with more cuts per machine per day.

The ins and outs of contouring

Vertical bandsaws can be designed with a very special capability—contouring. In this respect, the band machine can often do the work of a milling machine, a shaper, and even a broaching unit.

All contouring bandsaws can, of course, also be used for cutoff work, but their design generally makes them more efficient as straight contouring machines. This is because cutoff size is often severely limited by the throat design of the machine. Some contouring units have fixed tables that require the operator to push the workpiece physically through the blade. Hydraulic or air-operated tables allow the workpiece to be cut off in the normal manner.

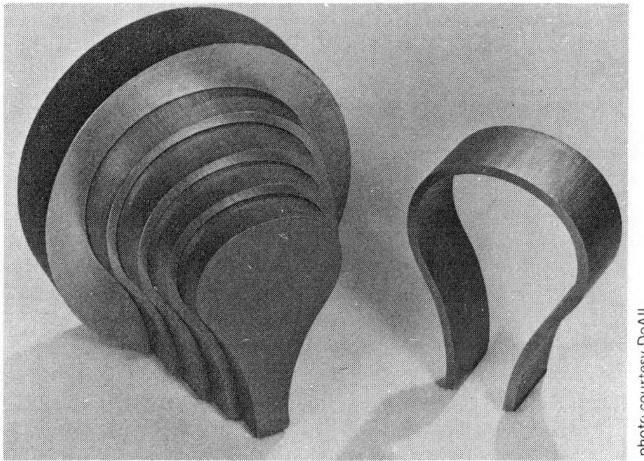

photo courtesy DoAll

Horseshoe-shaped examples show how a contour saw can cut various geometric shapes with minimum material loss

Contour bandsaws are available with horsepower ratings of 1½ hp to 15 hp or more, and some are equipped with tilt frames to permit mitering in the cutoff mode and three-dimensional cutting in the contour mode. These machines are often designed to accept the widest possible variety of bands.

Because of this wide selection of blade types, the typical contouring bandsaw is built with a wide selection of speeds, most often from 50 rpm to 5200 rpm. In addition, it usually incorporates a speed selector—a slide-rule-type device that shows what speed to use for different cutting operations on specific workpiece materials.

Material to be contour-cut can be fed throught the vertical saw in various ways. The most common method is simply to guide the workpiece by hand, and early bandsaws contoured this way. Precision capacities of this method depend largely on the operator. A good operator can usually maintain an accuracy of ±0.010 in.

Over the years, feeding has been improved to the point that, today, some models offer fully automatic tracing capability. In this "hands off" method, a thin metal or plastic templet is prepared and then taped or glued to the workpiece. An ultrasensitive stylus (photo, p53) accurately follows the templet contours for exact fidelity of multiple parts, or for repetitive sawing of cams, special tools, and similar work.

In fully automatic models, the sensing/control system is generally built in a closed-loop configuration, and thus can be made self-correcting. This virtually eliminates side loading of the blade and permits much better accuracy and finish than is possible with hand feed.

Sawing efficiency is optimal, since the saw automatically maintains the feedrate appropriate for the particular material throughout the entire cut, while even the most skilled operators will normally slow down for tight contours. Constant workpiece feed also lengthens blade life.

Flexibility of use is one of the most important features of the vertical contour bandsaw. Because of the wide selection of blade materials and designs, and of cutting speeds to use them, the machine can cut virtually any material, including glass, fiberglass, laminated materials, ceramics, marble, plastics, wood, asbestos, and paper. Metals, too, of course.

Hacksawing

photo courtesy of Armstrong-Blum

Probably the most natural extension of the human arm's sawing motion when using a hand tool is the powered hacksaw. The simple back-and-forth motion of the blade made the hack one of the first types of saws designed for power.

Blade-motion simplicity has also kept hacksawing machines' prices down, compared to other types of sawing machines. This low initial investment, plus the design's flexibility and adaptability, has made the hacksaw the industry's workhorse of cutoff.

Hacksawing is basically different from any other type of sawing in that the blade makes a noncontinuous cut. A single blade is tensioned in a bow, with pins or with groups of pin pairs, while it is drawn back and forth across the workpiece. Cutting is done only during half of the stroke cycle. The second half, the return stroke, does just that—it returns the blade to its starting position. It is this very motion that gives hacksawing its name. Feed pressure is applied downward at the onset of the forward cutting stroke, and the machine literally *hacks* at the workpiece.

The fact that the blade is not continually in the cut is often cited as a drawback of the system, an inefficiency. But many factors enter into an appraisal of cutting efficiency, including the pressure required to achieve an adequate chip load (based on workpiece machinability), the thickness of the blade, the number of cutting teeth in contact with the workpiece, cutting speed, and the total cut-cycle time.

Horsepower, too, is a partial measure of efficiency. Machine designers take average load and cutting capacity into consideration when providing a particular unit with a specific horsepower. Typically, a 2-hp hacksawing machine cuts up to 8-in. rounds; some 7½-hp machines are designed to cut mild steel in 30-in. rounds.

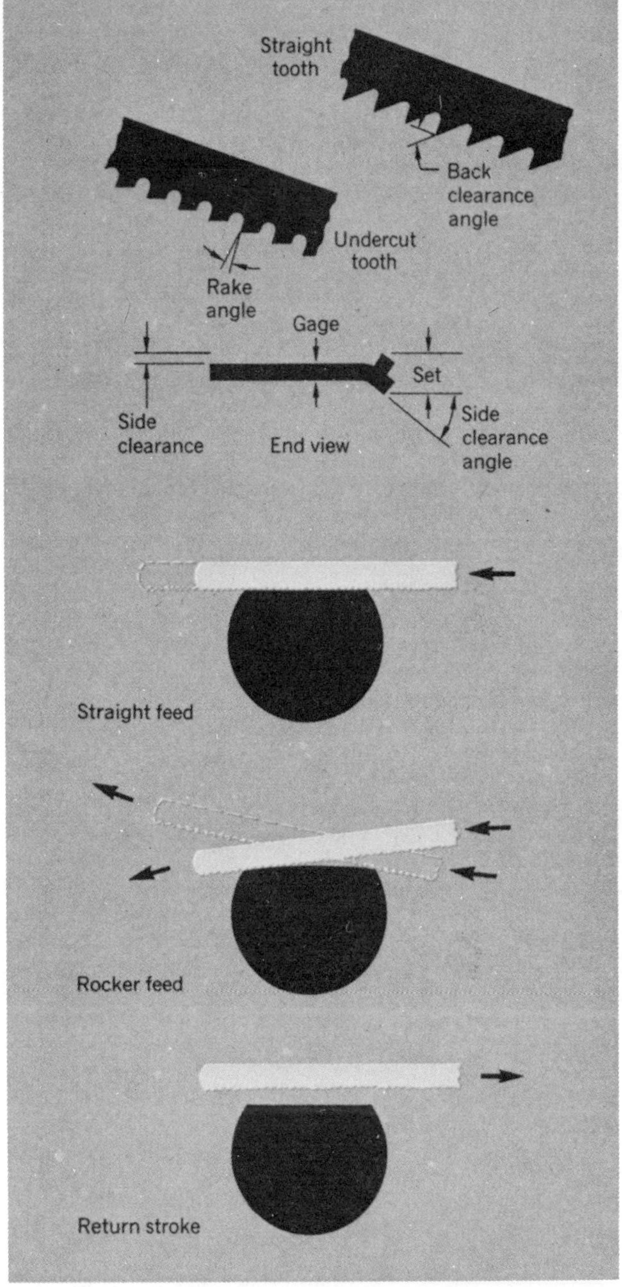

Another design basic is the construction of the machine's frame itself. Older designs use cast-iron frames to hold the saw rigid and stable. More-modern machines use welded-steel frames for higher breakage resistance, plus better adaptability (from the standpoint of the manufacturer) when a special design is requested. On the other hand, cast gearboxes and guideways in machine design provide excellent vibration resistance and good guidance characteristics.

Blade design

While the basic operating motion of the hacksaw is markedly different from that of other sawing machines, hacksaw teeth can readily be compared to those on a bandsaw blade. Further, hacksaw blades are constructed of materials similar to those used for bandsaw blades.

Blades range in size from 12 in. long, ⅝ in. wide, 0.032 in. thick (costing less than $5) for small, general-duty machines to 53 in. long, 2.7 in. wide, 0.15 in. thick for heavy-duty, 10-hp machines.

Large blades, those designed for machines that can cut 25-in. and larger rounds, are generally available in segmented styles. Segmentation of the hacksaw blade, a feature also found on large circular saws, allows the operator to replace only the badly-worn or broken section, and to regrind that section to match the rest of the blade. Segmented blades are regrindable up to seven times—a great cost saving in light of the fact that a 53-in.-long blade costs over $170. In machines that offer a choice of solid or segmented blades, the general rule is to use solid blades only when cutting tubes or sectional steel. For large-diameter solid bars, segmented blades are recommended.

Power hacksawing is most often done with hss blades. Tungsten steel is sometimes used if the workpiece is a hard alloy or a stainless steel; more often the blade will be made of molybdenum steel, which provides better toughness, wear resistance, and heat resistance. Standard tool-steel blades are sometimes used for soft metals, and very often for one-of-a-kind cutoff jobs in small shops.

Although hss blades provide the best overall cutting characteristics, they do have a safety drawback. Should a solid-hss blade—tensioned to, say, 380 in.-lb on the nut—break during cutting, the blade itself will have a tendency to shatter into sharp, flying pieces. Welded-edge hacksaw blades, invented by Armstrong-Blum Mfg. Co. and now available from many manufacturers, have largely overcome this shrapnel problem.

In a welded-edge blade, a strip of hss is first electrically welded to the cutting edge of a blade body made of steel alloy. Cutting-tooth profile is then milled into the hss edge. With such high-speed-edge blades, both heavier feed pressures and higher blade speeds—factors which might otherwise hamper safety—can be used. Yet the blade remains safe to use. Bandsaws are similarly available with welded-hss cutting edges.

What is the best blade for any particular job? For any particular machine? Many factors must be considered. Among these are workpiece hardness, workpiece cross section, strokes per minute, blade tension, and blade geometry. Some of these factors can be combined to represent a "cutting factor," which, like "machinability," directly relates to the material being cut.

Cutting factors can be calculated for each specific shape and type of metal (AM—Apr.17'72,p90) and are related to each other. If Inconel, for instance, has a cutting factor established at 1.00, SAE 4140 steel would have a factor of 0.35. This factor, when used to weight the square inches of material cut per blade, can give better comparability between performance of different blades.

But the basic analysis remains the same. To find what blade is best for your hacksawing application, figure out how many cuts (square inches of cut) you are getting out of each blade, then apply that figure to the price you pay for the blade. This simple exercise, using the basic recommendations made by your dealer as a starting point, can often mean considerable savings over your operating year.

Hacksawing terminology

Blades are tensioned in the bow by means of nuts; torque wrenches are often used to bring the blades up to the proper tension. The degree of tautness depends mainly on the width of the blade, until cutting results indicate that the blade should be either looser or tighter. Roughly, a 1-in.-wide blade should be tightened to around 96 to 120 in.-lb; a 2-in.-wide blade to 320 to 440 in.-lb.

One German manufacturer of hacksaws has devised a blade-tensioning guide that is built into the machine's frame and is further tied to automatic controls. On a line of automatic machines, a mark is inscribed on the face of a dial mounted beside the bow guide. In tensioning the blade, the nut is turned so the dial mark lines up with another mark on the bow guide, at which point the blade is properly tensioned.

A miniature switch in this system senses the tension of the blade and provides input to the saw's automatic control system. If the blade tension drops below a preset limit, indicating that the blade has broken, the control system shuts down the machine. Machine shutdown is also commanded by the sensor if the tension increases, indicating blade blockage.

The tensioning system on machines made by another builder incorporates a similar simple-to-use operation to obtain proper tension. The blade is placed in holders, and the hex tensioning nut is brought to finger tightness. From that point, the nut is turned a specific number of flats, corresponding to the rating of the blade.

When cutting, the hacksaw blade actually curves (bows away from the workpiece), because of the heavy forces it exerts on the stock. Some bowing is permissible, but if it results in any blade flexing, tension should be increased or a stronger blade should be installed.

Some manufacturers circumvent much of this bowing by designing their hacksawing blades with pin holes closer to the cutting edge—that is, below the vertical center of the blade. Tension is drawn closer to the teeth, and the rest of the blade absorbs bowing forces better.

This blade is slightly convex as it is positioned in the saw frame. (Blades with centered pin holes are straight, due to equal tension across the entire breadth.) But when work resistance is encountered, the flex of the blade is naturally greater in the unsupported center section. This special blade is bowed when it is not cutting and straight when it is cutting, and it has less of a tendency to wander during the cut.

Blades are held by the pins through their holes; if the pin diameter is much smaller than the hole ID, the ends of the blades may break outward. Proper matching of the blade to the machine is thus essential.

Design of many hacksaw machines permits the use of "back-up" support bars for the blades. These additions, teamed with high tensioning of the blade itself, assure cutting of adequate chips in difficult-to-saw materials.

Metal removal at the hacksaw tooth edge is similar to that of a bandsaw. Four major pitch sizes are available—4, 6, 10, and 14 teeth per inch. Tooth set may be conventional or wavy, with the wavy set generally used to widen the kerf in cases where the saw has a tendency to bind.

Depending on the metalcutting operation, blades may be designed with straight teeth, which provides relatively deep gullets for chip clearance and a positive rake angle, or with undercut teeth, having a shallow back-clearance angle and little or no rake, for greater stability in the cut.

Back-and-forth motion

The two component directions of the hacksaw blade determine its basic cutting efficiency.

The return stroke, sometimes called the "wasted" stroke, is the most obviously inefficient. No metal removal takes place while the blade is returning to its starting position in preparation for the next cutting stroke. Thus, from an efficiency standpoint, anything that can be done to shorten the time it takes for the blade to return to start would improve machine productivity.

Several manufacturers have applied this concept to their machines' basic motions. One builder incorporates a crank-sliding gear into the design of the driving mechanism. This arrangement reduces the time it takes to return the saw to the starting position, so that a greater number of strokes per minute is possible while maintaining the same cutting speed.

Another builder takes a slightly different approach: Using a crank/lever arrangement, as opposed to a simple crank, 212 deg of any single drive revolution are used for the cutting stroke, the other 148 deg return the blade. The result, once again, is more cutting strokes per min—33⅓% more.

How to increase the productivity of the second component direction of the hacksaw's basic motions—the cutting stroke—is not so obvious. At the heart of cutting-stroke efficiency lies chip load per tooth.

A quick look back at bandsawing will explain the principle. As the length of band contact increases, more teeth come into contact with the workpiece. Pressure must be increased, so each tooth will be forced to take a full chip load. When difficult-to-machine metals are to be sawed, this condition intensifies. Stronger feed must be applied with resulting problems in blade tension and stability.

These problems hold true for conventional hacksawing. The cutting stroke (usually a *pulling* stroke) meets with increased resistance as more teeth engage a larger workpiece, if optimum chip load per tooth is to be realized.

A different approach, which reduces the total number of teeth in contact with the workpiece at any given time but increases the chip load per tooth, is also available. This approach, developed shortly after World War II by KASTO, uses an arc-shaped cutting stroke and combines it with *pushing* motion during the cut.

The arc-shaped cut is produced by a rocking action of the blade and frame as they move across the workpiece; it can best be compared to hand-sawing a workpiece while maintaining very little elbow motion.

Because of the rocking action of the sawblade, relatively few teeth are actually cutting at any one time. By decreasing the number of teeth in the cut, the entire blade is able to absorb a higher cutting force from each tooth, and chip load is increased.

Care and feeding of blades

Feed techniques for hacksaw teeth are very similar to those used for bandsaws; in fact, the rules were originally established for bandsaw teeth:

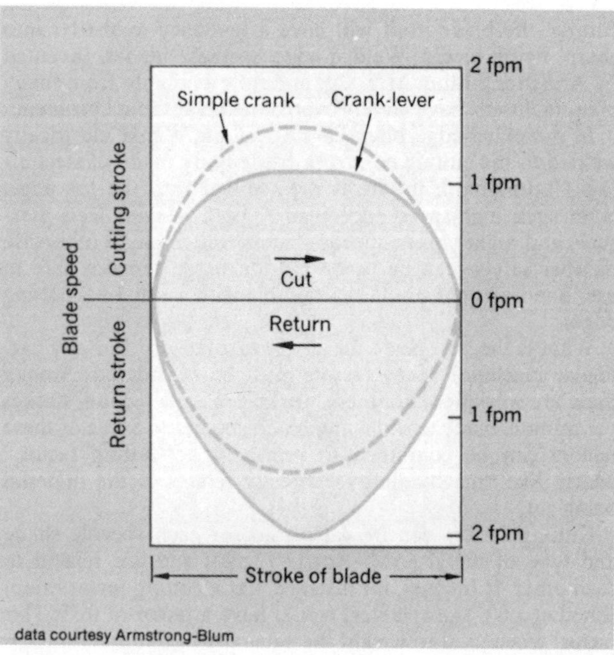

More cutting strokes per minute is the result of speeding up the blade return in a two-stroke cycle

- At least three teeth at a time should engage the work.
- A cut should be started with the blade against a long surface rather than an edge.
- The blade shouldn't be too much longer than the maximum width of the cut.

The rules are general, and a good deal of testing and tryout is needed to get optimum conditions for a particular job. One of the primary variables lies within the machine itself: the kind of feed mechanism used.

Basically, two types of feed can be built into a machine. Depth feed determines the minimum positive depth to which the blade is forced to cut; pressure feed controls the amount of effort that is exerted on the blade.

Both types have about the same effect when the blade is cutting straight through a workpiece of uniform width—especially in machines that employ a straight back-and-forth (rather than rocking) cutting motion. But in reality, little cutting is done straight through workpieces that have uniform width. In cutting a round bar, for example, the length of cut (corresponding to the workpiece width) increases as the blade moves toward the center of the circle and decreases as the cut is being finished.

Depth-type feed, whether it is achieved through controlled hydraulic volume or by spring controls, produces a comparatively slow cut in the beginning and end portions of a slice through round barstock (where cut lengths are short), and a comparatively fast cut in the middle of the slice (when the blade is sawing through very nearly the full diameter).

Pressure feed, whether provided by single-friction hydraulic pressure or by gravity feed, cuts relatively quickly through the shorter cut lengths and slows down through the wider-length-of-cut diameter.

A dual-power feed, available on some hacksaw machines, combines the advantages of both types of feed. This all-mechanical feed mechanism has a double-square-thread feedscrew, which rotates in bronze bearings in the upright. Mechanical gearing changes the method of feed from pressure (friction) type for shorter cut lengths to positive-depth type for longer cuts.

One of the most important ways of maintaining blade life is the proper use of coolants. It's important because the hacksaw blade, by the very nature of its stroke, tends to heat up nonuniformly, with the bow/blade joint acting as a heat sink. Coolant should be applied whenever possible—except when cutting cast iron, where coolant forms a sludge that retards cutting.

Cooling the production hacksaw operation is done by flooding the saw/workpiece interface. Oils and solubles both work well. Oils are typically mineral-based, with fatty oils added. These have added sulfur for antiweld characteristics and chlorine for film strength. Soluble cutting agents are used with fatty oils and are sulfurized for extreme pressure and antiweld properties. The water base aids in heat removal, an important consideration.

For an average speed of 100 strokes per min., soluble cutting fluids are mixed in a 1:5 ratio with water. The concentrate should be added to the water and stirred aggressively to permit full mixing.

A systems approach

Coolant and blade selection, machine capacity, automated auxiliary equipment, and materials-handling accessories make the modern hacksaw not just a machine but a system. All factors are interrelated. Operating efficiency for any particular machine is not only a factor of the basic machine design, but depends to a large extent on how the specific shape is handled and cut.

The way the workpiece is fixtured is one factor that affects productivity. Getting economy out of a hacksaw often means gripping as much stock as possible—bundle cutting.

But bundle cutting is not always practical on a production basis. There are occasions when small diameters must be cut one at a time on a large-capacity machine; only a small portion of the blade would then be used.

A variable-position vise solves this unequal blade-wear problem nicely:

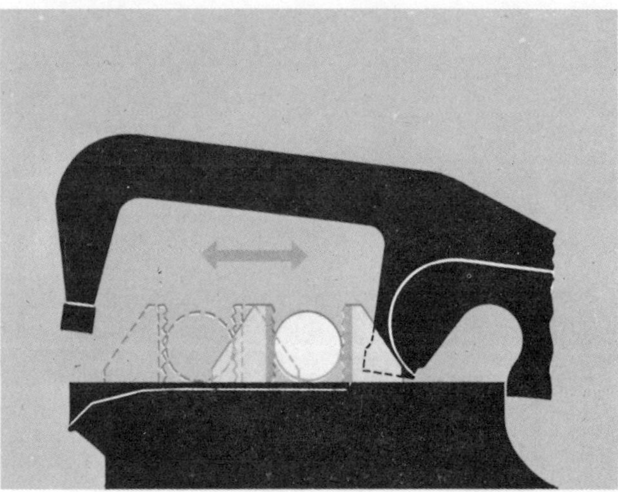

sketch after KASTO

The system works only on saws with nonautomatic material-feed systems. In automatic versions, the vise has a material-sensing function and cannot be moved without affecting accuracy.

Many other additions to the hacksaw machine are possible. Nesting equipment, including various clamps and extensions, permit efficient bundle cutting. Mitering can be done with special swivel-jaw vises. Automatic counters, when wired into machine controls, permit preselection of the number of parts cut off a piece of stock; a machine so equipped will shut down when the job is completed.

Small hacksaws, like some circular saws and some bandsaws, are portable. Retractable rubber wheels allow the operator to take the machine to the job. In such a situation, it is important to realize that workers other than the primary operator will likely use the machine. Operating instructions, consequently, should be simple and straightforward.

These instructions should include a list of things to look for if the machine starts producing poor cuts. Among them:

- Slow cutting—look for chip overflow in the gullets, too high a blade speed, or insufficient coolant, with consequent heat dulling the teeth, feed pressure too high, wrong tooth size, blade mounted in wrong cutting direction, or a new blade in an old cut with a kerf difference that causes the blade to bind.
- Tooth stripping—look for teeth too coarse for the stock thickness, teeth too fine with overloaded gullets, cut started on a sharp corner or through a thin section.
- Blade breakage—suspect worn blade, too heavy feed pressure, wrong blade tension, a new blade stuck in an old, narrow kerf.
- Crooked cutting—check for loose blade tension, hard inclusions in the workpiece, an out-of adjustment machine.

Producing straight cuts on the typical hacksaw is not much of a problem if you take a reasonable amount of care. But if all else fails, you may have to go to a larger machine, or even a circular saw.

Circular sawing

Photo courtesy of Hill Acme

Circular sawing, developed in Germany more than 60 years ago, offers advantages over other methods in many precision and production applications.

Circular sawing uses a continuous-cutting blade with many teeth, which can rotate through a large range of speeds. The process is often called "cold" sawing, although this is a slight misnomer. Cold sawing distinguishes other types of cutting from the kind often seen in foundries, where the workpiece is actually hot. All types of sawing commonly found in metalworking shops, including bandsawing and hacksawing, should thus be called "cold." Nevertheless, the term has come to mean the production cutoff of metal with a solid circular blade. Of course, some circular saws are often adaptable to other cutoff methods: Just as a bandsaw can be used for filing or friction sawing, some circular-sawing machines are designed also to handle hot sawing or abrasive cutoff.

But from a metalworking viewpoint, the circular saw has features that make it very different from other means.

If sawing is basically a milling operation, circular saws come closer to slot milling than any other type of machine. Any millable material can be cut on a circular saw, and chip loads per tooth are more on the order of magnitude of milling cutters than with other saw blades.

Since circular sawing is done at the fixed periphery of the wheel, machine design is affected. Surface speed at the blade's cutting edge varies directly with wheel radius. Cutting speed can thus be changed without an elaborate motor transmission by designing the machine to accept different blade sizes. Nevertheless, most circular-sawing machines do offer several spindle speeds, even if they are not designed with the special high-speed gearing needed for abrasive and friction cutoff.

Machines can be built in several basic feed configurations: horizontal, vertical, pivot (chop), and variations on these.

With vertical feed, the rotating blade travels downward in a straight line to engage the workpiece. Blade-feed motion is assisted by gravity, and the workpiece can be centered directly beneath the centerline of the saw blade for optimum cutting accuracy and minimum wobble. Some vertical-feed machines push a table-recessed blade up into the workpiece.

A rectangular-cross-section workpiece, loaded into the most convenient position (with a low profile), can be cut most efficiently in a vertical-stroke machine. If this rectangular stock measures 6 in. high and 12 in. wide, the blade would have to feed its way through only 6 in.

Other machines are designed for horizontal feed. The blade here is pushed into the workpiece from the side. Variations on basic horizontal-feed units include specialized plate-cutting saws, which traverse the blade along the length of solid plate. Most-efficient clamping of a rectangular workpiece in this type of arrangement would be in a "standing" position.

A third basic feeding configuration is pivot motion (chop stroke). It embodies much of the efficiency of a vertical-feed unit (the chop stroke is 90% efficient geometrically as the vertical motion), and it has a ruggedness advantage.

In this type of machine, there are only two pivot points. Friction, consequently, is limited to rolling friction in the bearings, which can be designed for maximum life.

The feed on the circular saw blade, whether it pivots or moves vertically or horizontally into the workpiece, can be supplied in various ways. The simplest chop-cut units are manually fed; air or hydraulic feed is used on almost all other machines.

Making chips

The deeper a sawing blade penetrates into the workpiece, the greater its tendency to jam in the cut. To reduce jamming, the kerf must be made larger than the width of the blade body. With bandsaw and hacksaw blades, this is done by designing set into the teeth. Some circular blades are also built with set, but most of them use a different approach.

Teeth in a circular saw blade get their rigidity from the entire blade, and, in many operations, most of the blade body's radius follows the teeth into the cut. This means that a circular blade must cut a wider slot into the workpiece than other types of blades to prevent jamming. In a typical operation, a bandsaw might need a 0.060 to 0.125-in. kerf to sever a bar; the same cutoff job on a circular saw might need a kerf of 0.250 in. This additional material loss may be of some significance, depending on specific cases.

Kerf is increased with a circular blade simply by designing it with teeth that are wider than the rest of the body. In solid-steel blades, a parabolic cross-sectional geometry is used; in carbide-tipped blades, the cutting tips are wider than the straight-sided body. Both designs increase the radial side clearance of the blade for free cutting into the workpiece.

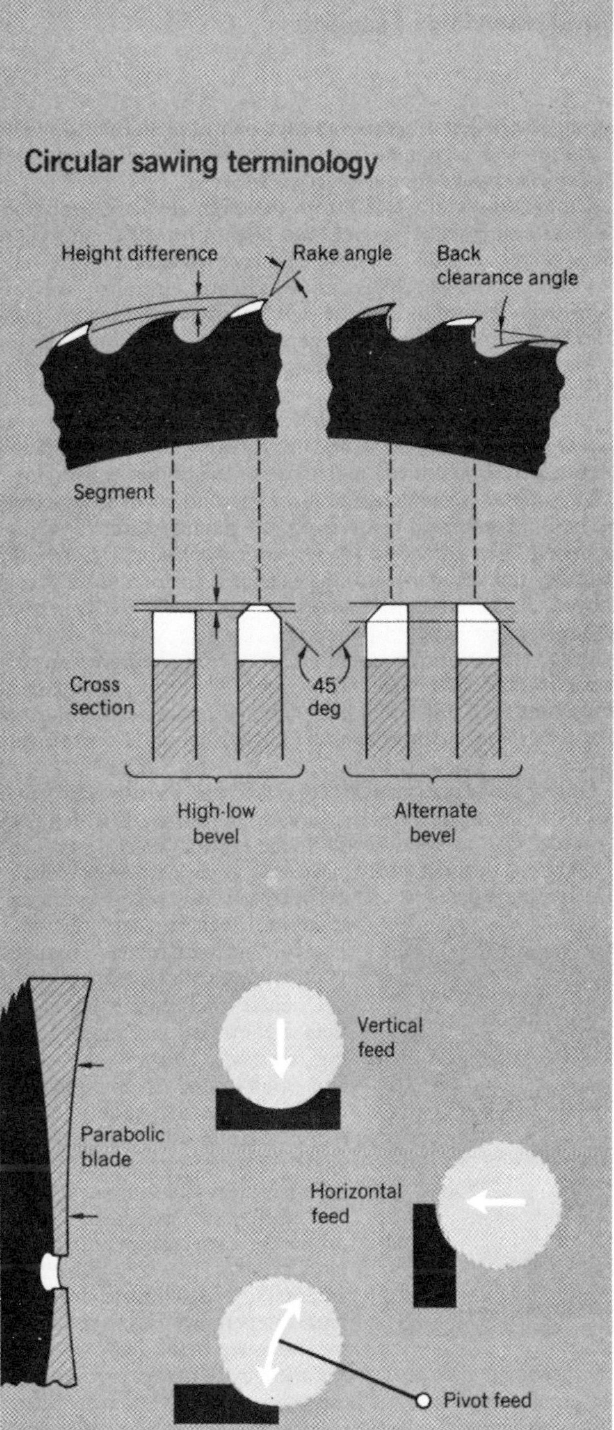

Two different types of tooth patterns are common in large circular saws: alternate-bevel patterns and high-low patterns.

Alternate-bevel tooth patterns use an adjoining pair of teeth to cut the kerf. Both teeth in the pair are the same height, and each has a side bevel—one left, one right. The first tooth bites a chip out of one side of the cut, the next tooth takes material from the other side.

In the high-low pattern, the first tooth is high and chamfered (usually to a 45-deg angle) on both sides. This tooth removes material from the center section of the cut. The following tooth is lower than the first (between 0.010 and 0.020 in. lower), and it opens up the cut removing material from both sides of the slit made by the preceding tooth. Each pair of teeth produces three cutting chips; for this reason the pattern is alternatively known as the "triple-chip" method.

This method places the greatest possible chip load on each

tooth. The height difference in each pair of teeth (called lead) is the major determining factor for the blade feedrate into the workpiece. Infeed should never exceed lead.

Other considerations in the tooth design are back clearance angles, cutting (rake) angles, and pitches for different materials. Increased side clearance and back clearance angles allow more room for chip removal. Greater clearance will reduce the tendency to load the saw, and thus will increase saw life. However, too much clearance will reduce tooth strength. A high rake angle reduces cutting forces, but also reduces the strength of the tooth.

Like larger hacksawing blades, circular saw blades are available in segmented styles, thus greatly reducing replacement costs, should teeth be damaged. As another cost saving, cold-saw blades are designed to be reground. For instance, a 32-in.-dia segmented blade from one manufacturer ($445) is designed to be ground (a $35 service not counting freight and segment replacement) until a total of 1 in. of radius is removed. Each grinding removes approximately 0.030 in., so about 35 regrinds per blade are possible.

Blade manufacturers often introduce controlled amounts of stress in blades by hammering them. This process is called tensioning, and it is not unlike pulling a hacksaw or bandsaw blade, in that it helps to hold the blade true under centrifugal force.

One of the primary advantages of a large circular saw is its ability to cut through materials with uneven grain structures. Because of the tooth-load characteristics, cold saws can quickly cut through such workpieces as railroad rails, which have perhaps the most uneven microstructure of any common material. Even in such jobs, accuracy and squareness remain good, often to ±0.002 in.

Photo courtesy of Bohle

Circular saws use carbide more than any of the other major sawing methods; carbon tool-steel and high-speed steel are also commonly used. Requirements for a circular-saw material include resistance to thermal shock, high hardness and toughness, and the ability to maintain these properties at cutting temperatures. Carbon steel is generally acceptable, but hardness drops at high temperature. Elevated temperatures are better handled with hss combined with proper coolant use. Tungsten carbide has the highest hardness and is less sensitive to heat. It is, however, sensitive to mechanical shock.

Cooling the circular saw-blade/workpiece interface is similar to the cooling used with other sawing methods. It depends largely on speed and the type of material being cut. Flood cooling with oil or water-soluble fluids is most often used, but some applications need special methods. One of these is traverse sawing—having a downfeed blade travel over a long length of plate. Mist cooling is often used this situation, and the blade is shrouded in a cover to contain the coolant.

Tote that load

Circular sawing machines, perhaps more than any others, make optimum use of large-scale material-handling and automated gaging systems. In fact, many cold saws destined for high-volume production situations are sold mostly on the merit of their conveying systems, with little emphasis on the virtues of the saws themselves.

This is for a good reason: The nature of the circular sawing operation permits to a high degree of automatic gaging.

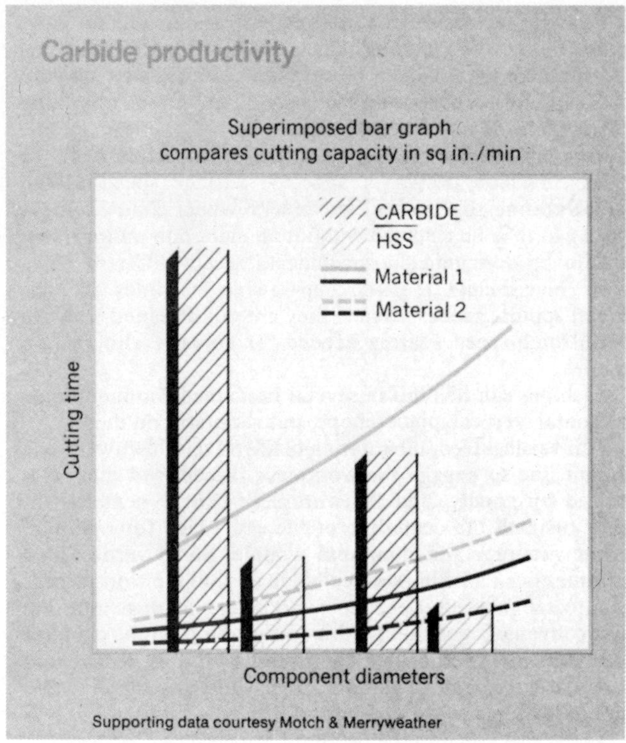

Supporting data courtesy Motch & Merryweather

In the simplest type of length gaging, a swing arm and a deadstop are provided. The operator controls vise opening manually, and pushes the material in toward the blade. From this basic configuration, automation can take over virtually the entire job. In the most highly automated structural fabrication plants, storage tables (magazines) cross-feed random-length material to automatic input tables, which feed the production saw. After gaging and cutoff, the sliced pieces are discharged toward an automatic beam punch, where the next operation begins.

The number of possible variations between the simple hand feed and the totally automated cut-to-length line are almost limitless. And once again, specialized situations call for unique solutions. One builder, for example, equips the feed tables on its plate-slicing saws with air-cushion jets.

Material clamping and vises, like feeding systems, can range from simple to highly complex. Factors to consider include mitering capability (some very large machines lift up and pivot on their circular baseplates for this function), double mitering, slotting capability, how many sides of workpieces can be clamped (a few machines offer descending clamping bars), the shortest length that can be fully clamped, and attachments for bundle cutting. A few machines on the market include head-swivel capability, in addition to mitering, so compound miters can be cut.

So the selection of the right kind of circular-sawing system—or, for that matter, any sawing system—depends on many factors. Determination of what criteria are important can only begin on the shop floor. ■

Fundamentals of broaching

There appears to be no end to the increasing size and productivity of today's new broaching machines or to the versatility of broaching tools. Here are the basics

Broaching—one of the most productive precision-machining processes known—is also a study in self-contradiction. It's a high-production, metal-removal process that sometimes is required to make one-of-a-kind parts. It's at its best when machining simple surfaces or complex contours. It's recent successes include such dissimilar items as high-precision computer parts and massive locomotive bull gears.

Broaching is similar to planing, competes with milling and boring, and gives turning and grinding stiff competition. Properly used, broaching can greatly increase productivity, hold tight tolerances, produce precision finishes, and minimize the need for highly skilled machine operators.

Understanding broaching

Tooling is the heart of any broaching process. The broaching tool is based on a concept unique to the process—rough, semi-finish, and finish cutting teeth combined in one tool. A broach tool frequently can finish-machine a rough surface in a single stroke.

In its simplest form, a broach tool resembles a wood rasp. It is a slightly tapering round or flat bar with rows of cutting teeth located along the tool axis. In advanced forms, extremely complex cross-sections and tooth designs may be found. However, the basic axial, multi-toothed tool shape remains.

For exterior broaching, the broach tool may be pulled or pushed across a workpiece surface; or the surface may move across the tool. Internal broaching requires a starting hole or opening in the workpiece so the broaching tool can be inserted. The tool, or the workpiece, is then pushed or pulled to force the tool through the starter hole. The final shape may be a smoother,

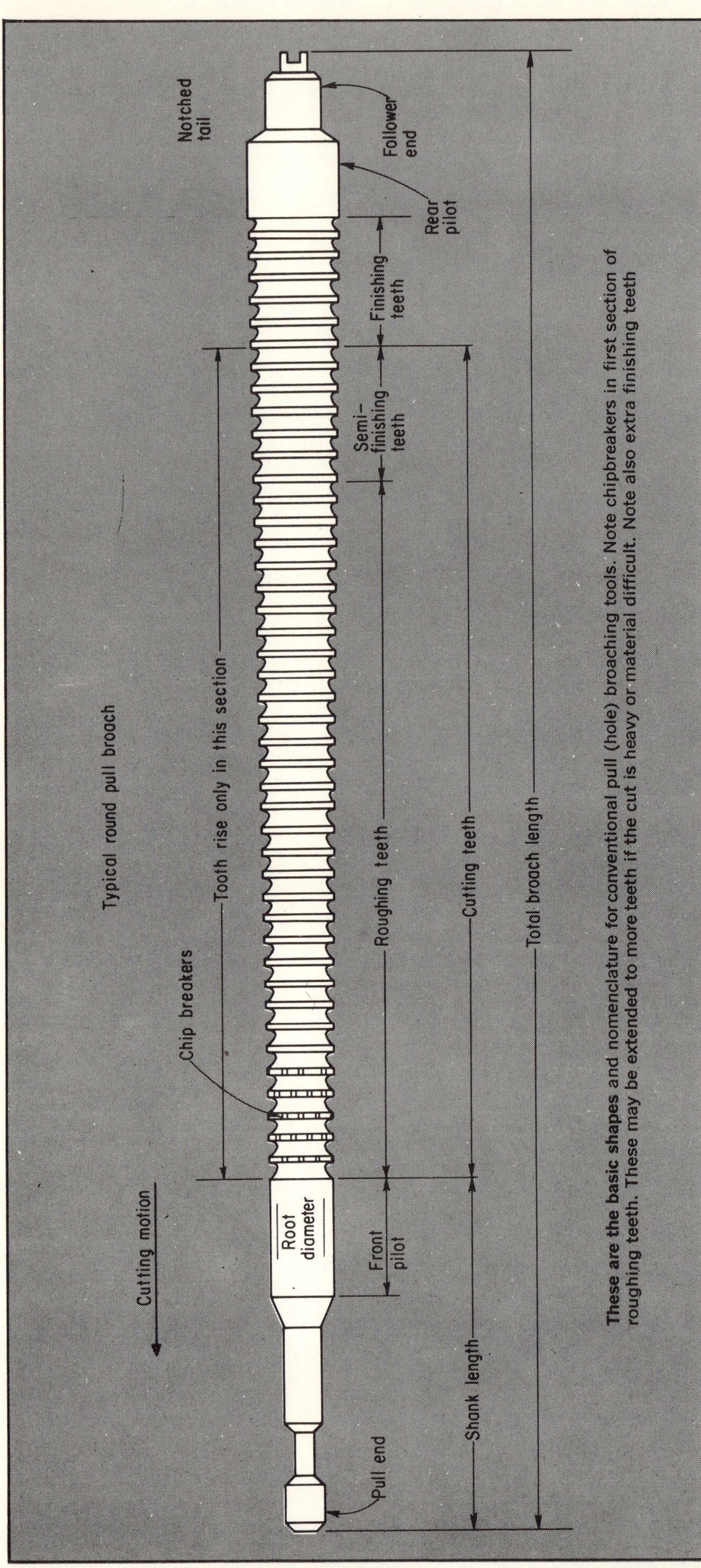

These are the basic shapes and nomenclature for conventional pull (hole) broaching tools. Note chipbreakers in first section of roughing teeth. These may be extended to more teeth if the cut is heavy or material difficult. Note also extra finishing teeth

flatter surface, a larger hole, or a complex splined, toothed, notched, curved, helical, or some other irregularly shaped section.

Almost any irregular cross-section can be broached as long as all surfaces of the section remain parallel to the direction of broach travel. The exceptions to this rule are uniform rotating sections such as helical gear teeth, which are produced by twisting the broach tool as it passes the workpiece surface. Blind holes can also be broached, and push broaches with limited travel are used.

The length of a broach tool is determined by the amount of stock to be removed, and limited by the machine stroke, bending moments (in a push broach), stiffness, accuracy, and other factors. Practically, the length of an internal push broach should not exceed 25 times the diameter of the finishing teeth, and a pull broach is usually limited to 75 times the diameter of the finishing teeth. Broaching tools can be as small as 0.050 in. or as large as 15 to 20 in. in diameter.

Whatever the actual tooth size and shape, standard nomenclature is used to describe the essential parts of a broaching tool. (See sketch at left.) When an internal pull broach is used, for example, the *pull end* and *front pilot* are passed through the starting hole. Then the pull end is locked to the pull head of the broaching machine. The front pilot assures correct axial alignment of the tool with the starting hole and serves as a check on the starting-hole size.

The *rear pilot* maintains tool alignment as the final finish teeth pass through the workpiece hole. Diameter of the rear pilot is slightly less than the diameter of the finish teeth. Often a *notched tail* is added to the tool to engage a handling mechanism that supports the rear of the broach tool.

Broach tooth terminology

Broach teeth usually are divided into three separate sections along the length of the tool: the *roughing teeth, semi-finishing teeth,* and *finishing teeth.* The first roughing tooth is proportionately the smallest tooth on the tool. The subsequent teeth progressively increase in size up to and including the first finishing tooth. The difference in height between each tooth, or *tooth rise,* usually is greater along the roughing section, less along the semi-finishing section. All finishing teeth are the same size.

Individual teeth (see sketch at left on page 88) have a *land* and face that intersect to form a *cutting edge.* The face is ground with a *rake* or *hook angle* that is determined by the workpiece material. Soft steel workpieces usually require greater hook angles; hard or brittle materials, smaller hook angles.

The land supports the cutting edge against stresses. A slight *clearance* or *backoff angle* is ground onto the lands to reduce friction. On roughing and semi-finishing teeth, the entire land is relieved with a backoff angle. On finishing teeth, part of the land immediately behind the cutting edge is often left straight so that repeated sharpening (by grinding the face of the tooth) will not alter the tooth size.

The distance between teeth, or *pitch*, is determined by the length of cut and influenced by the type of workpiece material. A relatively large pitch may be required for roughing teeth to accommodate a greater chip load. Tooth pitch may be smaller on semi-finishing and finishing teeth to reduce the overall length of the broach tool. Pitch is calculated so that, preferably, two or more teeth cut simultaneously. This prevents the tool from drifting or chattering.

Sometimes a broach tool will vibrate when a heavy cut is taken, especially when the cutting load is not evenly distributed. Vibration may also occur when tooth engagement is irregular. The greatest contributing factors to vibration are poor tooth engagement and extremely hard workpieces. Such problems must be anticipated by the broach designer.

Tooth rise is calculated so that the thickness of the chip does not impose too great a strain on individual teeth. A large tooth rise increases power requirements. When all teeth are simultaneously engaged in the workpiece, too large a tooth rise could cause an increase in power requirements beyond the rated tonnage of the machine. If the rise is too small to permit the teeth to bite into the workpiece, a glazed or galled finish will result.

The depth of the *tooth gullet* is related to the tooth rise, pitch, and workpiece material. The tooth root radius is usually designed so that chips curl tightly within themselves, occupying as little space as possible.

Chip load is a factor

As each broach tooth enters the workpiece, it cuts a fixed thickness of material. The fixed chip length and thickness produced by broaching create a 'chip load' that is determined by the design of the broach tool.

This chip load cannot be altered by the machine operator as it can in most other machining operations. The entire chip produced by a complete pass of each broach tool must be freely contained within the preceding tooth

Table 1: Commonly broached materials and typical results

AMS No.	Metal	Heat treat	Hardness (Rc or Rb)	Tolerance (in.)	Finish (mu-in.)
4132	2618-T61 Al	G	70 Rb	0.002	32-45
4135H	2014-T6 Al	G	70 Rb	0.0023	32
4928	Ti-6Al-4V	E	36-38	0.00075	24-32
5382B	Stellite 31	B	32	0.002	80
5613C	SAE 51410 (410SS)	H	32-36	0.002	63
5616C	Greek Ascoloy	I	32-38	0.001	35-42
5665C	Inconel	A	85 Rb	0.005	80
5668D	Inconel X	H	29	0.001	32
5727B	Timken 16-25-6	F	20-28	0.001	32-63
5735D	A-286	G	28-30	0.0024	32
			30-35	0.001	35
			32-38	0.0006	32
5765A	S-816	G	23-30	0.001	32-40
6250D	SAE 3310	E	20	0.010	63
6260E	SAE 9310	I	36-38	0.002	63
6302	17-22A(S)	H	29-34	0.001	60
6304	17-22A	H	35-40	0.003	——
6342B	SAE 9840	I	32-36	0.001	50
6370D	SAE 4130	I	32	0.0005	63
5382D	SAE 4140	I	25-29	0.002	32-63
6415E	SAE 4340	I	38	0.002	45-63
——	M-2 (tool)	A	24-28	0.0008	32
——	EMS 544	—	40-47	0.001	30
——	Inconel 901	I	32-36	0.0015	63
——	Rene 41	G	40-42	0.0024	32
——	WAD 7823A	—	28	0.0003	40-60
——	D-979	I	38-40	0.0005	60
——	EMS 73030	—	32-36	0.0028	63
——	M-308	—	36-38	0.0024	32
——	Chromoloy	—	31-32	0.004	32
——	PWA-682 (Ti)	—	34-36	0.001	32
——	Lapelloy	J	30-37	0.008	32
——	303 stainless	A	85 Rb	0.001	63
——	304 stainless	A	80-85 Rb	0.002	63
——	403 stainless	I	37-40	0.0006	63
——	SAE 1010	D	60 Rb	0.001	30
——	SAE 1020	D	3-12	0.002	60-80
——	SAE 1037	I	15-20	0.0003	30
——	SAE 1045	I	24-31	0.0005	——
——	SAE 1063	E	12-18	0.004	25-60
——	SAE 1070	E	5-10	0.002	28-60
——	SAE 1112	—	87 Rb	0.001	40-45
——	SAE 1145	C	13-18	——	50-100
——	SAE 1340	C	15-20	0.003	——
——	SAE 4047	C	8-15	0.002	60-80
——	SAE 5140	C	8-15	0.002	60-80
——	SAE 52100	D	25	0.0005	30
——	Gray cast iron	B	90 Rb	0.003	80-100
——	KP-7 cast iron	B	——	0.0005	125

Note: Treatment or condition

A	annealed	F	stress relieved
B	as-cast	G	solution and precipitation treated
C	as-forged	H	air quench, furnace temper
D	cold finished	I	oil quench, furnace temper
E	hot finished	J	salt quench, furnace temper

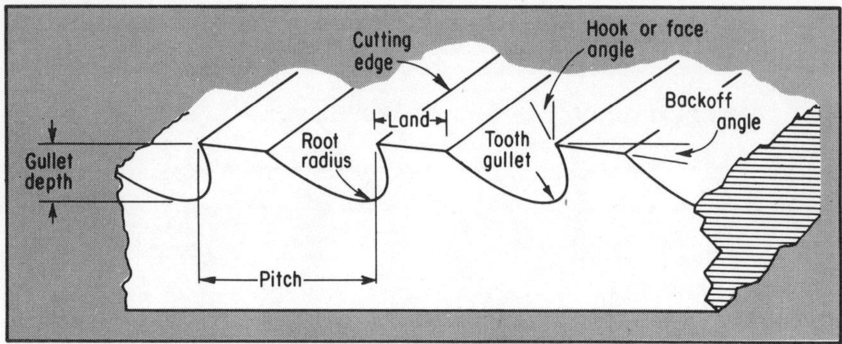

Teeth in a typical surface-broaching tool have standard nomenclature that corresponds to that of an internal broach. Gullets here are average size

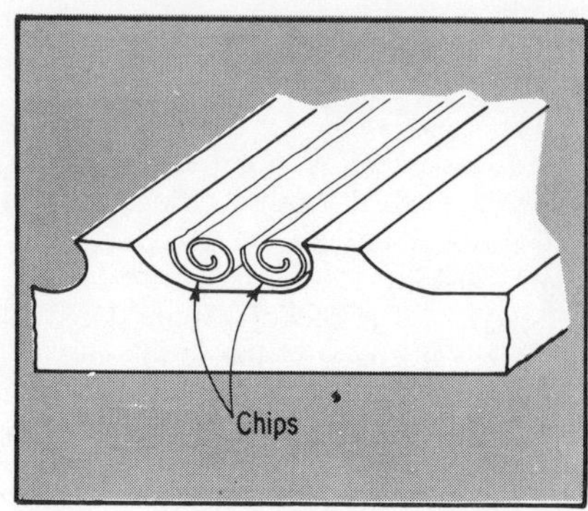

Flat-bottom gullet is an extended space between teeth for more chips

gullet. The size of the tooth gullet (which determines tooth spacing) is a function of the chip load and the type of chips produced. However, the form that each chip takes depends on the workpiece material; brittle materials produce flakes, while ductile or malleable materials form spiral chips.

Long cuts in ductile materials, or interrupted cuts producing two or more chips, would soon fill a circular gullet with chips. The solution is a flat-bottomed gullet with extra-wide spacing (above, right). This provides room for two or more spiral chips or a large quantity of chip flakes.

Broach designers may place broach teeth at a *shear angle* (upper sketch, page 89) to improve surface finish and reduce tool chatter. When two adjacent surfaces are cut simultaneously, the shear angle is a definite factor in moving chips away from the intersecting corner to prevent crowding of chips in the intersection of the cutting teeth.

Notches, called *chipbreakers,* are used on broach tools to eliminate chip packing and to facilitate chip removal. (See lower sketch on page 89.) The chipbreakers are ground into the roughing and semi-finishing teeth of the broach, parallel to the tool axis. Chipbreakers on alternate teeth are staggered so that one set of chipbreakers is followed by a cutting edge. The finishing teeth complete the job.

Chipbreakers are vital on round broaching tools. Without the chipbreakers, the tools would machine ring-shaped chips that would wedge into the tooth gullets and eventually cause the tool to break. Special chipbreaker designs can be used to increase the maximum tooth rise of a broach without overloading the machine. If deep slots are ground into the lands of the cutting teeth, the depth of cut can be increased on each tooth without fear of overloading.

The sections of the workpiece not machined by the first tooth are picked up by the next tooth, or the next, by staggering the array of slots along the tool axis.

Variety of broaching tools

All broaching tools are 'special designs,' in that they are generally made for a single user and a specific machining operation. However, certain types of broaches have become especially well-known for the type of work they do. Some examples are shown in sketches throughout the rest of this special report.

Rough forgings, malleable-iron castings with a hard skin, and sand castings with abrasive surface inclusions are cut with one of three types of *rotary-cut* broaches (sketches on page 90). The design idea is somewhat similar to that of a chipbreaking slot, but the cutting edge has been drastically reduced and the slots between the teeth have become much deeper. Rotary-cut broaching teeth are heavier, to withstand the heavy cutting load, and are spaced in staggered fashion along the axis of the broach to generate the entire circumference of the hole. The tools are designed to take deep cuts underneath a poor-quality surface. Once this surface has been penetrated, the balance of the broaching tool proceeds to semi-finish and finish the underlying metal in the normal manner.

The *hexagonal rotary-cut* broach, used for small diameter holes, removes little stock. Depth of cut is limited to the distance across the flats. The *radial rotary-cut* broach removes more stock than the hex-type tool because the cutting portions of the teeth are connected by arcs rather than by flats.

Spline rotary-cut broaches offer a greater degree of flexibility than either of the other tool types and also permit maximum stock removal. The amount of stock removal is governed primarily by the capacity of the broaching machine, rather than by any tooling limitations. Rise per tooth may be as much as 0.050 in. on such broaches.

Almost all keyways in machine tools and parts are cut by a *keyway* broach—a narrow, flat bar with cutting teeth spaced along one surface. Both external and internal keyways can be cut with these broaches. Internal keyways usually require a slotted bushing or *horn* to fit the hole, with the keyway broach pulled through the horn, guided by the slot.

If a number of parts, all of the same diameter and keyway size, are to be machined, an internal keyway broach can be designed to fit into the hole to support the cutting teeth. Only the cutting teeth extend beyond the hole diameter to cut the keyway. Bushings or horns are not required.

When several keyways are spaced around a hole, the resulting section is a multiple-spline cut. A single keyway broach can be used to cut all the splines by indexing the workpiece around a fixture. However, high production work usually requires a *multiple-spline* broach. This tool is equivalent to a series of keyway broaches combined in one tool, with the cutting teeth spaced around the tool diameter.

Internal helical splines, either straight-sided or involute, can be broached with a *spiral-spline* broach. The teeth are ground in a helical path around the tool axis. The helix angle corresponds to that required in the work.

The rifling of gun barrels is a special application of helical broaching. The work requires a special broach with straight-sided splines only a few thousandths of an inch deep.

Broaches have long been used to cut internal gears and splines with involute teeth. Use of seamless or welded mechanical tubing for the workpiece provides the broach tool with a ready-made starter hole. The work is usually done in one pass.

Burnishers are broaching tools designed to polish (by cold-working) rather than cut a hole. The total change in diameter produced by a burnishing operation may be no more than 0.0005 to 0.001 in. Burnishing tools, used when surface finish and accuracy are critical, are relatively short and are generally designed as push broaches.

Burnishing buttons sometimes are included behind the finishing-tooth section of a conventional broaching tool. The burnishing section may be added as a special attachment or easily replaced 'shell.' These replacement shells are commonly used to reduce tooling costs when high wear or tool breakage is expected.

Shell broaches can be used on the roughing and semi-finishing sections of a broach tool. The principal advantage of a shell broach is that worn sections can be removed and resharpened, or replaced, at far less cost than a conventional single-piece tool. Shells can also be used for the finishing teeth of long broaches; the teeth of the shell can be ground to far greater accuracy than those of a long conventional broach tool.

Spline-burring broaches are quite short and are generally designed as push broaches. They remove burrs created by machining work done after the splines have been formed. For example, a hole might be drilled and tapped into the spline for a grease fitting, leaving burrs that could create assembly problems. These broaches are made slightly undersize on the spline width and may be equipped with round teeth to remove burrs from inside the bore.

Special sizing broaches are pulled or pushed through a semi-finished hole to take out the last few thousandths of stock faster and more efficiently than a fine-feed boring tool can.

Pine-tree broaches cut the complex serrations used to lock turbine blades into their rotors. Common practice is to use a set of broaches; the first cuts a straight-sided V-notch in the rotor rim and is followed by one or more serrated broaches that progressively widen the notch to the full pine-tree configuration.

Sectional broaches are used to broach unusual or difficult shapes—often in a single pass. The sectional broach may be round or flat, internal or external. The principle behind this tool is similar to that of the shell broach, but straight sections of teeth are bolted along the long axis of the broach rather than being mounted on an arbor. A complex broaching tool can be built up from a group of fairly simple tooth sections to produce a cut of considerable complexity.

Carbide tool bits and the sectional-broach idea are combined into *heavy-duty* broaches for cutting deeply into heavily scaled surfaces. The carbide-tipped tool bits are arranged in a staggered pattern on the face of a tool holder. Each tooth is preset by means of an adjustable screw and locked in place on the tool holder by a setscrew.

Broaching tools with brazed *carbide broach inserts* are frequently used to machine cast-iron parts. Present practice, such as in machining automotive engine blocks, has moved heavily to the use of disposable, indexable inserts, and this has drastically cut tooling costs in many applications.

Spline punches, special types of broaches with only one tooth, are used for shaping holes through which conventional broaches cannot pass. One example is internal gear teeth in a blind hole. The gear teeth are rough cut by drilling and shaping, or milling, then one or more spline punches are forced into the work to produce the tooth form.

Blind-hole broaching violates two broaching principles: the tool does not pass completely through the workpiece, and it must be withdrawn backward over the broached surface. But it

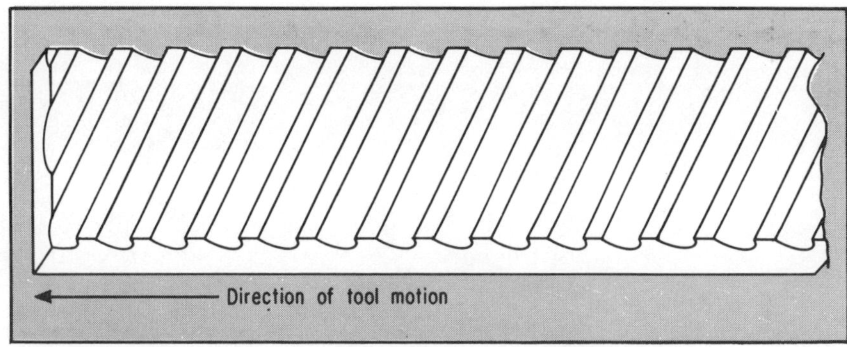

Shear-angled teeth are more difficult to regrind, and they produce side thrust on the machine ways, but a better surface finish is achieved

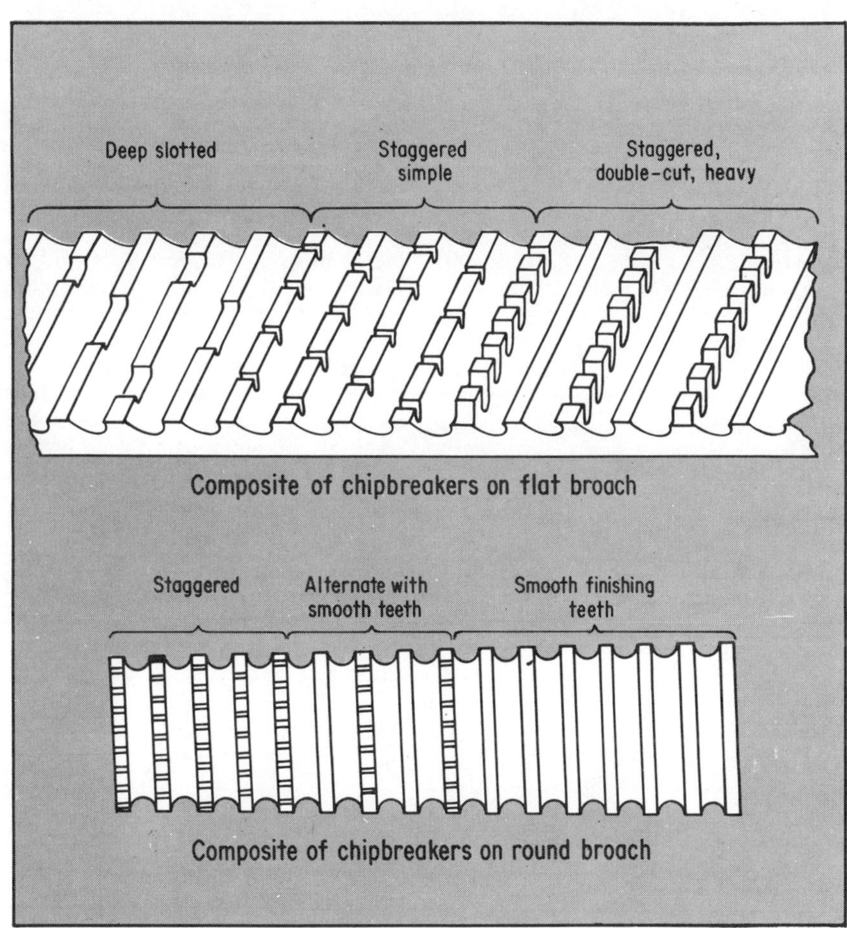

Chip breakers relieve the total load on the broaching machine by splitting the heavy chips in the roughing sections of the broach tool

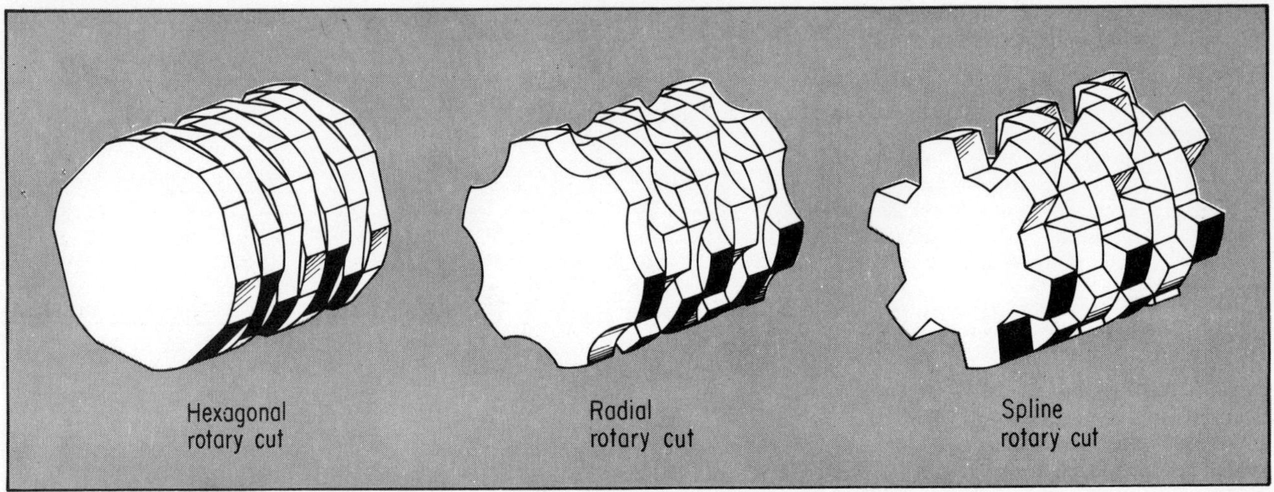

Rotary-cut broaching tools are designed to penetrate rough skins, as on castings and forgings, without exceeding hp ratings of a broaching machine. Following tools broach the surfaces more conventionally

can be done when necessary. The job usually involves a series of short push broaches, each slightly larger in diameter than the preceding tool. These short push broaches are mounted on a circular indexing table that rotates under the workpiece. As each tool stops under the workpiece, the broaching machine pushes the workpiece down over the tool, withdraws it, and then waits for the next broaching tool to index into position.

Strip broaching also violates the principle that a broach tool should not return through the workpiece, or else tool life will be reduced and the surface finish of the workpiece will be marred. In strip broaching, the broach tool is returned through the workpiece hole without stopping the machine to unload. Strip broaching is most commonly used for round-hole broaching of large quantities of low-cost parts when machining costs must be held to an absolute minimum. Strip broaches can be combined with burnishing buttons that slightly increase the hole diameter to provide a small amount of clearance, permitting the tool to be withdrawn without damaging the finished surface or dulling the cutting teeth.

Rotary broaches are special types of surface broaches. They are not commonly used, but they do offer advantages when producing work with external radial forms. In the most common setup, the broach tool is mounted on a rotating faceplate and the work is

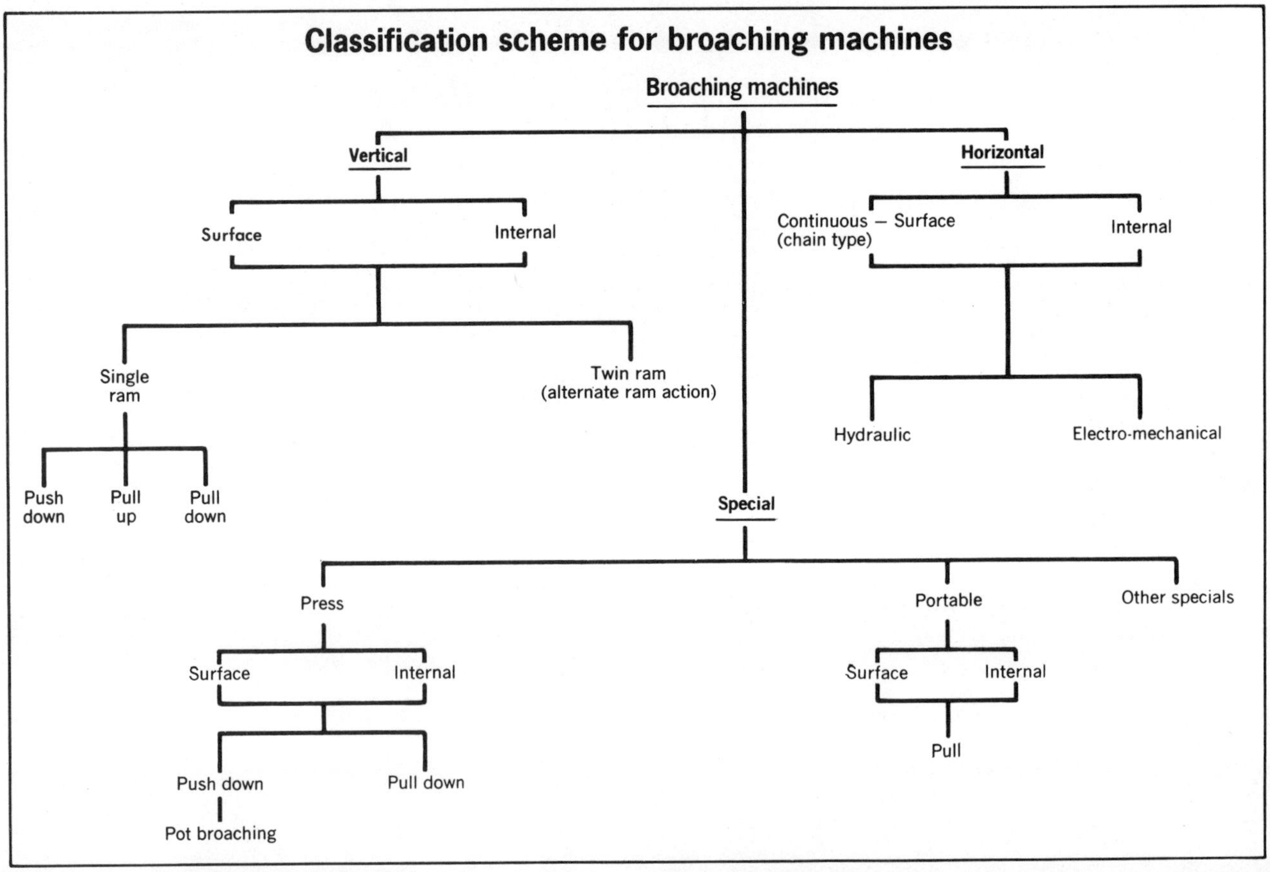

Classification scheme for broaching machines

Broaching machines

- **Vertical**
  - Surface / Internal
    - Single ram
      - Push down
      - Pull up
      - Pull down
    - Twin ram (alternate ram action)
- **Horizontal**
  - Continuous – Surface (chain type) / Internal
    - Hydraulic
    - Electro-mechanical
- **Special**
  - Press
    - Surface / Internal
      - Push down
        - Pot broaching
      - Pull down
  - Portable
    - Surface / Internal
      - Pull
  - Other specials

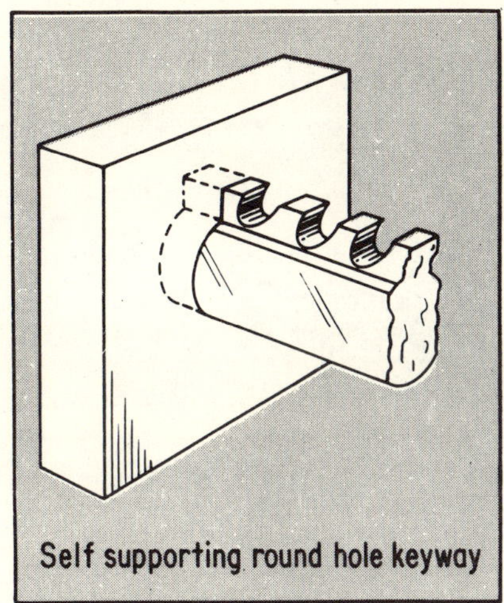

Keyway broach is a slotting tool that supports itself on backbone

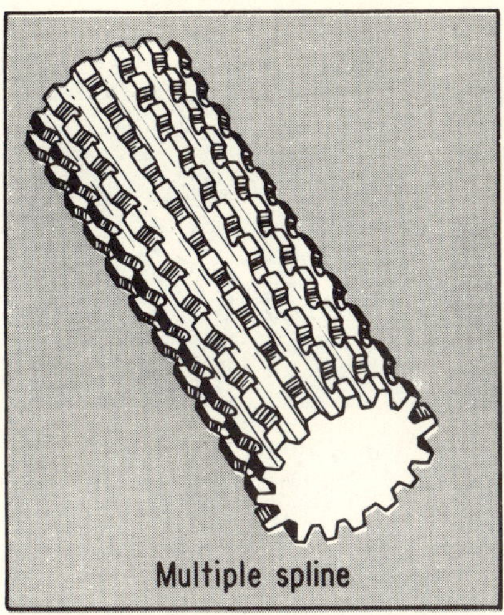

Multiple spline is a keyway broach taken to a higher degree

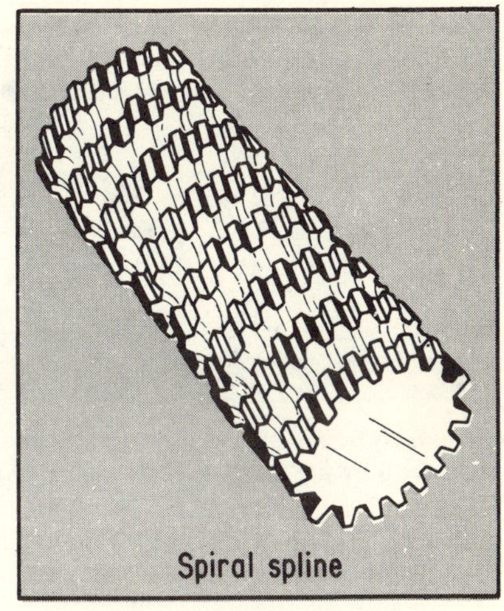

Spiral spline cuts internal helix, like several helical keyways

clamped into a hydraulic fixture. The tool makes one revolution to cut the desired shape. Circular slots can also be cut by a rotary broach that is turned around its own axis.

Ring or *pot* broaches can be extremely useful, and their productivity can greatly offset the initial tooling expense. Gears, for example, are broached by pushing them down through a series of cutting teeth held in a pot-type fixture. Power can be supplied by a conventional hydraulic press.

Broach tool materials

Almost all broaches are made of high-speed tool steels in monolithic construction. Brazed carbide or disposable inserts are sometimes used for cutting edges, most often on tools used for broaching cast irons.

In the early stages of broaching technology, broach tools were made from water-hardening tool steels. These tools were used on slow, screw-type broaching machines. With the introduction of new machines with higher speeds and greater production rates, high-speed tool steels became the principal materials for broach tooling.

Here is a list of tool steels and the materials that are commonly broached with these steels. (The list is only a sampling.)

M-2 tool steel: General use, including brass, aluminum, magnesium, and the following steels: 1020, 1063, 1112, 1340, 1345, B-1113, 4140, 4340, 5140, 8620 (Rc 26), 347 stainless steel, and 416 stainless steel (Rc 35-40).

M-3: Aluminum castings, cast irons, A-286 (Rc 32-38), Greek Ascoloy (Rc 32-38), M-252, D-979 (Rc 40), and the following steels: 4140 (Rc 32), 4337 (Rc 29-34), 4340 (Rc 32-38),

Table 2: Typical broach hook and backoff angles

Material	Hook angle, deg	Backoff angle, deg
Aluminum	6 to 10	. . .
Babbitt	8 to 10	. . .
Brass	–5 to 5	2 to 3
Bronze	0	½ to 2
Cast iron	6 to 10	2 to 5
Copper	15	2 to 3
Zinc	6	. . .
Aluminum bronze	15	2 to 3
SAE 1037	15	1 to 2
1112	15	2½
B-1113	15	2 to 3
1340	12	1 to 2
4140	8 to 15	1 to 3
4337	8 to 15	1 to 3
5140	15	1 to 2
5140 (Type 410 SS)	18 (roughing)	2
	20 (finishing)	2
9310	18 (roughing)	2
	20 (finishing)	1 to 2
303 Stainless	15	½ to 2
304 Stainless	15	½ to 2
403 Stainless	15 to 20 (roughing)	3
	30 (finishing)	5
431 Stainless	up to 28	. . .
M-308	15	3
N-155	20	2
Greek Ascoloy	15	2 to 3
Chromalloy	15	2
Lapelloy	12 to 15	2
A-286	10 to 15 (roughing)	2 to 3
	15 to 18 (finishing)	
Rene 41	15	3
Incoloy 901	15 (roughing)	3
	18 (finishing)	
Titanium 140A	5 to 15	2 to 4
Titanium 150A	5 to 9	2 to 5
Titanium PWA A682	12 to 15 (roughing)	3
	15 (finishing)	3

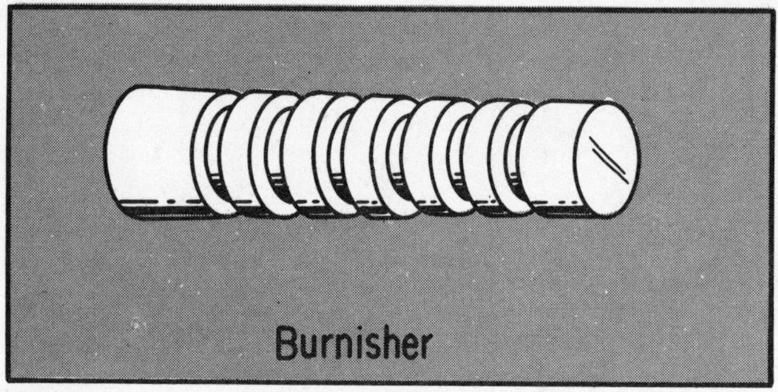

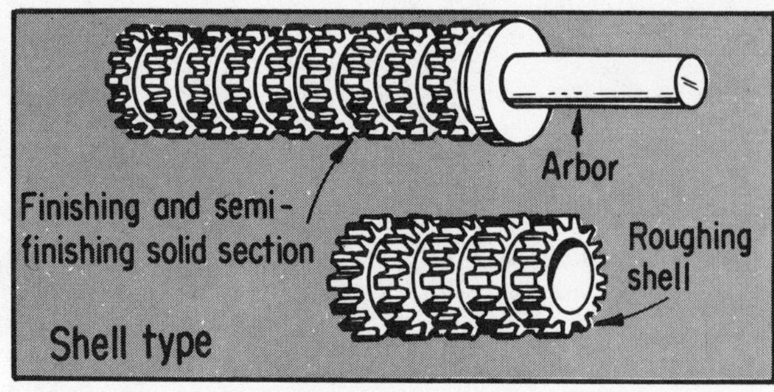

Burnishing tool increases diameter by only ½ to 1 thousandth, expanding by pressure, not by cutting away the metal

Shell broach, usually a section for roughing, is replaceable because it will wear more readily than finishing section

8617 (Rc 30-36), 8620 (Rc 32), 9310 (Rc 36-38), 9840 (Rc 32-36), 403 stainless (Rc 37-40).

M-4 (or T-5): Cast irons.

T-2: Steels: 1112, 4340 (Rc 35-40), 403 stainless (Rc 30-35); titanium alloys, PWA-682 Ti (Rc 36), Lapelloy (Rc 30-35), Greek Ascoloy (Rc 32-38), 19-9DL (Rc 20-27), Discalloy (Rc 23-32).

T-5: A-286 (Rc 29), Chromalloy (Rc 30-35), Incoloy 901, PWA-682 Ti (Rc 34-36).

T-15: Aluminum 2219, A-286 (Rc 32-36), Stellite, 17-22A(S) (Rc 29-34), N-155 (Rc 30-40), AMS 4925 titanium (Rc 32-40), Waspaloy, Incoloy 901 (Rc 32-36), and the following steels: heat-resistant steels, conventional alloy-steel forgings, 4340 (Rc 35-40), 52100, 9310 (Rc 26-30), and 17-4PH.

Most of the carbide cutters used to broach cast iron are used in flat surface broaching applications, although contoured cast-iron surfaces have been broached successfully. Surface broaching of pine-tree slots has been tried with carbides on high-temperature-alloy turbine wheels, but with little success. The carbide edges tend to chip on the first stroke.

Carbide-tipped broaches are seldom used on conventional steel parts and forgings. One reason is that good performance is obtained from high-speed-steel tools; another is the low cutting speeds of most broaching operations (from 12 to 30 fpm) do not lend themselves to the advantages of carbide tooling. The success of carbide tooling on cast irons is due to carbide's resistance to abrasion on the tool flank below the cutting edge.

Another problem with carbide-tipped tools is that a broaching machine work fixture must be exceptionally rigid to prevent chipping of the cutting edge. Experimental work with extra-rigid tools and workpiece fixtures, however, has shown that tool life and surface finish can be greatly improved with carbide-tipped tools, even when used on alloy-steel forgings.

Cast high-speed tool steels are almost never used in broaches. One property of the cast tool materials that prohibits their use in monolithic internal pull broaches is low tensile strength. Most cast alloys that can attain a hardness of Rockwell C 60 or higher do not have ultimate tensile strengths much in excess of 85,000 psi.

There are several practical ways of extending the life of a broach tool. One is to use a surface treatment, such as nitriding, oxidation, or hard chrome plating, to increase the surface hardness and wear resistance of the broaching tool.

Commonly broached materials

Broaches have been used on almost every material at one time or another —most of the known metals and alloys, some plastics, hard rubber, wood, composites, graphite, and so on. Metals and alloys are, by far, the most commonly broached materials. The products made from the other materials are not usually made to the stringent dimensional tolerances, or in the quantities, that make broaching economical.

The sample of broaching's surface finish and tolerance capabilities given in Table 1, page 87, does not define the limits of broaching technology; it simply shows what can be achieved in common practice.

In general, any material that can be machined can be broached. And the higher the machinability of the material, the easier it is to broach. In steels, machinability correlates closely with hardness. That is why workpieces with a high surface hardness, such as produced by previous work-hardening or scale, require that the first broach tooth cut beneath the scale or hard surface if possible.

The hardness of the workpiece material also influences the allowable cut per tooth. On harder metals, it is customary to take a relatively fine finishing cut; on softer nonferrous metals, a fine surface finish can be achieved with a heavier finishing cut.

Too heavy a cut, however, will tend to overload the broach tool—no matter what material is being broached. Too fine a cut, on the other hand, tends to interfere with free-cutting action and increases the tendency of the material to glaze, gall, or tear. Smaller steps can be used for finishing than for roughing.

The ductility of a metal has a considerable influence on the selection of an optimum hook angle for the broach teeth. In general, this angles decreases with decreasing ductility. Brittle materials, therefore, call for very small hook angles. (See Table 2, page 91.)

Free-cutting steel will allow a greater cut per tooth, or step, than will a hard or tough steel. However, a step of 0.0005 in. on a broach diameter is a practical minimum. Hook angles also vary with the material being cut as was mentioned previously. They range between 15° and 20° for the soft steels and between 8° and 12° for the hard steels. Backoff angles of 2° to 3° on the roughing teeth, 1° on the semi-finishing teeth, and 0.5° on the finishing teeth give good results when broaching steel. Chipbreakers should be used.

Stainless steels with hardnesses above Rockwell C 35 can be broached. Stainless harder than this, however, tends to dull broach teeth fairly fast, reducing the number of pieces produced between grinds.

The approximate cut per tooth (round broaches) runs from 0.001 to 0.005 in. This range will cover practically all types of stainless steel. Broaches with hook angles between 12° and 18° usually give the best results. Backoff should be held to a minimum; a 2° angle is preferable, but in no case should it exceed 5°. Chipbreakers should be used.

Cast and malleable irons permit a greater cut per tooth than even the free-machining steel. Brittle materials such as cast iron call for small hook angles, usually around 6° to 8°. Backoff angles are the same as for the general run of steels. Usually, a shorter pitch is permissible in broaching cast irons than in broaching steels because less chip room is required for the irons.

Brasses and bronzes allow a slightly heavier step, or cut per tooth, than steel. Too heavy a step, however, will tend to overload the broach. Hook angles usually range from 0° up to 10° and even higher, increasing with the ductility of the metal being broached. Brittle brasses call for smaller angles, from +5° to −5°. Backoff angles are usually 2° on the roughing teeth, 1° on the semi-finishing teeth, and 0.5° on the finishing teeth. Some form of chipbreaker is required.

Aluminum and magnesium can be broached with standard tool design, although special broaches give even better results. A hook angle of 10° to 15° and a backoff angle between 1° and 3° are recommended. Heavier cuts can be taken; even the finishing teeth can remove as much as 0.002 in. each. If trouble is experienced in maintaining proper tolerances, the size of the finishing cut can be increased, rather than decreased, to correct the situation.

Choosing a broaching machine

The type of broach cutting tool required for a given job is the single most important factor in determining the type of broaching machine to be used. Second in importance is the production requirement. Taken together, these factors usually determine the specific type of machine for the job.

The type of broach tool (internal or surface) immediately narrows down the kinds of machines that could be used. The number of pieces required per hour, or over the entire production run, will further narrow the field.

For example, a dual-ram machine with one operator may be chosen over two single-ram machines requiring two operators, to provide higher output per man-hour. The single operator can load one table of a dual-ram unit while the other ram is cutting. Even higher production requirements may dictate a continuous horizontal machine. The machine size in a particular model is a function of the tool size, workpiece size, broaching power requirements, and available production space.

For internal broaching, the length of a broach in relation to its diameter may determine whether it must be pulled rather than pushed through the workpiece, for a broach tool is stronger in tension than in compression. This in turn, helps determine the type

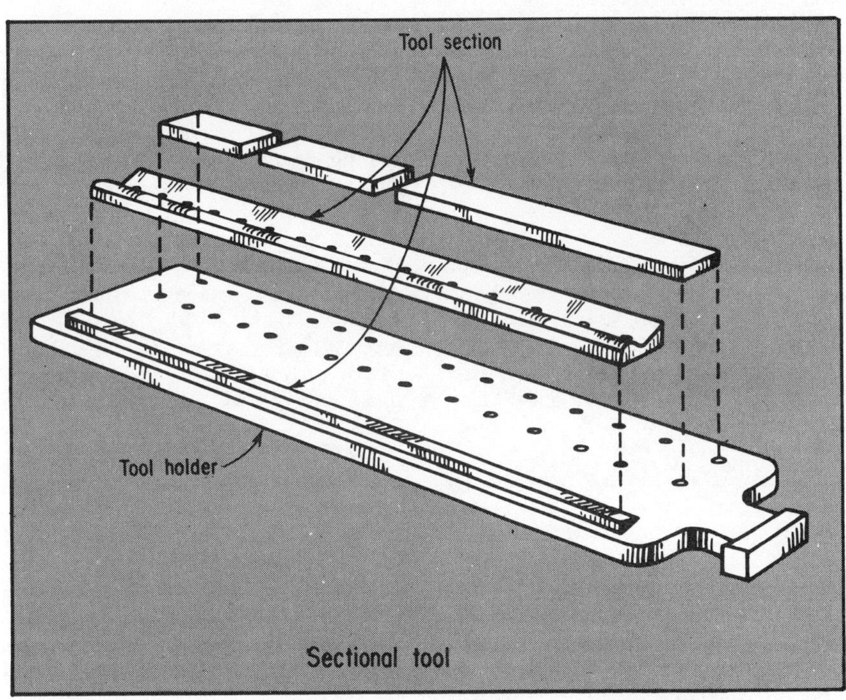

Sectional tool can be built up from several separately ground inserts, which reduces replacement cost. But location and clamping must be positive

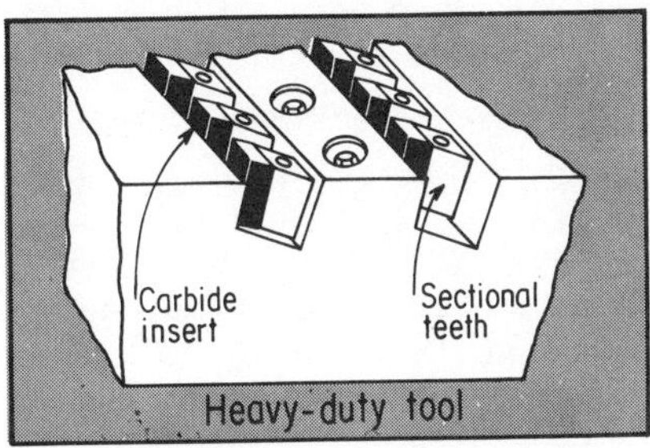

Carbide inserts are usually brazed because space in gullet is limited

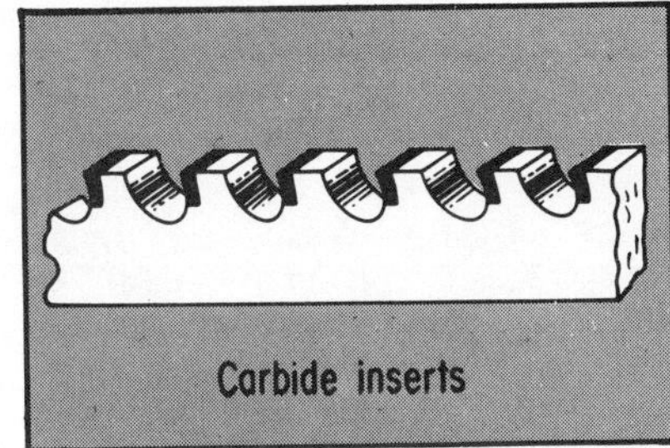

Carbide inserts in a heavy-duty tool are used for heavy stock removal

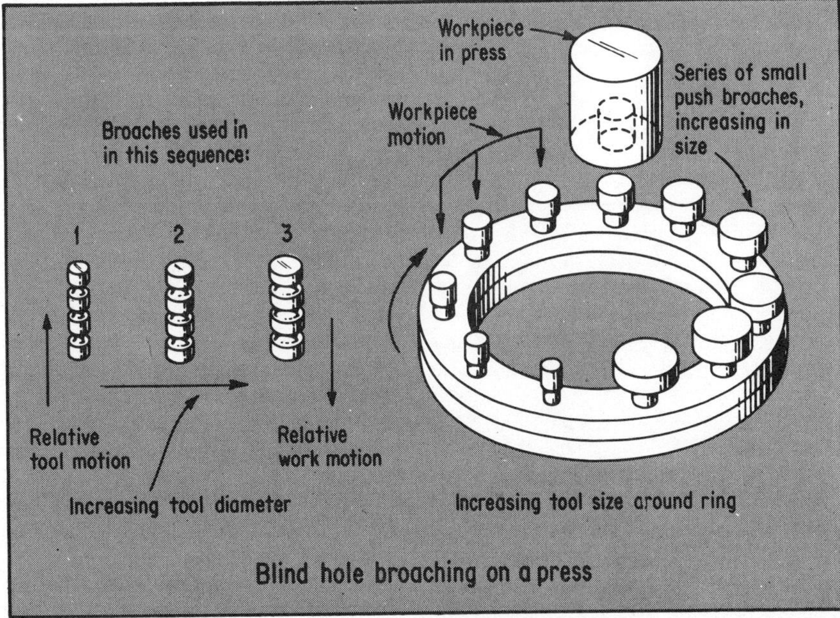

Blind-hole broaching is a special operation that requires separate tools for each step of enlargement. A rotary table is one way to do it

of machine for the job. A short push broach often is handled in a press instead of an expensive ram-type broaching machine. Presses, of course, can be converted to pull short broaches by the addition of a pull-down adaptor that converts push strokes to pull strokes.

Lubrication, workpiece size, chip-handling characteristics, and surface finish help determine whether a pull-up or a pull-down broach should be used. The trend is strongly to pull-down machines because gravity helps feed lubricant to the cutting teeth. Large workpieces are more easily handled in a pull-down than in a pull-up machine.

The type of drive—hydraulic or electro-mechanical—is another important factor in machine selection. So are convertibility and automation. Some machine designs allow for conversion from internal to surface work, for example. Some designs are fully automated; others are limited in scope and operate only with close operator supervision.

Here is a rundown of the major types of broaching machines:

Vertical broaching machines

About 45% of the total number of broaching machines in existence are verticals, almost equally divided between vertical internals and vertical surface or combination machines. Vertical broaching machines, used in every major area of metalworking, are almost all hydraulically driven. One of the essential features that promoted their development, however, is beginning to turn into a limitation. Cutting strokes now in use often exceed existing factory ceiling clearances. When machines reach heights of 20 feet or more, expensive pits must be dug for the machine so that the operator can work at factory floor level.

Vertical internal broaching machines are either pull-up, pull-down, or push-down, depending upon their mode of operation. The *pull-up* type, in which the workpiece is placed below the work table, was the first to be introduced. Its principal use is in broaching round and irregular-shaped holes. Pull-up machines are now furnished with pulling capacities of 6 to 50 tons, strokes up to 72 in., and broaching speeds of 30 fpm. Larger machines are available; some have electro-mechanical drives for greater broaching speed and higher productivity.

The more sophisticated *pull-down* machines, in which the work is placed on top of the table, were developed later than the pull-up type. These pull-down machines are capable of holding internal shapes to closer tolerances by means of locating fixtures on top of the work table. Machines come with pulling capacities of 2 to 50 tons, 15- to 90-in. strokes, and speeds of up to 80 fpm.

Push-down machines are often nothing more than general-purpose hydraulic presses with special fixtures. They are available with capacities of 2 to 25 tons, strokes up to 36 in., and speeds as high as 40 fpm. In some cases, universal machines have been designed which combine as many as three different broaching operations—such as push, pull, and surface—simply through the addition of special fixtures.

Vertical surface or combination broaching machines are found mainly in the automotive industry. These machines, produced in single- and double-ram versions (and even more rams occasionally), are hydraulically powered, with a few notable exceptions. Capacities range from 3 to 50 tons, with up to 120-in. strokes, and speeds of 120 fpm.

Electro-mechanically driven vertical surface broaching machines are available with either single or double rams and with strokes up to 120 in., capacities of 25 tons, and speeds of 80 fpm.

Horizontal broaching machines

The favorite configuration for broaching machines seems now to have come full circle. The original gear- or screw-driven machines were designed as horizontal units. Gradually, the vertical machines evolved as it became apparent that floor space could be much more efficiently used with vertical units. Now the horizontal machine, both hydraulically and mechanically driven, is again finding increasing favor among users because of its very long strokes and the limitation that ceiling height places on vertical machines. About 47% of all broaching machines are now horizontals. For some types of work, such as roughing and finishing automotive engine blocks, they are used exclusively.

Horizontal internal or combination machines, among the first used after the advent of powered broaching, have been driven hydraulically for many years. Hydraulic drives, developed during the early twenties, offered such pronounced advantages over the various early mechanical driving methods that only within recent years has any other method been used.

By far the greatest amount of horizontal internal broaching is done on hydraulic pull-type machines, for which configurations have become somewhat standardized over the years. Fully one-third of the broaching machines in existence are of this type, and of these, nearly one-fourth are over twenty years old. They find their heaviest application in the production of general industrial equipment but can be found in nearly every type of industry.

Hydraulically driven horizontal internal machines are built with pulling capacities ranging from 2½ to 75 tons, the former representing machines only about 8 ft long, the latter machines over 35 ft long. Strokes up to 120 in. are available, with cutting speeds generally limited to less than 40 fpm.

Horizontal surface broaching machines account for only about 10% of existing broaching machines, but this isn't indicative of the percentage of the total investment they represent or of the volume of work they produce. Horizontal surface broaching machines belong in a class by themselves in terms of size and productivity. Only the large continuous horizontal units can match or exceed them in productivity. Horizontal surface units are manufactured in both hydraulically and electro-mechanically driven models, with the latter now becoming dominant.

The older hydraulically driven horizontal surface machines now are produced with capacities up to 40 tons, strokes up to 180 in., and normal cutting speeds of 100 fpm. These machines, a major factor in the automotive industry for nearly 30 years, turn out a great variety of cast-iron parts. They use standard carbide cutting tools and some of the highest cutting speeds used in broaching.

But electro-mechanically driven horizontal surface machines are taking over at an ever-increasing rate for certain applications, despite their generally higher cost. Because of their smooth ram motion and the resultant improvements in surface finish and part tolerances, these machines have become the largest class of horizontal surface broaching units built. They are available with pulling capacities in excess of 100 tons, strokes up to 30 ft, and cutting speeds, in some instances, of over 300 feet per minute.

Larger machines have fully stress-relieved welded-steel frames, rather than gray-iron castings. Frequently two sets of cutting tools are attached to the ram so parts can be broached on both the forward and return strokes. A common operation on automobile engine blocks is broaching head surfaces on one stroke of the ram, and pan rail and bearing surfaces on the return stroke.

These machines can also be equipped with dual-speed controls, whereby the ram is driven at one pre-selected speed during one portion of the stoke and

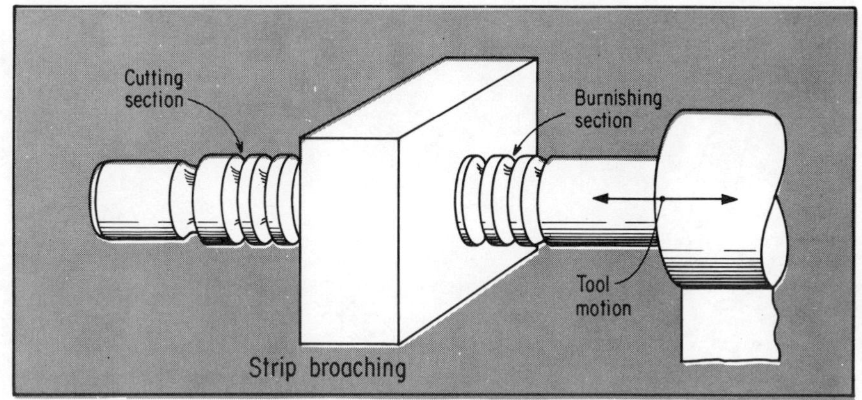

Strip broaching returns the tool over the already-cut work surfaces, which reduces tool life and means precision is limited

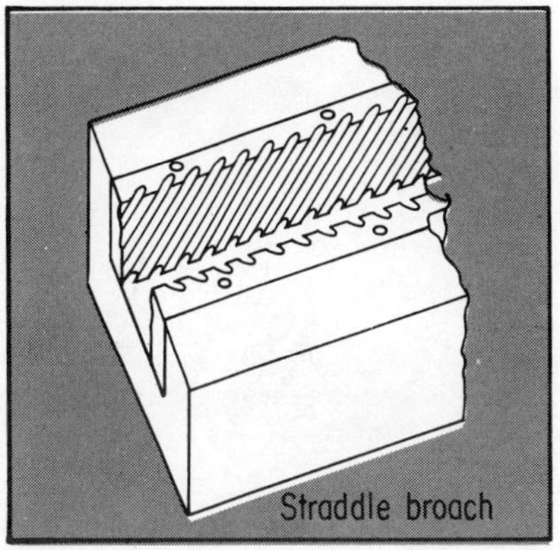

Straddle broach is like a double milling cutter, holds close tolerances

changed to a second pre-selected speed during another portion of the stroke. A typical application is the use of the high speed for the initial roughing cut on pine-tree slots in turbine wheels, and the slower speed while the finishing teeth are cutting.

Continuous surface horizontal broaching machines are rapidly becoming the most popular type of machine produced for high-production surface broaching. Recently in Detroit Broach & Machine Co's Rochester, Mich, plant, the largest unit ever produced neared completion next to one of the smallest broaching machines ever made. Both were continuous surface units.

The large continuous machine was a 42.5-ton giant with a 29-ft-long bed, a 220-in. stroke, and a 40-ton broaching capacity. It performs nine separate operations on 7½-lb, 12½-in.-long connecting rod-and-cap sets for farm-machinery engines. The pygmy of the pair was an 8.5-ton chain broach with a 2.5-ton capacity and a 20-in. stroke. It broaches 5-oz, 2.5-in.-long manual transmission shaft shifters in four different configurations for automobiles.

The key to the productivity of a continuous horizontal broaching machine is elimination of the return stroke by mounting the workpieces, or the tools, on a continuous chain. Most frequently, the tools remain stationary, mounted in a tunnel in the top half of the machine, and the chain-mounted workpieces pass underneath them.

Special broaching machines

Special broaching machines also fall under the general categories of internal or surface use, but beyond that it is difficult to classify the wide and often unique variety of special machines. Nevertheless, here is a sampling.

Sometimes it is impossible to bring the workpiece to the machine. This is particularly true in the marine, power-generation, construction, and airframe industries. Therefore broaching machine builders have designed portable machines that can be brought to the work.

A form of internal broaching called 'strip' broaching is used occasionally to effect large gains in productivity per machine and man-hour through reduced broaching time cycles. In strip broaching, the broach is returned directly through the hole just broached, immediately after the cutting stroke, eliminating the necessity for disengaging the broach tool from its pulling fixture. Broach life is reduced because the cutting edges rub against the work on the return stroke, but not to the extent where the overall saving derived from this technique is lost.

Internal broaching of helicopter rotor spar sections is an unusual special broaching application. In one instance a 24-ft-long workpiece had about ⅛ in. of 4153 aluminum removed around the periphery of the irregularly-shaped internal form by 35 progressively stepped broach sections. These were pulled through the workpiece one at a time by a special electro-mechanical horizontal machine with a 64-ft-long bed. Broach sections were semi-automatically loaded and unloaded from the pulling bar at the beginning and end of each stroke. One operator handled the entire job, riding from loading to unloading stations in an electric cart.

Fixturing is important

The forces on any type of broach fixture will probably exceed those encountered in any other machining process, simply because so many more cutting teeth are in contact with the work at one time in broaching. Fixtures are also important because of the tremendous cost savings they can produce by reducing work handling time and labor.

Nevertheless, the principal function of a fixture is to locate and hold a workpiece rigidly during the cutting stroke of a broach tool. Other functions—such as guiding the tool, speeding loading and unloading, or coordinating the broaching machine with other machines—are all secondary.

One big trend today in fixture design is the automation of fixture action to assist in integrating broaching machines into transfer lines and other automatic machining systems. A second trend is toward universal fixtures that can hold similar, but not necessarily identical, workpieces.

But fixture design is basically the job of the machine-tool builder; the user need only provide the necessary dimensional, machining, and production data for the job.

Cutting fluids for broaching

Cutting fluids are used to reduce cutting temperature, lubricate the cutting tool, improve surface finish, increase tool life, remove chips, or a combination of these purposes.

In broaching, temperature reduction is important because tool life is quite sensitive to temperature changes during cutting. And there are several benefits to be gained from reducing friction: heat generation and wear at the tool/chip interface are reduced, and the shear angle is increased, thereby lowering the shearing work done and the heat generated along the cutting shear plane.

An internal broach tool may receive a generous supply of cutting fluid upon entering the workpiece, but after the tool has entered the work, the flow of fluid is retarded. This has been noted in some horizontal broaching operations where surface finish and cutting tool life are good at the starting half of a horizontal internal shape but poor at the final half.

During horizontal internal broaching, the flow of cutting fluid into the interior of a workpiece is restricted by the cutting teeth. Fluid trapped between tooth spaces flows by gravity to the lower half of the tool; the upper

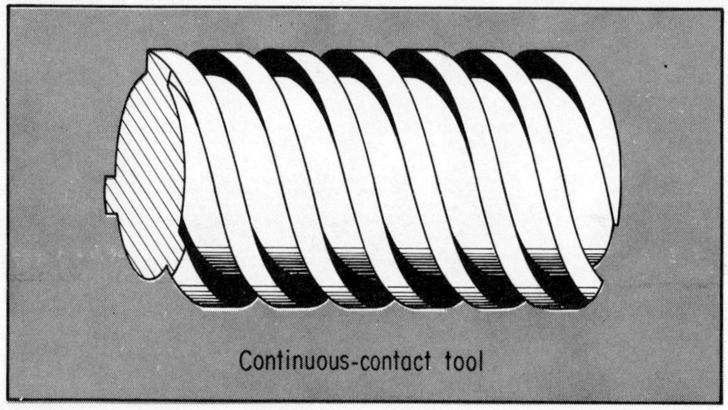

Continuous-contact tool is used on short lengths of cut and laminations when wide variations in tool pressure cause standard tools to chatter

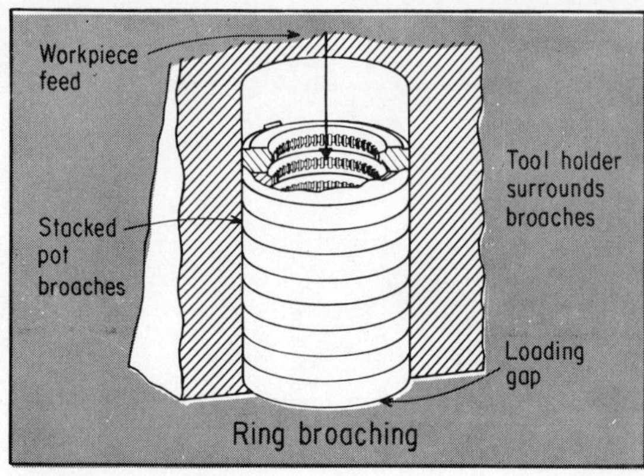

Ring broaches and pot fixtures combine to cut workpieces on a short time cycle

teeth may be cutting dry in a very short time. In some instances the problem can be solved by submerging the workpiece in cutting fluid during the entire broaching operation.

When broaching long internal shapes, such as rifle barrels, high-pressure streams of cutting fluid can be forced through the entire bore and around the broach. This also helps flush chips out of the cutting zone.

High-temperature alloys cause special problems, for cutting forces are higher and more heat is generated. The best approach is to reduce the cutting speed so that heat transfer by conduction can reduce tool temperature, and to use a cutting fluid that provides good lubrication at the tool/-chip interface to reduce heat generation. Most of the cutting fluids now used for broaching high-temperature alloys have a mineral-oil base with active chlorine and sulphur additives.

The usual approach to selecting a cutting fluid for a particular broaching operation involves trying several fluids to determine which gives the best performance. In most plants, 'best performance' is measured by a combination of surface finish and tool life.

The following list of materials and the cutting fluids used in broaching them should be considered as a starting point not as a recommendation:

Aluminum and alloys, aluminum and zinc diecastings: 9 parts kerosene to 1 part straight mineral oil; 7 parts kerosene to 5 parts straight mineral oil; a mixture of sulphurized mineral oil and sulphurized mineral lard oil.

Brass: 20-25 parts water to 1 part soluble oil; mineral lard oil.

Bronze: 5-15 parts water to 1 part soluble oil; mixture of sulphurized mineral oil and sulphurized mineral lard oil; 10-20 parts lard oil to 1 part straight mineral oil.

Copper, Everdur, Inconel, Monel, nickel: 5-15 parts water to 1 part soluble oil; mixture of sulphurized mineral oil and sulphurized mineral lard oil.

Wrought and malleable iron: 10-20 parts water to 1 part soluble oil; mixture of sulphurized mineral oil and sulphurized mineral lard oil; 10-15 parts lard oil to 1 part straight mineral oil; mineral lard oil.

Low-carbon and free-cutting steels: 10-20 parts water to 1 part soluble oil; mixture of sulphurized mineral oil and sulphurized mineral lard oil; 10-15 parts lard oil to 1 part straight mineral oil; mineral lard oil.

Medium-carbon and tough low-alloy steels: 10-15 parts water to 1 part soluble oil; mixture of sulphurized mineral oil and sulphurized mineral lard oil; 10-20 parts lard oil to 1 part straight mineral oil; mineral lard oil.

High-carbon high-alloy steels, including stainless steels: 5-10 parts water to 1 part soluble oil; mixture of sulphurized mineral oil and sulphurized mineral lard oil; 10-20 parts lard oil to 1 part straight mineral oil; mineral lard oil.

Broach sharpening

This section deals only with the sharpening of high-speed-steel broaches. Not only are they the most common broach types, but the other principal cutting edge, carbides, now are used in throwaway forms.

The original grinding is the responsibility of the broach producer. Resharpenings are the responsibility of the user, but the broach tool may be returned to the producer for resharpening in the producer's plant.

Most broaches will be good for at least a half-dozen resharpenings. However, the high-speed tool steels used in making broaches include some of the most difficult-to-grind steels known. For this reason it is not unusual for a second-choice tool steel (in terms of tool life) to be chosen over a slightly better steel if the first choice is extremely difficult to grind. This decision frequently arises with abrasion-resistant M-4 or T-15 high speed tool steel. The choice of a less-abrasion-resistant grade that costs less to resharpen often will more than offset any cost savings gained from the longer tool life of the M-4 or T-15 tool.

Internal broaches are sharpened by grinding them only on the face; metal removal on the top of the teeth changes the dimensions of the broached surface. Grinding on the tool face requires a small grinding wheel inclined at an angle greater than the face angle because of the geometry involved.

Surface broaches normally are resharpened on the top, but may be reground on the face of the teeth if excessive wear lands exist. When this is done the original dimensions may be re-established by shimming the broach on its holder. Care also must be taken to regrind the gullet space to the original tooth depth so that adequate chip space is maintained. Chipbreaker notches must be reground.

Grinding wheels suitable for broach resharpening are mostly of the vitrified aluminum oxide type, usually with grain sizes between 46 and 100.■

Fundamentals of grinding

It's not the same old grind anymore. Here's a brief review of the operating principles behind the more popular kinds of machines, and an update on the latest technology

Even though the process of grinding—the use of natural or artificial stone to cut metal—traces its origins back 6500 years ago to the Bronze Age, modern techniques didn't get started until just a little over a 100 years ago.

Rather than being created or invented, the forerunners of today's machines generally evolved from earlier methods. But remarkable progress, as illustrated by the emergence of basic designs and of patents issued, was made in the 20 years starting in 1860. Before that, the Barker & Holt machine of 1853 was used in machining balls required in steam-engine valves.

Even the early grinding machines—including the

By Joseph Jablonowski, associate editor

1867 Pratt & Whitney grinding lathe for external cylindrical work—came relatively late when compared to such other basic machines as the boring machine, lathe, miller, shaper, and drillpress. The kinds of parts that were machined and the rough tolerances and finishes required in the early days of the Industrial Revolution meant that machine builders weren't pushed into the manufacture of high-precision grinding units until the latter part of the 19th century.

Grinding machines in use, by SIC class

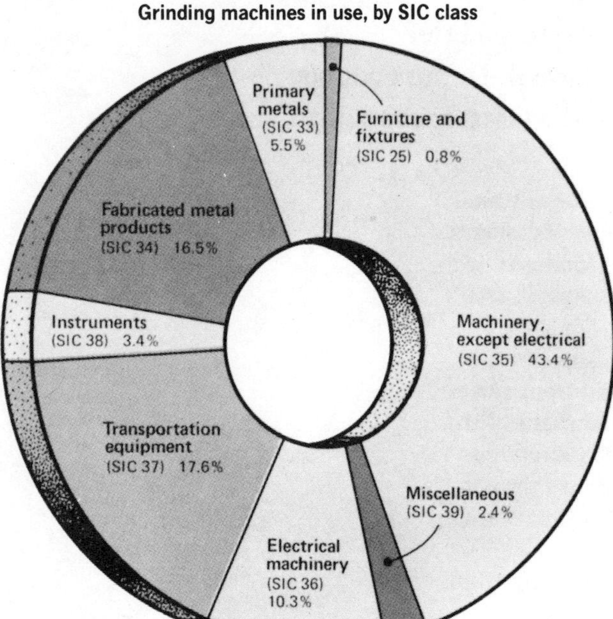

If the precision grinder was a latecomer among machine tools, the centerless grinder was even a later-comer among grinders. Though we can find evidence of such units as early as 1864—Englishman Henry Dyson's machine—centerless grinding as we know it today wasn't well developed until the 1910s and 1920s, when Heim, Detroit, and Cincinnati started marketing their machines. However, ball grinding, a variation of centerless grinding, was developed in the 1880s, by Fischer in Germany and by Richardson in the US.

The first US patent for a precision surface-grinding machine was issued in 1873. This was followed three years later—just 100 years ago—by the introduction of Joseph R. Brown's universal machine.

Since those early beginnings, much progress has been made in the fields of machine design, abrasive technology, and controls. But the early machines set the stage for the beginning of precision mass production, and especially helped the automobile industry get off on a firm footing.

The transportation-equipment industry in this country, in fact, continues to be one of the prime users of grinding machines. Car, truck, ship, and aircraft-building firms own about 34,500 grinding machines.

Where the machines are

The transportation segment of industry, however, is not where the bulk of the US's grinding machines are installed. Most of the units, some 190,000 of them, belong to companies in SIC 35 (nonelectrical machinery). A further breakdown of where grinding machines are installed comes from the American Machinist 11th Inventory (chart, above).

Within those overall Standard Industrial Classifications, heaviest users of grinding machines are firms within the group designated, "Machinery, except electrical." The aerospace and automotive companies followed (table, bottom).

The 11th Inventory, taken in 1973, showed that there were a total of 489,600 grinding machines of all types used by US industries. Of that total, 36% were less than 10 years old, 38% were bought 10 to 20 years earlier, and 25% were older then 20 years.

The Inventory further broke down the total population of grinders into types of machines. It identified external cylindrical, internal, disk, tool & cutter, bench, surface, belt, and other types of machines. (The proportion that each type of machine contributes to the total US grinder population is shown in the chart on the facing page.)

Numerical control does not figure importantly in grinding at this time. Only 180 units, of the total of 489,600 machines identified, had some type of NC. The bulk of those commanded by these more-sophisticated controls were reported to be surface grinders or in the "other (including jig grinders)" classification.

Grinding machines continue to be a major factor in the US machine-tool industry. In the past ten years, grinding and polishing machines annually accounted for over 18% of the total dollar value of all machine tools shipped, according to figures published by the Commerce Dept. Figures for shipments of grinding and polishing machines in 1974 show that some $248-million worth of the units were shipped. (The Commerce Dept does not separate grinding machines from honing, lapping, and polishing units, but it is estimated that grinders represent over 75% of that total.)

Production vs toolroom

Like most other types of machine tools, grinders can be used either for short runs or in high-production applications. Whether you classify a particular machine job for one type of operation or the other is probably bound up in the type of metalworking operation in which the shop is involved. To complicate this dichotomy further, many toolroom operations—diamond-wheel cutting is a good example—have moved into the field of production.

Production grinding encompasses a multitude of things.

Heaviest users of grinders, all types

Industry (SIC)	# machines
Special dies, tools & machine-tool accessories (3544, -45)	55,217
Misc machinery, exc electrical (3599)	43,480
General industrial machinery (356)	29,339
Aircraft engines & parts (3722, -23, -29)	27,585
Automotive parts (3714)	27,292
Fabricated structural parts (344)	25,521
Special-industries machinery (3551–55)	19,705
Machine tools (3541, -42, -48)	18,657
Complete motor vehicles (3711, -12, -13, 379)	15,799
Screw machine products, fasteners (345)	13,157
Ferrous & nonferrous foundries (332, 336)	12,523
Misc manufacturing industries (39)	11,734
Cutlery, hand tools, hardware (342)	11,542
Construction, mining & oilfield equipment (3531, -32, -33)	11,390
Heaviest-users total	**322,941**
% of total grinding machines	**66%**

Source: *American Machinist 11th Inventory*

According to Robert Houston, product engineer with Norton's development department, "If you can't walk away from it, it isn't production grinding." This implies both the capabilities and the potential problems of exploiting small-scale techniques on a large manufacturing scale.

More generally, this transfer involves automatic operations in the manufacture of tools or parts. A comparison of manufacturing and toolroom operations illustrates some of the differences. Toolroom work, is characterized by such things as a variety of short runs, a variety of setups, small shapes, small contact area, short contact time, infrequent truing, low metal-removal rate, small grit range, and dry (as opposed to coolant-flooded) grinding. Manufacturing-grinding operations, on the other hand, are likely to include such characteristics as long production runs, standardized setups, large and small shapes, large contact areas, long contact times, frequent truing, high metal-removal rate, full grit range, and the predominance of wet-grinding methods.

But because of the increasing crossover of toolroom techniques to manufacturing operations, and because many of the newer high-production techniques involve applications that cannot be simulated in the laboratory and must be proven in the field, it's becoming increasingly difficult to draw the line between the two. We simply cannot talk about one type of grinding as being limited to low-production applications and dismiss it out of hand. Similarly, some of the newest production techniques are initiated in the toolroom.

The grinder population, by type of machine

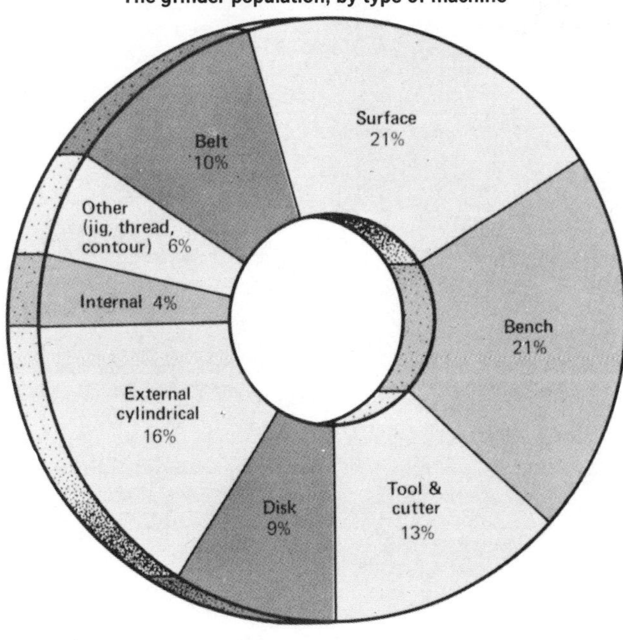

Traverse vs plunge

Just as grinding machines can be considered toolroom machines or production machines according to their use, they can also be separated into two types according to how their abrasive wheels are fed into the workpiece. Depending on how much of the workpiece stock is removed in a single grinding pass, machines operate by either the traverse principle or the plunge principle.

In traverse grinding—which is by far the most common type—the workpiece is carried in a fixture or a chuck, or between centers on the machine, and is presented to the cutting face of the wheel only one portion at a time. In order to grind the entire face of the workpiece, whether flat or cylindrical, the wheel must make several passes; each pass cuts only a fraction of the total material to be removed.

In a flat surface grinder, traverse grinding is typified by the reciprocating motion of the table. In cylindrical grinding, the wheel is moved back and forth across the face of the rotating workpiece, or the workpiece is oscillated across the wheel.

Plunge grinding, on the other hand, removes all or most of the stock in a single feed motion. The workpiece is held rigid with respect to the position of the wheel, and the wheel is continuously advanced at a constant rate until the desired amount of stock is removed. Limiting factor for the length of the workpiece section that may be ground in plunge grinding is the width of the grinding wheel itself.

Plunge grinding as a basic approach is normally used when the finished workpiece section is to have a profile or a series of diameters, or for workpieces with a variety of planes. The wheel is dressed, often with a crush roll that has a profile that is the inverse of the contour to be impressed on the wheel, and the duplicate of the desired workpiece contour.

In some situations, a given workpiece may be ground by either method. The guiding principle in the choice between traverse grinding and plunge-feed methods is that the former usually produces a better finish, while the latter produces completed shapes faster.

Flat vs OD vs ID

So grinding machines correctly can be thought to be either toolroom (limited-production) units of high-production units, and to operate either by the plunge-feed or the traversing method. These distinctions generally apply to all types of grinding machines. The approach that a particular grinding machine brings to cutting the workpiece depends more on the specific requirements of the individual part.

Rather than focus on the specific motions of individual machines, this special report will discuss various types of grinding machines according to the type of workpiece surface they are designed to handle. Basically, there are three types: Grinding machines can be made to cut flat surfaces, to cut external surfaces of rotating workpieces, or to cut internal diameters of holes.

This is a distinction that is not necessarily followed by the machine builders themselves. In some cases, a grinding-machine builder will specialize in only one type of machine; in other cases, one firm will produce a variety of machines to handle all surface shapes, and in rare instances, the grinding function will be incorporated into units designed primarily for other types of cutting—as in a recently introduced vertical boring mill that carries a separate grinding head.

So there are other categories on grinders, too. But it is beyond the scope of this report to discuss such specialized machines as contour grinders, thread grinders, and tool and cutter grinders. Similarly, snag grinders or simple bench wheels will not be included in this primer.

After discussing the basic operating principles behind the three general types of machines that perform the grinding function, this report will focus on the grinding wheels themselves, including some of the latest advances in standard abrasives and their bonding, and in the newer diamond and superabrasive wheels.

The final section of the report, "Fundamentals of Grinding," will cover the dynamics of the grinding interface itself, an area where much of the latest research is focused. The interplay of grain formation, chip size (for grinding wheels do indeed make chips, albeit small ones), and the role of coolants is crucial to increased productivity for grinding in the future.

Surface grinding

In a sense, of course, all grinding is surface grinding. But the term has come to mean the grinding of flat surfaces, as opposed to cylindrical ones.

Depending on the workpiece size and shape, many types of machines are available to grind flat surfaces. Since these differ in their basic orientation of grinding wheel to the workpiece surface, a review of operating principles is in order:

The most common type of surface grinder is the one that uses a **horizontal spindle** with a high-speed, **reciprocating table.** In this variety, wheel-spindle position is kept constant, and the table on which the workpiece is held—typically by a magnetic chuck—traverses longitudinally across the spindle centerline.

The other two travels in a horizontal-spindle, reciprocating table machine—the cross traverse and the vertical traverse (or wheel infeed)—can be achieved in a variety of ways. Cross travel—getting the narrow wheel to cut the full width of the workpiece—can be done either by moving the table on a slide or by moving the wheel through an overarm. Several manufacturers favor making this cross travel, or transverse movement, a function of the column's overarm; in this way, grinding forces are centered in the upright column. In any case, cross feed should never exceed one half the wheel face width per revolution.

Vertical travel, similarly, can be accomplished either in the table or in the wheelhead; in nearly all machines of this type, however, this feed function is performed by moving the wheel up and down. But the way the head is anchored into the machine's base is often a selling point of a particular machine design; some types of support are said to be more rigid than others.

With the cross travel of a horizontal-spindle, reciprocating-table grinder locked firmly into place, the machine can be equipped with a profiled wheel for creep-feed contouring. In this application especially, machine rigidity is of primary importance.

With a tilted headstock, machines of this type can be used with shaped cross-section wheels to cut sharp corners, shoulders, dovetails, and slots, with the angular face and the periphery of the wheel.

Horizontal-spindle, reciprocating-table machines are available in a wide range of sizes, from toolroom units with longitudinal travels of perhaps 16 in. to machine-way grinders with travels of 300 in. and more. Feeds can be provided by hand, hydraulic motor, or electric motor. In all varieties, the basic operating motions remain much the same.

A variant of the conventional reciprocating-table type of machine is the one with a **horizontally mounted wheel spindle** but equipped with a **rotary table** to carry the workpiece.

In this rotary-table version, generally used for small workpieces, the parts are transported in opposition to the direction of the cutting wheel, at speeds that generally run from about one-eighth to one-fifth of wheel speed. Parts are clamped by any one of a variety of methods, including being held magnetically (magnetic chucks are available with either radial poles or concentric poles) or being held with face plates or retaining rings.

Whereas a reciprocating-table machine generally has a head-tilt capability, rotary-table designs feature table tilt. This enables a flat-faced grinding wheel to cut dish-shaped workpieces to flat, concave, or convex cross-sections.

On one type of machine, the work is finished by a single pass under the wheel, the size being automatically controlled by devices that maintain the wheel face at a fixed distance above the rotating chuck. In another version, small workpieces are loaded on a separate rotating table, from which they are guided onto the grinding table. After the work passes under the wheel, other guides lead it off the table down a chute to a demagnetizer.

Just about all horizontal-spindle surface grinders are built to abrade the workpiece with the periphery of their wheels. Vertical-spindle machines, on the other hand, use the flat face of their wheels to cut the work. This basic fact has an important implication for speeds and feeds in this genre of machines: Since there is a greater contact area between the wheel's flat surface and the workpiece compared to peripheral-cutting wheels, surface speeds are generally lower.

In the **vertical-spindle, reciprocating-table** type of surface grinder, built specifically for rapid stock removal at high horsepower, the spindle is often designed to tilt, in order to keep table alignment true. One of the advantages of the vertical spindle is the ability to cut a cross-hatched pattern.

A variation of the vertical-spindle machine is the type that uses a through-feed belt, which traverses individual workpieces under the rotating cup wheel.

But the majority of **vertical-spindle** machines by far use **rotary tables.** Like the rotary tables used with horizontal-spindle machines, the ones on vertical-spindle units are able to clamp workpieces magnetically, with retaining rings, or with face plates.

Further, the rotary tables on vertical-spindle machines are often able to tilt. The specific methods used to mount the tilting table vary from one manufacturer to another, but most use a three-point column support for maximum rigidity.

In tilt grinding with a vertical-spindle machine, only the leading edge of the wheel is in contact with the workpiece. This results in a greater unit of force on the cutting abrasive, and causes two grinding conditions to change: First, the wheel will act softer. In cases where wheel loading or glazing occurred in flat-grinding operations, tilting the table enhances wheel breakdown. Second, tilting the table cuts power consumption, allowing higher feedrates. This is a useful feature for cutting ductile materials. But while the orientation of the rotary table itself and the method by which individual workpieces are loaded onto it are fairly standard, the orientation of the two rotating components offers some peculiar advantages to this method of grinding.

Work motions in the vertical-spindle, rotary-table grinder generally give good wheel action and offer maximum machining capabilities. Further, the work, whether in a single piece or a cluster of smaller pieces, revolves in a continuous motion; there is no "end" to the cutting operation. All parts chucked in the same setup receive equal treatment by the nature of the machine's design.

One factor to remember about rotary worktables is that not all work surfaces move across the grinding wheel with the same speed. Pieces mounted toward the outside edge of the circular table travel faster than those mounted closer to the center. In designing a vertical-spindle machine with a rotary table, the double-rotary motion tends to compensate for such uneven workpiece travel, depending on the path the wheel sweeps across.

In order to get proper cutting and wheel-wear characteristics, machines of this type are designed so the inside edge of the cylindrical grinding wheel is mounted above the rotary center of the circular table. As a workpiece is carried under the wheel, it first encounters the wheel portion that is moving from the outside of the rotary table to the inside, from right to left. The workpiece next passes through the empty section of the wheel, then encounters another section of the cup-shaped wheel, roughly a third. Here it is ground in the opposite direction: from the inside of the worktable to the outer periphery.

Finish grinding and sparkout with the double rotary motions of this type of machine provide a cross-hatched surface, because both the leading and trailing edges of the grinding wheel are in contact with the workpiece.

Wheels used in vertical-spindle machines may be made in segments, held in a special adapter, or made in the form of a shallow cylinder. Some of these have wire or steel strap winding around the periphery. These windings should be kept in place until the wheel has worn down to them. When they are removed, care should be taken to avoid damaging the abrasive surface of the wheel.

For high-volume applications, machines of this type are often equipped with automatically controlled tables and multiple spindles. The spindles are so arranged that the first one the workpiece encounters does roughing work, while the last ones do the finish grinding and sparkout.

A variation on vertical-spindle grinding machines is the type built to operate with a **swivel stroke.** In units of this type, the workpiece is mounted securely in a nonmoving table adjacent to the upright support for the counterbalanced swiveling wheelhead. The grinding wheel descends to, then feeds into the workpiece, while the headstock swivels back and forth across the surface.

Swiveling is usually done through a 170-deg arc, and machines of this type can be equipped with back-to-back worktables, for setting up a second workpiece while the first is being ground. After the first part is completed, the spindle is rotated around the support to grind the second part.

As with other vertical-spindle machines that use upside-down, cup-shaped wheels, coolant is introduced through the spindle into the grinding interface. Rotary tables are also available on swivel-action machines, enabling them to cut cross-hatch patterns.

One of the more interesting designs in vertical-spindle surface grinders is the **orbital version.** In this type, the cylindrical wheel (once again cutting with its face rather than with its periphery) is held to a central spindle by a gearing system. The spindle is centered on the workpiece surface; the wheel describes an orbital pattern as it revolves and rotates about the spindle. A normal feeding system is used, but it is the motion of the wheelhead itself that provides the unusual action of the wheel.

Disk grinding is the process of flat surface grinding in which the work is passed over the face of the abrasive member. It differs from most other types of surface grinding, in which the rim or periphery of the wheel does the cutting. Wheels used in disk grinding are commonly referred to as abrasive disks, and are usually made from bonded abrasive, although paper or cloth disks coated with abrasives and mounted flat on supporting plates are also used.

Disk grinders come in both single- and double-spindle types, and with either horizontal or vertical spindles.

The simplest versions of this type of machine uses a **vertical-spindle single disk,** which cuts a hand-fed workpiece. These single-spindle disk machines are provided with worktables mounted at a right angle to the spindle or with swinging fixtures to carry the work past the abrasive.

Double-disk machines are extensively used for production grinding on both sides of small workpieces. In these machines, the abrasive surfaces of the two disks are mounted in opposition, and flatness and parallelism can be closely monitored on a high-volume rate.

Feeding methods of double-disk grinders are classified by the geometry, accuracy, and stock removal of the workpieces.

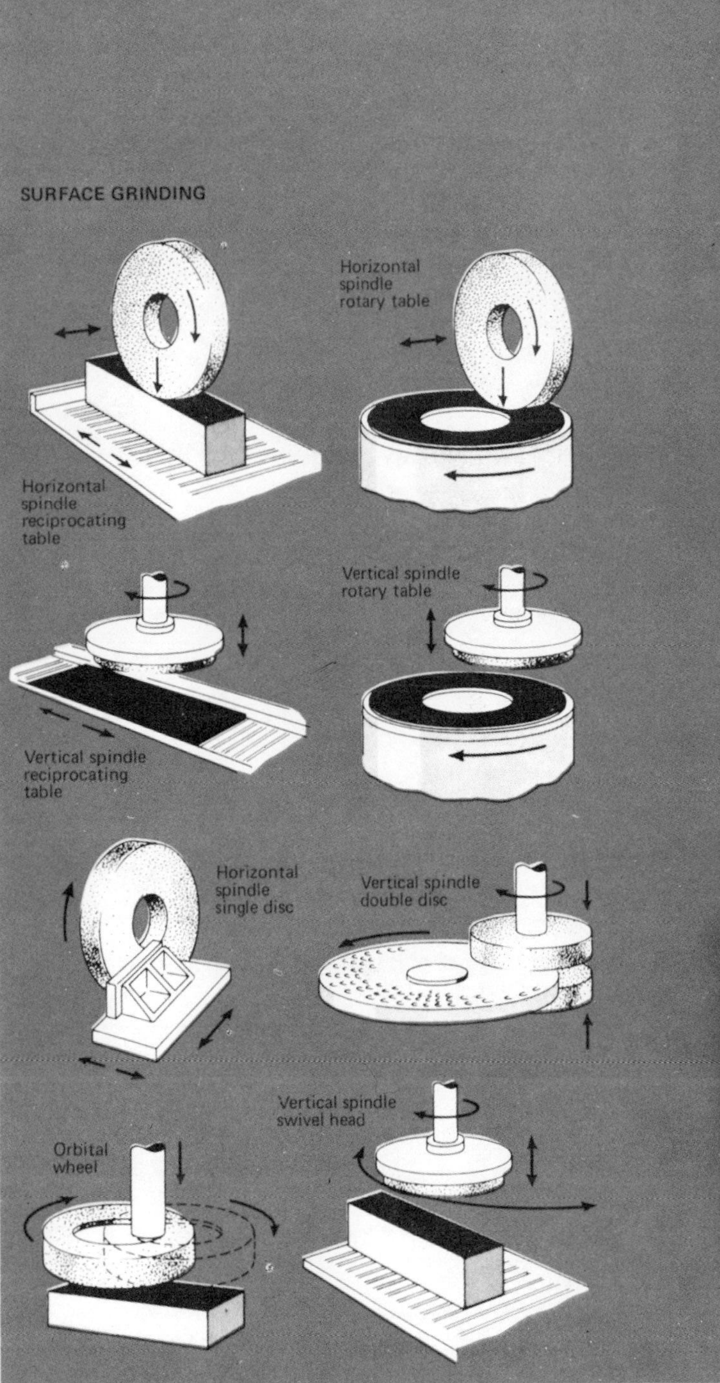

In the through-feed method, adapted for high-efficiency mass production of simple-shaped workpieces, individual parts are linearly fed through the center of the two spindles by a short, continuously running conveyor. The workpieces run through the disks—previously set to final dimension—unclamped, but guided between two rails. Parts fed in this manner are, typically, round or square shapes, whose length and width are at least three times the thickness to be ground.

A second feeding configuration for double-disk machines is the rotary carrier. This technique, used for small workpieces or ones of irregular shape, uses a carrier that is thinner than the workpiece and which carries the individual parts into the grinding interface. Automatic loading and unloading features are frequently seen in this type of macine.

Gun-feed and swing-arm methods are also used for feeding

parts into double-disk grinders. Both of these methods clamp the work in a thin holder, and slip it into and out of the double-wheel grinding area one piece at a time.

Grinding for flatness

One of the primary objectives in grinding with these machines is to achieve a flat surface, one that is free from waviness. All surface-grinding machines are capable of producing flat faces, but some units do it more easily than others.

Flatness is usually measured in the toolroom, with either a precision steel straightedge or a surface plate. When checking this characteristic with a straightedge, it's important to move the edge around all planes of the workpiece and check for daylight. Flatness measuring with a surface plate is generally done with a feeler-gage.

Many factors can contribute to lack of flatness in surface-ground parts. Stresses in the workpiece, caused by uneven cooling of castings and forgings, are often "unlocked" by grinding heat. Discontinuous surfaces, too, are more difficult to grind flat; as the wheel approaches a surface interruption, it tends to cut deeper. Thin and warped workpieces, similarly, are more difficult to grind to a good degree of flatness.

Often the solution is to feed the wheel into the workpiece with as light a cut as possible. Multiple passes under the wheel are often necessary to flat-grind discontinuous surfaces.

Thin sections that are warped and tend to spring up in the center may be pulled down by the magnetic chuck so that a false flatness is presented to the grinding wheel. When the workpiece is removed from the grinder, the section springs back up to its normal, bowed shape. These pieces are best ground by clamping them onto the chuck without magnetizing them. Or shims can be placed under the workpiece to support a bowed center section.

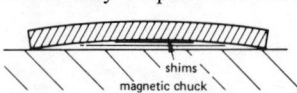

shims
magnetic chuck

Work that has been heat-treated should be rough-ground to relieve internal stresses. Should the part become misshapen after the rough grind, shims can be used.

Grinding for dimension

One of the most important dimensional characteristics usually sought in surface grinding is that of parallelism. Double-spindle disk grinders achieve this characteristic on a high-production basis with automatic size controls that provide continuous input to the two spindles. But with these, and with all other surface grinders, parallelism is a function of machine rigidity.

Various machine builders attack this rigidity problem in various ways, and such factors as cost, production rate, and basic component orientation play an important part. One midwestern machine supplier, for example, plans to introduce a concrete-based horizontal-spindle grinder this month in an effort to adapt this approach—already used in lathes to some degree—to grinding.

Sometimes lack of parallelism is caused by imperfections in the magnetic chuck itself. Between jobs, the chuck should be demagnetized, and the face rubbed lightly with a stone to remove any burrs. This will erase scratches and restore a flat surface.

Grinding to a particular dimension can be made easier or more difficult by controls on the machine itself. Surface grinders are available with a wide range of size-controlling devices, ranging from simple graduated handwheels to elaborate electronic positioning circuits with digital readouts.

Precision-grinding a flat surface to a specific dimension, then, is a function of the machine's rigidity and the controls it uses. The conventional test for accuracy uses a pencil mark on the ground surface. This simple check determines whether the machine is capable of accurate cuts. After the wheel has cut the surface and downfeed has stopped (when sparks are no longer produced), the pencil mark is placed on the surface. The wheelhead is raised, then lowered again to the last setting. The mark should be erased with a single pass of the wheel, indicating a 0.0001-in. repeatability.

Grinding profiles

Profile-grinding produces a form or contour rather than a flat or parallel face on the workpiece surface. This form may be a curve, an angle, or a tooth shape.

As with other types of machines, there are alternative methods for cutting forms on a grinder. Take the case of a V-shaped notch in the surface of a flat workpiece. It may be formed with a V-shaped wheel mounted in a horizontal-spindle machine, by mounting the workpiece on an angular workholder and cutting with the flat periphery of the wheel, or, in some machines, by tilting the wheelhead or the table.

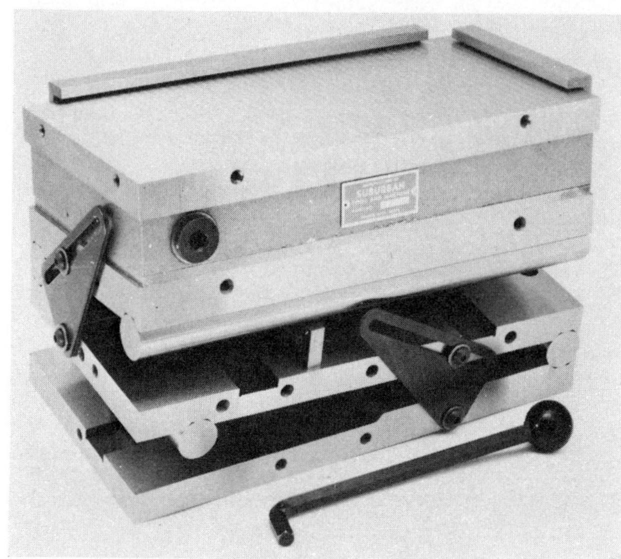

Magnetic chuck includes sine-plate angle grinding

More complicated forms, however, are normally cut with grinding wheels that are shaped—either by the wheel manufacturer or by the user.

Two very different methods are used to shape a flat wheel or dress a preshaped wheel: a single-point diamond tool or a crush roll. For simpler angles and radii, fixtures hold a diamond tool in a predetermined position and bring it across the wheel in the desired pattern. For more complicated cross-sections, a pantograph is used to describe the shape.

The crush roll is an increasingly popular method for shaping a horizontal-spindle grinding wheel to cut contours. The roll is shaped to the desired workpiece profile and is forced against the revolving wheel, crushing the corresponding shape into it. Crushing rolls may be of the idler type, in which the abrasive wheel drives the roll, or of the power-driven type, rotated by its own small motor.

Since the grinding wheel rotates more slowly while it is being crush-shaped than it does during actual grinding, the machine's spindle-speed control must be capable of at least two speeds.

Wheels trued by crushing cut faster and run cooler than those trued with a diamond. Crushing produces a wheel with many sharp, pointed grains, while diamond truing tends to produce many grains with flat surfaces. But the duller grains can often give a better surface finish.

Cylindrical grinding

Cylindrical grinders, as the name implies, are used for most external grinding of cylindrical parts. These units will cut straight cylinders, workpieces having more than one diameter, and tapered parts. Workpieces having irregular profiles are classed as cams or eccentrics, and are cut on a cam grinder, or a cylindrical machine with special attachments.

There are two distinct types of outside cylindrical grinders; those that hold the workpiece and rotate it about its center, and those that operate by the centerless principle. The former class includes plain between-centers-type machines, universal machines, and chucking-type machines. The latter class includes regular centerless grinders and those that operate on a throughfeed principle.

One of the more common cylindrical grinder types is the **between-centers** machine. In it, the grinding wheel is mounted on a cross slide perpendicular to the work axis. It is fed to or from the work either by hand or automatically, and sometimes is fed parallel to the workpiece. On most types, the headstock and tailstock are mounted on an auxiliary swiveling table carried by the machine table. This swiveling-table function permits taper grinding on the workpiece.

The workpiece is held between centers, as in a lathe, and steadyrests are sometimes used to support the work. These rests are located at suitable intervals—generally six to ten workpiece diameters apart along the work—to oppose the outward and downward pressure of the wheel. As grinding progresses, rests must be adjusted to take up reduction in work diameter. This is done by manipulating the rest's shoes, which in production grinders, are usually made of carbide.

A smaller motor rotates the workpiece in opposition to the wheel rotation, and cutting occurs when the separate feed motor is engaged. This feed may be either continuous to the full cutting depth—as in plunge cutting—or in incremental amounts for traversing the workpiece OD.

In addition to plunge grinding and traverse grinding, functions that are also found on other types of grinders, some cylindrical machines can be controlled to perform a combination of motions called "peel grinding." In this technique, which uses programmed cycles within the machine control, the wheel is plunge-fed almost to the finished size. The table is then traversed with the wheel still cutting at the depth to which it was originally plunged. This action peels a layer of stock away from the workpiece, and cutting is primarily done with the flat face of the wheel. The cycle is repeated as necessary until the finished size is achieved.

A variation on the between-centers type of machine is the one equipped for **chucking** grinding. Instead of centers, the machine is equipped with a single chuck, permitting greater workholding flexibility on shorter parts.

Chucking OD grinders can handle a variety of work, and like between-centers machines, may be used with steadyrests or center supports. Tubular-shaped workpieces ground with a chucking-type machine require careful mounting, however. A standard three-jaw chuck can distort the OD of a thin-walled tube and cause the wheel to grind flats into the diameter. Using a different chuck, a magnetic one perhaps, can eliminate this undesirable effect.

Of course, cutting flats into a workpiece is sometimes not so undesirable. One manufacturer of bearing races recently had to cut three "windows" into thick tubular parts and finish the OD at the same time. He did this by locking the chuck at 120-deg increments and plunging the grinding wheel into the workpiece until it cut holes into the shape. Unlocking the chuck rotation and adjusting feedrate then finished the OD.

Although some jobs must be held on a faceplate, in a chuck, or in a special fixture, most cylindrical work is held between centers. Wherever possible, two dead centers are used, but live centers are preferable when the work is such that it would score either center.

Variations on the basic external-cylindrical grinder that holds the workpiece at its center are numerous. One variation places emphasis on the control of motions: By equipping the machine with two independent variable-speed drive systems—one turning the workpiece, the other traversing the grinding head—one roll grinder can precisely adjust motions to suit individual workpiece materials.

Many grinding operations, particularly angular-approach plunge grinding, require the workpiece to be positioned axially relative to the grinding wheel. Axial positioning, featured on some machines, compensates for unequal center depths or for inaccurate pre-machining. The workpiece is displaced axially by a center that is mounted in preloaded bearings in the workhead spindle. This displacement can be made either by hand-operated feedscrew or by a servomotor. The correct position is determined by a dial indicator, a measuring instrument, or the indicator scale of a size-control unit. With the servomotor feed and a size-control amplifier, automatic positioning is possible.

Another variation changes the wheel-infeed orientation. In this roll grinder, the wheelhead tilts into, rather than slides into, the workpiece. The system is free of backlash, permitting more-positive wheel movement in either direction. Virtually free from sliding friction, the machine is readily adaptable to automatic controls.

But probably the most important variation on the basic de-

Photo: Jones & Lamson

Held between centers, part can be plunge-cut or traversed

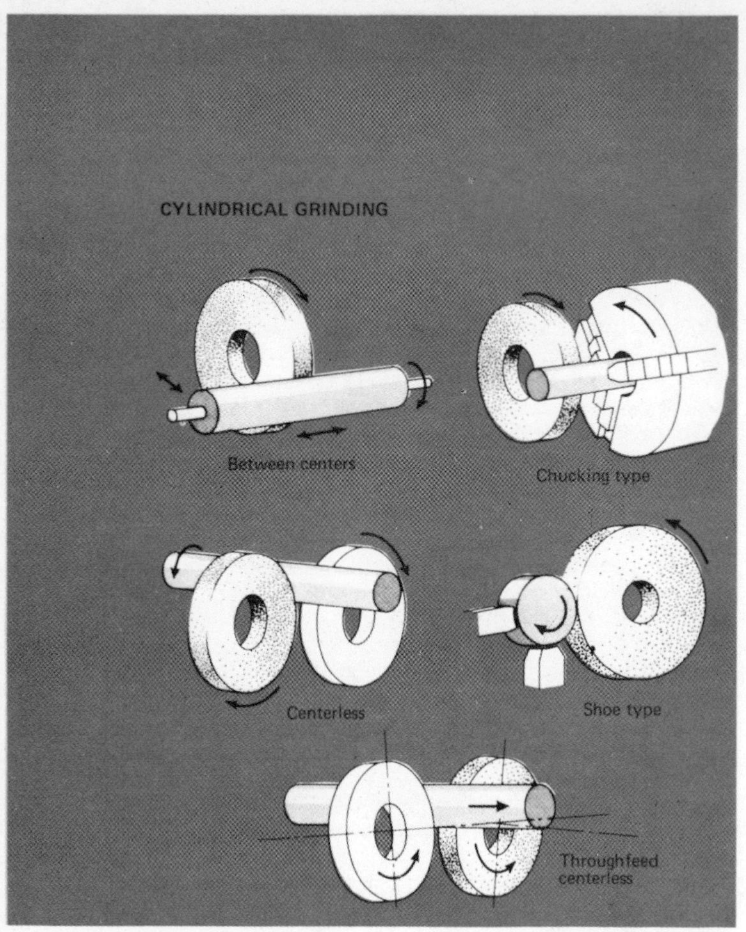

sign of the center-holding OD grinder is the **universal** machine. This grinder has a wheelhead that can be swiveled at an angle to the ways, and which can usually be arranged to do internal as well as external work.

Other universal machines use three spindles. One recent German import, for example, mounts all three—for cylindrical, face, and internal work—on a single grinding head that swivels 270 deg with positive 90-deg bayonet stops. One of the advantages of this type of grinder is that all machining operations can be done in a single setup.

The basic center-type external cylindrical grinding machine is adaptable to many types of production problems. By mounting two workpieces between centers, one below and one above the grinding-wheel centerline, multiple parts may be ground simultaneously. This, of course, shortens production time.

Multiple-wheel grinders are commonly found in use when a workpiece has multiple diameters. The workpiece is held between centers or chucked, and a corresponding number of wheels with preset diameters are fed into the part with no traverse movement. A typical application for this type of machine is in grinding the bearing surfaces of a crankshaft or camshaft.

Similarly, center-type grinders can be used for form grinding. A formed wheel, shaped to the negative image of the desired workpiece profile, is plunged into the part, producing the shape.

Like those used on surface form-grinding machines, the wheels on cylindrical form grinders can be shaped or dressed with a variety of methods: Crush-roll dressing uses a hardened, small-diameter roll that is either motor-driven or held in an idler, and is fed into the grinding wheel to shape it. Steel rolls are recommended for production runs of from 100 to 100,000 parts; for even higher production runs, machine builders recommend use of diamond rolls for dressing the wheel. Another dressing technique is the use of a templet and a single-point diamond tool. This method is usually recommended for low production runs because of the longer time it takes to form the abrasive wheel, and for parts that have relatively simple contours.

Still another variation on the basic between-centers-type machine is the polygon grinder. Although this is not strictly a cam grinder, which cuts asymmetrical lobes, the machine does use variable eccentrics and cyclical feed motions.

Polygon shapes can be used to advantage in load-carrying shafts whose torsional characteristics often exceed those of splines, keys, or flats. The polygon grinder, although limited to specialized applications, can produce these standard lobed profiles economically and to a high finish.

The machine supports the grinding wheel and workpiece spindle in parallel guides, which are moved in an elliptical path controlled by the horizontal movement of a connecting rod and by the vertical motion of the spindle carrier. The shape and size of the path are determined by an infinitely variable eccentric. Thus, motions are transmitted to the grinding wheel and to the work carrier to produce the regular-polygon cross sections.

While some workpieces must, by their very nature, be held at their centers to be ground, other shapes may be conveniently cut by the **centerless** method.

Principal elements of the centerless grinder are the grinding wheel, the regulating (or feed) wheel, and the workrest blade mounted between them. Pressure exerted by the grinding wheel forces the work against the regulating wheel and the workrest. The grinding wheel operates at about 6500 sfm. The speed of the rubber-bonded regulating wheel can be varied, but is usually around 50 sfm.

The slower regulating-wheel speed, combined with its direction of rotation, means that, at the point of contact, both the grinding wheel and the work are moving in the same direction. The work has a slower surface speed, however, and is therefore subjected to abrasion.

This is termed "down-grinding" in a centerless machine. The grinding wheel and the regulating wheel rotate in opposition to each other, and the workpiece is driven with a positive rotation.

Down-grinding is by far the most-widely used type of centerless action. But in certain variations on the process, notably in grinding threads, down-grinding imposes higher stresses in

Photo: Stoffel

In this universal grinder, multiple wheels are mounted in a swiveling head to cut ODs, IDs, and flat surfaces

the crest of the grinding wheel's thread-generating ribs.

In the "up-grinding" process, developed for certain types of threading applications, the workpiece is rotated in the same direction as the grinding wheel. (If the wheel rotates clockwise, the cylindrical part also rotates clockwise.) With this relationship, the surfaces of the workpiece and the grinding wheel will move in opposite directions at the point of their contact. Stresses imposed on the crests of the ribs crush-rolled onto the grinding-wheel periphery are reduced, enabling the use of work-surface speeds up to 30% greater than for similar grinding conditions in "down-grinding."

All forces in the up-grinding method act downward toward the workrest blade. The grinding wheel turns toward the blade, as does the regulating wheel. The part is held firmly on the workrest blade, and the tendency to produce a chattered finish under heavy cuts is reduced. Control of work rotation and of endwise travel is obtained by the regulating wheel in this situation; it is independent of friction between the grinding wheel and the workpiece. Spinning is practically eliminated, especially in coarse grinding.

Up-grinding has been in use as a basic method for some time. In more recent years, however, many manufacturers found it to have highly specialized application, and its use has almost died out.

In practice, the technique was sometimes applied to normal centerless grinding. More often, it was used for **throughfeed** grinding.

By tilting the rotational axis of the regulating wheel with respect to that of the grinding wheel, the workpiece is also given a longitudinal movement past the abrasive. This is called throughfeed grinding.

In this method, the workpiece is usually hand-fed into the grinding gap, although automatic loaders and stack-type feeders are also used. Once in contact with the grinding and regulating wheels, the part is pulled through to the far side of the machine.

Throughfeed grinding is applicable for cylindrical parts that have straight, parallel sides. For workpieces with shoulders, heads, or tapers, an adaptation of the throughfeed principle can be used. This adaptation can be either end-feed or in-feed grinding.

In **end-feed** grinding, tapered workpieces are ground with grinding wheels that have been trued to the proper taper. The workpiece is fed in from the front until it hits a stop. It is then withdrawn from the machine. On another type of end-feed machine, the regulating wheel is swiveled to obtain the taper, and the workrest blade is set at an angle.

In-feed grinding is much the same as plunge-cut grinding. It's used for work having a shoulder larger than the ground diameter. The workrest blade and the regulating wheel are set in a fixed relationship. Work is inserted, and either the regulating wheel or the grinding wheel is moved in—depending on the make of machine—until the proper diameter is obtained. The wheels are then separated and the piece is removed by hand or by an automatic ejector.

In centerless-grinding machines, whether plain centerless, throughfeed, end-feed, or in-feed, the workrest blade may be made of high-speed steel, Meehanite, gray cast iron, or carbide-tipped steel. On small machines, its maximum thickness is usually ½ in.; on larger ones, ¾-in. blades are used with 4-in.-wide wheels.

The workpiece supporting surface is usually ground at an angle 30 deg to the horizontal, although smaller angles are sometimes used to reduce the pressure between the workpiece and the regulating roll—on thin-walled tubing, for instance. Greater angles are sometimes used to obtain faster rounding-up action. (For an explanation of just how a centerless grinding machine rounds a workpiece, see next page).

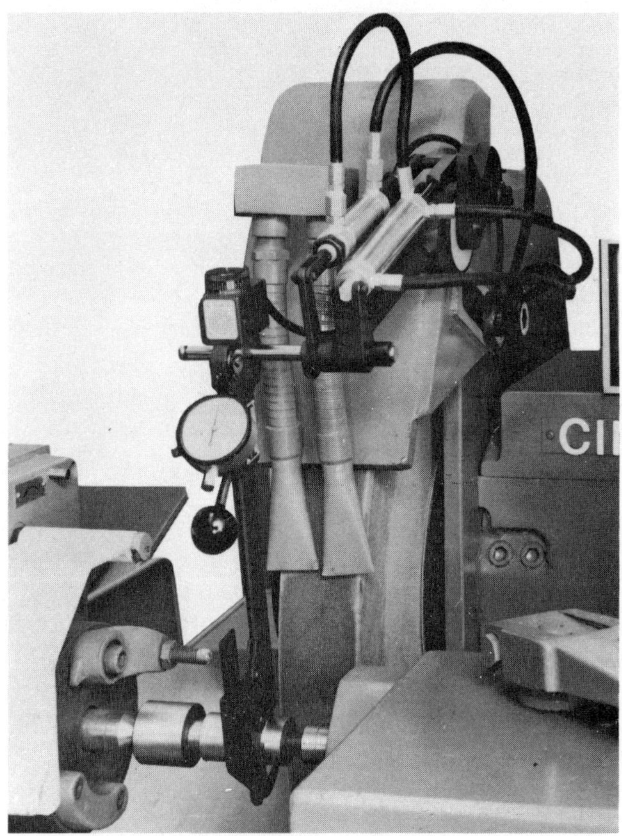

Workpiece size can be gaged in many ways. This bracket rests against the part and feeds back data to machine control

Within practical limits, the higher the work is mounted, the quicker the roundness will be produced. The limit is reached when the work begins to vibrate. Soft grinding wheels permit grinding higher above center than do hard wheels; for parts particularly difficult to round up, place the workpiece extra high and use an extra-soft wheel.

Another method for external grinding of cylindrical surfaces can best be described as being a hybrid between chucking-type grinding and centerless grinding. In this **shoe-type** grinding, characteristically used for bearing rings and similarly shaped workpieces, the parts are placed with their thrust face against, and driven by, an electromagnetic chuck rotating with the work spindle. The parts are also supported by stationary shoes, which are usually located at the bottom of the part, across from the grinding wheel.

Magnetic adhesion keeps the workpieces in place, but allows sufficient sliding play for their self-centering in the grinding process. This type of grinding avoids clamping the rings at their OD, and therefore also avoids the risk of deforming the workpiece circularity.

In shoe grinding, the work is located slightly eccentrically—i.e., below the work-spindle axis and toward the grinding wheel. The degree of eccentricity is determined by the grinding allowance of the work. Eccentricity is maintained throughout the grinding operation, thus assuring positive guidance of the workpiece.

The circular part, resting on the cemented-carbide shoes, turns with an opposite rotation compared to the wheel, and moves in the same direction at the point of contact with the wheel. This is the same motion used on other centerless-type machines. Work-rotation speeds are usually between 150 and 300 sfm. Speed of the grinding wheel is much higher, sometimes as high as 9000 sfm.

So circular parts may be ground any number of ways, including by methods that hold the workpiece and rotate it about its center, by techniques that do not fix the workpiece center but rather operate only by contacting the part OD, and by methods that tend to combine the two principles. In most cases, selection of a general type of method to grind outside diameters depends on a multiplicity of considerations like part size, part sturdiness, finish desired, production rate, final part shape, etc.

Because external cylindrical grinding is so commonplace, however, and because there are so many possible shapes that have to be cut, a wide choice of alternative methods and variations on these basic methods exists. Abrasive-belt stock removal is a typical example; this is enjoying increasing popularity lately. And one firm uses a four-headed machine that employs both belts and wheels to centerless-grind long lengths of barstock.

But even if there are no tremendous modifications made to a particular machine, some inventive process engineer is sure to dream up a way to slightly alter, say, a workrest angle, which will make remarkable improvements to productivity in his shop.

Rounding up in a centerless grinder

Centerless grinders, like most other external cylindrical machines, grind round workpieces from stock. But how?

Consider a hypothetical centerless with the workpiece riding on a level rest with its center lying in a straight line with the centers of both wheels. The workpiece has a high spot (shown in its lower left quadrant) to be removed.

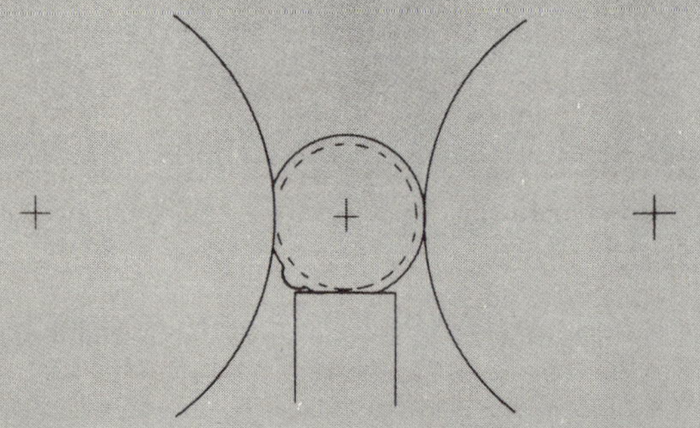

When the high spot hits the flat workrest, it raises the workpiece center, so the wheel misses part of its cut—the part that lies 90 deg behind the bump—and leaves another, smaller high spot there.

A quarter turn brings this second high spot to the workrest, again raising the workpiece centerline. But this time the original high spot is against the regulating wheel, shoving the workpiece against the abrasive to grind a concavity across from it.

The workpiece rests on this concavity after another quarter revolution, dropping its center below the grinding gap and producing another high spot, 90 deg behind it. By the time the original high spot reaches the cutting wheel, it is lifted out of the gap by this last high spot and is not given enough pressure by the regulating wheel, which now abuts a concavity.

Even though some rounding is achieved, the workpiece never gets perfectly round.

By moving the workrest up, so that the workpiece centerline is higher than the two wheel centerlines, and by tilting the face of the rest, so that its point of workpiece support is closer to the grinding wheel, the centerless process achieves a better degree of roundness.

This is primarily due to simple geometry. When the workpiece center is lifted, the extended tangents to the wheels at the workpiece meet at some point below the workrest. Research in Sweden by SKF, and in Russia, has shown that the optimum spread for this imaginary angle, drawn where the two tangents meet, is just under 10 deg.

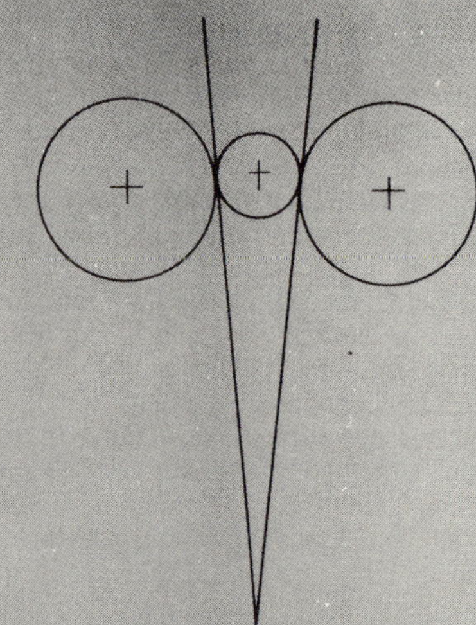

But moving the workpiece center up doesn't solve it all. A lump or bulge in the part still causes it to move up and down as it strikes the workrest. This vertical motion is greatest when the point where the workpiece touches the rest is horizontal and equidistant between the two wheels. By moving this contact point close to one of the wheels, the vertical component of the unround workpiece's "bounce" is reduced. A tilted workrest surface does this.

Also, when a high spot—of perhaps a few μin.—now reaches the regulating wheel, a low spot is still ground on the other side of the workpiece diameter, but not directly opposite the high spot.

High spots and low spots are ground into the diameter of the part, producing a polygon. But the highs and lows on the cylinder now do not complement each other, and the polygon—with a multitude of flats—approaches true circularity.

In theory, the angle of the workrest face is critical, but it also interplays with the workrest heights. Since it is often difficult to change workrests, general practice dictates raising or lowering the rest as part diameter and wheel wear dictate.

Internal grinding

The grinding of holes and internal shapes presents problems that are not always encountered with other methods–problems affecting both machine design and day-to-day operation.

Internal grinders are used to generate internal surfaces and to finish them accurately. They are used to bring holes to final size and shape and to eliminate any errors from previous operations, such as heat-treating or rough surfaces produced by roughing tools.

Straight or tapered holes, through holes, blind holes, or holes having more than one diameter may all be ground by this class of machine. In addition, internal shapes, and even flat sections such as bearing races (photo below), can be generated and finished with internal grinders. Machines are available to finish holes from a few thousanths of an inch in diameter up to 5 ft and more; dimensional tolerances are often held within a 50-μin. envelope.

Semiautomatic machines may be used in the toolroom or on a production line. They are equipped with power feed and traverse, but the cycle is manually controlled. Automatic machines are available, with fully automatic cycles in which the operator simply loads and unloads the machine, or fills a self-feeding hopper.

As in other basic types of grinding machines, internal grinders operate with a variety of different motions and functions, and are generally built to suit a wide range of workpiece sizes or shapes. As with external cylindrical grinders, internal grinding machines either hold the workpiece rigidly, or support it with wheels or shoes and allow it to rotate.

Of these two types of workholding configurations–chucking and centerless–there are, of course, variations. In the **chucking-type** machine, builders generally use one of three construction variations. The first has the wheelhead mounted on a compound slide so that, as the rotating grinding wheel traverses in the workpiece bore, it is simultaneously fed on a cross-slide to bring the hole down to size. The workpiece is chucked or mounted in a faceplate, which rotates in a fixed position. On the second type, the wheelhead traverses on one slide and the workpiece is mounted on a cross-slide. The part is fed to the grinding wheel as the wheel traverses. On the third type, the wheel spindle is suspended from an overhead bar, which acts as the slide. As the grinding wheel traverses, it is fed into the workpiece to bring the hole to size.

On the first two types, the workhead is swiveled at an angle to grind tapers. On the third type, the grinding wheel's path is governed by a control plate for grinding tapers or curved sections. One additional variation on this construction is the planetary-type machine, which holds an oversize or awkward workpiece, typically a casting, stationary in a faceplate. Rotation, feed, and traverse motions are all functions of the grinding-wheel spindle, which revolves in a planetary motion to provide the desired workpiece ID.

Thus, the chucking-type internal grinder can orient the workpiece and the grinding wheel in a variety of fashions. Some don't really chuck the workpiece; rather, they use a faceplate or some other holding device. Nonetheless, they operate with motions distinct from those of a centerless machine.

A good example of the chucking-type internal grinder is

Photo: La Salle

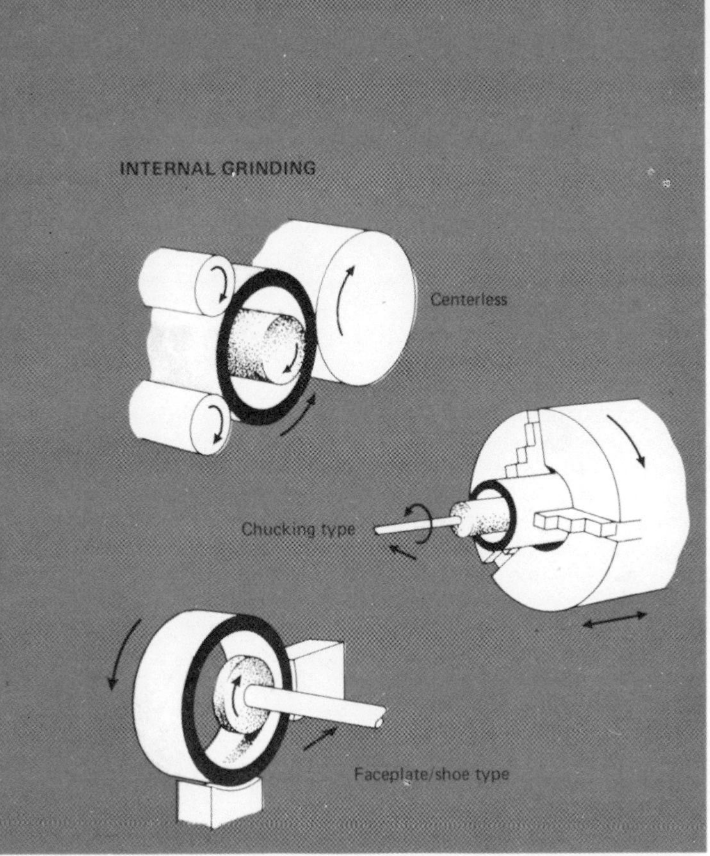

one we've already seen, the universal grinder (page G-8). In its ID-grinding mode, with its swiveling turret positioned and locked to present the smaller wheel spindle to the workpiece, this type of machine can produce internal holes with repetitive high quality. The universal machine is also flexible; it can handle outside diameters and flat surfaces as well. But if the shop deals in a particular type of internal-diameter work, or if parts have complicated shapes, a specialized ID grinder will prove more economical.

Chucking-type internal grinders can be equipped with any of several different workholders, to suit the part. Major styles in current use include round collet chucks, sliding-jaw collet chucks that pull the workpiece back against a plate in addition to holding it radially, diaphragm chucks with no sliding parts, jaw chucks, finger chucks that hold the part against a backplate, and magnetic chucks. All of these methods are found in OD grinders, also.

Choice of a particular chucking method is basic to the grinding of accurate IDs. Distortion caused by using the wrong type of workholder is perhaps the greatest single cause for scrapping internally ground workpieces. One frequent type of distortion is caused by using traditional three-jaw chucks on thin-walled tubing. In its free state, the ID very well may have been round, but with the jaw pressure applied, it becomes slightly triangular in shape. This distorted bore is then ground to roundness—but when the jaws are released, the part springs back to its original shape and leaves a reverse triangular hole in the bore.

Another type of distortion occurs when magnetic chucks are used in an internal grinder. Since the workpiece is not supported radially, walls are sometimes not thick enough to withstand the wheel's feeding force at a distance from the magnetic-chuck faceplate. Thus, the tubular part can deflect near the entry to the ID; the farther away from the chuck, the greater the deflection. The result is a bore of varying internal diameter.

Still another type of distortion is caused when tubular workpieces with nonparallel end faces are clamped under finger-type holders in a chuck. If each of the three fingers is designed to exert a certain pressure in toward the chuck faceplate, the single finger that rests on a lower portion of the face will exert less pressure than the other two, and will permit the workpiece to slide under it. Concentricity and circularity will be lost.

Centerless internal grinding alleviates some of these distortion problems under controlled conditions. Shoe machines and roll centerless machines depend on the roundness of the workpiece OD to cut round IDs.

In **internal centerless** grinding, the workpiece is rotated on its own OD between three rolls. These consist of a regulating roll to rotate the work, a support roll, and a pressure roll. The grinding wheel slips into the workpiece ID, and its feed pressure is directed against the regulating roll. One major advantage of this system is that, since the ID's circularity depends on the OD rather than the accuracy of chucking, perfect concentricity between OD and ID can be attained. The workpiece normally must be cylindrical in external contour, that is, shoulderless; irregular parts, however, can sometimes be handled if auxiliary cylindrical holders are employed. Auxiliary holders are normally made from the same material as the workpiece, for accurate holding even under the elevated temperatures of grinding.

A conventional type of internal-grinding spindle is used to traverse the wheel back and forth in the bore for centerless ID grinding, and small work can be loaded and unloaded merely by raising the pressure roll and inserting or removing the workpiece.

These machines can be arranged for either on-center or above-center grinding. On-center machines have the work, the grinding wheel, and the regulating roll all on the same horizontal centerline. This method gives maximum support and permits accurate grinding of thin-wall sections without distortion problems. Above-center machines have the rolls arranged so that both the workpiece center and the grinding-wheel center are above the horizontal centerline of the regulating rolls. Angular relationship of the support and the regulating rolls is such that parts with considerable variation in OD will remain on the same vertical centerline. Because of this, it is possible to grind different-size workpieces to the same tolerance, with little adjustment of the grinding geometry.

A slightly different variation on centerless internal grinding, **shoe-type** ID grinding, offers similar advantages. Internal shoe-type grinding uses an approach that is patterned after external cylindrical shoe grinders: The tubular part is held against a magnetic faceplate and rests on hard-faced shoe supports. The grinding wheel is inserted into the bore, and feed pressure is applied toward an external shoe, usually the one in the three-o'clock position. The grinding wheel rotates in one direction at high speed, while the faceplate drives the workpiece in the opposite direction at slow speed.

Since no pressure roll is used in shoe-type grinding, the waviness—sometimes called chatter—in the ID finish is reduced. The shoe that carries most of the feed load spans several wave bands and acts to smooth out spindle vibrations and unbalances.

Another advantage to the shoe-type holding system is the ease with which workpieces can be loaded and unloaded. This is especially useful in short-cycle production jobs.

Most modern production-type internal grinders, in fact, are highly automatic in operation. Little remains for the operator after the job is set up but to start and stop the machine and to load and unload the chuck or fixture. Traverse to the work, longitudinal and cross feeds, and dressing the wheel can all be carried out without operator supervision. A typical wheel cycle might go like this: Traverse quickly to the work, slow down to grinding speed, rough the bore, withdraw for automatic dressing to resurface the wheel for the finish cut, reenter the ID for finish grinding, withdraw again to the start position, and stop all motions.

Wheel-size considerations

Obviously, wheels used in internal grinding are much smaller than those that grind external surfaces; the wheel must fit into the ID to be ground. Abrasive tools for internal grinders range from wheels as big as 20 in. (or even larger) down to thin, diamond-charged wires only a few thousanths of an inch in diameter.

Wheel size in internal grinding has implications for the design of the machine spindle and wheelhead. In order to maximize the cutting rate at the workpiece/wheel interface, proper cutting speed—measured in surface feet per minute (sfm)—must be maintained. As in other types of grinders, optimum cutting speed depends on the type of wheel and the workpiece material, but generally is around 6000 sfm.

But to achieve that needed 6000 sfm with a ½-in.-dia grinding wheel, the spindle must deliver upwards of 45,000 rpm. With extremely small wheels, rotational speeds of 200,000 rpm are sometimes needed.

To accommodate this wide range of spindle speeds, wheelheads vary considerably in their basic design. Typically, grinding heads for internal machines are driven by flat belts. Ultra-high-speed heads use high-frequency, direct-drive motors or air turbines to reach speeds in the 200,000-rpm range.

(This relationship between spindle speed and cutting speed is stated in the formula:

$$\text{sfm} = \text{rpm} \times \pi \times D/12$$

where D is the grinding-wheel diameter in inches.)

Because of the larger area of contact between wheel and work surface in internal grinding compared with other types, optimum cutting speed is generally higher. This fact alone often requires the use of a different—faster—wheelhead on the machine.

But faster wheelspeeds aren't always the answer. Bigger, but slower, interchangeable wheelheads offer increased stiffness in the grinding system, and they still accommodate internal wheels small enough to fit the bore. With a stiffer system, less dressing is required, so, for production purposes, it is sometimes better to sacrifice some rotational speed—and consequently some cutting speed—for rigidity.

Another noteworthy relationship in discussing internal grinding is that of the area of contact between the workpiece and the working portion of the grinding wheel. In plain surface grinding, that area is relatively small; in external cylindrical grinding it is smaller still, because of the curved nature of the workpiece. But for internal grinding, the contact area increases significantly, because the arcs involved tend to be located around the same center.

Leading researchers have quantified this difference by deriving an "equivalent wheel diameter" (D_e), a research tool that makes it possible to equate grinding performances of different machine types. Comparison is based on conformity of contact areas.

In internal grinding, this increased area of contact, has, as noted above, implications for selecting the working speed of the grinding wheel. It also sets parameters for choice of wheel diameter.

A large contact area strongly affects wheel wear and the self-dressing action. It also affects cutting rate, because an abrasive grain must penetrate the work in order to cut. Too many grains in contact with the workpiece at a particular time will require too much pressure for optimum cutting. Thus, in internal grinding it is often necessary to reduce grinding-wheel size to reduce the contact area and improve machining.

As with most guidelines, a significant exception exists. In the case of interrupted cuts, where quill deflection alternately becomes large, then drops to zero, one way to avoid workpiece (and wheel) damage is to try to bridge the interruption. This is done by selecting the largest wheel diameter that will still permit adequate coolant delivery. Another alternative to cope with problems of interrupted cuts with an ID grinder is to dress the wheel to produce the sharpest possible cutting surface, in an effort to minimize quill deflection.

Controlling size

Workpiece geometry and tolerances, setup and maintenance requirements, and production needs determine the best sizing method. Grinding manufacturers recommend the simplest method consistent with the accuracy required.

One of the simplest forms of size control is **diamond sizing.** The truing diamond establishes a precise point of reference with respect to the finished workpiece size. By accurate control of machine feed, the grinding wheel is trued and then fed a precise distance, to establish finish size. Most high-production machines that use this form of size control feature automatic compensation for wheel wear. As the grinding wheel gets smaller, feed depth is correspondingly increased.

Other high-production machines, particularly those with automatic loading devices, use a **post-grinding gage** that provides feedback to the machine. This method gages parts as they are unloaded from the machine, and the gage unit, located outside the tooling area, corrects the process variables to hold size tolerances.

Both of these methods control the grinding size from variables observed outside the tool/workpiece interface. They are distinct from in-process gaging—or gage sizing, as it is sometimes called in ID operations—which measures the grinding action as it takes place.

In-process gaging improves operator control: Cycles can be varied, small bores can be gaged more accurately, the machine operator can control surface finish and taper more precisely, and the machine can accept a wider range of workpiece-size variation.

Depending on the tooling, size gaging can be either intermittent or continuous during the grinding cycle. A gage signal terminates the grinding when desired finish size is attained. The technique can also be applied to control infeed rates.

Four major versions of in-process gaging are used in internal grinding; the suitability of each depends on workpiece shape and production requirements. One version, the plug gage, consists of a plug, a disk shape with a specified OD, mounted within the workholding chuck and facing the bore. The plug reciprocates through the chuck and attempts to enter the bore each time the grinding wheel is withdrawn. Finish size is attained when the gage is able to enter the bore.

The second type of in-process gaging commonly used for internal grinding is the single air-jet. This technique uses an air nozzle mounted on a finger at right angles to a pivoting arm. The arm and the extension are pivoted into a straight bore between the workpiece and the grinding wheel, with the orifice directed toward the workpiece's ground surface. Size is determined by measuring the restriction of airflow (gaging back pressure).

A slight modification of this system, the air-fork gage, is the third commonly used in-process technique for ID grinders. It is used primarily for continuous gaging of tapered bores and uses two air jets, mounted in opposition, to measure the full diameter of the bore.

Electronic gages of the contacting type are the fourth common in-process measuring system. They can be mounted either outside the chuck or inside the workpiece holder, and they commonly use two diamond-tipped fingers, which are inserted into the bore. Operating like an ID caliper, electronic gages can be used to shut down the grinding operation or to adjust grinding conditions, depending on the specific type of control used.

Jig grinding

One type of grinder is very different from all other classes, even though it operates on some of the same principles as some internal grinders. The jig grinder was developed to locate holes accurately and grind them to size, particularly in hardened steels and carbides. This method uses the rectilinear positioning principles employed in jig borers, but spindle motion is different, and the wheel can be outfed accurately while grinding, so straight and tapered holes and contours can all be ground.

Cutting with a jig grinder uses three distinct motions: high-speed rotation of the grinding wheel held in the grinding head, slow planetary rotation of the grinding axis around the main spindle axis, and a vertical reciprocating movement of this entire planetary action through the length of the hole being cut.

Machines of this type require delicate positioning control, and many models are equipped with computerized numerical controls (CNC) as well as NC, optical measuring devices, and dial-input controls. Jig grinders are often equipped with high-resolution position-feedback systems that provide information to the control.

Great accuracy is required in setting up the workpiece for jig grinding. Workpieces that have several locations to be finish-ground are often best clamped to a rotary work table. The part need not be clamped or held so tightly for jig grinding as for jig boring, because there is less pressure tending to shift the piece. Care should be used to snug the clamps only enough to secure the workpiece and not to create appreciable workpiece distortion.

Plunge cutting on a jig grinder requires the same accuracy in setup and operation, since this roughing operation is usually followed by finish grinding in the same setup. Feed for plunge grinding should be hand-controlled, so that its rate can be sensitively felt to avoid glazing the wheel, yet grind freely. Too strenuous a feed pressure can cause excessively fast wheel breakdown, while too light a feed pressure causes glazing.

Jig grinders normally use either abrasive wheels or carbide burs. A good rule of thumb is to use burs for bores smaller than ¼ in., wheels for larger holes. In the case of very small wheels, diamond-charged mandrels are sometimes employed.

But while jig grinders can be partially classified with internal grinding machines, they are much more. The combination of planetary and reciprocating motions used with highly precise rectilinear workpiece positioning enables the grinding of contours. This fact gives the jig grinder its unique capabilities in the toolroom for cutting punches and dies.

The planetary motion of the jig grinder, which is also seen in the highly specialized grinders used for grinding the combustion chamber of trochoidal (Wankel-type) engines, also enables OD grinding. This application involves only a feed reversal: By reducing the radius of movement of the grinding spindle from the axis of the main spindle (or *infeeding* the wheel), the jig grinder can produce finishes on external faces or cylindrical parts.

Photo: Moore Special Tool

Wheels

It was a little over 100 years ago, when grinding-machine builders were just starting to introduce complex, high-production machines, that machinists began to realize the prime importance of the abrasive itself. Prior to those days, of course, several abrasive-grain types, notably emery, sandstone, and diamond dust, had been bonded into wheels. But even the most skilled machine operators could seldom get consistent results with those products.

Probably the first predictable wheels were made from corundum, first imported from India to Britain in 1825, or from emery, an impure aluminum oxide obtained in the early 19th Century from Asia Minor, and bonded with a variety of substances in trial-and-error fashion. Principal bonding agents were vulcanized rubber (developed in 1857), silicate of potash (1859), and borax for vitrified wheels (1872).

Robert S. Woodbury, in his *History of the Grinding Machine*, tells us, "By 1885, the elements necessary for a solid grinding wheel for production work were known, but much experimentation was required before really satisfactory wheels were available. The actual technical advances made from 1885 to 1910 in the manufacture of vitrified wheels were kept as trade secrets and are difficult to reconstruct today. Probably most of them depended upon the individual judgment and skill of foremen and key workmen."

By 1895, emery had been replaced for most grinding by the better cutting properties of natural corundum. The first artificial abrasive, silicon carbide, was developed in 1891 by Edward Acheson, who called it carborundum. It was, at the time, the hardest material available, except for diamond.

As reasonably uniform grades came onto the market, initial steps were taken to apply specific wheels to specific grinding situations. Many of the wheel-recommendation tables offered by abrasives manufacturers today are the result of early efforts by experts such as Charles H. Norton, who published a series of articles on applications around the turn of the century (AM—1895, !901, '03, and '04).

Woodbury continues: "By 1905 there was available reasonably satisfactory information on how to use the artificial grinding wheel in practice. It was no accident that Colvin and Stanley's *American Machinist Grinding Book* appeared simultaneously in New York and London in 1908, and James J. Guest's *Grinding Machinery* was published in London in 1915."

Grinding wheels today are available in an ever-increasing variety of abrasive types. Further advances have been achieved in the bonding system that joins the grains in the wheel or belt.

Wheels have been improved to the point where they are very specialized; a particular abrasive may be perfectly suited for cutting, say, bronze bearings, but would be uneconomical in machining a similar, but slightly different, material. Products are tailored to leave very fine surface finishes, to "hog" out metal at a fast removal rate, to operate as a replacement for a sawblade, etc.

Abrasive products, furthermore, are not always round; while a distinction may be made between bonded products and coated (belt) products, some of the technology is interrelated.

But before discussing the individual components of the various wheels, it's necessary to settle on some basic definitions used for all wheels.

The American National Standards Institute (ANSI), working with the Grinding Wheel Institute, defines an abrasive wheel as a cutting tool consisting of abrasive particles held together by various bonding materials. It further breaks down the bonds into two types: organic wheels, which are bonded by means of an organic material such as resin, rubber, shellac, or other similar agent; and inorganic wheels, which are bonded by means of an inorganic material such as clay, glass, porcelain, sodium silicate, magnesium oxychloride, or metal. Wheels bonded with clay, glass, porcelain, or related ceramic materials are characterized as "vitrified bonded wheels." The very new plastic-bonded products also fall into this type.

Use classification

ANSI also describes the different operations for which abrasive wheels are used. Wheels in different size ranges are required depending on intended use, and major categories include:

- Cutting off—for slicing or slotting of parts. Operation is usually performed with a thin abrasive wheel, usually with an organic bond.
- Cylindrical, between centers—for grinding the outside surface of a part that rotates about its center.
- Centerless—where the work is rotated by a regulating roll.
- Internal—for grinding the inside bore of the work.
- Offhand grinding—where the work is held in the operator's hand.
- Saw gumming—saw-tooth shaping and sharpening with a grinding wheel.
- Snagging—for removing relatively large amounts of material without regard for close tolerances or surface-finish requirements. Typically, snagging removes surface defects from billets and excess metal from welds.
- Surface grinding—where the workpiece is flat.
- Tool grinding—for grinding the shapening edge of such tools as drills, taps, reamers, etc.

In many cases, the classification of functions developed for grinding jobs coincides with the classification of machines already discussed in this report. Function dictates the shape of the grinding wheel to be used, and ANSI distinguishes between standard grinding wheels and mounted grinding wheels.

Grinding-wheel shape

Standard grinding wheels are made to cut either with the periphery of the wheel or with the side face of the wheel. So

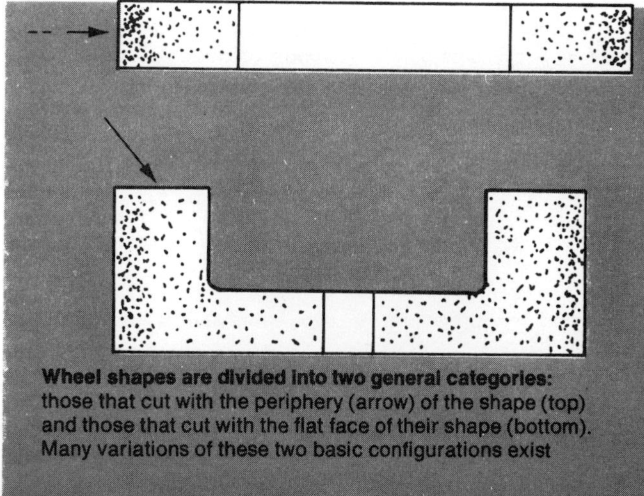

Wheel shapes are divided into two general categories: those that cut with the periphery (arrow) of the shape (top) and those that cut with the flat face of their shape (bottom). Many variations of these two basic configurations exist

standardized shapes for wheels may be broken down into two general groups (illustration pG-15), irrespective of abrasive grain or bonding agent. (For a description of wheel-composition designations, see p G17.)

The simple form of the periphery-cutting wheel (ANSI Type 1) and its specialized derivatives are coded with number designations, and carry recommendations for type of job (cutoff, snagging, cylindrical, etc.).

The second major classification includes all wheels that are made to grind with the flat face of the shape (ANSI Type 6 and its variations). Individual shapes in this class also carry primary-use designations (tool grinding, surface grinding, etc.).

Standard shapes all share well-defined dimensional tolerances. Specifications include definition of radial width at periphery flat, depths of blind hole, overall diameter, location of mounting hole, etc. A complete description of these dimensions is found in ANSI B74.2-1974.

Mounted-wheel shapes

Mounted wheels are usually 2 in. in diameter or smaller, and come in various shapes. They may be either organic- or inorganic-bonded, and are permanently secured to a steel mandrel. Shapes are classified into three groups (A, B, or W) according to geometry and size of abrasive mass. Further, an arbitrary number designates specific geometry.

Mandrels are classed by shaft diameter and amount of projection. The numbers 1, 2, 3, and 4 stand for ⅛-, ¼-, ⅜-, and ½-in. mandrel diameters, respectively. Thus, a mounted wheel designated:

B-52-D-1

shows a B-group abrasive shape with a "52-type" cross-section (the B-52 shape is one-half of an ellipse), mounted on a mandrel that has a 1½-in. projection and that is ⅛ in. in diameter. (For a full description of shape cross-sections, refer to ANSI publication B74.2.)

Diamond-wheel shapes

Diamond grinding wheels (and those that cut with the new superabrasives) are similarly classified by shape. Once again, these designations describe *shape* and *dimension*; they have nothing to do with abrasive *composition*—grit, concentration, bond. See p G-18 for diamond-wheel compositions.

The standard code for diamond-wheel shapes takes four components into account: basic core shape, shape of the diamond cross-section, location of the diamond section, and modifications. A four-position code is thus built; for example:

6-A-2-C

This indicates (in order) that this wheel has a No. 6 core (a shallow cup), uses an A-shaped diamond cross-section (rectangular), which is located at position No. 2 (on the side of the wheel from periphery to center), and that the wheel has a C modification (holes are drilled and countersunk into the core). For a complete description of diamond-wheel shape codes, refer to USAS B74.1, available through ANSI.

Wheel selection

The choice of a grinding wheel for a specific operation depends on many considerations. Principal factors—workpiece size, workpiece material, stock removal, machine setup, accuracy, and surface finish—can be evaluated to help narrow the choice.

The size and diameter of the workpiece in cylindrical and internal grinding largely determine the wheel hardness needed. Small-diameter work has a short arc of contact with the wheel, causing it to wear rapidly, so a harder wheel should be used.

Length has a similar effect. When a wider wheel is used—generally indicating a bigger workpiece—the wheel grade should be softer than normal. The unit pressure per grain on a wide wheel is lower than on a narrow wheel, since the total force available is distributed over a larger area. A softer-grade wheel in this situation permits proper wheel breakdown for free cutting without glazing.

In throughfeed centerless grinding, the angle of the wheel axis, which is determined by workpiece size and diameter, also affects wheel selection. A wheel acting too hard can be corrected, to some extent, by increasing the angle of the regulating roll. This increases the feedrate and causes the wheel to break down faster and act softer.

Workpiece material has the greatest impact on the choice of wheel. High-tensile metals, such as tool steels, alloy steels, and machine steels, are generally ground with aluminum-oxide wheels. Soft materials, and those that are both hard and brittle (such as cast iron, brass, bronze, aluminum, stainless, carbon, and sintered carbides), are generally ground with silicon-carbide wheels. This abrasive breaks down more easily than aluminum oxide and assures free cutting and rapid metal removal.

The new zirconia-alumina abrasive alloy, a proprietary product of Norton, is made to grind many of the same materials for which aluminum oxide is desirable. Made by fusing zirconium oxide, a white crystalline compound that is less brittle and very durable, with aluminum oxide, the new abrasive combination allows for a self-sharpening action. Individual grains, rather than being ejected when dull, are retained better and are continually fractured, exposing sharp new points to the work surface.

Choice of superabrasives, too, rests heavily on the type of workpiece material. Diamond wheels have virtually taken over in cutting tungsten carbides, and they are also often used to grind die steels, stainless steels, and combinations of steel and carbide. General Electric's cubic boron nitride (CBN) can grind ferrous metals, but debate continues over whether this fairly new product is cost-effective for mild steels. Similar debate may swirl around DeBeer's new amber-boron-nitride (ABN) abrasive when it comes onto the US market.

Stock removal is another consideration for grinding-wheel selection. Parts that are ground from rough, or that have a heavy stock allowance, require coarse-grained wheels. Deep cuts in grinding, as in other metal-cutting processes, require heavy cutting tools and a lot of chip clearance.

The kind of machine used and the way the workpiece is set up also help determine wheel choice, although to a lesser extent. In centerless grinding, for example, throughfed work may be ground with wheels that are softer than those used for normal centerless or infeed work. For multiple-diameter grinding with crush-trued wheels, differences in final part configuration will impose limitations on the type of wheel best suited for the job. Many cutoff wheels are reinforced with a single sheet of fiberglass bonded right into the resinoid wheel. Type 6 or similar wheels, which grind with their flat faces, are often reinforced around their periphery with wire or straps.

Desired accuracy and surface finish are other important considerations in wheel selection. Close accuracy and a high-quality finish are usually obtained with hard, fine-grained wheels. This means, however, that only a small amount of stock is removed per pass. But surface finish can be improved even when using a coarse wheel, if the spindle is equipped to reciprocate.

Many factors, therefore, go into the selection of a grinding wheel for a particular job—whether it is a conventional wheel, a high-performance tool, or a superabrasive.

Economic factors are becoming increasingly important in grinding-wheel selection, especially with the advent of the newest high-performance wheels, which modify abrasive

Conventional-wheel marking system

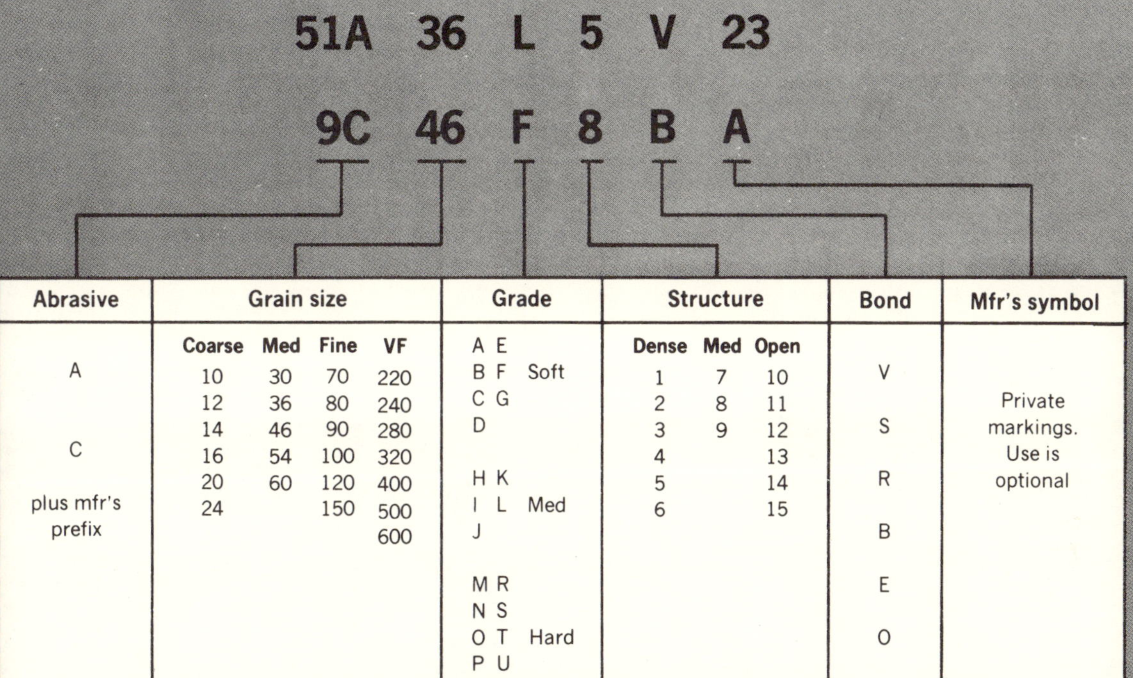

Abrasive	Grain size	Grade	Structure	Bond	Mfr's symbol
A C plus mfr's prefix	**Coarse** 10, 12, 14, 16, 20, 24 **Med** 30, 36, 46, 54, 60 **Fine** 70, 80, 90, 100, 120, 150 **VF** 220, 240, 280, 320, 400, 500, 600	A E B F Soft C G D H K I L Med J M R N S O T Hard P U Q V	**Dense** 1, 2, 3, 4, 5, 6 **Med** 7, 8, 9 **Open** 10, 11, 12, 13, 14, 15	V S R B E O	Private markings. Use is optional

Abrasive. The cutting agent used in the wheel. While natural abrasives—corundum, emery—exist, most wheel makers use manufactured abrasives. The A code stands for aluminum oxide (Al_2O_3); C stands for silicon carbide (SiC).The digit or letter prefix is a manufacturer's symbol indicating such other information as carbide state (black, green), or oxide friability, or a combination of abrasives.

Grain size. Corresponds to the number of meshes per linear inch in the sizing screen. Coarser grits cut faster; finer grits are used for finishing. Proper grit selection is further dictated by the nature of the workpiece material and by the cutting area.

Grade. Measures the relative hardness or softness of a wheel. The harder the wheel, the more securely the grains are held, and the greater the force required to break them out of the wheel. Grade is a relative measure, and some manufacturers offer only a limited range, or designate a particular number as "medium" or "hard" relative only to their own lines.

Structure. The density or openness of the grain spacing. Structure-number use is optional, and some manufacturers do not specify it in their labeling. Open-grain wheels provide the best chip clearance and are suited for heavy removal in softer materials; close-grain wheels are used in grinding hardened steel and exotic materials.

Bond. The type of material used to cement the grains together. Bonds can be of several types: **V**—Vitrified. Chief ingredient is clay fired in a kiln. This is the most-common type. **S**—Silicate. Uses silicate of soda as important ingredient. Releases abrasive grains more slowly than vitrified bonds, and is used where heat generated in the workpiece is to be kept to a minimum. **B**—Resinoid. Synthetic resin mixed with the abrasive grains and baked in an electric furnace. Resinoid bonds are generally stronger than other types and are used for high-speed grinding (upwards of 9000 sfm). **R**—Rubber. Uses natural or synthetic rubber in the bonding material. Used in thin cutoff wheels and for the regulating wheels in centerless grinders. **E**—Shellac. An organic bond used for high finishes on cutlery, machine rolls, and camshafts. Also used in thin cutoff wheels. **O**—Oxychloride. Used to a limited extend in grinding disks. Uses magnesium oxychloride. Bonding-material selection is highly important in adapting an abrasive to a particular job, and recent improvements in wheel performance have been attributed to the chemistry of the bonding agent.

Manufacturer's symbol. A marking system used to record batch production, to designate factory processes, and for coding. Typical symbols might show a W for a wire-wound wheel or a T for a tape-wound wheel. Blanchard uses a C to indicate additional porosity; Cincinnati Milacron tacks an RW onto the V (for vitrified) in the symbol for its latest high-performance product, and says the VRW stands for "very remarkable wheel."

Diamond-wheel marking system

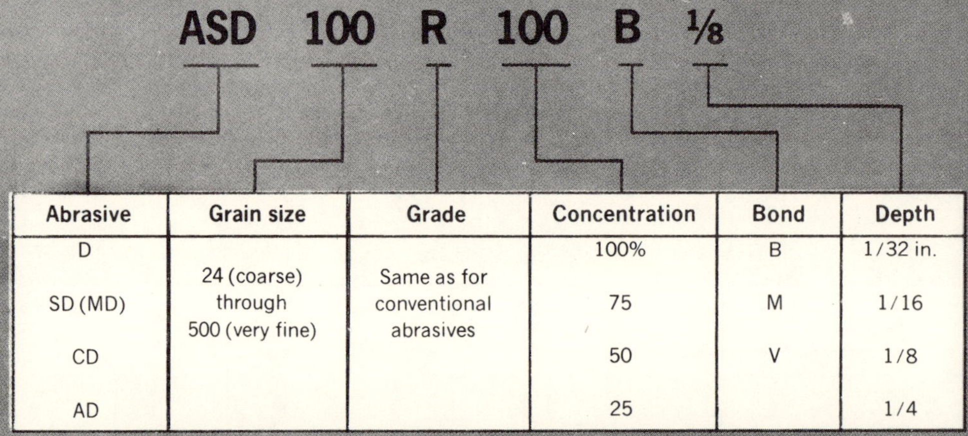

Abrasive	Grain size	Grade	Concentration	Bond	Depth
D	24 (coarse) through 500 (very fine)	Same as for conventional abrasives	100%	B	1/32 in.
SD (MD)			75	M	1/16
CD			50	V	1/8
AD			25		1/4

Abrasive. D stands for natural diamond; SD or MD means synthetic or manufactured diamond; CD, coated diamond; AD, armored diamond.

Grain size. Diamond grit is measured through a mesh, as in conventional abrasive grains. Grains coarser than 24 are rare (and probably fairly valuable).

Grade. Wheel hardness is measured with relative letter designations in much the same way conventional wheels are graded. Keep in mind that letter grades are strictly relative only within a particular manufacturer's line.

Concentration. The highest concentration of diamonds in matrix, 100%, contains 72 carats of grains per cu in. of conglomerate. Lower figures indicate smaller concentrations.

Bond. Resinoid bonds (B) are most common. Other bonding methods include metallic (M) and vitrified (V).

Depth. The thickness of the abrasive layer atop the wheel core; it is measured in fractions of an inch.

Note: Cubic boron nitride wheels and other "superabrasives" conform to the diamond-grading system in most cases.

structure, or the bonding method, or both, to achieve better durability. New factors are entering into the abrasive-cost equation—items such as wheel-change time, dressing time, and the cost of diamond-dressing tools.

This is particularly true in high-production methods. Cincinnati Milacron offers one case history about a changeover to its new high-performance product: An automotive part was ground on a chucking OD grinder. The conventional wheel gave 10 parts per dress, while the new wheel gave 20 parts per dress. Abrasive cost per 1000 parts with the conventional wheel was $5.26; with the high-performance wheel it became $3.58, because of lowered wheel wear. Wheel-change time with the former wheel cost 60¢ per 1000 parts; with the new wheel it cost half that. Diamond cost with the previous wheel was 80¢ per 1000 parts; the new wheel halved that expense also. Dressing time with the conventional wheel cost the manufacturer $6.30 per 1000 parts; with the new wheel, $3.10.

Superabrasives—such as diamonds and cubic boron nitride—can offer similar cost reductions in high-volume grinding. Improved wear rates over conventional abrasives more than compensate for the initial price differences in many cases. But those first-cost differences are considerable. Cost of superabrasives can run as high as 350 times that of a similar-size aluminum-oxide or silicon-carbide wheel.

The combinations involved in getting the most productivity out of a particular machine for a certain part and keeping quality high are almost endless. Grinding-wheel manufacturers themselves are the best source of application data in this rapidly changing field.

Before any wheel—conventional, high-performance, or superabrasive—can be put into service, it must be carefully inspected, then balanced and properly mounted to the machine.

Handling and mounting wheels

Handling of wheels is a prime consideration. Grinding wheels should be stored near the machine on which they are to be used, to minimize handling, and they should always be hand-carried or transported on a truck—never rolled on their periphery. Improper handling contributes to wheel cracks—and the slightest crack will *always* disqualify a wheel from use. Safety is the prime consideration here: A cracked wheel breaks with explosive force, enough sometimes to penetrate even the best of approved machine guards. Or as one AM reader puts it, "If you hear anything unusual during grinding, hit the floor with your face down; it may be the last chance you'll get."

Grinding-wheel breakup is the result of tremendous forces within the conglomerate of individual grains. Organic-bonded wheels are slightly more resistant to shock than inorganic ones, but any disturbance of the wheel structure is very hazardous. And while abrasive-wheel makers exercise extreme care in combining grains and bonds, the bonds themselves are necessarily limited in strength by the requirement that they release dulled grains during their life span.

The forces that try to shatter a wheel during its operation are mostly centrifugal in nature. Overspeeding the wheel increases these forces and accentuates the possibility of breakup: The centrifugal force on a wheel at 5500 sfm, for example, is nearly 50% greater than on the same wheel at 4500 sfm, even though the speed is only 22% higher.

Specific recommendations for use, care, and protection of abrasive wheels are published in ANSI B7.1-1970.

Visual inspection will disclose any obvious cracks in the wheel, but before being mounted, a bonded wheel should be given a "ring" test, in which the wheel is suspended from a hook and tapped lightly with a hammer. A sound wheel will give a clear ring. If it sounds cracked, return it to the toolcrib for more careful examination.

An operating grinding wheel is very sensitive to imbalance. Careful mounting and balancing not only will reduce the danger of breakup, but also will lessen harmful vibrations to the machine.

Wheels with small center holes are mounted directly on the spindle; those with large holes on a demountable sleeve. Two flanges, about ⅓ to ½ the diameter of the wheel, supply the real support. The inner flange is keyed, screwed, or pressed on the spindle. The outer flange should be an easy-sliding fit, so it can adjust itself to give uniform pressure on the wheel. Both saucer-shaped flanges should be relieved at the center so they bear evenly on their outer edges only. Place blotting-paper washers between the wheel and each flange. These are extremely important; if the washers are omitted, the wheel may crack where it is clamped or may slip during grinding.

Small-hole wheels are held by a single nut on the spindle end. This should be tightened firmly but not excessively, to avoid placing internal stresses on the wheel. Large-hole wheels are held by a series of flange screws around the sleeve. They should be tightened with finger pressure, then wrench-tightened one at a time, alternately across the center, as in tightening the lugs on an auto wheel. Continue tightening opposite screw pairs until all are uniformly snug, but do not overtighten.

Balancing the wheel

Wheels are sold in a perfectly balanced state, but can become out-of-balance after handling and mounting. Because of this, balancing is a necessary operation, particularly for medium- and large-diameter wheels. An out-of-balance wheel sets up excessive vibration, causing chatter marks on the workpiece and needless wear in the spindle bearings.

To balance the wheel, place it on a balancing arbor resting on a balancing stand. Two types of stand are in common use: One consists of a pair of parallel ways which are true and level; the other uses two Vs, formed by pairs of overlapping disks. Either type must be in good condition, level, and clean if it is to provide true readings.

The wheel will oscillate for a moment after it is placed on the stand and then come to rest. After a pencil mark is placed on the wheel at the exact bottom, it is rotated by hand and allowed to turn by its own weight. If the pencil mark once again falls to bottom, the wheel is unbalanced. Weights in the circular groove in the mounting flange are then displaced until a state of balance is achieved.

On one type of grinder, this conventional balancing procedure can be eliminated by an automatic mechanism built into the spindle. The wheel is balanced after a clamping lever is released and an indicator becomes stationary, at which point the lever should be reclamped.

Abrasives for disk grinders are mounted in several different ways, depending on their particular design. In general, however, the procedure consists of bolting the disk to its steel backplate. Both the mounting plate and the disk should be checked for flatness before mounting. Screws or bolts used must consistently be of the proper length. Longer bolts passing through the mounting plate to nuts embedded in the wheel will penetrate the abrasive; shorter bolts will not hold with the proper tension.

Mounting the disks should be done in a clean environment, to avoid foreign matter between the disk and its mounting plate. Dirt in the bottom of the noles has the same effect as bolts or screws that are too long. The bolts should be inserted in all holes before tightening, then tightened in opposite pairs.

Both disks and wheels must be handled and stored with care. After mounting and balancing, run the wheel at operating speed, with the guard in place, for at least one minute before contacting the workpiece. The guard should cover at least half of the wheel. After grinding is completed, turn off

The newest wheel

Research Abrasive Products Inc, Wickliffe, Ohio, has stepped beyond other high-performance grinding wheels with its newly patented (Dec 9, 1975) "plastic bonded grinding wheel."

Kenneth Dixon, Research Abrasive's president, explains: "It is not difficult to make a grinding wheel harder; you simply put more bonding material into the compound. Coupling agents also are an important consideration; these agents strengthen the bond at the interface between the abrasive grains and the bonding material. Thus, silane-treated abrasive grains yield a tougher grinding wheel.

"In our wheels, the organic-plastic-resin bonding system takes advantage of these mechanisms for stronger hold on the abrasive grains, but it is also microscopically resilient. This gives a very tough wheel, but one that also is freer-cutting."

Research Abrasive cites a case where its plastic-bonded centerless grinding wheel replaced a silane-treated grain wheel on tapered roller bearing race application. The RAP product not only gave longer wheel life, but also reduced burning, which the hard, silane-treated grain wheel had not been able to do.

"It is common industry practice to use J letter-grade wheels for grinding bearing races," says Dixon, "because the harder K, L, or M, grades burn the parts, yield poorer tolerances, and create intolerably wavy surfaces. We use grades that are two, three, or four letter grades softer than the J grade, because our plastic-bonded wheels are tough yet resilient.

"Our bonding system also enables us to tailor a grinding wheel for a specific application. We can take such variables as machine wear into account. At one major disc-brake manufacturing plant, we dropped down the equivalent of 14 letter grades to accommodate machine wear—and still gave the customer three times more useful life out of his wheels over those of our competitor."

The firm claims use-life improvements of five or six to one over the conventional resinoid-bonded and vitrified wheels, and two or three to one plus improved metallurgical results over other new high-performance wheels. Beginning in 1971 with steel-roll grinding wheels, the company now supplies disk and centerless wheels as well. Development work is underway on in-feed wheels.

With its plastic-resin bonding system, Research Abrasives can pigment its grinding wheels; it has used eight to ten different colors at various times. During the Christmas season, it has pigmented some orders red and green and hung a "Merry Christmas" sign on them.

the coolant supply before stopping the wheel, to avoid differential wheel cooling and an out-of-balance wheel.

Truing and dressing

Although truing and dressing a grinding wheel are similar operations, their purpose is significantly different.

"Truing" a wheel means to prepare it by removing enough of the cutting face to have it run true, or in perfect concentricity with its own spindle or mount. Truing also prepares a single wheel to cut contours, radii, or multidiameters.

When a wheel has been run for some time, it may become loaded or glazed. It is then "dressed" to restore the original sharp, clean cutting face, or to prepare it for a finish-grind cycle.

Three chief types of wheel dressers are used on precision grinding machines: those equipped with a diamond tool, with an abrasive wheel, or with a mechanical dresser. The most common is the diamond-tool type. All types must be held in firm mounts to do an accurate truing or dressing job.

Dressing attachments may be separate units that bolt onto the grinder, or they may be an integral part of the machine. One popular cylindrical grinder holds a truing tool in the tailstock; some surface grinders have it built into the wheelguard. In some automated machines, the dressing cycle is programmed to be part of the automatic cutting cycle.

A typical diamond dressing attachment is intended for truing or dressing the wheel to a straight cylinder for grinding plane surfaces. But it is often equipped with a profile slide, controlled by a cam, that regulates the distance of the diamond from the spindle axis as the tool moves across the face of the wheel. Cams may be interchanged for different wheel profiles.

The diamond truing tool consists of one or more industrial diamonds mounted in a nib at the end of a steel holder. The diamond is advanced to the wheel at its horizontal centerline, with the nib pointed down at an angle of 3 to 15 deg. If there is any doubt about the location of the centerline, lower the nib slightly; if it is brought to bear on the wheel above center, it may gouge the wheel. To make the diamond self-sharpening, rotate the nib at a 30-deg angle in the horizontal plane each time it is used.

The diamond is advanced to the highest part of the wheel and takes sufficient passes, truing at 0.001-in. depth, until the wheel is perfectly round. This will usually leave a surface that is satisfactory for rough grinding. If a finer finish is desired, take one or two cuts that are 0.0005 in. deep, and then make a few more passes with no infeed, to let the wheel spark out on the dressing diamond.

Dressing for finish

While the depth of cut during dressing has an effect on the surface finish that the wheel will leave in the workpiece, the rapidity of dressing is also a factor. As the diamond nib is traversed across the face of a rotating wheel, it describes a helical path. The proper lead, or the gap between the spiral grooves, is a function of both the rotational speed of the wheel and the traverse speed of the diamond. Lead distances of from 0.0005 in. per revolution to 0.005 in. are common, but the final choice depends on the job for which the wheel is being dressed. Too rapid a traverse will leave deep spiral marks on the wheel; too slow a pass will tend to glaze it. For roughing, a traverse at medium speed is best, with a somewhat slower speed for finishing.

Abrasive dressing wheels are mounted at a 5-deg to 7-deg angle to the grinding-wheel axis. They are brought into contact with the grinding wheel with an infeed of 0.002 or 0.003 in., and traversed across the grinding face. Care must be taken not to run the dressing wheel off the face of the wheel being dressed, as it may gouge the face when table travel is reversed. Similarly, the dressing wheel should not be traversed without any infeed; i.e., sparkout should not be attempted. Without feed pressure, the dressing wheel, the grinding wheel, or both will tend to become glazed.

If the grinding wheel is to be dressed true, the dressing wheel must be in good running condition. Runout in the dresser bearings will cause inaccuracies in the grinding face. True the face of the dressing wheel occasionally by canting it at a 10-deg to 15-deg angle and taking a light cut with the grinding wheel.

Diamond wheels and CBN wheels seldom require dressing, but when it is needed it is usually done by hand. A soft silicon-carbide or Al_2O_3-wax stick is held against the face of the rotating wheel. This cuts away some of the bond encapsulating the abrasive grains and exposes more grains to the workpiece. Diamond wheels that are cup-shaped are dressed with an abrasive wheel—usually made of silicon carbide of about 70 grit. Once again, dressing a superabrasive wheel cuts away the bond to expose new grains.

Dressing for contours

Contours, multidiameters, and multiple surfaces offer significant areas for cost-reduction in grinding. By combining operations with one machine it is possible to save time by eliminating secondary loading and unloading time, to improve accuracy by combining operations on a workpiece into one chucking, and to reduce machine investment.

Some machines in all types are designed for grinding multiple surfaces or contours; these have been discussed to some extent in earlier sections of this report. Contour-grinding machines, whether primarily surface machines, cylindrical units, or internal machines, must employ some method to true a wheel to the desired contour, and then to keep it dressed within that contour.

For simple angles and radii, fixtures hold a diamond dresser in a predetermined position relative to the wheel, and permit its travel along the desired plane or arc. Other contour dressing attachments use templets to guide the diamond nib to true a complicated shape.

Crush-roll dressing and truing uses a shaped roll that is pressed against the rotating wheel, crushing the corresponding shape into it.

High stock-removal rates are common when grinding with crush-trued wheels. The ability to remove the metal rapidly is primarily because of the freer-cutting or "natural" grain structure of crush-trued wheels. Because the wheel is freer-cutting, wheel wear is less than with diamond-trued wheels, less heat is generated while grinding, and the operation often uses less power.

It is probably these advantages that led General Motors engineers to begin designing a giant multistation grinding system that will use a 10- or 15-ft-dia wheel, cutting at 27,000 to 36,000 sfm. Such a high-performance system, it is expected, will be continually crush-dressed with an idler roll. The system may be built this year.

Crush-truing is best used when there is a production need to grind very accurate shoulders, radii, and slots. The method should be confined to vitrified wheels; other types tend to break out in crushed clusters and prevent duplication of form.

Wheels (and coated abrasives) are being constantly redesigned to improve cutting efficiency. The cutting action of a grinding wheel is, of course, closely tied to the design of the machine it's used on; this has been the pattern for the last 100 years. Much of the research into the principles of grinding, outlined in the next section of this report, will have significant impact on how both wheels and machines are designed and built in the future.

Grinding dynamics

Practical experience served the science of grinding from its first production beginnings in the latter half of the 19th century into the early 20th century. The gathering and application of empirical evidence and the creative genius of pioneers like Norton and Brown served the field well. But as the grinding process became a dominant machining method, as it moved from toolroom work to high-volume industry, new engineering principles had to be developed.

One of the first to apply engineering analysis techniques to grinding was James J. Guest. He analyzed the problem of strength and surface speed of the wheel as affected by varying diameters, and then he derived a formula. Workpiece distortion, the effects of various coolants, feeding systems, and the importance of time and costs were all considered by Guest. In many respects, these are some of the same variables that hold researchers' attention today.

But today's grinding theories are much more encompassing than those of earlier researchers. Grinding today is seen as the complex system of metal removal that it is. But many of the diverse–and often conflicting–principles of grinding held by scientists and engineers are based on the early theories of people like Guest and Alden, who introduced the principle of "grain depth of cut," accounting for wear losses.

The importance of primary research into the principles of grinding cannot be underestimated. Many of the newest–and more productive–specialized applications of abrasive technology can be attributed to experimentation and careful observation of various results. Techniques like "abrasive machining," use of superabrasives, "high-speed grinding," and electrochemical grinding grew up in the laboratory. Similarly, the basic design of machines is the result not only of simple refinements to a basic design, but also of the development of theories and mathematical models to study what goes on when an abrasive grain contacts a metal surface.

Grains cut chips

The traditional concept of how a grinding wheel removes metal holds that grinding is cutting. The process uses small abrasive grains that move along the surface of the metal with enough pressure to cause formation of a chip.

The fact that grinding produces chips has important implications: The principles that are applied to other types of metal cutting–milling, planing, etc–can also be applied to abrasive methods to a great extent. The terminology can be the same.

The purpose of efficient grinding is to maximize the effective cutting force of each abrasive grain while minimizing power losses due to friction and heat.

The chipmaking job of each abrasive grain used to be compared with the chipmaking job of a single-point cutting tool. Grains were just like the teeth of milling cutters but operated at very high speeds, had an indefinite geometry, and produced very small chips.

This comparison holds true only to a certain extent. When thermal research about 15 years ago showed that the energy remaining in the workpiece after grinding was substantially higher than that left by single-point cutting, the comparison had to be redefined.

That research measured calorimetrically the energy that ended up in the workpiece as 70-80% of the total expended. The result has been checked repeatedly against ordinary metalcutting–in which about 5% of the energy ends up in the workpiece–and has led to some new realizations about what a grain does.

An ordinary chipmaking tool is a hard, sharp instrument with a specific rake angle, or inclination of the cutting face. When it is drawn across a surface at a specified depth, it produces a shearing of metal directly in front of it. The workpiece metal ruptures along this shear plane, from the tip of the cutting tool to a point on the workpiece surface at some distance in front of the cutting tool, and consumes most of the energy applied by the tool. Much of the rest of this applied energy is used up in friction that occurs when the forming chip slides across the rake-face of the single-point cutting tool or milling-cutter tooth.

One theory about chip formation in grinding compares an abrasive grain to a ball. Certainly, individual particles do appear to be blocky crystals, multifaceted polyhedrons that roughly resemble a sphere with many sharp edges. As such, and with a depth of cut that is generally about only 5% of the sphere's overall diameter, it is easy to see why abrasive grains cut in a different manner from most single-point tools. The distinguishing feature of grinding is that, by and large, most of the cutting edges have negative rake angles. Much of the movement of metal is done, not by the sharp rake angle as in ordinary cutting, but by the overall shape of the spherical grain itself.

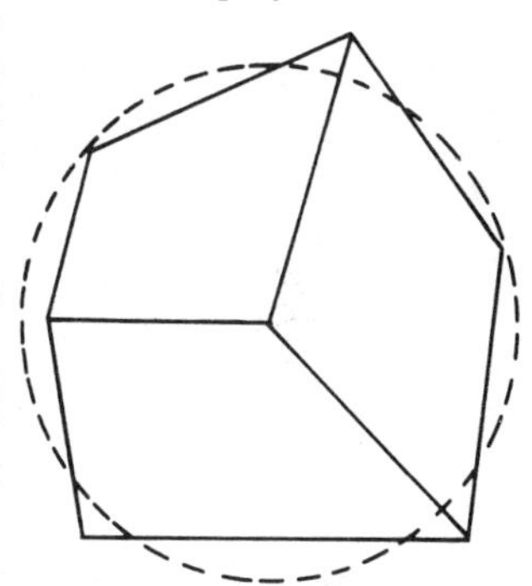

When such a grain acts on a metal surface, the action closely resembles the behavior of a Brinell ball in an indentation test. Pressure is applied in two directions: downward as the result of machine-feed force and forward as the result of grinding-wheel and workpiece motions. It is this pressure that resides in the workpiece and causes some chipmaking action.

As a Brinell ball is indented into a metal surface, the region immediately below it becomes plastic to some extent. But there is essentially no upward flow. The plastic material moves downward and outward. Much the same is true for the action of a single abrasive grain. But, as the force vector for the abrasive grain is tilted from the vertical, some of the outward flow of metal is directed upwards toward the surface of the workpiece.

The energy that grinding uses in relation to the amount of chips that are fractured is high. This is due to the large volume of metal that is deformed but not cut. Further, metal that lies beyond this deformed metal becomes plastic and undergoes a change in density; this in turn reduces the volume flowing upward in the form of a chip.

Keeping in mind that each individual abrasive grain is a sphere with multiple facets, it can be seen that the deeper a particular grain penetrates into the workpiece surface, the better the chances are that a sharper rake angle will present itself to the topmost layer of the workpiece metal. While the average rake angle of a shallow-cutting abrasive grain is a small negative-rake angle, the deeper-intruding grains will create sharper chips.

So the initial theory of how the grinding process removes metal can be slightly modified. Although it is true that grains cut chips, not all of the power that goes into the workpiece through the grinding wheel is used to make chips. The application of this horsepower has long-term implications for the pressure brought to bear on the workpiece, the size of the grain, wheel speed, workholding methods, and even the type of lubrication (if any) that is used.

Grains rub, too

Other researchers have delineated just when an individual abrasive grain or group of grains start to cut. In the grinding of metals, as force increases, three distinct processes take place: rubbing–the grain rubs on the workpiece and causes elastic and/or plastic deformation in the metal with essentially no material removal; plowing–the grain causes plastic flow of the workpiece material both in the direction of its sliding path and across this path, causing an almost extrusion-like movement of the material out from under the grain with small stock removal; and cutting–a fracture occurs just ahead of the rubbing grain causing chip formation and fairly rapid stock-removal rates.

These three distinct processes can be seen in a graph of "wheel depth of cut" plotted against force. The wheel depth of cut is the radial advance of the wheel per revolution of the workpiece in cylindrical grinding. The force measured is the

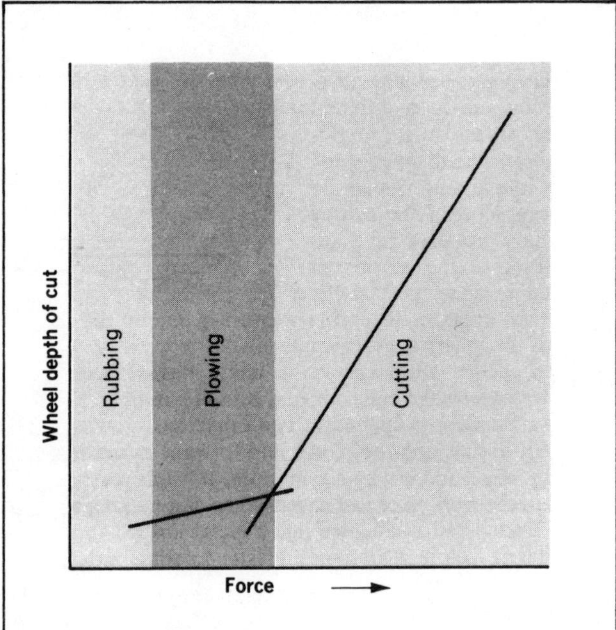

normal force per unit width of the grinding-wheel/workpiece contact, and it is corrected for variations in the stiffness of the machine-feed system. Rubbing takes place when the force is relatively low. Little or no metal is removed, and the wheel's depth of cut is shallow. As force is increased, plowing occurs, characterized by a deeper wheel depth of cut, and metal removal increases almost in direct proportion to increases in the force applied. As force is increased further, the grains shift to their cutting mode of operation, with significant increases in cut depth and metal removed.

Obviously, the cutting mode of operation is the most efficient for metal removal in grinding, but the other modes have importance, also. The plowing zone, for example, is a description of what happens during sparkout. Sparkout is the effect at the end of a grinding cycle, when relaxation of the elastic deformation causes wheel engagement to continue after the feed is disengaged. As such, it is useful in refining the finish at the end of the grinding cycle.

As both wheel depth of cut and force intensity increase, the relationship between them in any one zone remains constant. This graph function has the dimensions of the "volume of material removed per unit of time per unit of wheel/work interface," and it is called by research scientists the metal-removal parameter.

This parameter can be drawn for any given grinding application and permits a certain amount of predictability in changing the variables. (The chart illustrated is a highly stylized representation of what happens when the variables are changed. In many applications, there is no sharp delineation of zones along the metal-removal parameter.)

Wheel depth of cut, the vertical axis in the graph, implies two different phenomena: workpiece stock removal and wheel wear. In all forms of grinding, when the wheel and the workpiece are brought together with force applied, they begin to mutually machine each other–at different rates, of course, depending on a complexity of variables.

But for any given application, the two rates can be mathematically separated: Workpiece stock removal can be extracted from depth of cut, and so can the rate at which the grinding wheel wears.

When the rate at which the grinding wheel is worn down is plotted on the vertical axis of a similar chart (against applied force on the horizontal axis), a parameter of wheel wear is derived. This line, plotted on a graph, will begin at a very low rate and continue to be low as force is gradually increased. However, as force is increased well into the cutting zone, the wheel-wear parameter crosses a threshold and starts to climb rapidly. It is beyond this threshold that grinding begins to become uneconomical because of rapid wheel wear. Once again, however, the curve changes for different types of workpieces, wheels, coolants, etc.

Wheel speed

The slope, or steepness, of the metal-removal-parameter line in the chart of wheel depth of cut vs applied force represents a constant cutting speed. But speed, measured in surface feet per minute (sfm), is actually another variable in the grinding process–and an important one.

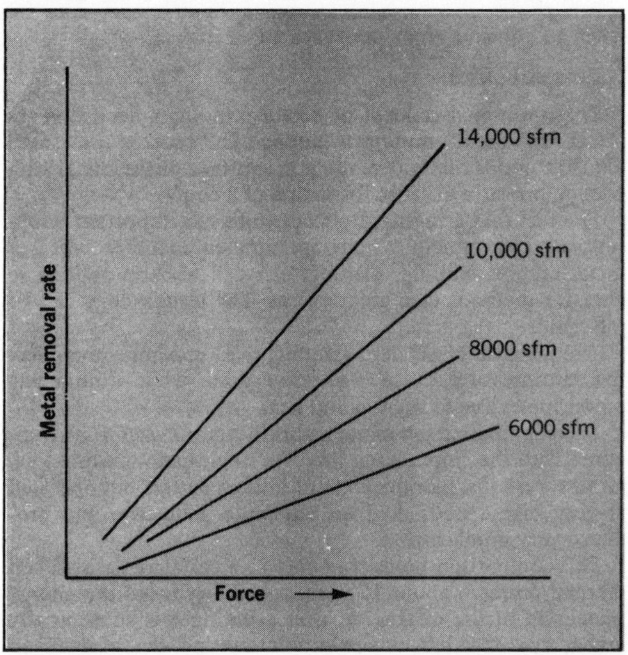

The effect of wheel speed on grinding is fairly well known. Normal operation entails decreasing force at a given feedrate as wheel speed is increased. If this change is plotted in a manner similar to that used to describe metal-removal and wheel-wear parameters, it will appear as in the graph above. As grinding speed increases, the rate of metal-removal per unit of applied force goes up also. Then again, so does wheel wear, but only after its threshold is crossed.

Grinding-wheel sharpness

Dull grinding wheels cut poorly. As a wheel becomes more dull, the metal-removal parameter falls in value; in other words, the force needed to remove a given amount of workpiece metal increases. Sharpness, therefore, can be—and has been—quantified for a particular wheel/workpiece/cutting speed condition.

Sharpness of the wheel is a term that is applied to the abrasive itself. A corollary condition, that of a "loaded" wheel, occurs when the structure of the wheel is too dense to permit free cutting. Wheel structure is the volume percentage of bond and grain in a wheel in relation to the total wheel volume. All wheels are composed of grains and bonds and are built with porous areas, voids. If the grains are too close together, if the voids are not large enough, there will not be enough clearance for chips. The wheel will load.

Abrasive sharpness, on the other hand, refers to the individual grains themselves and is a function of two attributes of the abrasive: hardness and friability. Hardness measures the ability of a particular abrasive to cut different types of material without dulling. Diamond is the hardest abrasive, to be sure, but it was no mean advance when very hard artificial abrasives were first offered at relatively inexpensive prices. (Hardness, measured on the Knoop scale, is being questioned as a valid measure. Some research has indicated that *softer* abrasive grains wear at a slower rate than do *harder* ones. It's suggested that the virtue of hardness of the grain itself undergoes some microscopic change during the dynamic grinding process and that softer grains easily take on a surface hardness that may be harder than that of "hard" grains. Obviously, more research is needed in this particular area.)

A second attribute contributing to abrasive sharpness is that of friability. This is the relative ability of a grain to fracture when its cutting surface becomes dulled. New, sharp edges of grains that have already been cutting the workpiece are thus exposed. The opposite quality of friability is toughness, the ability to withstand grain fracture. Even within the same abrasive type, one variety may be more friable than another. Green silicon carbide, for example, fractures more easily than the black variety. Aluminum oxide is graded all the way from heavy-duty (tough) to regular, to semifriable, to friable. Mixtures with aluminum oxide, e.g. zirconia alumina, are generally tougher.

But the sharpness of a wheel is also a factor not only of the abrasive but of the bond as well.

Under ideal conditions, a grinding wheel should resharpen itself as soon as the grains become dull, but not before. The bonding material is designed to eject dulled grains, thereby permitting new, underlying (sharp) grains to cut the workpiece. The wheel will do this while removing stock as rapidly as is consistent with the desired finish. In practice, a wheel that sharpens itself automatically all the time will wear away too rapidly for economical use. A properly selected grinding wheel will need occasional dressing.

Coolants and coolant delivery

Grinding, like all machining processes, can develop considerable heat. To keep this heat as low as possible and to flush out the wheel, many precision-grinding operations are performed with water-soluble cutting fluids. In some instances, such as thread grinding, straight oils are best. In others, dry grinding is preferred. And in still other operations, notably in abrasive cutoff, the best use of coolant directs the fluid flow at the abrasive wheel after it has departed the cut, but not at the wheel/workpiece interface itself.

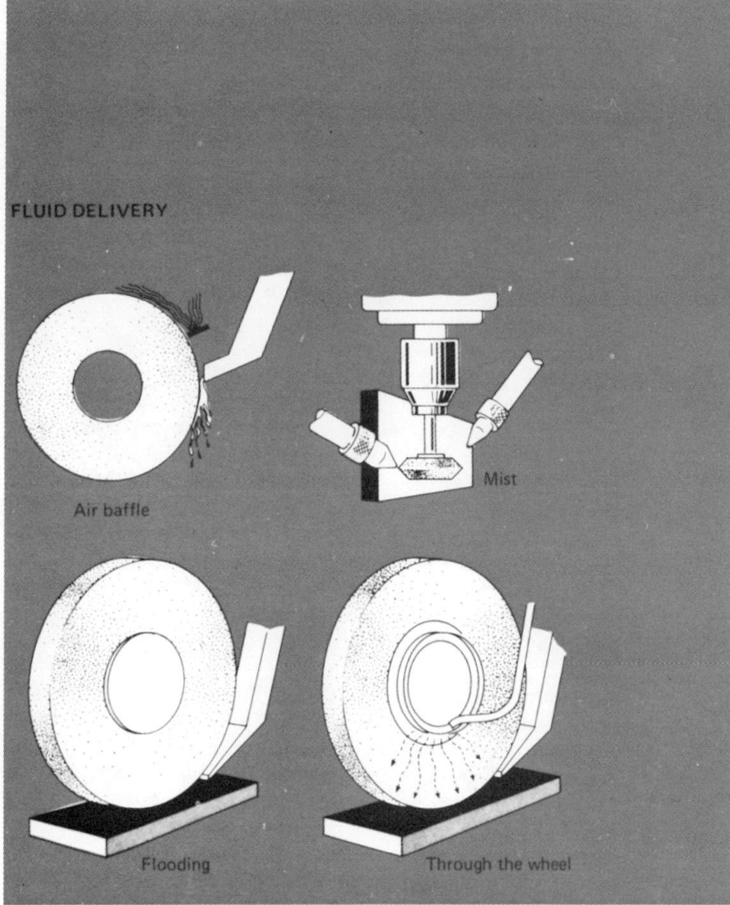

But the normal use of coolant serves the threefold purpose of keeping the workpiece cool, the wheel clean, and the chips lubricated. As such, the correct application of the fluid is extremely important. It should be applied in large quantities and under some pressure. Coolant delivery, furthermore, should be tailored to the type of operation and machine used.

There are a number of different techniques for applying coolant. The principal ones include flood, flooding with the use of an auxiliary air dam, through the wheel, and mist.

The most popular method of coolant delivery is by flooding the wheel/workpiece interface. A coolant pump, built into the grinder, provides moderate pressure to propel the fluid through a nozzle directed at the cutting edge. Large-enough quantities are delivered to effect heat dissipation, and the pressure of the fluid is usually enough to clear chips and swarf from the cutting interface.

On most center-type and centerless operations, application is easy, but this system is not totally without its problems. Hollow rolls and tubing sometimes present difficulties. With thin walls, it is often difficult to dissipate the heat quickly enough, and rapid grinding may burn the workpiece. Work of this kind can often be handled by filling the tube with coolant and plugging the ends; this will reduce grinding time and produce a more accurate part. Another problem sometimes encountered with flood cooling is that the pressure of the fluid is not enough to clear away chips. A higher-pressure pump may be installed, or, more simply, the nozzle can be narrowed, thereby increasing the venturi effect and increasing the velocity of the exiting fluids.

Still another difficulty with flood-cooling systems is that of the air barrier. As the grinding wheel rotates, it sets up a laminar flow of air around it. This airflow tends to rotate with the wheel and to cling to it. It becomes, in effect, a "second skin" that effectively prevents the coolant from reaching the wheel/workpiece interface.

Air dams have been successfully used to break up this laminar airflow. A blade at least as wide as the wheel is placed in radial fashion near the wheel but not touching it. This blade effectively breaks up this airflow and permits lower-pressure coolant to enter the cutting interface, where it can do its job.

Research is still being conducted to determine the proper distance from the dam to the wheel, the height above the cutting area, and specific means to have the dam become self-compensating as the wheel wears. But this relatively new airflow concept has already had good results for cases in which workpiece burning was a problem.

Internal grinding presents a more difficult problem. The coolant must flush the chips and wheel debris out of the hole and, because good practice calls for the largest wheel possible, there may be difficulty in getting sufficient fluid into the bore. The stream should be directed so the fluid will be dragged between the wheel and the workpiece by the wheel's rotation. For other than blind holes, the fluid is sometimes introduced from the rear through the spindle.

Through-the-wheel methods continue to be successful in applying coolant directly to the grinding interface. Fluid is pumped through tubes to holes in the grinding-wheel flanges; it flows through these flanges and through matching holes in the blotter and into the wheel itself. Centrifugal force literally pulls the coolant out to the periphery through the pores in the wheel structure. Often used in conjunction with flooding techniques, through-the-wheel cooling has the advantage that fluid emitted through the pores tends to wash away chips that might otherwise load the wheel. Its major disadvantage is associated with coolant cleanliness: unless the rest of the system is kept very clean and the pump filter is changed when necessary, the wheel itself can act as a filter and might soon become loaded from the inside out.

Mist coolants are becoming increasingly popular as grinder operators realize their heat-dissipation advantages. Coolant applied to the grinding interface in a mist tends to evaporate, thereby absorbing more than its share of heat. Modern mist-coolant systems are portable, and their nozzles can be adjusted to deliver either a mist, for maximum heat removal, or a stream, to flush away small chips.

Whatever method of delivery is selected, the coolant itself must be suited for the job. In cutting easy-to-grind materials at normal speeds, water-soluble fluids will suffice. But when grinding stresses increase or when, as in high-speed grinding, chips tend to weld themselves to the wheel, a better coolant is necessary. This may be a straight oil or a water-based material of decreased water content.

Grinding performance deteriorates with the decreasing quality of fluid. Experts agree that this is one area in which "you get what you pay for."

Use of better-grade coolant increases the degree of lubricity in the grinding process. It is interesting to note that lubricity also can be increased without the use of fluids. One manufacturer recommends the applications of microcrystalline wax products to the wheel for dry grinding. When the wheel heats up, the wax melts and lubricates the grain. This is similar to some machinists' practice of placing spots of oil on the workpiece during dry grinding: a practice, some insist, that started when operators were hesitant to crank up a coolant system whose bateriocidic odor-suppressing properties were known to have expired.

Specialized processes

The topics discussed thus far in this report have confined themselves to the types of methods most considered "conventional." In addition, there are the more specialized uses of abrasive tools, and some of these are gaining significantly in their importance.

Abrasive machining is one of these processes. Although it is currently fashionable to call any technique that removes material with a multitude of tiny tools "abrasive machining," the term, by definition, means a stock-removal process that brings a workpiece to size and shape within commercial tolerances and surface-finish specifications. Popularized by Norton some 15 years ago (AM—May 29'61,p58), abrasive machining implies the removal of moderate to heavy amounts of stock with bonded abrasives and has been used, to some degree, for over a generation. But the process is neither a roughing operation nor a finishing one. Emphasis is placed on high horsepower within the machine; the method takes advantage of the fact that metal-removal rate increases with feedrate up to the maximum machine rate.

The advantages of applying increased horsepower to the abrasive-cutting process particularly appears in decreased machining time. As such, the concept of abrasive machining has challenged other metal-removal processes in terms of efficiency, and it has had a decided effect on the way modern grinders are designed.

High-speed grinding, like abrasive machining, is a specialized technique that attempts to capitalize on the increase of one of the grinding variables to increase productivity. In this case, the increased variable is cutting speed. Whereas normal grinding occurs at approximately 6000 sfm, high-speed grinding pushes the abrasive velocity past 8500 sfm; 12,000-sfm production machines are in use, and experiments have been conducted on precision grinding at speeds up to 24,000 sfm. GM's theoretical multistation, superspeed continuous-dressing machine, mentioned in a previous section of this report, will be an attempt to push high-speed grinding to an upper limit of productivity.

In practice, though, higher cutting speeds have serious implications for the machine. It's usually necessary to set up the main drive motor on a solid, separate foundation beside the machine bed. The conventional feed method, similarly, is changed to a workpiece infeed, rather than a wheelhead infeed, to keep moving masses as rigid as possible. Additionally, these production grinders require special guarding, delicate wheel-balancing systems, and automatic gate closures. The problem is one of energy containment, for, if something goes wrong, potential energy will release broken wheel fragments at something like 0.5 Mach.

Still, high-speed grinding has great potential for specialized operations. Applications like groove grinding and the machining of exotic materials that still defy single-point tooling can definitely benefit from the use of high-speed grinding—or, if you will, "abrasive machining"—techniques.

Completely at the other end of the speed spectrum, several researchers are finding out that, in limited applications, notably the grinding of carbides with superabrasive wheels, the application of low speed but high torque is in order. Such tests have reduced the cutting speed from 6500 sfm (the speed that is generally considered normal) to about 1000 sfm and observed a gain in torque required. High material-removal rates without burning the workpiece are being found in these specialized operations. ■

The production man's guide to

Cutting fluids

If you're choosing cutting fluids merely on operator preference or cost per gallon, you're misusing the fluid and wasting money. Today's cutting fluids are tailor-made 'tools' that require careful selection. Here is a guide to what these 'tools' are and what they can do

The metalworking industry has played and continues to play a leading role in the recent revolution in production processes. Within the memory of many in the industry, the machining of metals has progressed from single-operation, belt-driven, hand-operated machines to multi-purpose automatic and semi-automatic machines that can produce one or thousands of parts to tight tolerances—under manual or programmed control.

Today, the metalworking industry is meeting demands for more parts, better parts, more complex parts through constant improvement in metalworking techniques: improved machine tools with higher speeds and feeds, better cutting tool materials, and greatly improved cutting fluids for wide varieties of cutting operations.

A premium cutting fluid as marketed today is far different from those considered to be the ultimate in performance only ten to fifteen years ago, both in the components that go into modern fluids and in the manufacturing techniques employed in making them.

By John J Dwyer, Jr

Today's metal cutting fluids contain a wide selection of special chemical agents designed to supply a definite degree of lubricity, surface activity, stability, and anti-weld properties.

Modern fluids are quite remarkable in what they can do. On jobs where machining performance was entirely satisfactory, a simple change in fluid has boosted production or otherwise improved machining performance by 20% to 30%. For many jobs, particularly severe, troublesome ones, choice of the proper cutting fluid has more than doubled performance.

History of cutting fluids

Cutting fluids are not new. Liquids have been used as aids to metal cutting since 1883, when F W Taylor first proved their value. Now regarded as the father of cutting fluid research, Taylor in 1883 showed that a heavy stream of water, flooding the tool-chip-workpiece contact area during cutting, permitted cutting speeds to be increased 30% to 40%.

This early discovery led to the development and use of fatty oils for all types of metal cutting operations. Next, mineral oils were formulated for machining brass and other nonferrous alloys, and for light cutting on steel. For more severe operations, blends of mineral oil and lard oil proved highly effective.

With the need for higher production came the development of better cutting tool materials; this in turn created a demand for improved cutting fluids. At this point, extensive research brought forth chemical additives that could be added to cutting fluids to impart the properties needed in heavy-duty machining.

A fluid for every job

Continuing research and development has brought us to the point where today there is no cutting operation that need be hindered by the lack of a proper cutting fluid. We have the straight cutting oils, with all their possible combinations of blends, and we have additives such as sulfur, chlorine, and/or phosphorus and other chemicals to increase the working range of these fluids.

In an effort to make use of the excellent cooling ability of water, the chemical and petroleum industries came up with emulsifiable, or soluble, oils which mix with water to form emulsions of high cooling ability. These emulsions are widely used for high-speed machining and grinding operations.

The newest members of the cutting fluid family are the chemical cutting fluids. These fluids, blends of a number of chemical agents in water, are basically coolants, though some are also lubricants. Strictly experimental only a short while ago, these chemical fluids are being used in hundreds of up-to-date metalworking shops and are finding ever-increasing applications in the industry.

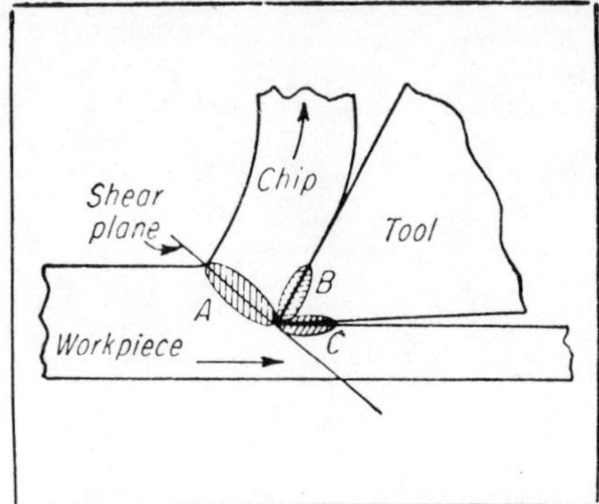

1. Heat sources in cutting: 60% or so comes from shear zone A, 30% from chip-tool zone B, 10% from tool-work zone C

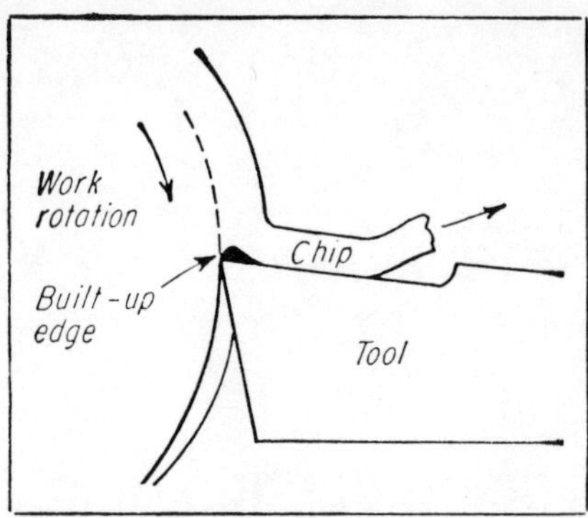

2. Built-up edge, a wedge-shaped mass of hardened metal, is formed by chip fragments pressure-welding to the face of the cutting tool

Functions of a cutting fluid

Basically, a cutting fluid should accomplish five things:

- Cool the work piece and tool
- Reduce friction
- Protect work against rusting
- Provide anti-weld properties
- Wash away the chips.

The relative importance, of course, depends upon the material being machined, the cutting rate, the cutting tool used, and the finish required on the part.

No matter how one looks at it, however, the prime function of a cutting fluid is the control of heat in the machining process. It does this in two ways: by straight cooling action (carrying away heat generated in cutting), and lubricating action (reducing the heat generated by reducing the amount of friction).

Before we can discuss the function of a cutting fluid in detail, we must look closely at how a cutting tool cuts. And it's not as simple as it sounds.

How a cutting tool cuts

Just how does a cutting tool function? Originally, it was thought that chips formed in metal cutting were created in much the same way that wood is split by an ax, that is, that the work split ahead of the tool. Investigators reported seeing such a crack in the work metal ahead of the tool, giving birth to the notion that cutting fluids get into the cutting zone by entering a large crack ahead of the cutting tool. (This may well be the case in machining brittle materials like cast iron and bronze, but this theory does not hold true for the majority of metals.)

That chips are formed by plastic deformation was first mentioned by Rosenhain, of the Staffordshire Iron and Steel Institute, in 1906. In this shearing process, the metal in the area immediately ahead of the cutting edge of the tool is severely compressed, resulting in temperatures high enough to allow plastic flow. This action requires tremendous power; it has been said that approximately 97% of the useful work done during metal cutting is released as heat.

As the atoms in the metal ahead of the tool are disturbed, the friction involved in their sliding over one another in the shear zone is thought to be responsible for 60% or more of the total heat generated (Fig. 1). This "internal" friction and the heat it generates can be compared to the friction and heat caused by bending a paper clip back and forth until it breaks.

As the tool continues to push through the workpiece, a chip eventually slides up the cutting face of the tool. This sliding creates an "external" friction, releasing heat (about 30% of the total heat generated in cutting). As the flank of the tool slides over the cut surface, it too produces friction and releases heat (the remaining 10% or so). As the tool wears, these percentages of heat liberated in the various zones are likely to vary considerably.

Built-up edge on a tool

As the cutting tool continues to cut into the workpiece, the tool face and flank are chemically "clean," and so is the metal rubbing against these surfaces. Investigations have shown that freshly machined ("clean") surfaces brought into rubbing contact in the absence of any contaminants undergo metal-to-metal seizure at the points of actual contact.

In other words, pieces of the chip, given sufficient time, will tend to weld to the face of the tool. (This is not surprising when you consider that the pressures involved may range from 50,000 to 300,000 psi and possibly higher.)

The pressure-weld so formed is known as a built-up edge (Fig. 2), and it can be either helpful or detrimental. The built-up edge is a transient, wedge-shaped mass of metal that is much harder than the workpiece (actual hardness varies with the strain-hardening characteristics of the metal being machined).

If the built-up edge is large and flat along the tool face, it decreases the effective rake angle of the cutting tool, making it necessary to exert more power to cut the metal. If this built-up edge keeps breaking off and reforming, it mars the surface of the part with rough spots, and results in excessive flank wear and cratering of the tool face. In fact, it has been authoritatively stated that "almost all of the roughness on a machined surface consists of tiny fragments of metal which have been left behind by a built-up edge from the nose of the cutting tool."

If, however, the built-up edge is small and sharp, it increases the effective rake angle and lowers the required cutting force. It also helps protect the cutting edge of the tool.

As we will see later, one of the jobs of a cutting fluid is to control the amount of built-up edge.

Action of the cutting fluid

We mentioned earlier that cutting fluid must provide both a cooling and a lubricating action. For a liquid to be most effective in dissipating heat it must have a high specific heat, a high heat of vaporization, and a high thermal conductivity. Oil, unfortunately, is a poor liquid in these respects.

Actually, water is the best coolant, but it promotes rust and has

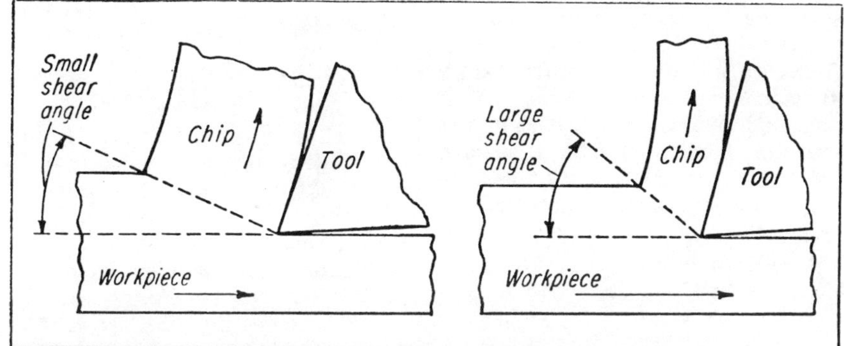

3. Small shear angle (left) produces short, thick chip and considerable heat. The application of a cutting fluid reduces friction between the chip and tool, thereby increasing the shear angle (right) and reducing the amount of heat generated during cutting. Cutting fluid also controls built-up edge

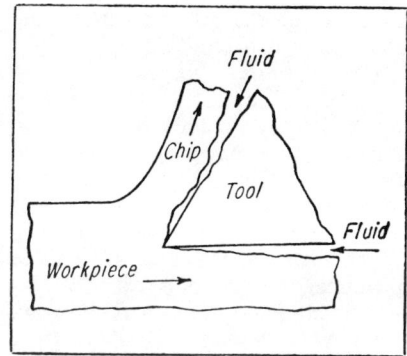

4. Cutting fluids have two paths to the cutting edge: between the chip and the tool or between the work and the tool. (See text for details)

no lubricity at all. (The chemical coolants and water-based emulsions overcome these drawbacks and are widely used in industry for just that reason.)

The second way in which a cutting fluid can reduce tool temperature in machining, that of reducing the amount of heat generated, is not as widely understood as is the relatively simple heat transfer effect.

An analysis of the mechanics of cutting reveals two interesting things about the shearing process responsible for chip formation.

First, for a given material, cutting tool, and metal removal rate, the amount of heat produced in the shear zone depends essentially upon the size of the shear angle (Fig. 3). If we think of the plastic flow, or shear, as taking place along a single plane, the shear angle is the angle between that plane and the direction of tool travel.

If this shear angle is small, then the plane of which deformation takes place (the shear plane) runs out a considerable distance ahead of the tool, so that the layer of metal removed is deformed into a short, thick chip. This severe cutting action generates considerable heat.

But if the shear angle is large, the shear path is short. The result is a longer, thinner chip, and considerably less heat.

But what can a cutting fluid do to influence the size of the shear angle? Remember that we said that an analysis of cutting mechanics revealed *two* interesting facts. The first was that the amount of heat generated depended upon the size of the shear angle (small angle, much heat; large angle, less heat). The second fact is that the size of the shear angle is controlled directly by the coefficient of friction between the sliding chip and the tool face. The lower the friction, the larger the shear angle. Here is where the cutting fluid exerts its influence.

If the friction between the chip and the tool face is reduced by application of a lubricating-type cutting fluid, then the heat coming from this rubbing friction is also reduced. But, at the same time, this reduction in friction causes the shear angle to become larger and, consequently, heat coming from plastic deformation (the major source of heat) is also reduced.

The cutting fluid also controls the built-up edge on the tool by reducing the friction between the tool and the chip, thus minimizing the chip-to-tool welding responsible for the built-up edge.

Getting the fluid to the cut

In lubrication practice, it often happens that sliding parts cannot be separated by a fluid film of lubricant because of a combination of high pressure between the parts, rough surfaces, and low viscosity of the lubricants. This is the case with most metal cutting operations. Under such conditions, metal-to-metal contact can be prevented by using lubricants containing additives that encourage the formation of an adsorbed film. Such lubricants are called extreme-pressure lubricants.

This adsorbed film is composed of molecules of the lubricant which have become chemically combined with the surface atomic structure of the metal. It is a tough film and resists removal even at high pressures and high sliding velocity. This type of lubrication is referred to as boundary lubrication, for the lubrication is provided at the boundary or interface of the two contacting surfaces. (Such films are often only three or four molecules thick—a few millionths of an inch.)

But how does the cutting fluid penetrate to the cutting edge of the tool? In Fig. 4, where the roughness of the surfaces is greatly exaggerated, we see that the fluid can get to the cutting edge by either of two routes: between the chip and the tool face, or between the workpiece and the tool flank.

(Investigations have shown that when turning with high-speed-steel tools, tool life can be doubled by applying cutting fluids at the tool flank rather than by conventional flooding, and carbide tool life could be increased by flooding at the chip surface and also from the bottom up along the tool flank.)

Obviously, not too great a flow of fluid will run down either of these two paths to the cutting edge, for both the chip and the workpiece are moving in directions opposite to the fluid flow. Furthermore, the extreme thinness of the labyrinth formed by the surface irregularities discourages copious fluid flow.

It is probable that the cutting fluid penetrates part of the way (some may go all the way) to the cutting edge by capillary action, then, because of the high temperature at the chip-tool interface, the liquid vaporizes and reaches the cutting edge as a gas. This gas provides a pressure separation between the chip and the tool face and also carries to the edge the additives that provide the adsorbed film.

Proper choice of additives will promote formation of adsorbed films on the freshly cut metal. These films are more easily sheared than the parent metal, thereby protecting the tool face from pressure-welding.

For example, suppose that the cutting fluid used in cutting a steel workpiece has a chlorine-containing additive of proper chemical reactivity. Then, at the chip-tool interface, the chlorine (whether it gets there as a gas or liquid) will react with the freshly cut steel to produce an iron chloride film.

Iron chloride is a relatively weak solid with a fairly high melting point. As such, it provides an excellent lubricant to keep the chip and the tool apart, thereby reducing chip-tool friction.

Tailored to the job

Thus, the action of a cutting fluid in reducing friction in metal cutting depends upon its chemical properties. A cutting fluid must be tailored to contain just the right amount and type of additives. Too weak a reaction, and the lubricating film may not form; too strong a reaction and both the tool and the workpiece may be chemically attacked.

It has been shown that both the cooling and the lubricating actions of a cutting fluid are important in metal cutting. But it is also necessary to understand how the relative importance of each is affected by the machining process itself.

At low cutting speed, cutting fluids vary greatly in their ability to reduce chip friction. In other words, at low cutting speeds very considerable benefits can be obtained from chemical action. However, as cutting speed increases, there is less time for the fluid to get to the cutting edge—and less time for it to react—and friction-reducing ability, therefore, falls off with increasing speed.

At speeds of about 100 sfpm, chemical action is quite a bit less effective than at lower speeds. And at speeds of about 400 sfpm, such as are used with carbide tools, chemical action is practically nil. Thus, at high speeds good cooling ability is of prime importance, while at low speeds good friction-reducing ability is the prime requisite. An all-round cutting fluid should combine both these properties to the highest degree possible.

Tailoring of cutting fluids, therefore, is a problem requiring full knowledge of the chemistry of various fluids and chemicals, and a knowledge of the physics of the metal cutting process.

For just these reasons, this report should be considered an introduction or a guide to cutting fluids. It will not attempt to make specific recommendations. With this in mind, let's look first at the various cutting fluids that are available, then at how the machining operation and the material being cut influence the selection of the cutting fluid.

Types of cutting fluids

There are so many cutting fluids on the market today that it is easy to overlook the fact that almost all of them fall into one of three types: cutting oils, emulsifiable oils, or chemical (often called synthetic) fluids.

First, let's list the various types available, bearing in mind that this is just one of many possible breakdowns. This will become evident as the individual formulations are discussed later. Here are our three types:

A. Straight cutting oils (not mixed with water)

1. Inactive oils
   - a) Straight mineral oils
   - b) fatty oils
   - c) fatty oil-mineral oil blends
   - d) sulfurized fatty-mineral oil blends
2. Active oils
   - a) sulfurized mineral oils
   - b) sulfo-chlorinated mineral oils
   - c) sulfo- or sulfo-chlorinated fatty oil blends

B. Emulsifiable oils ("soluble" oils)

- a) emulsifiable mineral oils
- b) super-fatted emulsifiable oils
- c) extreme-pressure emulsifiable oils

C. Chemical (synthetic) cutting fluids

- a) true-solution type
- b) wetting-agent type
- c) wetting-agent type with extreme-pressure lubricant

Straight cutting oils

Straight cutting oils are "straight" only in that they are not mixed with water for use in metal cutting. They are available in many formulations, ranging from straight mineral oils to highly compounded blends specifically formulated for severe operations.

Straight cutting oils, as can be seen in the list, are classified either as "active" or "inactive." These terms relate to the oils' chemical activity, or ability to react with the metal surface at elevated spot temperatures to protect it and improve machining.

An active oil is defined as one that will darken a copper strip immersed in it for three hours at 212 F. An inactive oil, under the same conditions, will not darken the strip.

Both types of cutting oils are used for machining, but in somewhat different applications. The difference between the two types of oils is that in the active oils a given amount (usually about 2%) of sulfur has been added to the oil to produce an active sulfurized oil, which upon reheating (in the cutting zone) will release some of this added sulfur to react chemically with the metal work surface.

On the other hand, the entire content of sulfur in an inactive oil is natural sulfur contained in that oil. Because it is tightly bonded within the hydrocarbon structure of the oil, the sulfur is not released and has no value chemically in the functioning of the cutting fluid during machining.

Also included in the inactive oil group are such fluids as fatty oils, fatty oil-mineral oil blends, and sulfurized blends of these oils. These are considered inactive because the sulfur is so firmly attached that little is released to react with the work surface.

Active cutting oils are generally used in machining steel. These oils may be dark or transparent, straight sulfurized or sulfo-chlorinated, mineral or fatty compounded. The dark oils generally contain more sulfur than the transparent sulfurized oils, and are usually considered better for heavy-duty jobs. Today, however, the newer additives and concentrates make the transparent oils just as suitable for tough machining applications.

Chlorine compounds, if chemically active at tool temperatures, act similarly to sulfur compounds, but usually at lower temperatures, to form a metallic chloride film at the tool-work interface. Both sulfur and chlorine, among other chemicals that are added to cutting oils, help to provide extreme-pressure lubrication and anti-welding properties under the conditions usually found in metal-cutting operations—high unit pressure, low sliding motion, high spot temperature.

Let's look now at the various oil products available.

Inactive cutting oils

Straight mineral oils are occasionally used as cutting fluids today. Because of their low viscosities (usually 40 to 200 SSU at 100 F), straight mineral oils have faster wetting and penetrating factors and, therefore, are used for light-duty machining of nonferrous metals like aluminum, magnesium, and brass, where the lubricating and cooling requirements are not severe.

These oils are also effective in machining sulfurized and/or leaded ("free-cutting") ferrous metals, for these materials have the lubricant built right into them. They machine easily without significant built-up edge, so a boundary lubricant (one that reduces friction by minimizing the minute weldments that tend to form between the chip and the tool) is not needed in the oil itself.

If these free-cutting steels were to be machined with an active sulfurized oil, the sulfur in the oil might react chemically to wear or erode the tool surface.

Straight mineral oils are also recommended for certain difficult operations, such as tapping and threading of white metals.

Greatest use today for a straight mineral oil is for blending with cutting oil bases. The latter contain concentrated amounts of anti-weld agents, oiliness agents, or a combination of both. With the proper base oils, a number of cutting fluids can be prepared to meet the requirements of many individual cutting operations.

Fatty oils, though widely used at one time, find limited use today as cutting fluids, mainly because of cost. In general, blends of fatty oils and mineral oils perform almost as well as straight fatty oils, except in some special operations like forming gear hobs on machines using high-speed-steel cutting tools.

The most used fatty oils are lard oil and sperm oil. They are used in severe operations on tough nonferrous alloys, where a sulfurized oil might produce undesirable discoloration.

Fatty oil-mineral oil blends are combinations of one or more fatty oils (usually lard oil) blended into straight mineral oils. The percentage of fatty oils in the blend may vary from 10% to 40% depending upon the depth of cut, cutting speed and feed, and the type of chip in the machining operation.

Fats in mineral oils reduce the interfacial tension of oil to metal, in a manner similar to the action of soap or wetting agents in water. Fatty-mineral oils have better wetting and penetrating properties than straight mineral oils and provide a better machined finish on ferrous and nonferrous metals.

These oils are used extensively in automatic screw machines where the operations are not exceptionally severe and where high surface finish and precision work is required. Noncorrosive to both ferrous and nonferrous metals, fatty-mineral oils are also widely used for machining copper and its alloys.

If machined parts are to be subsequently case-hardened, and it is inconvenient to wash the parts thoroughly before hardening, a fatty-mineral oil is often used in place of a sulfurized or sulfurized-chlorinated oil (even if one of these latter types would have provided better machining characteristics). The reason: The sulfur from the sulfurized oil, if not washed off, could result in surface softening rather than hardening.

Sulfurized fatty-mineral oil blends have excellent lubricity and stain less than sulfurized mineral oils. They are made by chemically combining sulfur with fatty oils (usually lard or sperm oils) under carefully controlled conditions, then "cutting" the resulting product with mineral oils of selected viscosities to make a finished product with controlled concentrations of fat and sulfur.

Such oils will be inactive (will not react chemically) at lower temperatures, but will be active above 700 F and under pressure conditions at the chip-tool interface.

Many sulfurized fatty-mineral oils can be used in machining nonferrous metals, especially where a high surface finish is required. These oils are commonly used in job shops where nonferrous and ferrous metals are worked in the same machines. They are also used as dual- and tri-purpose oils in multiple-spindle bar and chucking machines as the cutting fluid, the lubricant, and the hydraulic fluid combined.

The excellent anti-weld characteristics and lubricity of the sulfurized fatty-mineral oils are essential where cutting pressures are high and where tool vibration is likely to be excessive.

Sulfurized-chlorinated fatty-mineral oil blends are interchangeable with sulfurized fatty-mineral oil blends but have better anti-weld properties at lower temperatures and pressures. Chlorine's primary function as a cutting oil component is to act as an extreme-pressure agent, but it increases anti-weld effectiveness as well.

Active cutting oils

Sulfurized mineral oils, because of their added sulfur have increased cooling, lubricating, and anti-weld properties. They are useful for general and severe machining of tough, highly ductile metals. The most commonly used oils of this type are light-colored transparent oils with sulfur contents of 0.5 to 0.8%.

Of high chemical activity, these oils are effective in machining low-carbon and plain steels and other highly ductile metals with low machinability ratings, because of the lower temperatures generated in cutting these materials. Highly sulfurized mineral oils are not suitable for machining copper and its alloys, for the sulfur would stain these metals.

Sulfo-chlorinated mineral oils usually contain up to 3% sulfur and 1% chlorine. They can be made by adding a chlorinated wax to a sulfurized mineral oil, or by combining a sulfur chloride with mineral oil. (The chlorine in oils made by the latter method is more reactive and may promote corrosion in the presence of moisture.

Superior anti-weld characteristics of sulfo-chlorinated mineral oils prevent excessive built-up edge, and make longer tool life possible. These oils are excellent for threading soft draggy steels.

Because the chlorine in sulfo-chlorinated mineral oils takes effect at a lower temperature than the sulfur, this combination is employed whenever anti-weld and pressure-resisting properties must be effective over a wide temperature range.

These oils are more effective than the sulfurized mineral oils for machining tough low-carbon steels and chrome-nickel alloys.

Sulfo- or sulfo-chlorinated fatty oil blends, because of the fats present, contain more sulfur than the sulfo-mineral blends and are much more active at lower temperatures than the inactive-type sulfo- or sulfo-chlorinated fatty-mineral oil blends.

With their high oiliness and high extreme-pressure properties, these oils are a most effective cutting fluid for a wide range of machining operations, particularly heavy-duty machining.

Emulsifiable or soluble oils

To be most effective as a coolant, a liquid should have a high specific heat, a high thermal conductivity, and a high heat of vaporization. Neither mineral oils nor blends of mineral oils are among the best fluids in these respects; consequently, they are not the most effective fluids for removing heat once it has been generated.

Actually, water is about the most effective cooling medium known, but it has several obvious drawbacks: It promotes rust and has little or no value as a lubricant.

Lubricating and rust-preventing properties, however, can be combined with water's excellent cooling properties in what are called soluble oils—oils compounded so they will form stable emulsions when mixed with water.

Soluble oils, more properly called emulsifiable oils (for they form emulsions not true solutions), are mineral oils containing a soap or soap-like material to make them "soluble" in water. These emulsifiers break the oil into minute particles and keep the particles dispersed in water for long periods of time.

Because of their cooling/lubricating properties, emulsions are used for metalworking operations with high cutting speeds and low cutting pressures, accompanied by considerable heat generation.

Manufacturers of cutting fluids supply emulsifiable oils as concentrates that the user prepares by mixing with water. Mixtures can range anywhere from one part oil in 100 parts water to a 1:5 oil-water ratio. The leaner emulsions are used for lighter machining operations and where cooling is the essential requirement. Lubricating properties and rust prevention increase with higher concentrations of oil.

Emulsifiable mineral oils are composed of a light-bodied (usually 100 SSU at 100 F) mineral oil made emulsifiable with water by introduction of petroleum sulfonates, amine fatty acids, resin condensates, chromium oleate, and coupling agents such as glycol.

Emulsifiable mineral oils are by far the most widely used of the soluble oils, primarily because of their low costs. They have good rust inhibition and adequate lubricity for the ordinary cutting applications. Dilutions usually range about one part oil in 20 parts water.

Super-fatted emulsifiable oils are similar to the emulsifiable mineral oils just discussed, but they are oilier because of the added fatty oils (lard, rapeseed, sperm oils) and, therefore, can be used for tougher machining operations. They are used richer—something like one part oil in eight to 15 parts water. Quite often they are used to machine aluminum.

Extreme-pressure emulsifiable oils contain sulfur, chlorine, phosphorus, as well as fat to impart boundary-lubrication properties for even tougher machining operations than those handled by the previously discussed emulsifiable oils. These extreme-pressure oils are usually diluted one part oil in five to 20 parts water.

In some cases, these oils, commonly known as heavy-duty soluble oils, have replaced oil-type cutting fluids for broaching, gear hobbing, gear shaping and shaving, and turning operations. Because water-based coolants cool better than oils, the smoking and fogging usually associated with heavy cuts are eliminated, or at least minimized, when soluble oils are used.

Some emulsifiable oils are formulated specifically for grinding operations and are used in concentrations of one part oil in 25 to 60 (and sometimes even higher) parts water.

Chemical cutting fluids

Chemical cutting fluids, sometimes called synthetic fluids, are the newest members of the cutting fluid family. In a few short years they have come out of the experimental and only-for-the-real-special-job category and, today, are being used in hundreds of up-to-date metalworking plants.

These fluids are blends of a number of chemical agents in water. All of these fluids are coolants; some are also lubricants. Chemical cutting fluids contain no mineral oil; they form either true solutions in water or extremely fine colloidal solutions.

Chemical agents that go into these fluids include amines and nitrites, for rust prevention; nitrates, for nitrite stabilization; phosphates and borates, for water softening; soaps and wetting agents, for lubrication and reduction of surface tension; phosphorus, chlorine, and sulfur compounds, for chemical lubrication; glycols as blending agents and humectants; and germicides, to control growth of bacteria.

True solution fluids contain mostly rust inhibitors (inorganic and organic nitrites), sequestering agents, amines, phosphates, borates, glycols or ethylene or propylene oxide condensates. These fluids are best used as grinding solutions, where they prevent rust and permit rapid heat removal.

They have a tendency, however, to leave hard or crystalline deposits when water evaporates from them, and these may interfere with the operation of chucks, turrets, slides, or other moving parts.

These agents form clear solutions in water (but often carry a dye to color the water) and are used one part in 50 to 250 parts water. These chemicals can also be added to emulsifiable oils or to other chemical coolants to enhance their rust-inhibiting properties.

Wetting-agent types contain one or more wetting agents to improve the wetting action of the water, thereby providing more uniform heat dissipation and more uniform anti-rust action. This type of chemical fluid may also contain mild lubricants (organic or inorganic), water softeners, anti-foaming agents, and humectants.

The wetting-agent type of chemical fluid is a very versatile product; it has an excellent lubricating effect on moving machine parts and permits very rapid heat dissipation. Used one part in 10 to 30 parts water, it has replaced cutting oils on many operations. It can be used with high-speed-steel tools as well as carbide tools.

Wetting-agent types with extreme-pressure lubricants are similar to the plain wetting-agent type fluids but have chlorine, sulfur, or phosphorus additives to impart extreme-pressure or boundary lubrication effects. They are excellent for tough machining jobs and can be used with both high-speed-steel tools and carbide tools. Usual dilution is one part in five to 30 parts water.

Advantages of chemical fluids

The high heat-removal rates of the chemical cutting fluids permit higher machining feeds and speeds and also boost tool life. Barring substantial contamination by lubricating oils and hydraulic fluids, the chemical cutting fluids have a long useful life.

The detergent action of these fluids helps keep the coolant system open by preventing clogging of the pipes. And chemical fluids with wetting agents are highly compatible with other processes, such as washing. Often, such intermediate processing steps can be eliminated.

There are problems, too

Before switching to a chemical cutting fluid, consult the machine tool manufacturer. The design of some machines permits coolant to get into the machine's lubricating system. Some machines can use special water-resistant greases and oil, and this solves the problem; but other machines can't. Then, too, the lubricating and hydraulic oils could leak into the coolant system, contaminating the coolant.

Where lubricating, rather than cooling, is the prime consideration, oil-base fluids are still superior to chemical fluids.

The material being machined should also be considered. Generally, chemical cutting fluids work best on ferrous metals, though many aluminum alloys can be machined with them. Most chemical cutting fluids are not recommended for machining alloys of magnesium, zinc, cadmium, or lead.

Another problem is that some of the paints used on machinery may be removed by these chemicals; this not only mars the appearance of the machine, but it allows the paint to get into the coolant.

Cutting operations

One of the most important factors in selecting a cutting fluid is the nature of the cutting operation itself. When we consider the number of different cutting operations and then consider the possible number of combinations of these operations, it becomes clear that no one could possibly list them all (someone would always manage to combine two that had never been used together before), no less make specific cutting fluid recommendations for each of them.

But general recommendations, or suggestions, can be made for the basic operations and the materials being worked on. Such suggestions should be regarded as just that; they represent years of cutting fluid research and shop experience and can be regarded as good starting points in the search for the best cutting fluid for a particular metal cutting job.

The various machining processes differ, naturally, in metal-removal characteristics. Rated in order of decreasing severity, here is how the machining processes rank:

1. Internal broaching
2. External broaching
3. Tapping
4. Threading
5. Generation of gear teeth
6. Deep drilling
7. Boring
8. Screw machining with form tools
9. High-speed light-feed screw machining
10. Milling
11. Drilling
12. Planing and shaping
13. Turning, single point tools
14. Sawing
15. Grinding

Here, in somewhat the same order, are detailed discussions of each of these processes. (Some operations are taken out of order because of their similarity in nature or because of the similarity in cutting fluid recommendations.)

Broaching

Internal broaching is used to cut holes, grooves, or slots in machine parts, especially when the shape, size, and/or length of cut make other types of machining impractical. A broaching cutter consists essentially of a series of equally spaced teeth which take progressively deeper cuts as the tool makes its pass through the opening.

External broaching, while first used only for cutting keyways, square, hexagonal, or splined holes and other irregular shapes, is now being used for machining all sorts of external surfaces, both flat and curved. Surface broaching often replaces milling when speed and good surface finish are important.

Horizontal broaching of steel requires a heavier-bodied, more-chemically-active oil than vertical surface broaching under comparable conditions. The heavier oil clings to the horizontal broach better, and chemical activity aids in efficient cutting. For vertical surface broaching of mild steels, emulsions or solutions may be used, but the usual choice is an oil.

Here is a list of recommended cutting fluids for broaching various materials:

Steels: emulsifiable oils, sulfo-chlorinated mineral-fatty oils.

Stainless steels: emulsifiable oils, sulfo - chlorinated mineral - fatty oils.

Cast iron: sulfurized mineral-fatty oils, emulsifiable oils.

Monel, nickel: sulfo-chlorinated mineral-fatty oils, mineral-fatty oils.

Copper: inactive mineral-fatty oils, sulfo-chlorinated mineral-fatty oils, emulsifiable oils.

Brass, bronze: inactive sulfurized fatty-mineral oils, emulsifiable oils.

Aluminum and its alloys: inactive sulfurized fatty-mineral oils, emulsifiable oils.

Magnesium and its alloys: inactive mineral-fatty oils.

Tapping and threading

Although tapping and threading are essentially turning operations, the cutting speeds used are much lower than normal turning speeds. Both tapping and threading involve many small cutting edges in continuous contact with the work throughout the cut. Because of the design of the tools and the nature of these operations, the edges of the tools are shielded from the flow of cutting fluid and its cooling effect, particularly in tapping.

Here are the recommended fluids for tapping and threading various materials:

Steel: sulfo- or sulfo-chlorinated fatty-mineral oils.

Stainless steel: sulfo-chlorinated fatty-mineral oils.

Cast iron: sulfo-chlorinated fatty-mineral oils, emulsifiable oils.

Monel, nickel: sulfo-chlorinated fatty-mineral oils, emulsifiable oils.

Copper and its alloys: inactive mineral-fatty oils, sulfo-chlorinated mineral-fatty oils, emulsifiable oils, sulfurized mineral-fatty oils.

Aluminum and its alloys: inactive sulfurized fatty-mineral oils, emulsifiable oils.

Magnesium and its alloys: inactive mineral-fatty oils.

Generation of gear teeth

All of the gear-making operations—gear shaping, gear hobbing, and gear shaving—are about equal in severity.

In gear shaping, the cutting tool is shaped like a gear with cutting edges. Motion of the tool is back and forth past the gear blank, taking a cut with each forward stroke. After the first tooth space has been cut, both the blank and the tool are indexed before each pass, so that metal is removed to allow cutter and work to mesh as two mated gears.

Gear hobbing is a continuous gear-cutting operation. The cutting tool and the workpiece bear the same relationship to each other as a mated worm and gear. The tool meshes as if it were a gear itself. Essentially, a gear hob is a worm or screw that has been slotted to give it a large number of cutting edges. In the cutting operation, the gear blank turns slowly, the hob rapidly.

In rotary shaving, the tool is gear-shaped and has teeth that are closely gashed to form many cutting edges. The tool and gear mesh and rotate together. The cutting edges have a sliding motion that shears off thin, threadlike chips. In rack shaving, the tool is a rack that reciprocates in mesh with the workpiece. Straight teeth are shaved with a helical-tooth rack; helical teeth, with a straight-tooth rack.

In all of these gear-cutting operations, the teeth are cutting only during a small part of the total machining time, and each cutting edge is adequately cooled by ample exposure to air and cutting fluid. The prime requisite here, then, is lubrication.

The most often used fluids are active mineral and mineral-fatty oil blends compounded with sulfur and chlorine. For metals with extremely low machinability, highly active sulfurized oils are usually recommended.

Automatic screw machining

Automatic screw machines combine a variety of operations at various speeds—turning, drilling, reaming, tapping, threading, knurling, etc. Different operations, occurring simultaneously, are in sequence for a single part being machined. For example, on a six-spindle machine six parts are in process at the same time. A different operation in the sequence necessary to complete the job occurs at each of the six spindles.

Through automatic speed, feed, and indexing devices, each spindle carrying a part moves successively to each station without operator supervision. Each time a part is completed, the other five parts move up one operation step and a new piece starts at the first spindle station.

This multiplicity of cutting operations makes selection of a cutting fluid somewhat complicated, but for general usage with steel, a low-viscosity, chemically active mineral-fatty oil is usually recommended. In many instances, sulfurized fatty-mineral oils are used as tri-purpose fluids: cutting fluid, machine lubricant, and hydraulic fluid, all in one. Soluble oils are also used widely on automatic screw machines.

Here are the recommended cutting fluids for multi-spindle automatic screw machines and turret lathes (for drilling, turning, reaming, cutting-off, tapping, and threading operations):

Steel: sulfurized mineral oils, sulfo chlorinated mineral oils, emulsifiable oils, mineral-fatty oils.

Stainless steel: sulfo-chlorinated mineral oils, sulfo - chlorinated mineral-fatty oils, emulsifiable oils (regular and EP types).

Cast iron: emulsifiable oils.

Monel, nickel: sulfo-chlorinated mineral oils.

Copper and its alloys: inactive mineral and mineral-fatty oils.

Aluminum and its alloys: inactive sulfurized fatty-mineral oils.

Magnesium and its alloys: inactive mineral-fatty oils.

The high-speed light-feed automatic screw machines use basically the same cutting fluids, except that there is seldom any need for extreme-pressure (or heavy-duty) emulsions.

Plain milling

In the milling operation, a circular tool having a number of teeth revolves on its own axis, bringing the teeth into the work one at a time as the work feeds into the cutter.

Cutting speeds depend mainly upon the material being cut and the rates at which heat is generated and dissipated during cutting. Generally, the harder the material being cut, the lower the cutting speed. If both the work and cutter are flooded with a suitable cutting fluid, the cutting speed can be increased up to 1½ times that possible in the absence of a cutting fluid, all other conditions remaining the same.

Both light-viscosity, additive type oils and emulsifiable oils are used for milling steel; the former usually with HSS cutters, the latter with carbide tools.

Here are the recommended cutting fluids for plain milling operations on various materials:

Steel: emulsifiable oils, sulfurized mineral oils.

Stainless steel: emulsifiable oils (regular and EP types), sulfurized mineral oils, fatty mineral oils.

Cast iron: emulsifiable oils, dry.

Monel, nickel: sulfurized mineral-fatty oils (light), emulsifiable oils.

Copper: inactive mineral-fatty oils, emulsifiable oils.

Brass, bronze: emulsifiable oils, sulfo-chlorinated mineral oils.

Aluminum and its alloys: inactive mineral-fatty oils, emulsifiable oils.

Magnesium and its alloys: inactive mineral fatty oils.

The above recommendations apply to plain milling only. Multiple-cutter milling of ferrous metals is usually done with sulfurized mineral oils or mineral-fatty oils. Emulsions, sulfurized oils, and mineral oils can be used on the nonferrous materials, always keeping in mind that water-based fluids should never be used on magnesium because of the fire hazard.

Drilling

Basically, drilling is a process for making or enlarging holes. It can be done on lathes, turret lathes, and automatics, as well as on drill-presses. Drills, like turning tools, are in continuous contact with the metal they are cutting, but the cutting edges of drills are shielded from the flow and beneficial cooling action of the cutting fluid.

For this reason, drilling speeds are generally somewhat slower than those used for other operations. For conventional drilling operations, the preferred fluids are emulsified oil, and sulfurized or chlorinated mineral oils. These provide some degree of lubricity to prevent chatter and frictional heat, and they carry away the heat generated by chip formation.

Here are the recommended cutting fluids fór drilling operations on various materials:

Steel: emulsifiable oils, sulfurized mineral oils, sulfurized mineral-fatty oils.

Stainless steel: emulsifiable oils, sulfurized mineral oils, sulfo-chlorinated mineral-fatty oils.

Cast iron: emulsifiable oils, dry.

Monel, nickel: sulfurized mineral oils, emulsifiable oils.

Copper: emulsifiable oils, inactive mineral-fatty oils.

Brass, bronze: emulsifiable oils, inactive mineral-fatty oils, dry.

Aluminum and its alloys: emulsifiable oils, inactive mineral-fatty oils, kerosene and lard oil compounds.

Magnesium and its alloys: inactive mineral-fatty oils.

Deep hole drilling presents the problem of maintaining a sufficient flow of cutting oil to the cutting edges. One solution is the use of oil-groove, oil-hole, or oil-tube drills, which utilize drill flute space for cutting fluid passages.

Another approach to making deep holes is gun-drilling. Here the tool is essentially a single-point end cutter similar to a boring tool, except that it has an internal passage for fluid.

Trepanning is a hole-making operation that cuts a cylindrical path into the metal, leaving a solid core. This core passes through the hollow cylindrical cutting head as the tool feeds into the metal. Once properly started, a trepanning tool is self-piloting and cuts a straight, accurate hole in a fraction of the time needed to do the same job by boring.

In trepanning, the cutting fluid is pumped around the outside of the tool under pressure, forcing chips back through the center. Cutting fluids for trepanning must have good extreme-pressure and anti-weld properties, must be low enough in viscosity to flow freely around the tool, and must have good oiliness.

Because carbide and other new type tools are used in these operations, some shops prefer to use emulsions in dilutions from one part oil in five to 15 parts water to secure good cooling as well as lubrication. (It has been reported that some cutting oils with active chlorine and sulfur can react with the cobalt binder in carbide tools, leaching it out and weakening the cutting edges.)

Turning

Turning is simply the cutting of a cylindrical surface as the workpiece revolves in a lathe, turret lathe, or automatic. The tool feeds into the metal parallel to the axis of revolution of the surface being cut, or in a facing operation, along a line perpendicular to the spindle axis.

No matter what type of turning operation is involved—simple, single-point cutting; facing the ends of stock; cutting and squaring shoulders; cutting-off; complex turning with form tools—considerable heat is generated because the cutting edge is in constant contact with the work during cutting. But both the work and the tool are open to the cooling action of air and receive a free flow of fluid.

Here are the recommended cutting fluids for turning operations:

Steel: emulsifiable oils (both regular and EP types), sulfurized mineral oils, mineral-fatty oils.

Stainless steel: emulsifiable oils (regular and EP types), sulfurized mineral oils.

Cast iron: emulsifiable oils.

Monel, nickel: emulsifiable oils, sulfurized mineral oils.

Copper and its alloys: inactive emulsifiable oils.

Aluminum and its alloys: inactive emulsifiable oils, inactive mineral-fatty oils.

Magnesium and its alloys: inactive mineral-fatty oils.

Boring

The discussion of turning operations is fully applicable to simple, single-point boring operations, and the same cutting fluid recommendations can be made.

Planing, shaping

Normal planing and shaping operations require no cutting fluids, but for heavy planing work an emulsion may be brushed on the work surface ahead of the tool to provide moderate cooling and facilitate the cut.

Sawing

Reciprocating hack sawing, continuous circular sawing, and band-sawing are usually done with emulsions or sulfurized mineral oils. The fluids serve to clear the saw teeth, prevent chip adhesion, carry away heat, and minimize vibration of the blade by cushioning the cutting action.

Thread rolling

Thread rolling appears to offer no problems in selection of a fluid, even though this is more of a forming than a cutting operation. Consideration must be given to die life and to possible discoloration of bronze bushings in the machines. For these reasons, an inactive oil is sometimes preferred over the more commonly used light sulfurized mineral-fatty oils.

Thread grinding

Thread grinding is essentially cylindrical grinding with a wheel shaped to the thread form.

The primary aim in applying cutting fluids to thread grinding is prevention of high heat generation, and the most commonly used fluids are active mineral oils. High-sulfur mineral-fatty oils and medium-sulfur medium-chlorine mineral-fatty oils are used for thread grinding on all ferrous metals.

For thread grinding of aluminum, magnesium, and copper and its alloys, sulfurized fatty-mineral oils of the inactive type are recommended.

The oils used for thread grinding perform better with the fine-grit, dense wheels than the emulsifiable oils do; the latter tend to load up the wheel and prematurely glaze the grinding surfaces. These oils are usually higher in viscosity (300-350 SSU at 100 F) than other cutting oils, to minimize fogging of oil thrown off the high speed wheel. Also, the flash point of these oils is higher, precluding some fire hazards.

Plain grinding

Prevention of heat build-up is not as important in plain (cylindrical, centerless, and surface) grinding as in thread grinding; physical cooling is sufficient for good performance.

An opaque, milky white emulsion is the most commonly used cutting fluid for plain grinding. Generally, it is made from emulsifiable oil, although paste-type compounds are also used. As coolants, these emulsions are inexpensive, efficient for many grinding jobs, and, if properly prepared, cope with difficulties such as water condition and the usual run of contaminants.

Translucent grinding emulsions, prepared from highly compounded grinding oils, and the newer chemical solution coolants are particularly well suited for fine finish grinding. Good coolants, these emulsions and chemical coolants permit the operator to see at all times the line of contact between the wheel and the work.

When using an opaque emulsion, the operator cannot see through the fluid and, occasionally, he may have to shut off the flow temporarily, risking damage to the work surface.

Paste-type compounds high in "soap" content are also used in preparing grinding fluids. These compounds may also contain fatty oils which impart good wetting and lubricating properties. But an excess of fatty oils in an emulsion leads to loading of the grinding wheel and formation of deposits in the coolant system.

Materials being machined

Another of the major factors to be considered when selecting a cutting fluid is the material being machined. Unless we know its properties and its machining behavior under specific conditions, we can't make an intelligent selection of a fluid. What we really need to know is the machinability of the material.

But what is machinability? Definitions are hard to come by. If you are primarily interested in cutting tool performance, you're likely to say that a machinable metal is one that permits long tool life on a given job. But if your interest is in high production rates, you would probably say that it is a metal that can be cut at very high speeds at reasonable tool life.

Then, again, if your primary interest is some quality of the finished product, such as its surface finish or dimensional accuracy, your choice for a machinable metal would be one that yields these qualities at reasonable speeds and tool life. And one wit has claimed that a metal is machinable if it does not raise the ire of the machinist.

Though it is dfficult to define machinability, it is possible to rate metals according to their relative machinability, that is, the ease or difficulty of machining them.

The Independent Research Committee on Cutting Fluids of AISI has prepared tables of relative machinability that serve as a guide or framework for selecting operating conditions to achieve a desired result.

These machinability ratings are based on a value of 100% for cold drawn B-1112 steel. This value involves turning at a cutting speed of 180 surface feet per minute for feeds up to 0.007 inch per revolution and depths of cut up to 0.250 inch, using a suitable cutting fluid and high-speed-steel (18-4-1 composition) tools hardened to Rh 63-65C.

Various factors influence machinability and, therefore, the figures in these tables must be considered as averages only. For example, within the composition range permitted for B-1112 steel, it was found that unintentional variations in carbon, sulfur, and, principally, silicon contents cause the machinability index of B-1112 to vary as much as 20% below or 60% above the normal value of 100 assigned to it.

The amount of cold reduction, mechanical properties, grain size, and microstructure can, and do, influence machinability ratings to varying degrees. Bear in mind that these values are based on steels in a cold-drawn condition. Metals that have been heat treated or processed in any other way will, of course, have different ratings.

Hot rolling affects machinability in different ways: In steels containing up to 0.30% carbon, hot rolling decreases machinability; in steels containing from 0.30% to 0.40% carbon, hot rolling has little effect; but in steels containing over 0.40% carbon, hot rolling markedly increases machinability.

These machinability tables are used by most cutting fluid manufacturers as a starting point in

Machinability ratings

Class 1: ferrous

(70% machinability or higher)

A.I.S.I.	Rating %	Brinell
C-1016	70	137-174
C-1022	70	159-192
C-1109	85	137-166
C-1110	85	137-166
B-1111	95	179-229
B-1112	100	179-229
B-1113	135	179-229
C-1115	85	143-179
C-1117	85	143-179
C-1118	80	143-179
C-1120	80	143-179
C-1137	70	187-229
A-4023	70	156-207
A-4027*	70	166-212
A-4119	70	170-217
Malleable iron (ferrite)	100-120	120-200
Malleable iron (pearlite)	70	190-240
Steel casting (0.35% carbon)	70	170-212
Stainless iron (free cutting)	70	163-207
Cast iron (soft)	80	

Class 4: ferrous

(40 machinability or lower)

A.I.S.I.	Rating %	Brinell
2315*		
2350*	35	196-235
A-2515*	30	179-229
E-3310*	40	170-229
3141		
6140*	40	187-228
E-52100**	30	183-229
E-9315		
Ni-Resists*	30	
Stainless 18-8 (austenitic)*	25	150-160
Manganese (oil hardening steel)**	30	
Tool steel (low tungsten, chromium & carbon)*	30	200-218
High speed steel	30	
High carbon high chrome tool steel	25	

Class 2: ferrous

(50% to 70% machinability)

A.I.S.I.	Rating %	Brinell
C-1020	65	137-174
C-1030	70	170-212
C-1035	65	174-217
C-1040*	60	179-229
C-1045*	60	179-229
C-1141	65	183-241
A-2317	55	174-217
A-3045*	60	179-229
A-3120	60	163-207
A-3130*	55	179-217
A-3140*	55	187-229
A-3145*	50	187-235
A-4032*	65	170-229
A-4037*	65	179-229
A-4042*	60	183-235
A-4047*	55	183-235
A-4130*	65	187-229
A-4137*	60	187-229
A-4145*	55	187-229
A-4165	65	174-217
A-4615	65	174-217
A-4640*	55	187-235
A-4815	50	187-229
A-5045*	65	179-229
A-5120	65	170-212
A-5140*	60	174-229
A-5150*	55	179-235
A-8620	60	170-217
A-8630*	65	179-229
A-8720	65	179-229
A-9440	60	187-235
Cast iron (medium)	65	
Cast iron (hard)	50	

Class 5: nonferrous

(Above 100% machinability)

A.I.S.I.	Rating %	Brinell
Magnesium alloys Dow "J"	500-2000	58
Magnesium alloys Dow "H"	500-2000	50
Aluminum 11-S	500-2000	95
Aluminum 2-S	300-1500	23-24
Aluminum 7-S	300-1500	100
Brass, leaded F C, C D	200-400	50
Brass, red	180	50
Bronze, leaded	100	55
Zinc	200	75
Bronze, silicon	100	

Class 3: ferrous

(40% to 50% machinability)

A.I.S.I.	Rating %	Brinell
C-1008	50	126-163
C-1010	50	131-170
C-1015	50	131-170
C-1050*	50	179-229
C-1070*	45	183-241
A-1320	50	170-229
A-1330*	50	179-235
A-1335*	50	179-235
A-1340*	45	179-235
A-2330*	50	179-229
A-2340*	45	179-235
A-3115	50	174-217
3230*	45	184-235
A-3240*	45	183-235
A-4140*	50	184-235
A-4150*	45	196-235
A-4340*	45	187-241
A-6120	50	179-217
6130*	50	179-217
A-6141	50	179-217
A-6152**	45	183-241
Ingot iron	50	101-131
Wrought iron	50	101-131
Cast iron	50	161-193
Stainless 18-8 (Free mach'g)	45	179-212

Class 6: nonferrous

(Below 100% machinability)

A.I.S.I.	Rating %	Brinell
Aluminum bronze	60	140-160
Brass, yellow	80	50-70
Brass, red	60	55
Bronze, manganese	60	80-95
Bronze, phosphor	40	140
Copper, cast	70	30
Copper, rolled		
Everdur	60-120	200
Gun metal	60	65
Nickel (hot rolled)	20	110-150
Nickel (cold drawn)	30	100-140
Monel metal,* regular	40	125-150
as cast "H"	35	175-250
as cast "S"	20	280-325
rolled	45	207-224
"K"	50	215-265
Inconel, temper B cold drawn	45	130-170

*Annealed
**Spheroidized annealed

selecting a cutting fluid. The ratings are useful for they give us another way of measuring the severity of a cutting operation. (We already have listed the operations according to severity; now we can refine this because we know the relative difficulty of performing any given operation on any material.)

Keeping in mind that these figures were developed only for turning, that they are averages, and that they vary considerably, we could say that a metal with a machinability of 50% (C-1010, for example) would have, roughly, one-half the machinability of the standard, B-1112. Putting it another way, C-1010 steel would be twice as difficult to machine.

And if it is more difficult to machine, it calls for a cutting fluid formulated to meet the higher pressures and temperatures that may be encountered. This is how the machinability ratings help in cutting fluid selection.

Based on B-1112 steel as 100% machinable, metals are grouped into six machinability classes:

Ferrous metals:
- Class 1: 70% or higher
- Class 2: 50% to 70%
- Class 3: 40% to 50%
- Class 4: 40% or lower

Nonferrous metals:
- Class 5: 100% and higher
- Class 6: below 100%

Note that Brinell hardness ranges are given for the metals in the machinability tables. These ranges represent the most desirable hardness limits for normal machinability.

Machining properties

In this section we will discuss briefly the nature and machining characteristics of various commercial materials and how these factors influence the choice of a cutting fluid. Near the end of this section (p 118) you'll find a Cutting Fluid Recommendation Chart based upon the machinability groupings just discussed.

Let's look first at the properties of the most commonly used material: steel.

Steels

Plain carbon steels vary in their machinability depending upon carbon content. In general, the low-carbon grades are soft and "draggy," while the higher-carbon grades require higher cutting pressures and greater work input.

With steels of carbon content lower than 0.10%, cutting speeds may be quite low because of extreme ductility and heat generated. At the other end of the scale, extreme hardness due to high carbon content reduces machinability and, consequently, cutting speeds.

Free-machining grades are, basically, low-carbon grades modified by additions of various elements to improve their machinability. Sulfur content above 0.055% and phosphorus above 0.045% make steels more clean cutting. These resulfurized and rephosphorized steels facilitate machining by permitting faster cutting than do plain carbon steels, longer tool life, better surface finish, and lower costs.

A content of 1.00% to 1.90% manganese also produces a free-cutting steel. But hardness of these manganese steels increases with increased carbon content, with the result that a medium-carbon steel may, depending upon its manganese content, be hard and strong enough to have undesirable machining properties.

Leaded steels are actually sulfur, phosphorus, or manganese free-cutting steels to which 0.15% to 0.35% lead has been added for increased machinability. This addition does not affect the basic mechanical properties of the steel. Selenium is the most recent addition to steel in the search for increased machinability.

Alloy steels contain small concentrations of chromium, cobalt, nickel, molybdenum, tungsten, vanadium, and other elements, either singly or in combination. Naturally, these steels are more difficult to machine than are the ordinary steels. Cutting pressures are necessarily higher, speeds lower, cutting tools stronger—and the cutting fluids used must reflect these more severe conditions.

Stainless steels, like alloy steels, have a lower machinability rating than the carbon steels. Straight chromium stainless steels, containing from 12% to 18% chromium and 0.12% to 0.35% carbon, can be machined at speeds of 80 to 100 surface feet per minute; the same grades containing small amounts of sulfur, phosphorus, or selenium cut clean at speeds of 120 to 150 sfpm.

High - chromium, high - nickel grades, such as 18-8 stainless machine with considerable difficulty because of their ability to work-harden. Here again, additions of selenium improve machinability.

Several types of cutting fluids are used for machining stainless steels. For free-machining grades, a sulfo-chlorinated mineral oil is usually recommended. For the harder-to-machine grades, a sulfo-chlorinated mineral-fatty oil is often preferred. Emulsifiable oils (both regular and extreme-pressure types) are often used.

Cast iron

There are several classes of cast iron; they differ from each other in total carbon content (somewhere between 1.5% and 4.5%), in the relative amounts of the two forms in which the carbon appears (free or combined), and in silicon, sulfur, phosphorus, and manganese content.

The wide variation in structure of the cast irons makes for a wide divergence of machining properties, ranging from readily machinable gray cast irons to hard white irons that are hard to machine.

Gray cast iron casts into almost any shape and machines quite easily. Free carbon exists as graphite flakes, making for easy machining and giving fractured specimens a dark gray color.

White cast irons contain carbon in the combined form (cementite), as shown by the silvery-white color of fractured specimens. Hard, strong, with little or no ductility, white cast irons machine with extreme difficulty.

Malleable iron is produced by annealing white cast iron, thereby transforming the cementite to free iron and spheroidal graphite. The resulting ductility causes some dragginess, but malleable iron machines quite easily.

Nodular cast iron, whose spheroidal graphite and iron carbide structure results from the addition of small amounts of magnesium or cesium to molten high-carbon iron, is both strong and ductile. Nodular iron, as cast, machines about as well as gray cast iron, but its higher hardness requires greater cutting pressures.

Wrought iron is a two-component material; it consists of high purity iron with iron silicate (slag) added and worked in to give corrosion resistance. The glass-like siliceous slag is entrained in the iron as long, evenly distributed fibers. This structure results in rather poor machinability: the ductility of the pure iron makes the metal soft and draggy, and the slag makes it difficult to achieve a fine surface finish.

Machining of cast irons is most often performed dry, but emulsions are often used as coolants and as a means of keeping down the dust. Because cast iron is quite susceptible to rusting, any emulsion that is used must have good anti-rust properties.

Aluminum and its alloys

Compared to the ferrous metals, aluminum alloys machine easily. Cutting pressures are low; cutting speeds, feeds, and metal removal rates are high.

Chief factors affecting relative machinability of the various alloys are composition, physical form (cast or wrought), and temper. The stronger, copper-containing alloys permit rapid cutting, long tool life, and smooth finish. Alloys containing small lead additions permit clean cutting and good machinability.

Pure aluminum and its low-alloy grades are the least machinable; in some cases the metals are gummy, in other cases they produce long, stringy chips.

Because of its high coefficient of thermal expansion, aluminum must be kept cool during machining if dimensional accuracy is to be maintained.

Aluminum and its alloys can be machined dry (depending upon machinability and the severity of the operation), with oils, or with emulsions. In the past, mixtures of kerosene and lard oil (or straight lard oil) have been used successfully for general machining operations on aluminum, but emulsions and sulfurized fatty-mineral oils now provide superior results on almost all jobs.

Where oils are used, a light-viscosity oil is preferred, for it will provide good cooling when the flow is continuous and plentiful.

Magnesium and its alloys

When alloyed with other metals, magnesium develops properties suitable for many applications. Aluminum (3% to 10%), manganese (0.1% to 0.3%), and zinc (1% to 3%) are the most common elements alloyed with magnesium, but beryllium, cesium, copper, silver, tin, and zirconium are also used at times.

Almost all magnesium alloys have excellent machinability. In some respects, machining behavior of magnesium alloys is directly opposite to the behavior of aluminum alloys. Because there is no tendency to drag and tear, magnesium alloys finish to a fine, smooth surface.

Magnesium is a chemically active metal which, in finely divided form, will burn in air and react readily with water and acids, producing heat and hydrogen gas. For this reason, cutting fluids used with magnesium must contain no more than 0.2% free acid by weight, and must be free of water. In addition, they should have low viscosity for adequate cooling to reduce the fire hazard.

Magnesium is machined dry or with mineral oils, mineral-fatty oil blends, or inactive sulfurized fatty-mineral oils. Remember that emulsions and water-based fluids should never be used on magnesium.

A few other tips: A supply of powdered asbestos should be kept handy to smother a fire if one should start. Sharp tools minimize rubbing and build-up of frictional heat. Frequent collection and disposal of magnesium fines and chips is a good practice.

Copper and its alloys

Copper and copper-containing alloys run the machinability range from hard-to-machine to free-cutting and all the steps in between.

Hard-to-machine materials include the commercial grades of pure copper, and the essentially single-phase alloys like copper-nickel, copper-silicon, copper-tin, commercial bronze, phosphor bronze, and beryllium-copper. Typically, these soft metals have poor machinability, they are draggy and yield long, springy coil-type chips.

Medium-machinability alloys include leaded bronzes, Muntz metal, naval brass, tin bronzes, and aluminum bronzes—all of which include an appreciable amount of the alloying constituent. When machined, these alloys usually produce coil-type chips that are fairly brittle and break up into short pieces.

Free-cutting alloys have the same basic compositions as the two previous groups but they contain substantial amounts of lead, sulfur, selenium, tellurium or bismuth—elements that improve machinability.

Each of the three types of copper alloys require a different type of cutting fluid. For the first group (hard-to-machine alloys and pure copper), a mineral oil base with 10% to 20% fatty oil is recommended. If the alloy contains nickel, sulfurized mineral oil or sulfurized mineral-fatty oil should be used. Emulsifiable oils are also used for some of the less severe machining operations.

The intermediate copper alloys are usually machined with a mineral-fatty oil, but in this case the fatty oil content need only run 5% to 15% (the higher fatty oil content for the alloys that produce stringy chips). Emulsions can also be used.

A light, unblended mineral oil, emulsions, or mineral-fatty oils are used to machine the free-cutting alloys of copper.

In most cases, inactive oils are used for copper and its alloys to prevent staining. If a sulfurized oil is used and staining does occur, the stain can be removed by immersion in a 10% sodium cyanide solution for about 20 minutes. (Be careful, this is a deadly poison.)

Nickel and its alloys

Pure nickel, like pure iron, tends to be difficult to machine because its ductility results in dragginess. But many nickel alloys, like their steel counterparts, machine more readily than their parent metal.

Nickel silver, actually a nickel-copper-zinc alloy, ranges in nickel content from 5% to 30%. These alloys are comparable to plain carbon steels in hardness and strength, and have reasonably good machinability and clean-cutting properties.

Other nickel alloys are commonly used where corrosion or heat-resistance are important. Inconel (nickel-chromium-iron) is hard to machine because it is draggy. The Hastelloy family (nickel alloys with various combinations of copper, iron, molybdenum, or chromium) generally exhibits poor machinability. The Monel family (nickel-copper alloys) ranges in machinability from good to poor.

Nickel and its alloys can be machined with active oil-type cutting fluids. Sulfurized mineral oils (with or without fatty oils) and sulfo-chlorinated mineral oils (again with or without fatty oils) are recommended for use with high-speed-steel tools and cobalt-base alloy tools. Soluble oils are often used when machining with cemented-carbide tools.

Heat-resisting alloys

Many of today's industrial applications require tougher metals with greater resistance to high temperatures. This demand has been met successfully with stainless steels and a great variety of other heat-resistant alloys composed of nickel, chromium, molybdenum, cobalt, tungsten, titanium, iron, and other metals in varying proportions.

The recent improvements in metal cutting have made some of these "unmachinable" materials machinable, and some of the improvement is attributable to development of superior heavy-duty cutting fluids.

Machining of such alloys places a heavy burden on the fluid's ability to provide extra lubrication, remove heat, or both. For the relatively lighter, more moderate-speed operations, cooling is the most important factor, and an extreme-pressure emulsifiable oil fluid in a rich concentration is recommended. Here the water content of the emulsion removes heat effectively, while the EP additives supply the special lubricating properties required for the heat-resistant alloys.

For extra-severe, low-speed operations where lubrication and the prevention of build-up on the tool are the prime requisites, heavy-duty cutting oils (sulfo-chlorinated fatty-mineral oil blends) are recommended.

These recommendations apply to the materials designed for use at temperatures over 1200 F, including the precipitation-hardening stainless steels, iron-nickel-chromium-molybdenum alloys (such as Discaloy and A286), nickel-base alloys (Inconel X, Hastelloy C, Udimet 700), cobalt-base alloys (S-816 and HS-25), and titanium alloys such as Ti-6Al-4V. (This is by no means a complete listing of the superalloys and other heat-resistant materials.)

An important point to remember in dealing with these materials is that where the nickel content is high, as in the Inconel or Nimonic alloys, the cutting fluid must be thoroughly removed before the part is exposed to extreme high temperatures. This is to prevent the intergranular corrosion that might otherwise be caused by the sulfur in the extreme-pressure additives.

If sulfurized oils cannot be used at all, mineral oils with a high fatty oil content are recommended, but these cannot equal the performance of the heavy-duty oils and emulsions on these materials.

Nonmetallics

The principal nonmetallic materials that are machined are the thermoplastic and thermosetting plastics. Here, there is some need for lubrication, but the chief function of the cutting fluid is cooling.

Cutting fluid recommendations by machinability groups

		Machinability ratings					
Severity	Operation	Class 1	Class 2	Class 3	Class 4	Class 5*	Class 6
1	Internal broaching	A	A	A or J	A or K	D	C
2	External broaching	A	A	A or J	A or K	D	C
2	Pipe threading	A or B	A or B	A or B	A or B	D or G	D or H
3	Plain tapping	A or B	A or B	A or B	A or B	H to K	H to K
3	Plain threading	B or C	B or C	B or C	B or C	D or G	D or H
4	Gear shaving, cutting	B	B	B	A	G or H	J or K
4	Reaming	D	C	B	A	F	G
5	Deep drilling	E or D	E or C	E or B	E or A	E or D	E or D
6	Plain milling	E, C or D	E, C or D	E, C or D	C or B	E, H to K	E, H to K
6	Multi-cutter milling	E, C or D	E, C or D	E, C or D	C or B	E, H to K	E, H to K
7	Multiple-head boring	C	C	C	C	E	E
7	Multi-spindle auto	C or D	C or D	C or D	C or D	F	G
8	Hi-speed, lo-feed auto	C or D	C or D	C or D	C or D	F	G
9	Drilling	E or D	E or C	E or B	E or A	E or D	E or D
9	Planing, shaping	E	E	E	E	E	E
9	Turning	E	E	E	E	E	E
10	Sawing, circ or hack	E	E	E	E	E	—
10	Surface grinding	E	E	E	E	E	E
10	Thread grinding	A or B	A or B	A or B	A or B	—	E

Legend:
A—High sulfur mineral-fatty oil
B—Medium sulfur, medium chlorine mineral-fatty oil
C—Low sulfur, medium chlorine mineral-fatty oil
D—Chlorinated mineral oil
E—Water-miscible fluids (light to heavy duty as needed)
F, G, H, J, K—Mineral-fatty oil (F has highest fatty oil content, K the lowest)
*Never use water-base fluids on magnesium

Thermoplastic materials soften and melt even at intermediate temperatures, so cooling of the workpiece and the tool is essential. The most commonly used coolant is an aqueous emulsion or solution; a normal dilution is one part soluble oil in 20 to 40 parts water. Chemically active constituents in the coolant should be checked to prevent any surface effects on the plastic. Often, thermoplastic materials are cooled simply with an air blast during machining.

Thermosetting plastics also show the effects of local heating; while they do not melt at the point of work, they may char. Consequently, cooling with an aqueous cutting fluid is good practice. The thermosetting plastics may contain fillers of paper, wood pulp, asbestos, graphite, or other substances. While graphite is obviously beneficial in assisting machining, other fillers cause premature dulling of tools. Tools, therefore, should be kept as sharp as possible, preferably honed or lapped, to prevent overheating.

Cutting fluids other than emulsions may be used for threading plastics provided they have no chemical effects on the plastics (staining, swelling, softening, or fading). Light-viscosity fatty-oil blends and soft soap in a heavy solution are often used as machining lubricants for plastics.

Some thermosetting plastics, such as the urea formaldehyde type, are fairly resistant to the effects of oils and solvents, providing some latitude in the choice of a cutting fluid.

Using and handling cutting fluids

The manner in which a cutting fluid is applied has a considerable influence on tool life and the machining operations in general. Though there are many good, extremely effective devices and systems for supplying fluids to the cutting area, special equipment is not generally necessary for good results.

Increased tool life and production, better finish, and lower power consumption are possible with conventional low-pressure application methods if the following basic rules are observed:

Apply the fluid in a copious stream so that the cutting tool edge and the work are completely enveloped. In addition to supplying an adequate amount of fluid to the cutting zone, a copious flow provides a cooling action that prevents undue temperature rise. An empirical rule for operations with lathe-type tools is that the coolant supply nozzle should have an inside diameter of at least three-quarters the width of the cutting tool.

Fluid should be directed at the zone of chip formation and the portion of the tool producing the chip should be completely enveloped. For heavy-duty turning and boring operations, an additional nozzle supplying fluid from below (along the flank of the tool) is desirable.

For horizontal drilling and reaming operations, application of the fluid through hollow tools is preferable to external application because it provides an adequate flow at the cutting edges and flushes chips out of the hole.

In grinding, a copious flow of cutting fluid at low pressure will generally provide optimum results. Where application of a large volume of fluid results in undue splashing, it is better to install splash guards on the machine than to reduce flow of the fluid.

For slab milling, the cutting fluid should be applied in a copious quantity to both incoming and outgoing sides of the cutter by fan-shaped nozzles with widths about three-quarters of the width of the milling cutter. For face milling, use a ring-type distributor that directs the fluid at all cutting edges and keeps the cutter completely bathed in fluid.

Maintaining cutting fluids

Cutting fluids, like any other fluids that are used over and over again, must be cared for properly. There are several precautions that should be observed:

Cutting oils

Oil-type fluids perform satisfactorily if applied in full flow to the tools and the work, and if sufficient volume is maintained in the system to hold oil temperature around 70-75 F.

Cutting and grinding oils become contaminated rapidly during use. Extraneous materials, chips, dirt, etc, should be removed continuously or at periodic intervals by filters, strainers, centrifuges, or settling tanks. The mechanical edge-type filter, incorporating metal strip or disks as the filter element, acts principally as a strainer. The absorbent type filter uses paper disks, cotton waste, or cloth bags as the filtering element. Magnetic filters are suitable for separation of ferrous particles.

Centrifuging is used for removing heavy contaminants and particles from oil. Centrifuging, together with a heating unit and settling tanks, is often used for extracting oil from chips.

All cutting oil systems should be drained at intervals, manually cleaned, flushed and replenished with filtered or new cutting oil. Frequency of cleaning depends upon individual conditions

Emulsion fluids

Emulsions generally require more maintenance and care than cutting oils do.

In preparing emulsions, always add the oil to the water. This initial mix should be agitated thoroughly while the oil is being added, otherwise soap combining with the water may cause mineral oil to separate out. If not enough water is used, or if the water is added to the oil, an invert emulsion will result, in which water particles are dispersed in the oil phase. Though suitable for certain metal drawing operations, invert emulsions are undesirable for metal cutting.

Water used in preparing emulsions is very important. Soft waters, where available, present no problems. But hard waters containing various minerals and salts often hinder or impede emulsification. It is not uncommon for emulsions made with hard water to "break" readily, that is, to separate into a stratified condition, with a layer of oil or creamy emulsion floating on the surface. Such separation is detrimental.

Use of a specially formulated hard-water soluble oil with soft water may lead to the formation of a bluish-black stain on freshly ground ferrous parts.

Pretreatment of hard water is often desirable and sometimes quite necessary. Water conditioning agents consisting usually of polyphosphate combinations are readily available, as is trisodium phosphate. The general rule for using these materials is to add about 1.5 ounces per 100 gallons of water per grain of hardness. Chemical treatment of extremely hard water is far more economical than the purchase of a specially formulated soluble oil.

Microorganisms in the water shorten the service life of a soluble oil emulsion. Microorganisms of three types—bacteria, algae, and fungi—are often encountered in soluble oils, and all three have a detrimental effect on the stability of the emulsion. Many soluble oils are compounded with a bactericide, but the amount that can be added is limited by its solubility in the oil. And when the bactericide is further diluted when the emulsion is made up, its concentration is even further reduced.

Rancidity, the term applied whenever a cutting fluid gives off a bad odor, is usually caused by the growth of bacteria. The rotten-egg stench emanating from the sump of a machine that has been shut down, say, over a weekend, is caused by bacteria that attack inorganic sulfates found in all natural waters. A quality cutting fluid is the best insurance you can get against rancidity.

Emulsion concentration is not always given the attention it deserves. In heavy-duty cutting operations, heat at the tool causes water to evaporate at a rate faster than the carryoff of oil on the machined parts, resulting in an increased oil to water ratio. Carried too far, this causes an invert emulsion. In grinding, oil carryoff is higher, and the emulsion becomes increasingly dilute with use. This may cause rusting unless the concentration is checked and controlled frequently.

Length of time an emulsion is kept in service varies widely; it ranges anywhere from one week to six months. Cooling an emulsion by aeration, mechanical circulation, or refrigeration is useful in extending its service life and in producing better finish. In use, emulsions should be held at a temperature between 55 F and 70 F.

Before putting an emulsion into use, wash and flush the system thoroughly. Deposits of all kinds must be removed. Emulsions are extremely susceptible to contamination. If there is any reason to suspect that bacteria are present in the system, flush it with a germicidal solution before putting the new emulsion in.

Occupational dermatitis

Dermatitis is not an affliction that must be tolerated. It can be closely controlled, or eliminated, by observing the simple principles of cleanliness that the average person carries out at home.

Dermatitis is frequently, though mistakenly, associated with the handling of petroleum products, particularly in the machine shop, where the operator's hands and forearms are in prolonged contact with soluble oils and cutting oils. Combined with dirt, these fluids form grimy compounds that may become embedded in the skin, often blocking the pores and hair follicles.

In many cases, these areas become infected, and dermatitis sets in. But this is a condition that may occur with almost any material that is allowed to remain on the skin for a long period; it's not limited to petroleum products.

A contributing factor may sometimes be the solvent action of the cutting fluid. Left on the skin, the fluid can dissolve the natural skin oils, inflaming the skin and causing it to crack.

The bactericides used in cutting fluids have no known effect on incidence of dermatitis.

Women, with their thinner skin, are more susceptible to dermatitis than men are. And fair, blond-complexioned people are usually more susceptible than are dark, oilier-skinned people. For hypersensitive workers, the only answer is a switch to another job, one in which they will not be exposed to cutting fluids.

Personal cleanliness, keeping the fluid clean, use of commercially available hand creams, use of protective clothing, and installation of splash guards on the machines —all these can help to control dermatitis in the shop. ■

Acknowledgements:

We wish to thank the following companies for their cooperation in the preparation of this Special Report:

Adam Cook's Sons, Inc
American Oil Company
American Oil & Supply Company
Anderson Oil & Chemical Co
Atlantic Refining Company
Cincinnati Milling Products Div
Cities Service Oil Company
Gulf Oil Corporation
E F Houghton & Company
Humble Oil & Refining Company
Johnson's Wax
Kerns United Corporation
Macco Products Company
Magnus Chemical Company
Master Chemical Corporation
Mobil Oil Company
Monroe Chemical Company
Nopco Chemical Company
Pure Oil Company
Quaker Chemical Products Corp
Rust-Lick, Inc
Shear-Speed Chemical Products
Shell Oil Company
Sinclair Refining Company
Sun Oil Company
Texaco, Inc
Tidewater Oil Company
White & Bagley Company

Cleaning metalworking fluids

Cleaner cutting fluids, quenching oils, and hydraulic fluids cut machine downtime and boost profits. This report tells you how to do it and what it will cost

Even if the government weren't looking over your shoulder to see what kind of contaminants you might be discarding, there is a perfectly valid reason to recycle and reprocess all the different kinds of oils and fluids used in your plant. The reason, of course, is economics. It costs money to use dirty cutting oils and contaminated hydraulic fluids. Tools wear out faster, machine elements wear out or rust, workpieces must be reworked or scrapped, and the health of machine operators is threatened.

Thorough, well-planned filtration systems, such as those described in this report (and these are but a sampling), can make a substantial contribution to a metalworking plant's profit picture. It will pay you to examine all the units available and see which fits your operation.

The economics of filtration

Naturally, it costs money to filter cutting fluids, but it can cost more if you don't. Here are two case histories and an economic analysis that let you figure the potential savings

Long before pollution control became a national concern, economics dictated the need for filtration of grinding and cutting fluids. This is proved by the hundreds of filtration units currently operating in metalworking plants of all types and sizes.

Many of the companies that are adding filtration equipment today to meet stiffening pollution-control regulations are finding, to their surprise, that the move makes sense economically. And, as this article will demonstrate, the benefits—ecological and economic—are not restricted to large companies.

Before we get into our cost analyses, let's take a look at the two most common plant routines for handling coolants. The first approach, and the most basic, involves a maintenance man who, on a regular or irregular basis, shuts the machine down, drains the tank, mucks the sludge, and recharges the machine with fresh coolant. The regularity depends on the sophistication of the maintenance group; often, the frequency depends upon the number of trouble calls from the production department.

A more technical approach to coolant changing is one tied to production. When the production supervisor notes a drop in either the rate or the quality of output, coolant is one thing he checks immediately. If necessary, the machine

By James J. Joseph, manager, filtration systems
Hoffman Air & Filtration Div.
Clarkson Industries, Inc., Syracuse, N.Y.

Hoffman Vacu-matic cutting fluid filtration system at Sossner Tool, div. of Mite Corp., Melville, N.Y., is one of two handling swarf generated by grinding of taps

Large, centralized Vacu-matic system costing $58,000 was installed at Rollway Bearing Co. to handle fluid for 75 machines. Previously, fluid for each machine was changed weekly; now it is done once a year

is shut down and cleaned, then filled with clean fluid.

In both approaches, there are elements of unnecessary cost: production loss, machine downtime, maintenance labor cost, cost of new coolant, and scrap or rework expense.

The cost equation

These items can all be converted into simple factors that become part of the equation we will use to determine coolant "operating" cost with and without filtration.

For a typical grinding or machining operation without filtration, the equation is:

$$\text{Cost} = A(CG + LN + P + Q + T)$$

where A = the number of times the coolant is changed within a given period of time. Preferably, the period should be a year, for ease of calculation and for leveling unusual circumstances. Seasonal surge, if a factor, must be qualified. By definition, a coolant change is when a machine is shut down, cleaned, and recharged with fresh coolant.

CG = the factor for coolant costs. C is the cost in dollars per gallon of the coolant, and G is the number of gallons lost per change. If it is known, include the cost of removing the dirty liquid.

LN = the factor for labor costs involved in draining and dumping the coolant, mucking and cleaning out the sludge, and recharging with fresh coolant. L is labor cost in dollars per man-hour, and N is the total number of man-hours required.

P = the factor for cost of lost production. If a machine is cleaned during a normal downtime, this factor is zero. If, however, production must be halted, this factor should include the cost of lost production parts, the cost of idle man-hours, and costs of other operations affected by the shutdown. Usually, these entries can be estimated as the pieces per hour times the standard cost.

Q = the factor for the loss in quality or the increase in rejects. Quality-control reports are the best source of this data. If no record is available, the operator usually has a good idea of the number of rejects. The total rejects for the given time period (see factor A) should be recorded. If this is impossible, then a smaller interval just prior to shutdown can be used. This shorter interval will be accurate enough, for the ultimate calculation will be comparing "before" and "after" costs. Rework costs should be added.

T = the cost of tools involved. This is usually an expendable item, but it can get out of control. This factor can be determined from the number of pieces produced per tool or from the frequency of wheel dressing.

As the equation directs, all the factors are lumped and then multiplied by A to get the total cost over the time period defined in A. This cost can then be extended to an annual figure.

Case histories

Let's look at how some actual examples work out. Rollway Bearing Co. has a large, central Hoffman Vacu-matic filtration system that cost about $58,000. As the following data will show, Rollway's savings through addition of the filtration system were large enough to pay off the cost of the equipment in less than two years. But first, let's calculate the cost of coolant handling before filtration equipment was installed.

$$\text{Cost} = A(CG + LN + P + Q + T)$$

where A = 50; coolant was changed in each machine 50 times a year.

G = 5000 gallons; total system capacity for 75 individual machines.

C = $0.09 per gallon; cost of coolant mixed, added to machine, and discarded.

L = $3.50 per man-hour (direct labor only).

N = 75 hours; 1 hour per machine.

P = 0; machines cleaned during normal downtime with no production loss.

Values of F, no. of batches of coolant used per year with filtration

Capacity (gpm)	Water-based coolants: Grinding, Cast iron	Water-based coolants: Grinding, Other	Water-based coolants: Machining, Cast iron	Water-based coolants: Machining, Other	Straight oils: Grinding, Cast iron	Straight oils: Grinding, Other	Straight oils: Machining, Cast iron	Straight oils: Machining, Other
0-50	2-4	1-3	2-4	2-3	2-3	1-3	2-3	1-2
51-300	2-3	1-3	2-3	1-3	2-3	1-3	2-3	1-3
301-1000	1-2	1-2	1-2	1-2	1-3	1-2	1-3	1-2
1000 and up	1-2	1-2	1-2	1-2	1-3	1-2	1-3	1-2

Values of Y, filtration operating costs, cents per operating hour

Capacity (gpm)	Cast iron: Grinding	Cast iron: Machining	Other materials: Fine grinding	Other materials: Rough grinding	Other materials: Machining
Water-based coolants:					
0-200	0-5	0-5	5-10	0-10	0-5
201-1000	0-15	0-15	15-25	0-25	0-10
1000 and up	0-50	0-50	25-40	0-50	0-50
Straight oils:					
0-200	0-10	0-10	10-15	0-15	0-10
201-1000	0-25	0-25	20-50	0-40	0-25
1000+	0-75	0-50	50-75	0-60	0-50

Q = Privileged information.
T = Privileged information.

Substituting these figures into the equation, we find that the yearly cost before filtration =

$$50[(.09)(5000) + (3.50)(75) + 0] = \$35{,}625$$

Here is an example for a smaller plant. The Lagoe Oswego Corp., a sizeable contract job shop, has two grinders requiring 200 gallons of coolant. Here are the calculations to determine the cost of coolant handling prior to the installation of filtration equipment.

$$\text{Cost} = A(CG + LN + P + Q + T)$$

where A = 50; coolant changed in each machine 50 times a year.

G = 200 gallons; total coolant-system capacity for two grinders.

C = $0.20 per gallon; cost of coolant.

L = $3.83 per man-hour (direct labor only).

N = 5 hours; 2½ hours per machine.

P = 0; machines cleaned during normal shutdown.

Q = $12; estimate based on time lost in correcting rejects.

T = $30; estimate based on wheel changes.

Substituting these figures into the equation, we find the yearly cost before filtration =

$$50[(.20)(200) + (3.83)(5) + 0 + 12 + 30] = \$5{,}057.50$$

We have now calculated the coolant-handling costs at both Rollway and Lagoe Oswego before either firm installed filtration equipment. The next step, obviously, is calculate the savings resulting from use of filtration equipment. We can do this by using the following equation:

$$\text{Saving} = A[(1-K)(CG + LN + P) + R + Z] - Y$$

This equation develops "before" and "after" costs and subtracts them to get the savings. There is really no need to use the cost equation (developed earlier) alone, unless one wants to determine the magnitude of the costs prior to filtration. Note that the "saving" equation uses most of the earlier factors; but now a coefficient (1-K) has been introduced that will alter the values according to filtration performance. This factor takes into consideration the extension of coolant life as a result of filtration.

The factor K in this new coefficient is the ratio of the number of batches of coolant used in a year *with* filtration (a figure we will call F) to the number of batches used *before* filtration was added (already determined and called A).

Variations in coolant composition, water characteristics, toolroom atmosphere, and many other conditions prevent us from assigning a single value of F for all cases. The values of F in the table above are based on actual field experience, but there are many installations where coolant is changed more often. Even so, filtration is economically sound.

The factor R in the "saving" equation is defined as the dollar value of savings attributable to improved quality and lowered rejects as a result of filtration. If a study were made of machine performance in a period just before cleaning and also in the period just after cleaning and recharging with fresh coolant, the difference in quality could be correlated to what the improvement (and its dollar value) would be if continuous filtration were used. The duration of the test periods is not important; a half hour before and a half hour after cleaning might be sufficient.

The next factor in this equation is Z, which represents the dollar value of the saving attributable to longer tool life. This can be determined in much the same way as factor R.

The final factor, Y, is the cost of the filtration. In the table above, we offer a guide to values of Y, based on flow volume and type of operation. (Electrical expense for motors and other components has been neglected.)

Now that we have all the factors, we can calculate the savings produced by continuous filtration. Since the values of F and Y are given in range form, it is probably wise to run the calculations through, using both extremes. This will give you an idea of the range of potential savings.

Let's get back now to our case histories and calculate the savings offered by introduction of continuous-filtration equipment. At Rollway Bearing, the annual cost of coolant handling without filtration had been calculated to be $35,625. We can figure the savings by using the equation:

$$\text{Saving} = A[(1-K)(CG + LN + P) + R + Z] - Y$$

where A = 50 (as before).

K = 1/50; centralized system dumped once a year during major shutdown.

CG = As before.

LN = As before.

P = As before.

R = Privileged information.

Z = Privileged information.

Y = $5000; media cost per year for filters.

Substituting these figures into the equation we get:

$$\text{Saving} = 50[(1 - 1/50)(712.50)] - 5000 = \$29{,}912.50$$

Recall that the initial cost of Rollway's Hoffman Vacumatic system was $58,000. The annual saving of nearly $30,000 means that the system pays for itself in less than two years.

At Lagoe Oswego Corp., we had determined a coolant-handling expense of $5,057.50 a year before filtration. Now here are the factors that will be used to determine the annual saving:

A = 50; as before.

K = 1/50; firm reports that it never dumps coolant, but we have assumed one change per year.

P = As before.

R = $8.00 a week; saving from reduction in rejects.

Z = $6.00 a week; saving from improved tool life.

Y = $112.00 a year; cost of filter media.

Substituting these figures into the "saving" equation, we get:

$$\text{Saving} = 50[(1 - 1/50)(40 + 19.15 + 0) + 8 + 6] - 112 = \$3{,}486.35$$

In this case, we see that the saving comes to nearly $3500—this as a result of equipment which, if installed today, would cost only $1675.

Other factors involved

We obviously have not covered all the factors involved in a complete economic analysis of filtration costs. The refinement of interest rates for the time value of money; factors of maximum utilization of manpower; and the costs of electricity, water and sewage, makeup coolant, scrap, and rework have been omitted for simplicity, as have all overhead costs.

There are other factors, not given dollar values here, which are "plus" factors for filtration. These include reduced rancidity and dermatitis hazards, reduced machine wear, increased salvage value of chips, lower toolroom inventory costs, and improved housekeeping.

Although the examples reported in this article deal only with metalcutting operations, continuous filtration of this type can be applied (and analyzed) in much the same way to wire drawing, metal rolling, honing, metal etching, and electrical-machining operations.

Cleaning metalworking fluids

Squeezing dollars out of oils

You win—economically and ecologically—if you can reclaim metalworking fluids. One way to do the job is with a centrifuge

Economy and ecology—both words provide perfectly valid reasons for re-processing lubricating oils, quench oils, cutting oils, hydraulic fluids, or any other fluids which are normally discarded after certain periods of use. Indeed, why throw them out when, at very low cost, you can reclaim them?

One way of reclaiming these fluids is a centrifuge, whose operating principles have been known to every engineer and technician since the days of the cream separator. Centrifugation permits removal of most solid contaminants, as well as water or other supernatant liquids, from the oil, restoring it to practically as-new condition.

Once the impurities have been removed, the oil may be reused, adding only a small amount as makeup. Not only does this reduce the cost of plant operation (economy), it also largely eliminates the problem of dumping used oil (ecology).

The cost picture

Let's take a closer look at the economics. In a typical heat-treating operation, such as that at Milwaukee Gear Co., 2400 gallons of marquenching oil in each of five quench tanks had to be replaced once a year, and the tanks had to be thoroughly cleaned out. At 55¢/gal for oil, that represented a yearly outlay of $6600, not including cleaning labor.

By investing in one $4000 Model AS-14V Sharples purifier, which is periodically shifted from one tank to the next, the company has completely eliminated oil-change-over expenses. Not only that, the quality of quenched parts is now more uniform and predictable, with fewer rejects, simply because oil-quench conditions are more stable. Continuous removal of carbon, scale, and dirt from the quench prevents fouling of heat-exchanger tubes and allows more precise temperature control. Another heat treater reports large savings because parts quenched in centrifuged, carbon-free oil do not require subsequent sandblasting or cleaning.

There's money in chips

Don't let those chips leave your plant! At least, not until you've wrung the last ounce of oil from them. Experience shows that a basket-type centrifuge, also known as a "chip wringer," can recover up to 40 gal of cutting oil from every 1000 lb of chips of the type produced by single-

By William F. Gilliland, field engineer
Sharples-Stokes Div., Pennwalt Corp., Warminster, Pa.

Sharples centrifuge is rolled from one furnace to another at Milwaukee Gear Co. to purify marquenching oil

Sludge from dirty quenching oil (above) is trapped in bowl of centrifuge. After bowl is cleaned and relined with paper (below), impeller is added and cover is put back on

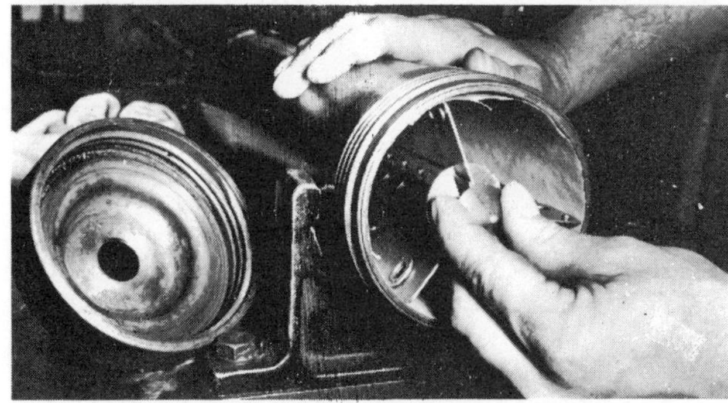

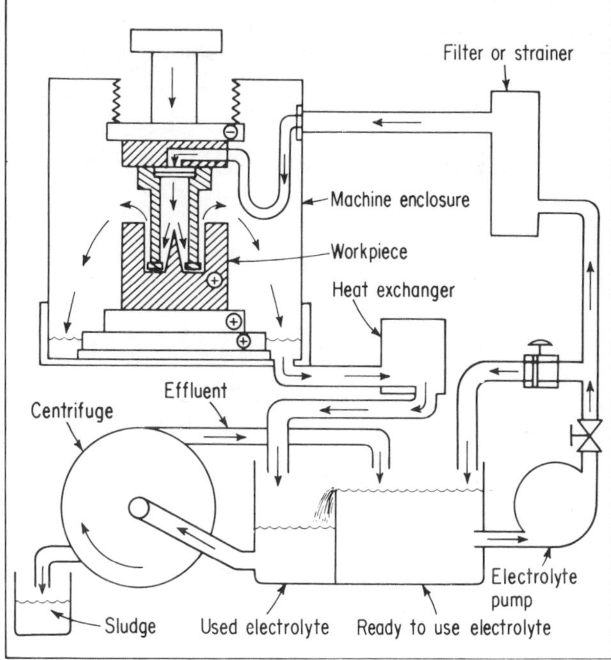

Cincinnati electrochemical-machining unit (photo, top) is equipped with Sharples centrifuge that removes sludge from the recirculating electrolyte (see sketch)

or multi-point tools (turret lathe, automatic screw machine, milling machine, shaper, etc.). Even larger chips, such as those from multi-spindle automatics, will yield up to 20 gal of oil per 1000 lb of chips

What does this come to in dollars? Assume a plant that produces four tons of chips a week, not an unusual amount for a high-production operation. At the minimum recovery of 20 gal per 1000 lb, the centrifuge will salvage 160 gal; multiplied by a typical oil cost of 50¢/gal, the saving amounts to $80 per week. In other words, the centrifuge could pay for itself in one year. Additional savings stem from reduced oil inventories, longer tool life, better finish, and lower reject rates.

An extra bonus

Actually, the savings are even larger, for the relatively dry chips command a higher price in the scrap market. And dry chips help minimize fire hazards and improve housekeeping.

Of course, the oil delivered from the chip wringer is not ready for immediate reuse. It should now be subjected to the purifying force (13,200 x gravity) developed by a solid bowl-type centrifuge to remove solid contaminants and water.

When does a chip wringer become economical? Generally, 2000 lb or more of oily chips a week is enough when ferrous metals are involved. Even lower quantities, 600 to 750 lb a week, are adequate when brass or other costly metals are being machined, because of the higher scrap value of dry chips.

Other benefits

In addition to the savings in replacement oil, there are several intangible, but important, benefits from the installation of a centrifuge:

- Because abrasive particles have been removed, purified oil does not act to dull tools, thereby reducing tool-changing time.
- Consistent oil quality leads to more consistent cutting action, easier maintenance of close tolerances, and less wear on machine parts.
- Elimination of water by centrifuging of cutting oils keeps machine parts from rusting.
- Continuous reprocessing of cutting oil in a complete system also sterilizes the oil, killing and removing bacteria that could lead to dermatitis. In addition, there is the aesthetic advantage of working with cutting oil that also *looks* clean.

Cleaning metalworking fluids

Filtering away machine downtime

Ultrafine filters on 19 NC machine tools remove small silt particles from the hydraulic fluid and cut tool downtime

When your yearly tab for hydraulic-system downtime on 19 numerically controlled machine tools is $142,500, you should take a hard look into the reasons for that high cost.

The Lubricants Technical Subcommittee of the Maintenance Standardization Committee at the Convair Aerospace Div. of General Dynamics Corp., San Diego, the firm saddled with that healthy cost, took just such a look and found that the hydraulic-system failures were caused by worn parts, and that the wear was caused by 1- to 5-micron silt particles in the hydraulic fluid.

The problem was solved by replacing the 10- to 25-micron nominal filters used on the machines with 3-micron absolute filters. That project was started in September, 1969, and completed in April, 1970. Since the new filters have been working, there has been no hydraulic-component replacement and no hydraulic-system downtime.

Just how do the silt particles cause wear? In the first place, particles in the 1- to 5-micron size range are "awkward" because many of the components in modern hydraulic systems have critical clearances of this magnitude. Larger particles are blocked by conventional filters, and, when they aren't, the particle is too big to fit in the critical clearance between moving parts anyway. Particles that are substantially smaller than the tightest clearance between moving surfaces tend to pass through the clearance without eroding the parts much.

However, the silt particles are just the right size to scrape along the surfaces of the parts and, at times, become partially wedged. They gouge out some hard metallic particles, which join the silt particles already in the hydraulic fluid. Without ultrafine filtration, this buildup will be rapid —and damaging.

At Convair, tests done with a portable contamination laboratory developed by the Industrial Hydraulics Div. of Pall Corp., Grand Rapids, Mich., showed that the hydraulic fluid used in the machining centers, profilers, skin/spar mills, and a plank mill contained many silt particles in the 1- to 5-micron size range—too fine to be trapped by the 10- to 25-micron filters. These particles were, in effect, lapping or grinding the surfaces of critical clearances. The new filters prevent this.

Actually, in switching to the ultrafine filters, the company took a leaf from the aerospace specification books. There is a trend in that industry to 3-micron absolute (0.9-micron mean pore size) filtration to remove particles down to 1 micron in size from hydraulic systems in aircraft and in the latest generation of machine tools. Convair's first step was to retrofit three of the 19 NC machine tools with 3-micron absolute filters. Each filter features an industrial-type housing, a differential-pressure indicator, and a disposable filter element. Initial tests showed that these filters could maintain a Class 1 hydraulic system as defined by the tentative SAE contamination classifications.

Before the retrofit, the three machine tools used for the evaluation had this 30-month history:

Cost to rebuild 6 hydraulic pumps	$17,772	(32%)
Cost to rebuild 14 servo valves	7,349	(13%)
Cost to replace 6 hydraulic motors	3,450	(6%)
Labor cost for troubleshooting and installation time at $6/hr	4,600	(8%)
Average cost of downtime at $25/hr	23,000	(41%)
Total cost	$56,171	(100%)

Interestingly, most of the cost, which was due to hydraulic-system malfunction, is for hydraulic pumps and hydraulic motors. These components are common to most hydraulically operated machine tools, including the non-NC

Checking a 3-micron absolute filter on a large metalcutting machine are W.J. Stanley, manager of plant engineering at Convair (standing) and G.W. Oliver, lubrication engineer

types. Consequently, the benefits of hydraulic-system-contamination control should apply to all types of machines.

Because of the results obtained with the three machine tools, Convair decided to equip the remaining 16 machines with the 3-micron absolute filters.

Quick payout

Initially, Convair figured that the cost of the retrofit, based on an assumed reduction in hydraulic-system downtime, would be recovered in less than two months. Since there has been no downtime, that payout period is more like one month. This excludes the cost, which can be substantial, of scrapping or reworking parts that are damaged by erratic machine tool operation due to hydraulic-system malfunction.

Also, hydraulic-fluid life has been extended because of the ultrafine filters. Before the retrofit, Convair changed the fluid every two or three months; now the fluid is changed when the machine tools are taken down for their annual maintenance work. But, instead of dumping the fluid, the company uses it in less critical applications. These savings weren't figured in the payout period, either.

A general rule of thumb to follow before making this kind of filter switch is first to determine what the critical clearances are in the hydraulic-system components and then select filters with an absolute rating as fine as the smallest critical clearance to be protected or with a mean pore size rating as fine as the smallest critical clearance to be protected.

The filter with the absolute rating will prevent any circulation of abrasive particles; the mean pore size filter will limit the initial number of particles circulating and, after multiple passes, will remove substantially all abrasive particles big enough to cause wear.

Cleaning metalworking fluids

Cyclonic filtration whips up a storm

Centrifugal force and the principle of the free vortex combine to produce an effective filtration system for grinding fluids

When the makers of a household detergent claimed that it cleaned "like a white tornado," everyone was skeptical. But, as a matter of fact, "tornadoes" can clean. For perhaps two decades or more, the free vortex has been applied in a great variety of industrial applications to separate and remove undesirable solids from liquids.

The free vortex is a natural phenomenon all of us have seen many times. It causes the swirl in the wash basin as the water rapidly drains out, and we have seen it in the form of whirlpools in swiftly flowing rivers. The phenomenon is created when a given tangential velocity is put into a reducing radius of rotation. The result is rapid radial acceleration of any material put into such circumstances.

Recently, these principles have been "discovered" as a way of cleaning the metal chips and bits of abrasive and grinding wheel from water-based and chemical grinding fluids.

The essential element of the process is a conical device approximately a yard long (it could be longer or shorter) inside which a cyclone is created—a genuine liquid tornado—of the liquid to be cleaned. The liquid enters the top of the cone at a tangent, which starts the fluid spinning. As the liquid spins downward, the forces of the free vortex created increase its speed of rotation so that, near the bottom, centrifugal forces of between 7000 and 7500 times the force of gravity are generated.

It's this centrifugal force, of course, that does the cleaning. All solids larger than a few microns are thrown to the wall of the cone and spiral down to be ejected in a flying spray through an opening at the cone's lower end. Very little of the fluid leaves with the ejected solids.

By William M. Fulton, president
Cyclonics Inc., Babson Park, Mass.

While generating its centrifugal forces, the swirling liquid cyclone develops a rising column of air at its core; this, in turn, starts an upward flow of the cleaned liquid adjacent to it. Just as in an atmospheric cyclone or tornado, there is low pressure at the "eye" of the storm, causing the material in this eye to rise. In a properly functioning cyclonic coolant-filtration system, the contaminated coolant enters the cyclone at about 40 psi, while the rising column of clean coolant returns to the grinder at about 10 psi. These pressure differentials are an important consideration in cyclone filter design. To illustrate, let's consider the extremes.

With reference to the sketch on page 83, if the input pressure at A is 40 psi and the output at C is vented to atmosphere, then a vacuum is created at the reject orifice B. The air rushing in at B will not permit any outflow of liquid or contaminant at all; the cyclone is worthless.

A condition at the other extreme is created when we increase the back pressure at outlet C to roughly 20 psi. The vacuum at opening B disappears, and we get a heavy flow of liquid from this opening. But filtration performance is very poor. What we've done by applying this back pressure is throttle down the centrifugal action—the liquid spins lazily—and we have lost nearly all of the cyclonic-filtration capability.

Optimum pressure relationship

There is a relationship between input and output pressures that is optimum for every cyclone. There is some room for adjustment, but it's narrow; an increase or decrease in back pressure on the output side of as little as 5 psi can decrease filtration performance from removal of 10-micron particles to removal of 20 microns, at best.

The shape and size of the cyclone must be tailored to the size, shape, and specific gravity of the material being filtered and the flow requirements of the system.

The beauty of a cyclonic filter is that there are no moving parts inside the cyclone itself and there are no filtering media to be cleaned or replaced. The fluid simply cleans itself.

The cyclone's centrifugal force is capable of separating particles of very small size. In a typical grinding-fluid

Cyclonics filtration system uses six ceramic cyclones to clean 120 gpm of coolant for a large surface grinder. Sketch at right shows basic elements of typical system

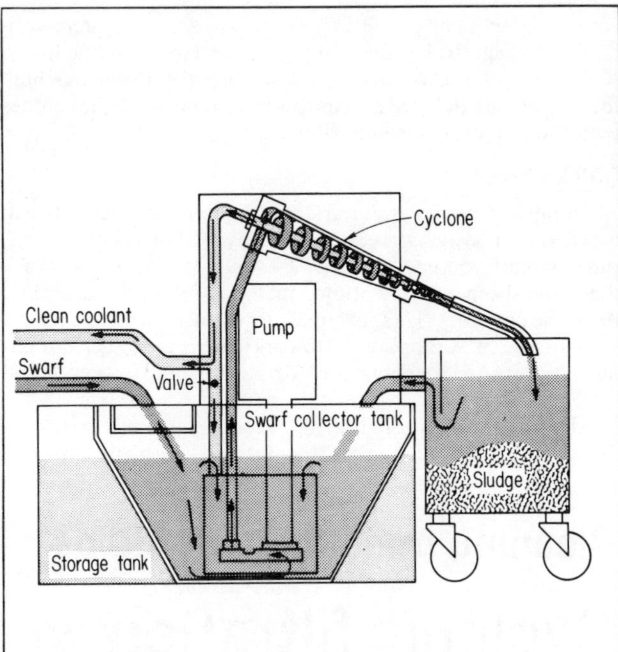

Filtration system consists of cyclone (or cyclones), swarf-collection tank, sludge truck, and immersible pump. Principle of the cyclone is explained in the text

application, approximately 98% of the particles 5 microns and larger are removed, approximately 66% of the particles of 2 microns and larger, and roughly 50% of the particles 1½ microns and larger. Particles as small as sub-micron size are also taken out. If the fluid were re-circulated, the maximum size of the residual particles could be continously reduced.

There is a practical limit, however, to the smallest particle size a cyclonic filter can remove. The reason for this is that there are, in addition to the centrifugal forces at work, two other forces being generated by the cyclonic action: the force that would keep the particle spinning around the axis of the cyclone and the force that would pull the particle into and up the rising column of clean fluid. As the mass of the particles is reduced, there is an increasing tendency for these other forces to overcome the centrifugal forces, and, at some point, the mass of the particles could be so small that it escapes the centrifugal forces altogether.

Under these circumstances, then, we can see that the specific gravity of the particles being removed has some bearing on the minimum particle size a cyclone can be capable of removing completely. Because most grinding fluids will have a mixture of particle types—metals, bits of grinding abrasive, bond—it is practically impossible to be precise about the size of particles that might remain in the grinding fluid after cyclonic filtration.

Surface finish and integrity

An increasing concern for the finishes developed in grinding operations has focused attention on coolant filtration in recent years. It has been clearly demonstrated that surface integrity can seriously influence the integrity of the whole part under conditions of high stress. Therefore, when ground parts are being specified for critical application, the tendency is to call for extraordinary finishes. These finishes simply cannot be achieved with unfiltered grinding fluids.

But finish hasn't been the only reason for growth in the application of filtration to grinding operations. There has also been an increasing awareness that the grinding process is a system comprising the operator, the machine, wheel, dressing tool, workpiece, and the grinding fluid. In this system, the grinding fluid may be the least expensive component, but it has an important effect on the performance of each of the others and, therefore, on the cost and effectiveness of the system as a whole.

What the fluid does

A properly applied grinding fluid performs four important functions:

- It effectively cools the workpiece to prevent damage to its surface.
- It lubricates the cutting action. This also reduces heat and facilitates grinding, thereby improving production rates.
- It prevents rust and corrosion of both the machine and the workpiece.
- It washes away chips of metal and abrasive which, if allowed to build up, would foul the machine and damage specified finishes on the workpiece.

There are two principle degraders of the grinding fluid: metal particles in the swarf and bacteria.

As swarf builds up in the sump of the settling tank, the metal particles that make it up attract and hold increasing amounts of the grinding fluid's lubricating and rust-inhibiting ingredients to their surfaces. They literally "sponge up" these valuable properties of the grinding fluid. Separating the swarf from the fluid and then keeping the clean fluid and swarf apart will greatly improve grinding-fluid life.

Bacteria of several types thrive in grinding fluid and tend to degrade it in three ways:

- They feed on the ingredients that provide lubricating and corrosion-resistant properties.
- While removing these properties, they leave behind corrosive acids which they produce as they multiply.
- They generate a variety of objectionable odors.

"Monday morning odor"—the smell of rotten eggs—is perhaps the most prevalent and obnoxious. It's a result of

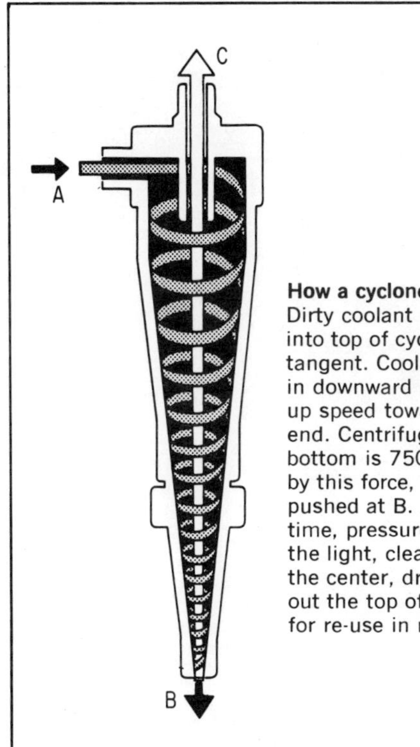

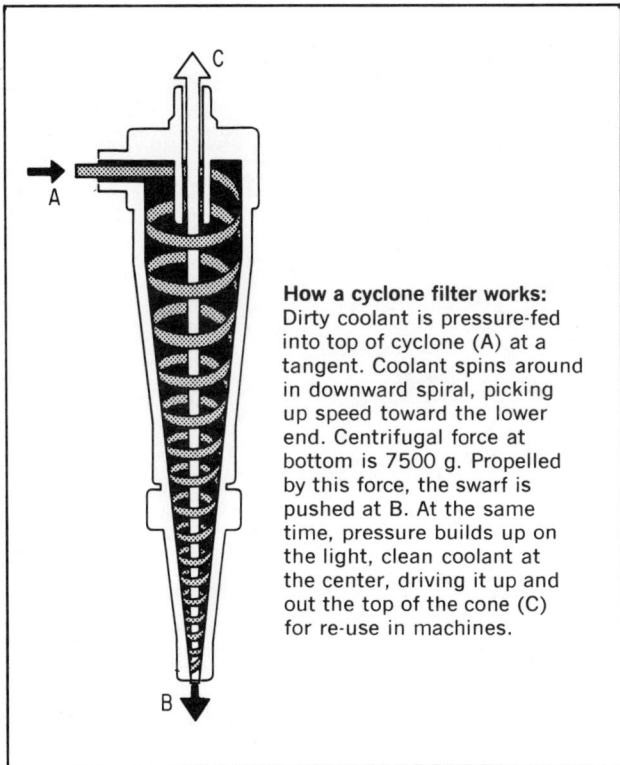

How a cyclone filter works: Dirty coolant is pressure-fed into top of cyclone (A) at a tangent. Coolant spins around in downward spiral, picking up speed toward the lower end. Centrifugal force at bottom is 7500 g. Propelled by this force, the swarf is pushed at B. At the same time, pressure builds up on the light, clean coolant at the center, driving it up and out the top of the cone (C) for re-use in machines.

the presence of anaerobic bacteria, a variety which cannot develop in the presence of oxygen. The action of the cyclonic filter permits the fluid to take on large amounts of oxygen, making cyclonically filtered fluid a very bad environment for the development of anaerobic bacteria.

As for the other types of bacteria, they can be kept under control by keeping the fluid clean and separated from the swarf.

A grinding fluid is inexpensive to begin with, but it gets increasingly expensive as it recirculates the swarf and loses its lubricating and corrosion-inhibiting properties. If the swarf is not quickly and completely removed, it is deposited on machine parts and builds up to a point where machine wear occurs, precision is lost, and maintenance costs begin to rise.

Effect on the grinding wheel

The grinding wheel has an important stake in clean coolant. One of the most important functions of the fluid is to wash away chips of metal and abrasive from the face of the grinding wheel, keeping it clean and sharp. As the wheel "loads," chip clearance disappears, and the wheel stops cutting. As the cutting action slows, the energy that should be going into chip removal goes into the workpiece in the form of heat. So, in a sense, dirty coolant progressively defeats its principal job of keeping the workpiece cool.

Clean, filtered grinding fluid is so influential in the working of the grinding wheel that, in many cases, a cyclonic filtration system permits the use of a harder grade wheel and reduces the frequency of wheel dressings. In some cases, higher infeed rates are possible, too, and this means faster stock removal.

Diamond-tool-dressing costs sometimes get lost in more important considerations, but it is simple arithmetic that, if clean grinding fluid can reduce the frequency of wheel dressings by one-half, it will reduce dressing-tool costs to the same degree.

The operator, the human element in the grinding system and its most important one, is definitely interested in clean grinding fluid. It is easier on his nose, and it does away with sump cleaning.

Capacity is no problem

Cyclonic filtration systems are completely adaptable to any capacity situations; it's simply a matter of having enough cyclones and tank capacity. Each cyclone unit can process about 20 gallons per minute; this would be enough to handle one grinder, perhaps two small ones. But, if you have a large grinder or central system requiring 100 gal per min or even 1,000 gal per min, just divide the flow requirement by 20 to determine the number of cyclones needed.

A standard cyclonic-filtration system consists of the conical filtration unit, an integrated swarf-collector tank and clean-coolant-storage tank, the coolant pump, and a separate sludge truck. Where additional pressures are required at the grinder, a booster-pump-and-tank system is available, simply stacking on top of the basic unit.

The cyclonic filtration system for a single grinder requiring a fluid flow of 20 gpm is priced at under $1500. This system has a single cyclone filtration unit. The price drops rapidly on a per-gallon-per-minute basis as the systems get larger. A 40-gpm system employing two cyclone units is priced at $1700.

Minimizes pollution problems

One more point, but one that is becoming increasingly important these days, is the matter of water pollution. The size of the water-pollution problem for any manufacturer is, of course, a matter of how much he has to dump. A cyclonic filtration system will drastically reduce the need to dump grinding fluids. It uses a great deal less to begin with, but, more important, the frequency of dumping will be cut from perhaps once a week now to once or twice a year. And, when the fluid is ready to be disposed of, it is ready to go into a neutralizing process because the solid contaminants have been removed.

There are no less than nine manufacturers of cyclone filtration systems offering units for grinding-fluid filtration today. Most systems are basically the same; with some, the difference is in the design of the cyclone units themselves. Most cyclone units are made of plastic, but Cyclonics Inc. employs cyclone units made entirely of alumina ceramic, a highly abrasive-resistant material that can withstand the abrasion involved in rapidly moving mixtures of abrasive grains and metal chips under high centrifugal forces.

Cyclonics Inc. offers kits to convert to cyclone filtration those settling tanks supplied by most grinding-machine manufacturers. It can also add filtration units to existing drag-out systems.■

Electrical-discharge machining

Spark erosion is solidly established in toolmaking and some types of production. This report explains the fundamentals of EDM, plus some new tricks that expand its capabilities

ELECTRICAL-DISCHARGE MACHINING has been with us for more than 30 years. When it was first introduced, it was hailed by some as the ultimate way to machine metal, and some of the more optimistic sages predicted that the process would make all other metalcutting methods obsolete within a few years.

They would like to forget they said that, no doubt. But it wouldn't be fair to evaluate EDM's successes in light of those extravagant predictions. The process of cutting with a spark has carved out a few significant niches for itself in the metalcutting field, particularly in tool-and-die making, where its hold is getting tighter all the time.

EDM is well-established in production small-hole drilling, too. Automobile and aircraft engines and their fuel-supply systems, particularly, are being filled with more and more holes in an effort to keep up with emission standards. Many of them are being drilled with special EDM units that produce at least as fast as conventional metalcutting machines and that have greatly reduced the tooling cost associated with drilling small holes.

Aside from drilling small holes, EDM has made a few other inroads into production metalcutting. Small, intricate parts made of hard materials, such as watch and computer parts, are sometimes made with EDM. But EDM has not been accepted for jobs that can effectively be done with conventional machine tools.

The reason for this, and EDM's great handicap, is its slow cutting speed. Metal-removal rates are limited by factors inherent in the process. It should be pointed out that early supporters of the process did not foresee this and, thus, cannot be blamed for their wild predictions. Much better theories of what happens at the cut have been developed since those days; but applying the best theories with the most sophisticated electronic circuitry has helped specific metal-removal rates only creep slowly up from those obtained 20 or 25 years ago.

Although cutting rates do continue to climb, most of the development work in EDM today is concentrated on getting more out of the rates that can be attained now.

By Edward A. Huntress, assistant editor

Orbiting, wirecut, and multiple-lead circuits are some of the steps in that direction that we will examine in this report.

Simple setup and operation are also among the top priorities of EDM manufacturers. Trends in both the conventional, ram-type EDM and its offspring, wirecut, are leading toward completely automatic machining. Some wirecut units have achieved this goal already and are often left to run by themselves overnight or over a weekend.

Progress in EDM seems to make a leap whenever a new idea removes one of the bonds imposed by thinking of it along the lines of conventional machine tools. To get a perspective on the present state of EDM, we'll briefly examine its short history.

Early history

For a technology so young, EDM's beginnings are surprisingly cloudy and are still the subject of debate. Part of the reason is the time in history during which the initial research and development of EDM took place: World War II and the years immediately afterward. Researchers in both the US and the Soviet Union were doing work on the process almost in parallel in their experiments and discoveries. Largely due to the general atmosphere of rivalry and technical secrecy that prevailed at the time, development had been going on for some time before information became available on the Soviet machines.

American Machinist carried the news in its May 8, 1947, issue that two Soviet scientists, B.R. and N.I. Lazarenko, had been awarded the coveted Stalin prize for their work on a new electric-erosion machining method, work they had been doing for ten years. Two months later, the Lazarenkos bylined an AM story in which the process was explained and drawings of the machines and power supplies appeared.

The machines showed many similarities to tap disintegrators that had been in use in the US at least since 1942; they used a dc power supply and a bath of liquid dielectric, and one of the three machines described in the Lazarenkos' article used a vibrating electrode that apparently started and stopped the spark. Such machines were available at the time from the recently formed Elox Corp (then located in Michigan). In fact, Elox had sold a similar machine to the Russian Purchasing Commission in 1945.

No doubt, this is the root of the debate. What is sometimes overlooked in the argument, though, is that one of the other units described in the Lazarenko article was a much different type of machine: it used a servo-controlled electrode holder that positioned the electrode at a relatively constant distance from the work. Spark initiation and termination were controlled by a resistance-capacitance (RC) relaxer circuit, not by movement of the electrode. This is the basic principle of modern EDMs, and this machine—not the vibrating-electrode model that some consider a copy of the US-built tap disintegrator—represents the technology upon which EDMs have been built in the years since.

But the debate isn't over yet. Another US researcher discovered that EDM performance improved when the electrode was maintained at a constant distance from the work and discharged through a liquid dielectric with a dc source. At least a few units were built on this principle during the mid-'40s.

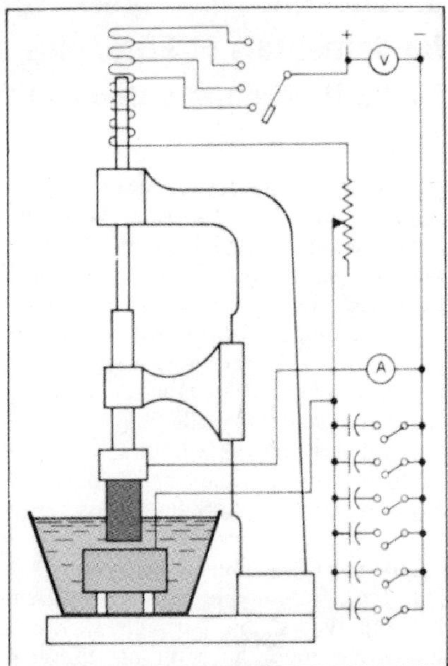

1. First EDM unit to feature a feedback-controlled electric servo was this model illustrated in a 1947 AM article. The Lazarenko brothers, who developed the machine, usually get credit for inventing the first practical EDM

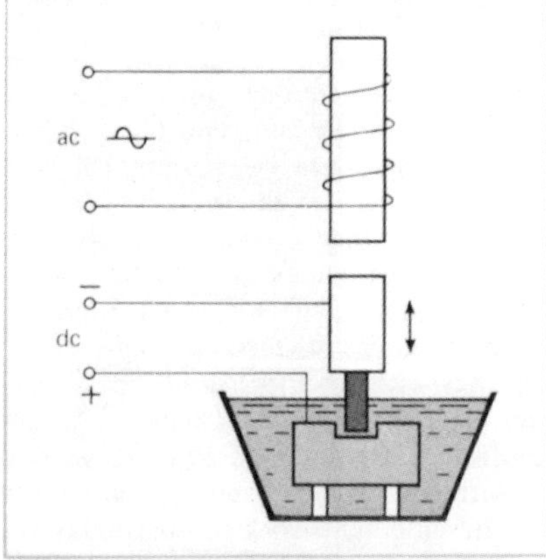

2. The electrode vibrator was a carryover from tap disintegrators. Although it solved the electrode-positioning problem, it sacrificed cutting speed

But, according to the records available, the problem of positioning the electrode as the workpiece erodes from under it was not solved until after the Lazarenkos' announcement. And it's not clear that an efficient self-oscillating power supply was developed before that time, either. The debate could no doubt go on—not on documented facts, perhaps, but on interpretation.

The five years following the revelation of the Lazarenkos' work saw a rapid development of the electrical-discharge-machining process. The idea of using pulsating dc was universally adopted. Some used the RC circuit, others used rectified ac line current, and still others used a high-frequency motor-driven alternator, but all accepted the idea of controlling the spark rate at the source. Machines appeared looking very much like the toolroom EDM units in use today, arranged like a vertical mill, with a servo-controlled quill holding the electrode above the workpiece, a bath on the machine table for the dielectric, and a pressure system for delivering fresh dielectric to the cut. Many of the early toolroom EDM units were indeed converted milling machines.

Even with the relatively slow machining rates common in the early days of EDM, the process was beginning to show some real savings in tool-and-die making. A report from General Motors in the mid-1950s described some dramatic savings the company had realized by using EDM to modify existing dies that would have otherwise had to be remade from scratch. One job, for instance, took 48 worker hours with EDM. The company estimated that 4200 hours would have been required to remake the dies, and there was no other way to salvage the existing ones.

Experimentation with EDM as a production tool was going on at the same time. Benefits of the system, it was soon discovered, made it an excellent choice for certain types of jobs that were difficult, or even impossible, with conventional machine tools. It could drill precise holes through hardened steel or carbides—without burrs—to 0.0002-in. tolerances. It could drill long, small-diameter holes without runout (one early machine drilled 0.080-in.-dia holes 28 in. deep in a production part, using two heads to drill two such holes simultaneously). And it could cut nonsymmetrical internal shapes with tight corner radii in any electrically conducting material.

As experience with the process accumulated, it became apparent that EDM had unique abilities worth cultivating. But it also had some troublesome idiosyncrasies. Cutting performance was unpredictable: two jobs that looked identical could differ in cutting time by a factor of two, three, or more. Tuning the power supply to specific jobs required more art than science, and experienced EDM operators guarded their secrets. Consistent, predictable performance had to wait for better theories to explain what was going on at the cut and for better application of those theories.

How it works

Electrical-discharge machining is a simple process to understand—or a difficult one, depending on how much you want to know about what actually happens at the cut. Essentially, it's a process of eroding an electrically conductive material with a series of electric sparks.

How the sparks do the eroding has been the subject of several theories and a considerable amount of research, and most researchers would probably agree that some elements of the process are not yet fully understood. But what is known has served the equipment manufacturers well in their efforts to build reliable, predictable machines that are fairly simple to operate. And, from the users' point of view, the present theory explains the results they're getting and enables them to exercise control over the cut.

We'll explore the theories later, but, first, let's look at a simplified EDM system. The basic elements of any system, conventional or wirecut, are the electrode, the dielectric, the power supply, and a servo system that determines when the electrode should be advanced in the cut and then moves the electrode accordingly. For the moment, we'll discuss the conventional, or vertical-ram-type EDM, which also has a tank and a filter for recycling the dielectric fluid.

Servo positions electrode

The workpiece is mounted inside the tank, connected to a terminal of the power supply, and covered with a dielectric liquid, usually a light oil. The electrode is lowered until it is within a few thousandths of an inch of the workpiece, the dielectric pump is turned on, and the power supply and servo control are switched on. Throughout the cutting operation, the electrode's vertical position is controlled by the servo mechanism, which maintains approximately a constant distance between the electrode and the workpiece.

The power supply then establishes a potential voltage between the electrode and the workpiece that, if the distance between the two is correct, should be enough to penetrate the dielectric and cause a spark to jump. If the spark doesn't jump, the servo drives the electrode closer for another try. If the electrode and workpiece touch and cause a short circuit, the servo retracts the electrode.

Lots of sparks, one at a time

The sparks occur at a rate of tens of thousands, or even hundreds of thousands, of times each second; but only one spark at a time. Each spark erodes a tiny bit of metal from the workpiece and leaves a small crater. The metal bits, cooled into balls, become suspended in the dielectric and tend to interfere with the cut. Washing them away in a stream of dielectric can prevent this, and so most EDM systems use a pump to supply a flow of the liquid to a nozzle or through the tool or workpiece itself. The dielectric liquid is then filtered and recycled.

Erosion of the workpiece occurs in advance on the electrode position; the two should never touch. The gap, or "overcut," is on the order of 0.001 in., and exists on the sides of the cut as well as on the bottom. When the cut is finished the hole in the workpiece will be about 0.001 in. larger than the electrode in all dimensions.

The workpiece is an essential part of the machine: once it is connected to the power supply it becomes the other electrode. Since the machine can't tell which electrode is the tool and which is the work, the tool (which we'll call the "electrode" from now on, even though both tool and work are electrodes) and the work will wear away equally as the cutting progresses unless the materials for each are chosen properly and the power supply is controlled in a way that directs its efforts in the desired direction.

This was the unhappy result with some of the pioneer eroding machines that used unrectified ac current for power. Electrode wear was not a problem in the application for which those machines were intended: blasting out broken taps and drills jammed in holes. But, before the process could be used to machine intricate shapes, a way had to be found to minimize the wear of the electrode while maximizing the wear of the workpiece.

One-way electrical current

A major step was taken in this direction when the polarized-dc power supply was introduced. Researchers found that, depending on the materials chosen for the electrode and the workpiece, discharges that polarized in one direction or the other would change the proportion of wear between the two.

So power supplies were developed to provide sparks polarized in one direction at a time, usually switchable from one polarity to the other for different materials. Modern power supplies work on the same basic principle. [Continued]

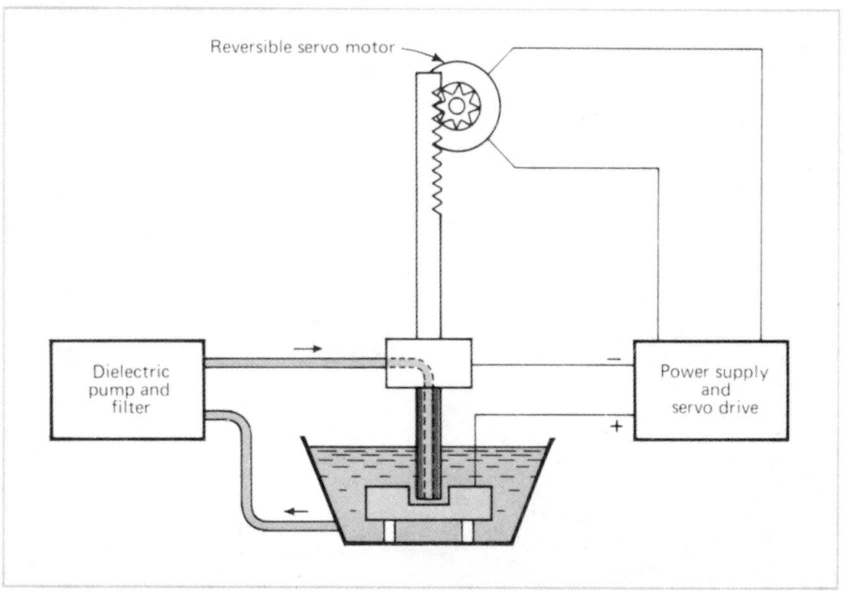

3. The basic elements of a ram-type EDM. Dielectric fluid, which controls spark-gap size, flushes away chips, and cools work and electrode, often is pumped or drawn through the electrode or the workpiece to improve flushing

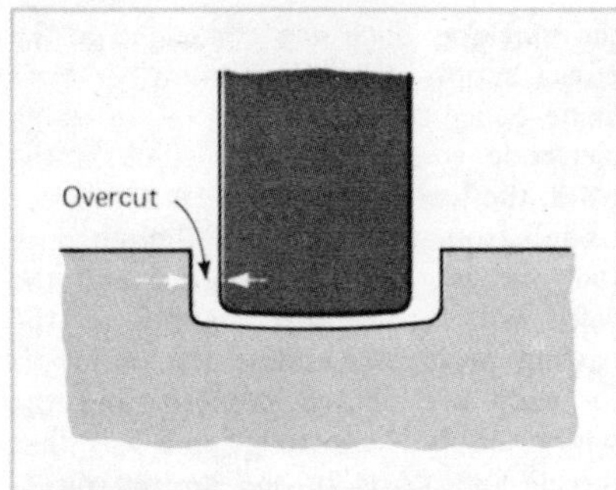

4. Overcut, the gap between electrode and work, is determined by spark energy, dielectric, work and electrode materials. Typical gap size ranges from 0.0005 to 0.002 in.

Tools, or electrodes, have been made of many different electrically conductive materials. The choice of one material over the other is based on wear rate (a factor of the spark polarity, workpiece material, and other elements of the process), cost, machinability, and the nature of the finish to be left on the workpiece. Copper and brass were the first materials used with success, but tungsten, molybdenum, graphite, steel, and other conducting materials are also in use today.

Several different schemes have been used over the years to control the electrode position relative to the workpiece. The first tap disintegrators (which were also used for drilling holes in hard steels) used a hand feed. No attempt was made in those early days to differentiate between sparking and continuous arcing, and so, as long as light was coming out of the gap, the machine was assumed to be doing its job. The operator simply cranked the handwheel back and forth until he got maximum light and smoke.

Naturally, this was a slow process. Automatic electrode-position control appeared next, long before EDM evolved into its present form. Most of the early controls used some type of vibrator that dropped the electrode until it contacted the work. The vibrator then lifted the electrode a predetermined distance, stopping the spark, and then dropped it again. Machining occurred only during a short portion of the electrode's travel, but the vibrating action provided two benefits: the pumping action of the electrode in its hole tended to flush the chips out of the cut, and, by making and breaking the circuit mechanically, the vibrator provided the intermittent sparking that would later be provided by the power supply itself. Even after the development of more-advanced power supplies, the vibrating feature was maintained on some machines, primarily to improve flushing. A few present-day EDMs still offer it as a way to solve difficult chip-removal problems.

The RC relaxer circuit developed by the Lazarenkos eliminated the need to vibrate the electrode, at least for the purpose of turning the sparks on and off. Faster cutting, it was found, is obtained when the electrode is maintained at a constant distance from the workpiece and sparking is controlled at the power source.

So the next stage in the development of EDM was based on the RC circuit in combination with a servo system that would respond to the constantly changing gap between the electrode and workpiece. Since it measured only thousandths or even ten-thousandths of an inch, the gap couldn't be measured directly. An analog of the gap was provided by the voltage that appeared at the power supply output terminals each time it initiated a spark: the closer the electrode and workpiece are when the charge is delivered by the power supply, the lower the voltage required to penetrate the dielectric and pass current through it.

The Lazarenkos and other early EDM developers took advantage of this indirect measurement of the gap and used it to control the forward and reverse drive of a servomotor they had connected to the machine quill. A simple circuit established an equilibrium—represented by zero voltage at the motor-drive terminals—when the spark voltage was at a level that corresponded to the proper gap.

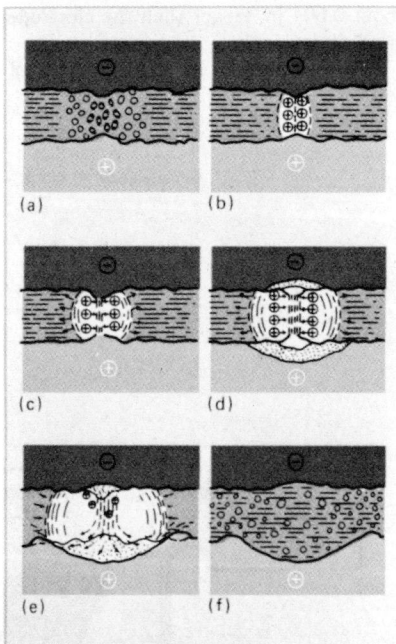

(a) (b) (c) (d) (e) (f)

5. The thermoelectric theory, which this model illustrates, is the most commonly accepted explanation of what causes metal erosion in electrical-discharge machining. Details of the six different steps are given in the text

Increases and decreases in the spark voltage appeared as different polarities of current at the servo-drive terminals, turning the motor in one direction or the other.

One of the keys to improved EDM efficiency in recent years has been the development of more-responsive servo mechanisms. The reversing dc motor system just described suffered from poor sensitivity and response rate because of the inertia of all of the moving parts required: a gearbox, the motor's armature, the gear driving the quill, and the quill itself. Reversing the motor required bringing all of the moving parts to a halt before they could be reversed.

Maintaining a sensitive gap

Another electromechanical system that evolved improved the situation somewhat. It consisted of a constant-speed, constant-direction motor and a pair of electrically actuated clutches that engaged the motor drive through forward and reverse gearboxes. Inertia of the armature and motor shaft was eliminated with this system, and it responded more quickly to changing conditions at the cut.

An even better system, the one in use on the majority of present-day EDM units, is a hydraulic actuator combined with electromechanical or electrodynamic servo valves. Inertia of the hydraulic fluid and the piston it works upon are much less than for a comparable motor-drive mechanical unit. Response rate and sensitivity are correspondingly better.

Each of the components of the EDM system—electrodes, power supplies, dielectrics, and servo mechanisms—have been developed in an attempt to make EDM faster-cutting, more consistent in its results, more efficient in terms of electrode/workpiece wear, and more predictable in its cutting time and accuracy. Development proceeded despite the lack of an adequate model of what was happening at the cut. Theories of how sparks eroded the workpiece (and the electrode) have never been completely supported by the experimental evidence, but one theory, which proposes that the eroding action is a complex combination of thermal and electric effects, is the best one at present.

It's called the thermoelectric theory, and it displaces two others: the electromechanical theory, which states that erosion takes place because of a "ripping away" of bits of the workpiece by a strong electrical field, and another theory, which claims that various electrical effects of the discharge generate jets of flame that melt the workpiece.

The electromechanical theory had assumed that there were no effects of heat involved in the cutting, and that the material properties of workpiece surface were unaffected by the EDM process. This proved to be wrong. Tools and production parts made by EDM sometimes cracked or broke for no apparent reason; a situation that created some strong arguments between EDM's adherents and its detractors. The question was settled in the early '60s when high-power magnification, combined with etching and other metallographic investigation techniques, showed the presence of a thin layer of melted and rehardened metal on the workpiece surface. Testing showed that this layer was usually harder than the parent metal and that it was highly stressed. This reduced the fatigue life of the parts in question and explained their cracking and breaking.

Heat, electricity combine

Without question, heat is generated at the cut. But it was some time before a satisfactory theory was developed to explain the nature of the chip-removal process, changes in cutting rates with changes in spark time and frequency, and the extremely high temperatures (tens of thousands of degrees C) that appeared to be developed at the cut.

According to latest thermoelectric theory, the discharge occurs in several stages, illustrated in Fig 5. First is the ionization stage, shown in the drawing at upper left.

The charge induced on the two electrodes by the power supply creates a strong electric field, which is strongest where the electrodes come closest to touching. This is where the discharge takes place. Molecules and ions of the dielectric fluid are polarized and oriented between these two peaks, forming a narrow, low-resistance channel. Actual current flow is initiated as current "streamers" organize (second drawing) and open the way to the main flow of current.

Electrons through a plasma

Ionization continues through the next two stages, even though current is already flowing between the two electrodes. In the next stage (second left), the resistance of the channel continues to decrease while the current is increasing. At this point the ionized path consists of plasma formed from positive ions and free electrons, mixed with gas formed by the chemical decomposition of the dielectric. Metal vapor is expelled from both electrodes.

Current intensity in the channel is very high at this stage, perhaps 10^7-10^8 amp/sec. The high current continues to further ionize the channel and to produce a strong magnetic field, which attracts the ions towards the axis of the discharge channel. The magnetically attracted ions compress the channel of current and increase its temperature to tens of thousands of degrees C, which melts a portion of the workpiece and vaporizes some of it. Vapor provides some of the force that expels the molten metal from the workpiece, and the rapidly collapsing ionized channel assists. Another effect—the mechanical impact of the discharge itself—contributes to the expulsion. The path of expulsion leaves marks on the edge of the crater, shown lower left.

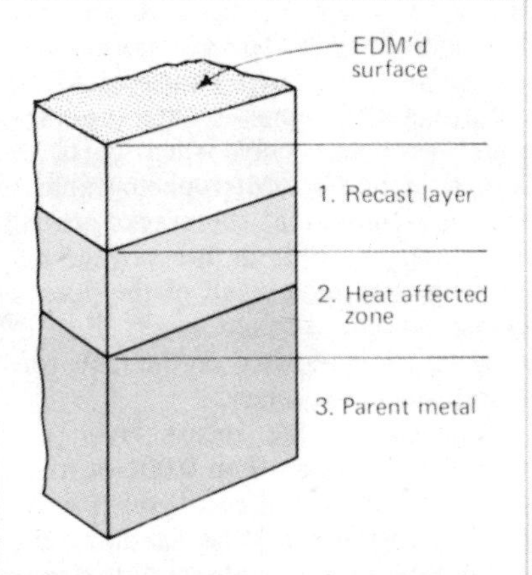

6. Contrary to old theories, EDM causes metallurgical changes in the workpiece surface. Layers are explained in the text

In the last drawing, the result of the discharge is shown: a large crater in the workpiece and (hopefully) a smaller one in the electrode. Large and small, of course, are relative. The range of crater diameters varies considerably, but a typical figure might be 100 μin. for the workpiece crater.

The cycle is now complete. Molten metal removed from the electrode surfaces (the tool and the workpiece) cools quickly in the dielectric liquid, forming tiny spheres that are flushed away.

Other elements of the EDM process have been analyzed in light of the thermoelectric theory, including power efficiency in the metal removal, heat absorbtion in the dielectric and the electrode, effects of pulse duration, and characteristics of the EDMed surface.

Experiments show that more than 90% of the energy of the spark is wasted in heating the electrode surfaces. Most of this energy is absorbed in electrode material that is melted but not expelled. This material quickly cools and remains on the electrode surface.

Different pulse durations, it was found, tend to alter the relative importance of the different electrical and mechanical effects of metal removal. Experiments also show that longer pulses are less efficient, in terms of metal removal for a given amount of energy expended. But the metal-removal rate is higher for longer pulses, up to a certain point. Longer pulses also tend to increase surface roughness (see Figs 7-10) and decrease relative electrode wear.

An unusual surface finish

Figure 7 also reveals the nature of the EDMed surface. Unlike almost any other type of machined surface, it has no texture or pattern that extends beyond the individual craters. The edges of the craters overlap to create a random pattern of irregular pock marks. Small globs of resolidified metal cling to the edges of each crater, but they are much smaller than the craters themselves and therefore have little effect on surface roughness.

Nondirectional surface patterns have proved to be one of EDM's major advantages over conventional cutting methods in several applications, particularly in the making of punches and dies. Punched parts don't cling to the surfaces of dies that have an EDMed finish as easily as they do to ground or milled surfaces. Surface roughness can be twice as great for an EDMed punch or die than for a conventionally machined one, as a rule of thumb, and EDM-equipment manufacturers are consequently searching for a roughness-measurement system appropriate in comparing EDMed tools to conventionally machined ones.

Fluid-control parts, such as fuel jets and injectors for auto and truck engines, also benefit from the nondirectional surface texture. Drag on the stream of fuel over an EDMed surface is consistent and predictable, and there is no tendency for fuel to be directed by tool marks as there is when conventional tools are used to drill the holes in these parts.

All of the metal exposed on the surface of an EDMed part is melted and quickly rehardened by the cooling dielectric, so that thin layer (typically 0.001 in.) is bound to be affected in some way. Usually, it becomes harder than the parent metal. Sometimes, though, it becomes softer; when high-energy pulses are used for fast cutting, the mass of melted metal that remains cools more slowly than if the pulses were weak and the melted pools were small. Slower cooling may leave the surface softer than the parent metal, depending, of course, on the type of metal being cut.

Whether the surface becomes harder or softer, though, it always develops internal stresses that can reduce the strength of the finished part. When high-energy pulses are used, these stresses can lead to cracks large enough to see under a microscope. They are random in direction, and often branch into a maze of smaller cracks.

Tests of fatigue life of EDMed parts indicate that cracks are present in all EDMed surfaces, even those machined with low power. The cracks are too small to see in many cases, but their presence is revealed by the greatly reduced fatigue strength (typically 25% of original strength) exhibited by the test pieces.

This evidence contradicts the experience of tool-and-die makers who have been using EDM for years. They have found that EDMed tools are often stronger and give longer life than conventionally machined tools. The generally accepted explanation is that punches and dies are loaded in compression and the cracks cause problems only when parts are loaded in tension. Microphotographs of used dies show that the cracks are still there and that bits of the surface may even flake off as a result of the maze of cracks in the surface layer. But the working loads imposed on the tools rarely pull the cracks open.

Adjacent to the recast layer is a region, usually less than 0.001 in. thick, known as the heat-affected zone (see Fig 6). This region, too, is heated and cooled in the EDM process and contains stresses (grain-boundary stresses are suspected) that reduce fatigue strength. Tests show that removing the recast layer increases fatigue strength only 4-5%, but removing both the recast and the heat-affected zones restores the strength of the part to 95% or more of its original value.

Power supplies

The job of the power supply is to provide a series of electrical charges to the electrode and the workpiece. If all of the variables affecting the cutting process could be controlled, the power supply's task would be simple enough: to provide a certain flow of current at a particular voltage for a predetermined period of time and then to shut off and start the cycle over again. But the variables are never completely under control. Each spark the power supply delivers requires a slightly different voltage to initiate, and the time required to ionize the channel in the dielectric varies from spark to spark. If consistent, efficient cutting is to be achieved, the power supply must deliver a consistent charge.

The quantity of metal removed with each spark is dependent on the amount of energy dissipated; that is, the product of the current and the time over which it flows. A 10-amp current that flows for 100 μsec will remove about the same amount of metal as a 5-amp current that flows for 200 μsec. Long, high-amperage pulses, therefore, achieve the fastest cutting rates.

But high-energy pulses also generate the roughest surfaces and the thickest recast layers. For a fine finish, short, low-amperage pulses are required. These, however, create a higher percentage of electrode wear than longer, medium-amperage pulses.

The power supply has quite a balancing act to perform. Consistent cutting, fast metal removal, fine finish, and low electrode wear seem to work against each other. And, although the power supply might be adjusted to give a satisfactory compromise for one particular electrode/workpiece combination, there is no reason to expect it to work for other electrodes or workpiece materials.

It's no wonder that operators of the crude EDMs of 25 years ago were regarded as practitioners of a black art. Controlling those units demanded an experienced hand, usually guided only by a vague idea of what the adjustments actually did.

As knowledge of the cutting parameters increased, power supplies were made more sophisticated and easier to control. Active electronic elements—first vacuum tubes, then transistors, and recently digital integrated circuits—have been

Photomicrography reveals effects on work surface

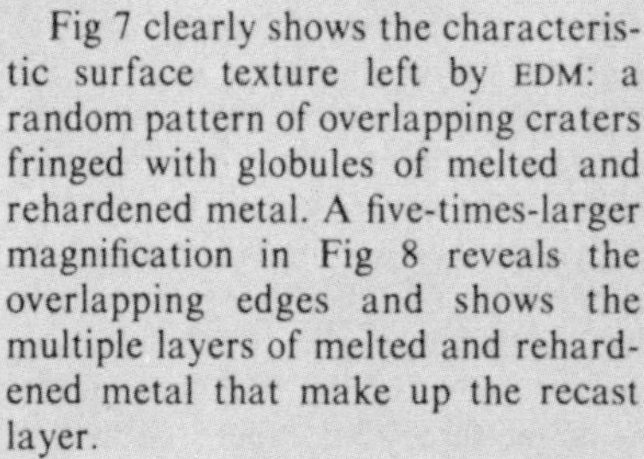

Photographs taken through the scanning electron microscope reveal the pock-marked surface left by EDM and the effects of different spark-energy levels on surface roughness.

These four photos, taken by Associate Professor C.H. Kahng and A.H. Bekkala of Michigan Technological Univ (Houghton), show the EDMed surface of AISI 4140 steel. In Figs 7 and 8, pulse duration was 2.5 μsec; a short, fine-finishing pulse. Figs 9 and 10 show the effects of a 2800-μsec (coarse-roughing) pulse.

Both tests were made on a conventional ram-type EDM at a 19-A peak-current setting.

Fig 7 clearly shows the characteristic surface texture left by EDM: a random pattern of overlapping craters fringed with globules of melted and rehardened metal. A five-times-larger magnification in Fig 8 reveals the overlapping edges and shows the multiple layers of melted and rehardened metal that make up the recast layer.

Figs 9 and 10 show the effects of increasing pulse duration by more than 1000 times: large, cracked globules, large craters, and a very thick recast layer. Closer inspection might reveal a random pattern of microcracks.

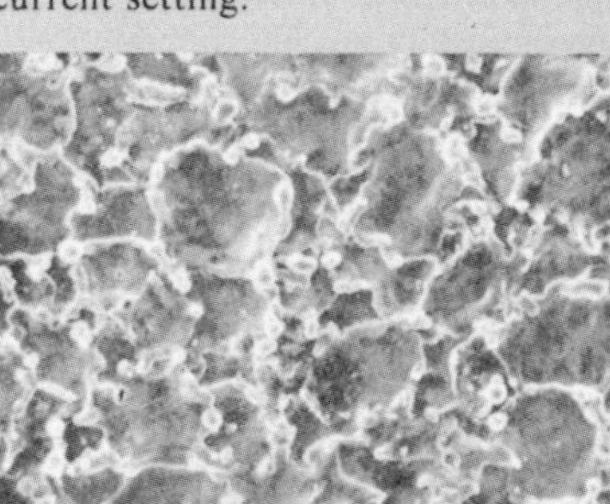

7. 2.5-μsec pulse. 100X

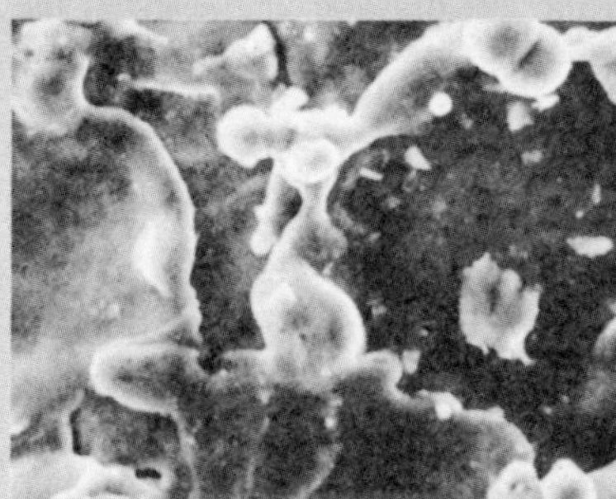

8. 2.5-μsec pulse. 500X

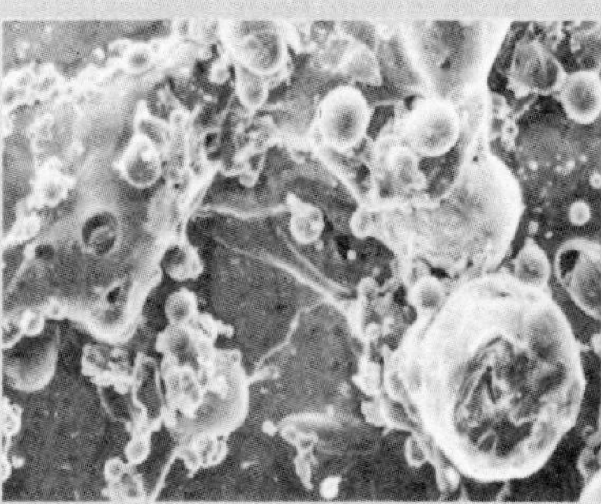

9. 2800-μsec pulse. 100X

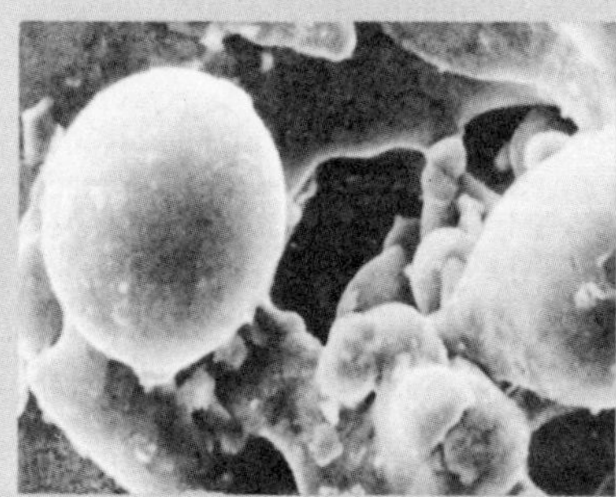

10. 2800-μsec pulse. 500X

designed into the power-supply circuits for two primary reasons: to control the pulse generated within the power supply and to see that the pulse is actually delivered to the workpiece. Achieving the first doesn't guarantee the latter.

The RC circuit, which is still used for finishing, does very little of either. The amount of energy delivered can be changed by varying the capacitance of the circuit, and the frequency of discharge can be varied by increasing or decreasing the circuit's resistance. But the ratio of "on" time to "off" time (that is, the ratio of the time of current flow to the pauses between those periods) can not be controlled independently. It is a function of the other circuit characteristics and, by today's standards, is very low: from less than 1/1000 to about 1/10. The RC circuit doesn't produce consistent pulses, because it lacks a way to control spark initiation. It is vulnerable to changes in cutting conditions.

For fast cutting with this circuit, high rates of current flow are required. The current limit is determined by the point at which the capacitor no longer stores the charge. Intermittent sparks stop, and continuous arcing takes over when this point is reached.

Vacuum tubes have been used to over-

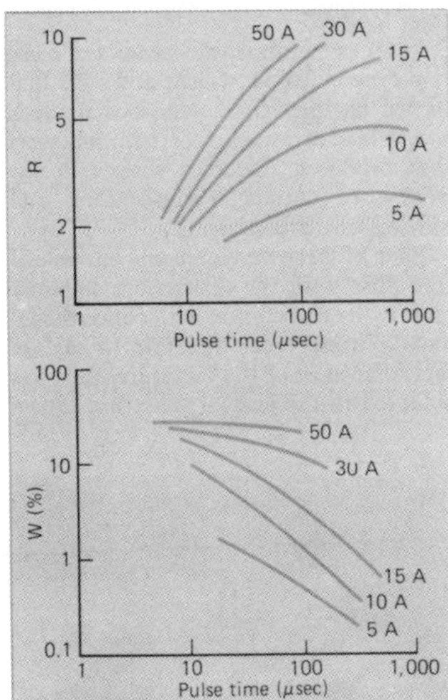

11. Surface roughness (R) and electrode-wear ratio (W)% are plotted against pulse duration in these graphs.

In general, longer pulses produce rougher surfaces and better ratios of electrode/workpiece wear. Roughness and wear are tradeoffs: if a relatively rough surface can be tolerated, electrodes can remove a much greater amount of workpiece material before wearing out or losing accuracy

Six basic types of EDM power supplies

These illustrations represent, in chronological order, the major steps in the advancement of EDM power-supply design.

The RC circuit is the one that got EDM started. It is one of several basic electronic circuits known as "relaxers," which means that they store a charge for a period of time and then "relax" by discharging through a resistance. In this case, the charge is stored in a capacitor and relaxed through the low resistance of the ionized discharge path. The tool and workpiece electrodes actually form another capacitor themselves, connected in parallel with the circuit capacitor, which is chosen so that the electrode-workpiece capacitance breaks down (ionizes a path in its dielectric) first.

The value of the resistor, which determines charging rate, and the parallel capacitance determine the discharge frequency of the circuit.

The RC circuit is simple and dependable, but it is a slow cutter. It can't deliver long or high-amperage pulses.

The rotary impulse generator, a motor-driven generator with rectified-dc output, delivers long, intense pulses. But it can't produce fine finishes.

Vacuum tubes have been used in RC circuits to increase the amount of energy they can deliver. The tube, though, doesn't mate well with the power and voltage characteristics of EDM and thus is not very effective when used to control the pulse directly. Its function in most vacuum-tube supplies is to intermittently turn off the current, leaving the resistor and the capacitor to determine spark energy.

Transistors are ideal for directly switching the pulse both on and off. Controlled by feedback systems, and most recently through adaptive control, they are the basis of all modern power supplies.

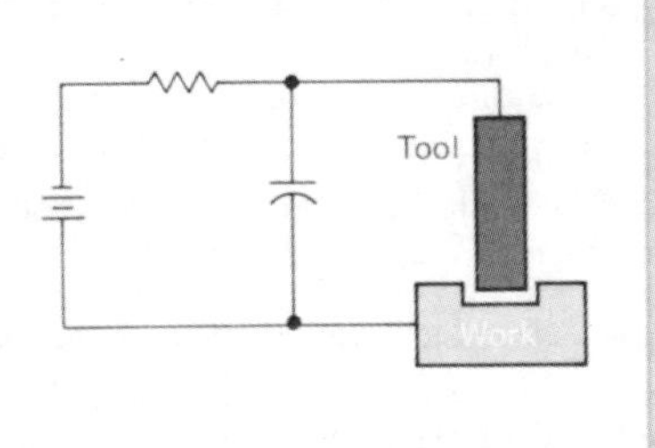

12. RC relaxer circuit

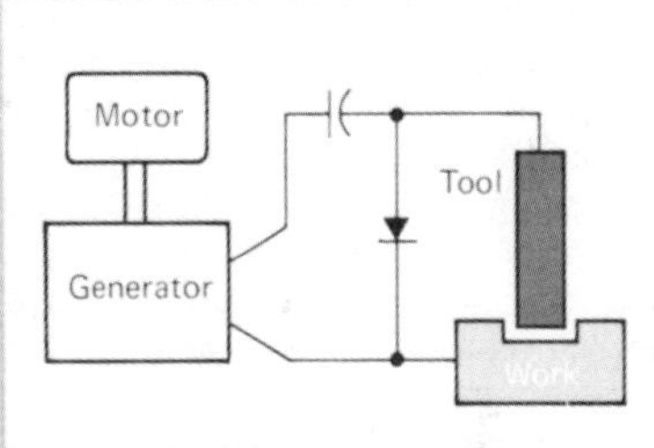

13. Rotary generator

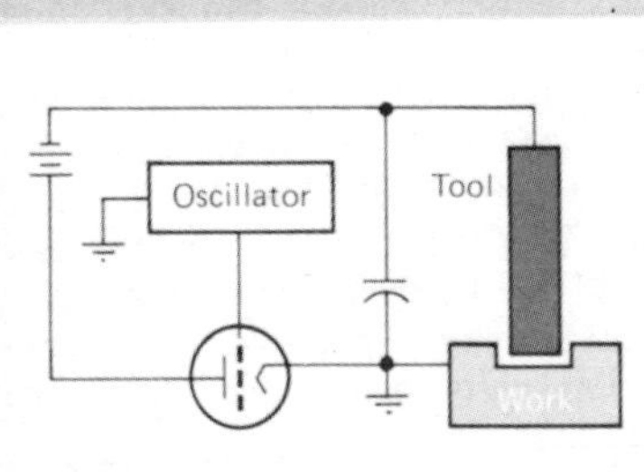

14. Vacuum-tube generator

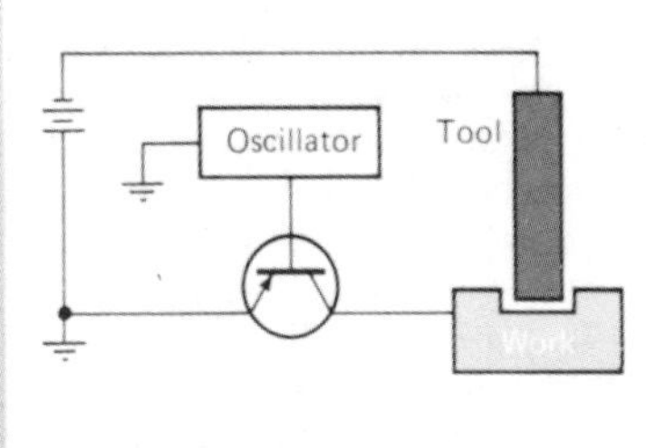

15. Transistor generator

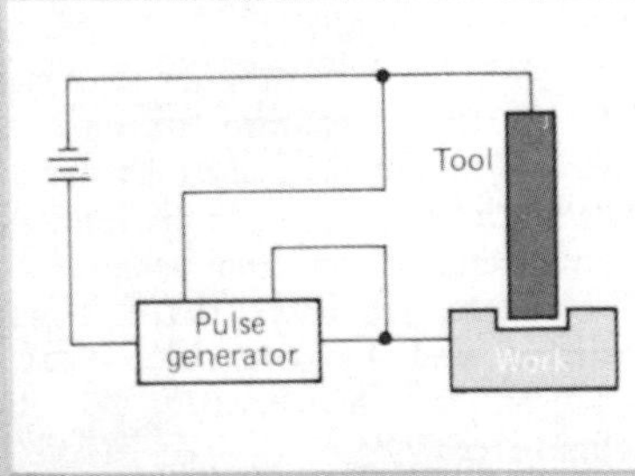

16. Pulse gen. feedback

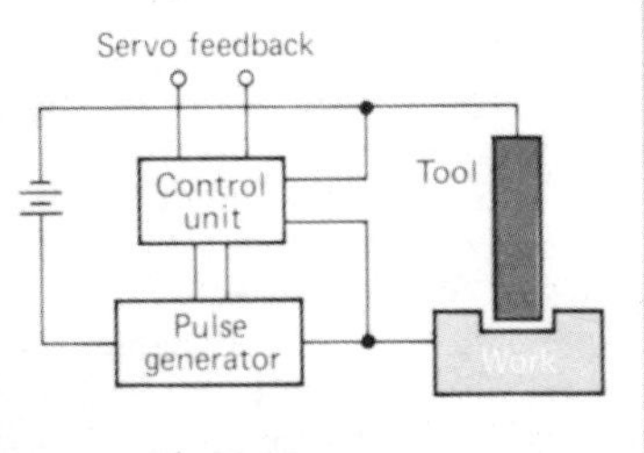

17. Adaptive control

come this limitation by acting as electronic switches that cut off the power to the capacitor for a short interval after the discharge occurs. The arc stops as a result, allowing time for the ionized column of dielectric to collapse. When the electrical path between the electrode and the workpiece is broken, the capacitor has an opportunity to recharge. Tubes allow direct frequency control, too, by means of a variable-frequency oscillator that turns the tubes on and off.

The vacuum-tube circuit, however, still suffers from the other limitations of the RC circuit: lack of control over the on-time/off-time ratio and inconsistent sparking.

The value of greater on/off ratios was proved in the 1950s by a power supply used on some Russian machines: the rotary pulse generator. This was nothing more than a motor-driven generator that produced a fluctuating dc output. Output frequency of this machine was low, but the amount of energy it could deliver in a given period was high, largely because its on time was longer than its off time. It was a fast rougher that left a very uneven surface on the workpiece, and it demonstrated that long pulses improve the electrode-wear ratio as well as the cutting rate.

When high-power transistors became readily available in the late '50s, power-supply designers finally had a control element that would enable them to exercise direct control over the on time, off time, peak current flow, and spark initiation. Transistors allow high current flow at low voltage and so can be used as a switch to simply turn the power on and off. A variable-frequency oscillator can be used to control the rate of the transistor's switching, as on tube-type power supplies. But, on the transistor circuits, the signal from the oscillator controls not only the termination of the pulse, but also the time at which the pulse is delivered to the electrode and workpiece.

This is a great advantage. With this circuit, the generated pulse can be completely controlled. The oscillator is often in the form of two independent timing circuits that allow on time and off time to be adjusted independently.

This is still not a guarantee, however, that the pulse will be delivered to the workpiece. One major factor that can interfere with the delivery is variations in the time it takes for the ionized channel to become conductive enough for the main flow of current to pass.

At present, this is a variable that cannot be controlled. But a feedback circuit has been devised to distinguish the ionization period from the main current flow, and this is used to control the start of the on-time-measuring timer. The timer starts only when it is signaled by the sensing circuit that the current flow is starting. On time and peak current remain constant, but frequency varies as the ionization time changes.

A new generation of power supplies with much more elaborate feedback and self-control circuitry is just beginning to emerge. One of the events the new units will monitor is electrode advance, which is a controlling parameter for the power supply. Interaction between the power supply and the servo mechanism is made by means of adaptive control.

Digital circuits are serving several functions on the new power supplies. One unit that will soon be introduced uses a semiconductor memory to store machine-control parameters for repeat jobs.

EDM applications

The first thought that comes to mind when EDM is mentioned is tool-and-die making. That's where the process has enjoyed its greatest success and greatest acceptance.

EDM offers clear advantages over other processes for making complex shapes, particularly internal shapes, in hard steel and carbide: dies made with EDM can usually be made in one piece, no matter how complex their internal form; material hardness has no effect on cutting rate or tool wear; several, perhaps even dozens, of separate machining steps are reduced to one or two with EDM, and even those can often run without operator attention; workers can learn to operate an EDM machine in a relatively short time, compared with the years needed to train a tool-and-die maker to work with conventional tools. The list could fill a page if it included the benefits realized at all levels, including tool design (it's often simpler), floor-space requirements, tooling costs, and floor-to-floor production time.

Often, EDMed tools are better than the conventionally machined tools they replace: tungsten carbide can be substituted for steel, where appropriate, for a 2- to 10-time improvement in die life; single-piece dies are stronger than those built up from separate sections; the finish left by EDM is usually harder and holds lubricant better than conventionally machined surfaces; and, because tools can be EDMed from steel that is already hardened and from carbide that is already sintered, heat-treating distortion is eliminated.

EDMed punches and dies are presently stamping parts that range in size from microelectronic connectors to large auto and aircraft panels. Molds and die-casting dies made with EDM are casting tiny plastic toy parts, gears, appliance parts, automobile grills, and boats—often with a textured finish left on the mold by the EDM process itself. Forging dies for small fasteners, hand tools, aircraft engine parts, and even large structural beams for off-road equipment are made with EDM, as are extrusion dies, header dies, and compacting dies.

All of these punching, drawing, and forming tools are being made on the vertical-ram-type machine described in the "How it works" section of this report. Massive EDMs, based on multiple-column structures and sometimes on large presses converted to the task, handle the larger workpieces.

But all of these machines represent only one approach to EDM: sinking a single tool into a workpiece along a straight line. Step by step, machine designers have applied new ideas to the design of electrical-discharge machines, and, with each new development, a broader range of potential applications has been revealed. Turning the quill was the first step: controlling the turning rate with a leadscrew permits thread-cutting with EDM. An early application of this idea was cutting threaded holes in carbide tool inserts to anchor them in their toolholders.

Later came attachments for the basic ram-type machine: wheels and rolls that turned on their horizontal axis to work the surface of a workpiece; cam followers that produced undercut shapes in the work; arc cutters; reciprocators; and moving-wire attachments.

Each of these developments has helped free EDM from the restrictions imposed on it by its association with conventional metal-cutting tools. It's fair to say, in fact, that most of the versatility EDM has achieved in the last 20 years has grown

18. EDMing the die and stripper in this tool set took 2 hours each; same as the time needed to mill the punch. Two holes required 3 electrodes, right

out of thinking that emphasizes the differences between EDM and conventional machining.

Comparing the two reveals several disadvantages inherent in EDM:

- Metal-removal rates are low.
- Workpieces must be electrically conductive.
- An EDMed surface is highly stressed.
- Cavities cut by EDM may be slightly tapered from the point of electrode entry.
- Electrode wear can cause cross-sectional cavity inaccuracies.

EDM does have many advantages over metalcutting methods, though:

- Hard and tough metals can be cut.
- No machining loads are imposed on the machine, electrode, or workpiece.
- Hardened steels can be cut without the need for later heat treating.
- No burrs are produced.
- Machining is automatic, even when many separate planes or three-dimensional shapes must be cut.
- Intricate shapes can be cut on both the internal and external surfaces of a part.
- Secondary finishing operations may be eliminated.

EDM's advantages are attractive for many metalcutting operations besides tool-and-die making. For a few specialized applications, EDM has been used as a production tool from its very beginning, mostly for making small, burr-free holes.

Most of the limitations that would prevent EDM from finding wider use as a production tool have been overcome: electrode wear can be completely eliminated with new technology; taper can be controlled and even eliminated with electrode-orbiting accessories; stressed surfaces can be honed or polished off. But slow cutting remains a major handicap.

The rate of metal removal however, is not necessarily a measure of how fast an EDM machine can produce parts. If a machining operation involves cutting completely through a workpiece, only enough metal must be removed to allow the electrode to circumscribe the part. The electrode might be a tube or a wire that is only 0.003 in. thick; allowing for overcut, only 0.020 cu in. of metal would have to be eroded away to cut out a 1-cu-in. plug.

This is one approach to maximizing the production rate of EDM: cutting around the part, instead of eroding away everything that isn't wanted. Wirecut EDM, which we'll examine more closely in a later section, works on this principle. So does the magnet-making system shown in Fig 19.

Another way to make efficient use of EDM is to produce as many parts as possible in a single step. Ten individual electrodes working on ten separate workpieces might be accommodated by a single machine. If the power supply can deliver the maximum usable power to each electrode and if sparks are supplied independently to each one, the machine's production will be ten times its normal rate. The magnet-making system also uses this principle, as does the machine that uses the cluster of hole-drilling electrodes shown in Fig 24.

A variation on this idea is to make a single electrode in several sections, separate each section with an insulating material, and supply each electrode section from its own power supply. The

19. 28 Miniature alnico watch magnets are produced simultaneously by the electrode block on right. Five blocks are mounted on each machine

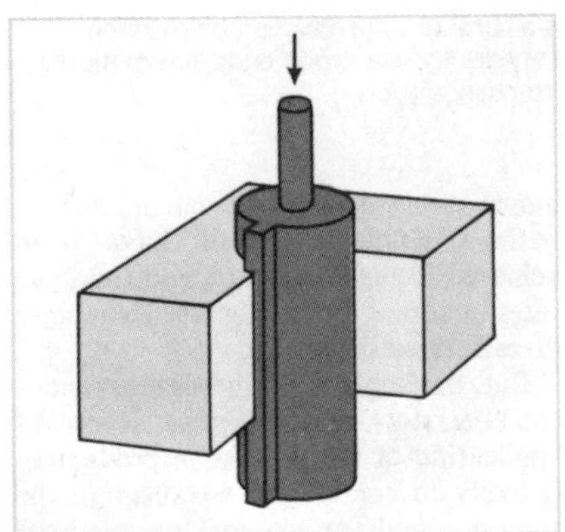

20. Straight sinking

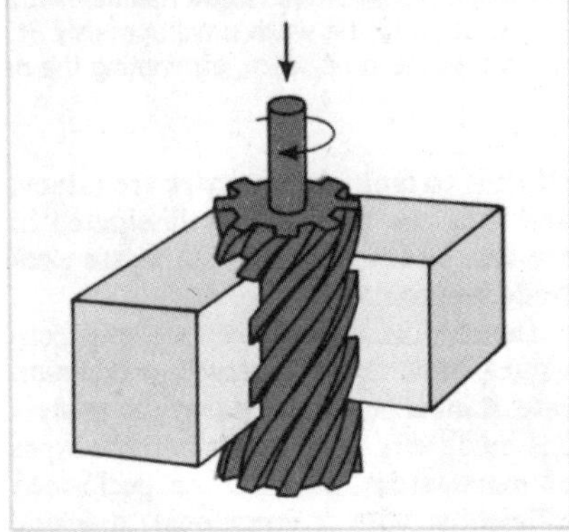

21. Threading

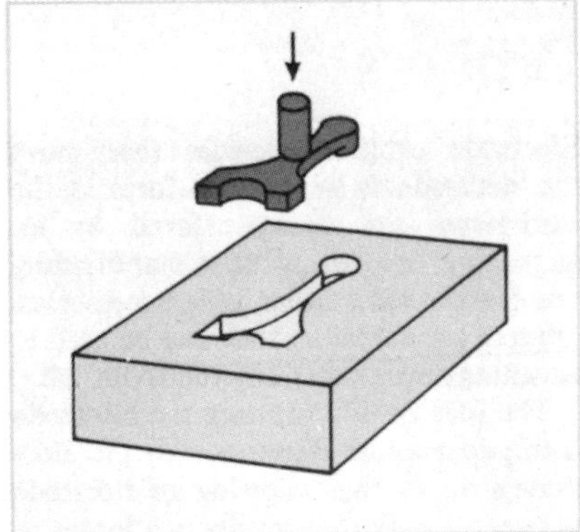

22. 3-D diesinking

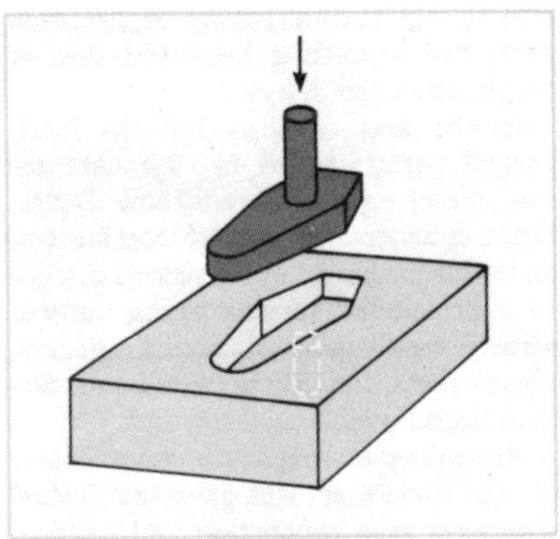

23. Other blind sinking

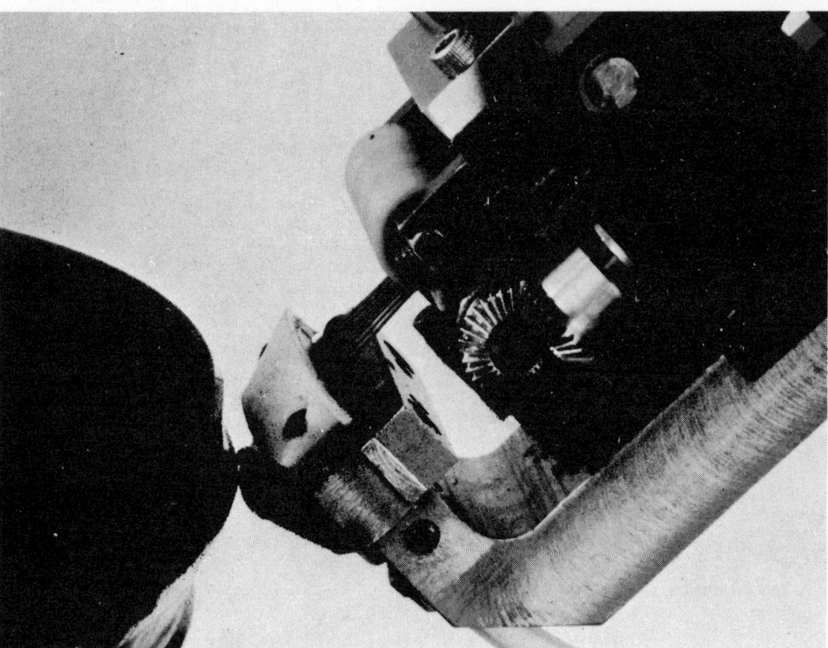

24. Raycon's patented refeed mechanism feeds a number of production hole-drilling electrodes into the work simultaneously. It also shears the electrode ends automatically when they become worn, eliminating the need to change tools

effect is to multiply the spark frequency, and thus the total energy dissipated in the cut, by the number of separate electrode sections.

Despite the improvements these techniques have made in EDM's production rate, it must be admitted that the process has made very few inroads into the types of metalcutting that can be performed efficiently with conventional machine tools. Most of EDM's success as a production tool has been in drilling small-diameter (under 0.10-in.) holes in precision parts and in cutting holes and slots in tough steels and alloys.

In the first category fall the fluid-control parts: orifices in auto carburetors, diesel injectors, jet-turbine blades, brake cylinders, gas-turbine combustors, and valve bushings. In the second category are included turbine vanes, turbine wheels, watch magnets, textile spinners, digger teeth for earthmovers, and carbide-tipped punch-card sensors.

According to some EDM manufacturers and some users who have had success with EDM as a production tool, further applications are limited by imagination as much as by technology. The time to consider EDM, they say, is before a product is designed. A deep hole that is designed round might be better made square, but it is unlikely to be designed that way unless EDM is considered as a production tool for the job.

Technical advances in the EDM field will tend to make it more attractive as a production tool. Cutting rates have crept slowly upward since its inception. Removal rates of over 100 cu. in. per hr with a single electrode have been achieved in the laboratory, and roughing rates in actual production use sometimes exceed 20 cu in. per hr.

But, barring a breakthrough in practical EDM stock-removal rates, increased application of the process in production is likely to come from advances in the several simultaneous-sparking methods and from increased automation.

Orbiters

Electrode orbiters—devices that move the electrode to generate a form in the workpiece—are being offered by an increasing number of EDM manufacturers. Unlike most other EDM accessories, orbiters are versatile; they can be used to advantage on almost any toolroom job.

The idea behind orbiting the electrode is to add another dimension to the electrode's travel, thus allowing an electrode to sweep a path that results in a larger or more intricate cut than would result from sinking the electrode in a straight line. Several different sizes of holes can be machined with one size of electrode, or a single electrode can be used both for roughing, which requires a large gap, and for finishing, which demands a smaller one.

Orbiters also produce a number of other benefits. Flushing is greatly improved, for example, because of agitation caused by the moving electrode and because more clearance exists, on the average, between the electrode and the workpiece.

Finishing is faster because the amount of electrode displacement can be adjusted in steps; instead of one roughing and one finishing step, an orbiter can produce a number of intermediate steps. The final, slow finishing cut removes much less material than it would if it followed the coarsest roughing cut.

Perhaps the greatest advantage of orbiters is that they spread the wear over the electrode surface. By increasing the size of the orbit, either in steps or in a continuous path, orbiters present the sides of the electrode to the sides of the cut. A cylindrical rod or a straight-sided electrode sunk straight into the work does all of its cutting with the leading edge. Corner sharpness holds up better of orbited electrodes, therefore, further increasing their life. And, since wear is evenly distributed over the electrode, it can be used even after a considerable amount of material is worn off, by increasing its orbit. Total electrode saving from all of these effects is on the order of 70%.

Still another benefit provided by orbiters is increased cutting accuracy. Reduced corner wear is one reason: rounded corners on straight-sunk electrodes tend to produce a tapered hole with a rounded bottom. And improved flushing, because it helps maintain constant sparking distance and energy, keeps the overcut uniform. Orbiters also compensate for variations in electrode size and spark-gap size.

Jig 'grinding' with EDM

The ability to adjust the size of the cut opens up another range of possibilities for EDM. Add a rotating spindle and an orbiting unit, and an EDM can stand in for a jig grinder. An X-Y table can be added to increase the machine's versatility. Copper or graphite tubes and rods replace the grinding wheels used with a jig grinder, and combining the vertical and horizontal movements of the EDM permits small disk electrodes to be used.

Disk electrodes present another opportunity: making undercuts in the workpiece. Grooves, double tapers, and reverse tapers can be made with this setup.

Direct the undercut to one side, instead of in a circle, and complex undercuts, steps, and slots can be cut. Or a very difficult sharp corner can be picked out.

All of these operations can be performed with some orbiters presently on the market. It's likely that they will soon become part of the standard repertoire of EDM.

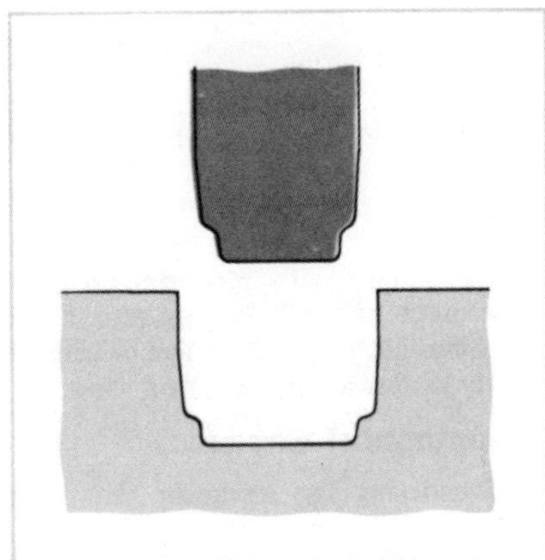

25. Corners wear first on electrodes that are sunk straight into the work, causing irregular taper in hole

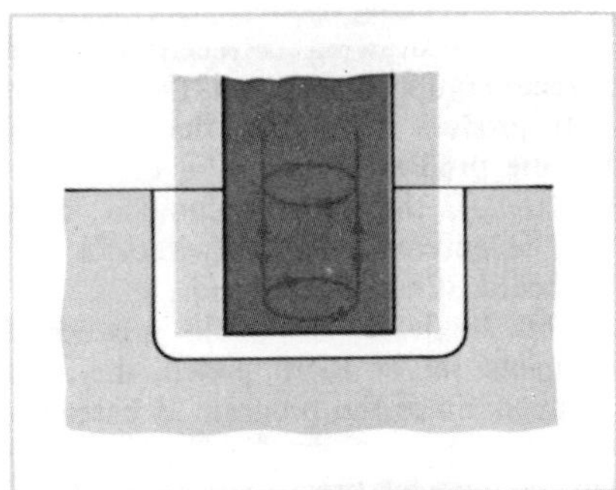

26. Wear is evenly distributed by orbiting electrode in ever widening path. Flushing and electrode life are improved

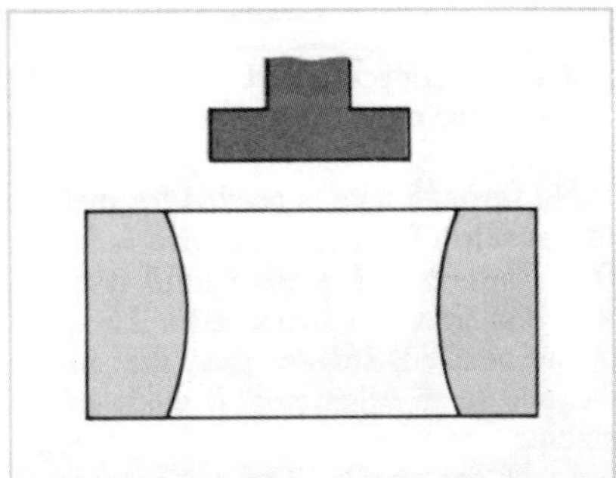

27. Reverse tapers, curved shapes, and undercuts can be generated with some orbiters that feature adjustable paths

Wirecut EDM

Wirecut EDM, traveling-wire EDM, or the "electric bandsaw," as it has been called, started out as an attachment for conventional ram-type units, much like the orbiters. But it took off on a tack of its own about ten years ago and has become established as a separate technique, with its own specialized machines, numerically controlled positioning, and different approaches to cooling and flushing.

The operating principle of wirecut is the same as for any EDM. The fundamental difference is the electrode; instead of a shaped electrode that is sunk into a workpiece, wirecut uses a constantly moving strand of copper, brass, or occasionally molybdenum wire that passes through the workpiece in the vertical axis and cuts a contour in the two horizontal axes.

Since the wire is used only once and is then discarded, electrode wear is not a factor. There is one less element that must be entered in the various compromises of the EDM process, therefore, which allows for simpler control of the power supply.

Cutting conditions are also more straightforward with wirecut EDM: flushing remains the same throughout the cut, unless the workpiece thickness changes, and the area of exposure between electrode and workpiece is constant.

Overall, wirecut is simpler to operate than conventional EDM.. Once the machine is set up, operation can be completely automatic. Tool-and-die shops using the process commonly let their machines run unattended overnight or over a weekend (see box on page 95). Training of operators takes only weeks, sometimes only days.

Wirecut has some advantages over conventional EDM in the job for which they're both suited: through-hole cutting. But weighing against wirecut is its high price. Only recently has a CNC wirecut unit been offered for less than $60,000, three times the price of some contemporary ram-type toolroom EDMs.

A complex machine

A wirecut machine is much more complex than a ram type. First, there is the wire-feeding mechanism itself. It consists of a spool for the wire, several rollers to direct the wire through the machine, a sensor to stop the machine when the wire runs out, a metal brush to conduct power to the wire, hard (often sapphire) guides to keep the wire straight in the cut, pinch rolls that provide drive and proper wire tension, a level-wind roller, and a take up spool.

Then there is the X-Y positioning system. This can be a ballscrew-driven machine table or a mechanism for driving the whole machine head. Either way, it must be precise, since the machine's positioning accuracy must be comparable to a jig grinder's. Typical positioning accuracy is ± 0.0002 in. over the full length of travel, with leadscrew compensation supplied by the system's controlling computer.

Computer control is a fairly new addition to wirecut; some of the earlier units used a line-tracer to control the wire's

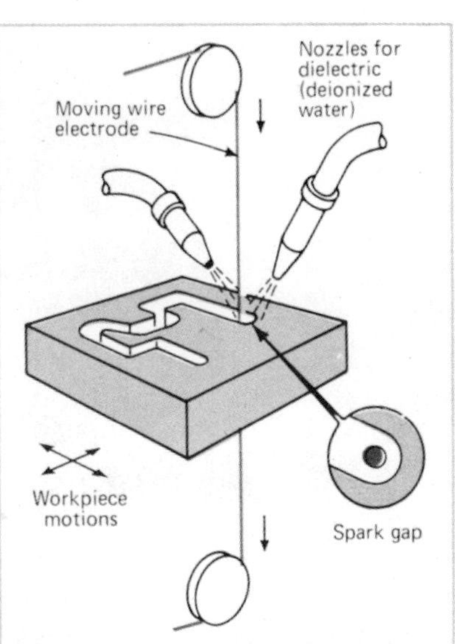

28. Wirecut circumscribes the desired shape instead of eroding it away

path. But CNC adds greatly to the machine's versatility and, no doubt, will be a standard component on new wirecut EDMs.

Next is the servo system, whose job is more complex on wirecut machines than on conventional units: it must retrace the wire's path, whether it's straight or curved, to back away from the cut.

Dielectric systems, on the other hand, are simplier on wirecut machines. A single jet, or sometimes two, can direct the dielectric fluid into the cut and flood it. There is no need to completely submerge the workpiece. Flushing is consistent because the electrode is always the same shape.

Water is almost always used as the dielectric. It produces rapid electrode wear, which is why it is not suitable for conventional EDM, but its low viscosity and rapid cooling rate make it ideal for wirecut.

Power supplies for wirecut are not much different from those of conventional EDMs, though they are of smaller capacity and are oriented toward high-frequency pulses with short on-times and long off-times. Metal-removal rates are higher with this type of pulse under the cutting conditions of wirecut.

Low power is used

Power supplies are smaller because the average cutting power of wirecut is low: power is typically 2 or 3 A, compared with 5-100 A or so average power for conventional EDM. That's all the power that can be delivered by the thin wire that serves as the electrode.

Metal-removal rates are correspondingly low, compared with those for other EDMs. But it's the time it takes to cut

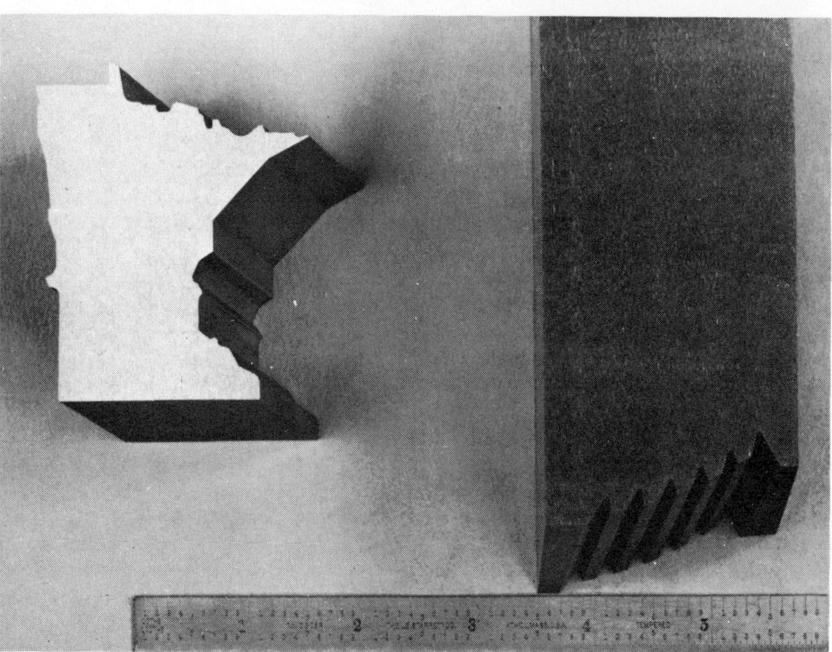

29. **Intricate punch shapes,** like the one shown at left, and form tools (right) are jobs for which wirecut is best suited. Almost any two-dimensional shape can be cut in workpieces up to 4 or 6 in. thick, in single pieces or in stacks

30. **Tilting the wire opens up additional possibilities** for wirecut. Controlling the tilt with a CNC is an intriguing idea, but unsolved problems remain, such as how to accommodate variations in feedrate that occur during simultaneous tilting and cutting

around a shape that counts. Travel per hour, measured in inches, is the only meaningful way to measure cutting rate. Contemporary machines cut about 2-4 in./hr.

If the power supply can deliver as much power as the electrode can handle, the thickness of the material does not have much effect on cutting rate. Some machines cut 6-in.-thick steel or carbide nearly as fast as they cut thin sheet.

Advantages of wirecut

Considering the advances that have been made in conventional EDM, including sophisticated power supplies and orbiting devices, it's clear that wirecut must offer important advantages to justify its higher cost. That it does, and they are best illustrated in the manufacture of a set of punching tools.

When a drawing is received in the wirecut-equipped toolroom, the first step is to produce a program that describes the die profile. This can be done on a digitizer, if the shape is complex, or it can be entered into the wirecut's MDI keyboard, if it's so equipped.

This single program will produce a complete set of tools: punch, die, and stripper. Since the program is based on the dimensions of the die, it's easiest to cut the dieblock first.

A through-hole can be drilled in the die workpiece before it is heat treated, or a starting hole can be drilled on a conventional EDM. The workpiece is then fixtured on the wirecut's table, the wire threaded through the hole, and the unit started. Cutting proceeds automatically, allowing the operator to work on another job.

No through hole is needed for making the punch, of course; the wire is fed in from the workpiece edge until it intersects the desired cutting path. The path in the punch is smaller than that in the die, and so an adjustment is made on the machine or on the CNC to decrease the scale of the profile. The basic program remains unchanged.

Scale is then enlarged to cut the stripper, which completes the set.

These tools are assumed to be straight-sided, but we could taper the sides just as easily. Most wirecut EDM machines built today have a taper-generating provision built in, either mechanically controlled or operated from a third control axis of the CNC.

Compared with making these tools with conventional metalcutting machines or even with an advanced ram-type EDM, the process just described takes a fraction of the time and separate steps. Each part is made in one piece, with one cutting pass, and with only one electrode.

Wirecut reduces 32-section die to 4 sections

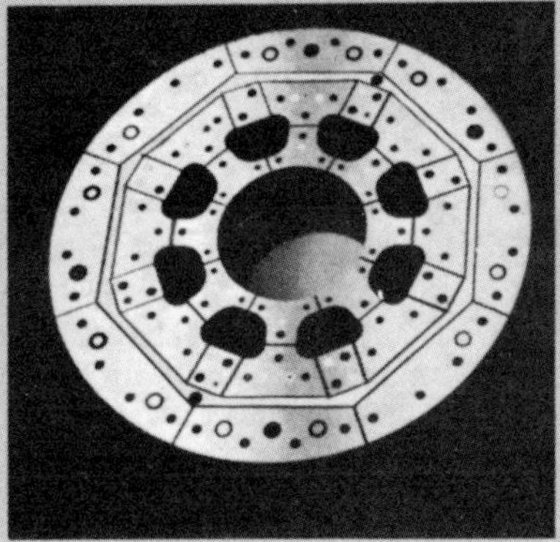

31. Conventionally cut

32. Wirecut

Experienced die-shop managers often find it hard to contain their enthusiasm for wirecut EDM, since it can solve many problems at once. The most obvious benefits stem from the ability of that thin wire to get into places that would be impossible for any other kind of tool.

The bulk of conventional grinding wheels often makes it necessary to build dies in many sections, which requires a lot of precise grinding of mating surfaces, long assembly times, and, inevitably, accumulated errors. This is not to mention the time and skill required to grind the shapes in the dies and punches themselves.

Time and skill were in short supply at GE's Large Motor & Generator Dept (Rochester, NY) toolroom in early 1976, when an influx of new business ran up a 40-week backlog for grinding blank dies, rotor- and stator-slot dies, and other tools for manufacturing the motor parts. Alternatives for solving the problem were considered: farming out some of the work, expanding the department's own grinding capacity, installing conventional EDM, and introducing wirecut were the possibilities. The evidence favored wirecut: it could be implemented quickly and promised a simple way of achieving the high accuracy required on the company's punches and dies. So two CNC units were chosen and quickly installed.

Savings were realized almost immediately. A short time after the units were installed, the department received an order for a replacement rotor-blank separating die that produced parts with a 26-in. OD. The eight kidney-shaped vent holes in the tool (see Fig 31) were difficult to produce with conventional grinding techniques, to put it mildly, and required making the die in 32 separate sections, each requiring extensive grinding after heat treatment. A wirecut machine was put on the job and produced the die in only four sections—which could have been reduced to one, had material of the proper size been available. The only grinding required was for final sizing of the ID and OD.

Die design is much simpler, in GE's experience, when the dies are to be cut with wirecut EDM. Fewer parts mean less assembly time and less opportunity for errors to accumulate. Material is saved, too: the scrap piece from a die block can often be used as a riser or stripper without any additional work. And one NC tape can be used to cut a complete punch, die, riser, and stripper set.

GE currently operates the machines on a three-shift basis. After the wire machines are set up and running, the operators do bench work or operate the company's conventional EDM machines. But, weekends, the machines work a few more shifts on their own. They are simply programmed to run through a long cycle and to shut off about ⅛ in. from the end of the cut. After 18 months of experience with the two machines, GE reports labor reduction on the order of 15% for simple dies and up to 40% or more on complex ones.

Ed Doust, manufacturing engineer, Ellis Forbes, manager of the department's tool-and-die unit, and Chet Mohr, manager of manufacturing services, recently filed a machine-performance report that sums up the reaction of some experienced tool-and-die managers to wirecut: "Wire EDM," says the report, "is one of the biggest forward steps made in the die-making industry in many years. . . . Additional applications will be limited only by imagination."

The same program could have cut prototypes of the parts to be stamped. Dozens or, perhaps, hundreds of sheets of stock could be stacked together and cut in one pass, each with the precise dimensions of the production parts that the dieset will cut.

For a short run, the punches and dies can be dispensed with: wirecut EDM is a short-run production process in its own right.

Electrode materials

Any electrical conductor can be used as an electrode, and just about every one has been tried. Years of experience have shown that a few materials give good performance on a variety of workpiece materials at a reasonable cost, and these have become the standards: copper, copper-tungsten, and graphite. But a long list of other materials are used in certain applications because of their formability or machinability, low cost, or good performance with one particular workpiece material.

Only one generalization can be made about how to choose an electrode material, which is that it is risky to generalize. So many variables are involved in electrode performance that the universal advice is to start with the machine manufacturer's recommendations and begin experimenting from there. The literature on electrode materials is of little help; it's filled with contradictions at every turn.

At the risk of adding fuel to this fire, we'll make a few cautious statements about the advantages and disadvantages of different electrode materials.

Copper remains a popular electrode material because it is easy to machine and because it can be stamped, drawn, extruded, or coined to mass-produce electrodes with fine detail. It gives fine finishes and has low electrical resistance, but it wears quickly when used with high power. This is due to the metal's low melting point; high-intensity sparks, such as those used for roughing, erode copper quickly, reducing workpiece/electrode wear ratios to 2:1 or less. With some power-supply settings, though, ratios climb to 10:1.

Copper is the choice when the wear ratio is acceptable in relation to the cost of making electrodes from the material. It has been used successfully with steel, carbide, and other hard workpiece materials commonly machined with EDM. It is also the most commonly used electrode material for wirecut.

One way to improve copper's performance is to alloy it with a material that increases its melting point. Tungsten is a popular choice, and various alloys (or pseudo-alloys, to be accurate) of the two are common choices for electrode materials.

Copper-tungsten has an excellent wear ratio, both for roughing and for finishing, and is a first choice for cutting deep holes and slots that require high accuracy. It is more difficult to machine than other copper alloys, however; this is the main strike against it. Cost is higher, too, but is offset by longer life.

One way around the machining difficulties is to cast copper-tungsten in a mold (we'll examine one molding technique in the next section). If the cost of the mold can be justified, molded copper-tungsten electrodes can be competitive with electrodes made of any other material.

Wire electrodes for sinking small, deep holes present no machining problem; so copper-tungsten is commonly used for these.

Like plain copper, copper-tungsten is suitable for all workpiece materials. It gives especially good performance on carbides.

Graphite is most versatile

If any electrode material comes close to being all-purpose, it is graphite: bits of petroleum coke bonded together with coal-tar pitch. Graphite comes in so many grades that it is difficult to generalize about its performance, but, considered together, the different grades of graphite can be used for coarse roughing to fine finishing, in any workpiece material.

Graphite's big advantages are that it is easy to machine (though it's highly abrasive) and gives exceptional resistance to wear. Wear can be completely eliminated with some power settings (the "no-wear" operating mode, commonly used for rough cutting).

Graphite wears slowly because it doesn't melt; it sublimates, or vaporizes, instead, and only at a temperature over 5000F.

Special-purpose electrodes

Several other materials continue to be used for special purposes, despite graphite's dominance. Tungsten, TiC, and silver tungsten are used for drilling small slots and holes where electrode wear must be minimized and graphite's brittleness gives it a disadvantage.

Brass is a fast-cutting electrode material, but it wears quickly, even faster than copper. It was commonly used in the early days of EDM. Graphite has replaced brass for most applications because it wears longer and shares a major advantage of brass: easy machining. Some wirecut units are made to be used with brass wire, and at least one wirecut manufacturer believes that brass will eventually replace copper as the basic wirecut electrode material.

Aluminum, because it is easy to work, is sometimes used for roughing large forging dies.

Occasionally, steel electrodes are used to cut steel. A punch, for example, can be sunk into a steel block to make its mating die.

Electrode making

Making electrodes can be the most difficult and the most time-consuming step in electrical-discharge machining. An electrode can sink a cavity with complex contours, curved slots, and odd angles with comparative ease; producing the inverse of those shapes in the electrode isn't that easy.

Traditional machining methods are used to produce most electrodes. If the shape is projected in two dimensions, as for a punch or stamping die, making the electrode is straightforward; standard drilling, milling, turning, grinding, lapping, and honing are used.

Three-dimensional machining requires mold-making techniques or at least a three-axis NC mill. Time savings of the EDM process itself can go out the window if the electrodes are complex, since it may take a few, or maybe even a dozen, electrodes to cut each cavity. And they may be of two or three different sizes for roughing and finishing.

Computer control can be a great time-saver in these circumstances. Once the basic electrode shape is programmed, any number of electrodes can be run off. And different dimensions for roughing and finishing electrodes can be produced with a few editing steps.

Before the days of orbiters and graphite, using five or ten electrodes to cut a single cavity was not uncommon. A lot of thought was applied to techniques for making electrodes in quantity.

Molding was an attractive choice. Easily cast metals, including aluminum and zinc, can be diecast in practically any shape. Electrodes made of these materials wore quickly, but they could always be replaced as many times as was necessary.

Techniques for spraying and plating copper on the inside of cavities were also tried, and used with some success. Different-sized electrodes were made from the basic molded pieces by etching them to size with acid.

Another approach is to stamp or coin the electrodes to shape: a technique that works well for mass production of finely detailed electrodes.

All of these mass-production methods, however, start with a tool-steel mold or die. Since the problem often is making the first mold or dieset, they don't help everyone.

Graphite and orbiters have helped to

Electrode material applications

	Forms	Uses	Common applications	Workpiece materials
Copper	solid, tube, bar, wire, stampings	rough, finish	all	all metals
Copper-tungsten	bar, sheet, wire, tube, castings	rough, semi-finish, finish	holes, slots, molds, dies	all metals
Brass	solid, tube, bar, wire, stampings	rough	holes, wire-cut	all metals
Tungsten	rod, wire, flat strips	semi-finish, finish	deep holes, small slots	refractory metals
Silver-tungsten	compacted shapes	semi-finish, finish	deep holes, small slots	all metals
Tungsten-carbide	sintered	semi-finish, finish	holes, small slots	refractory metals
Molybdenum	tube, wire	rough	holes, wire-cut	refractory metals
Aluminum	cast, forged	rough	forging dies	steel
Graphite	solid, small rods	rough to fine finish	all tooling	steel
Copper-graphite	solid	finish	all tooling	steel
Steel	all	semi-finish	holes, stamping dies	steel

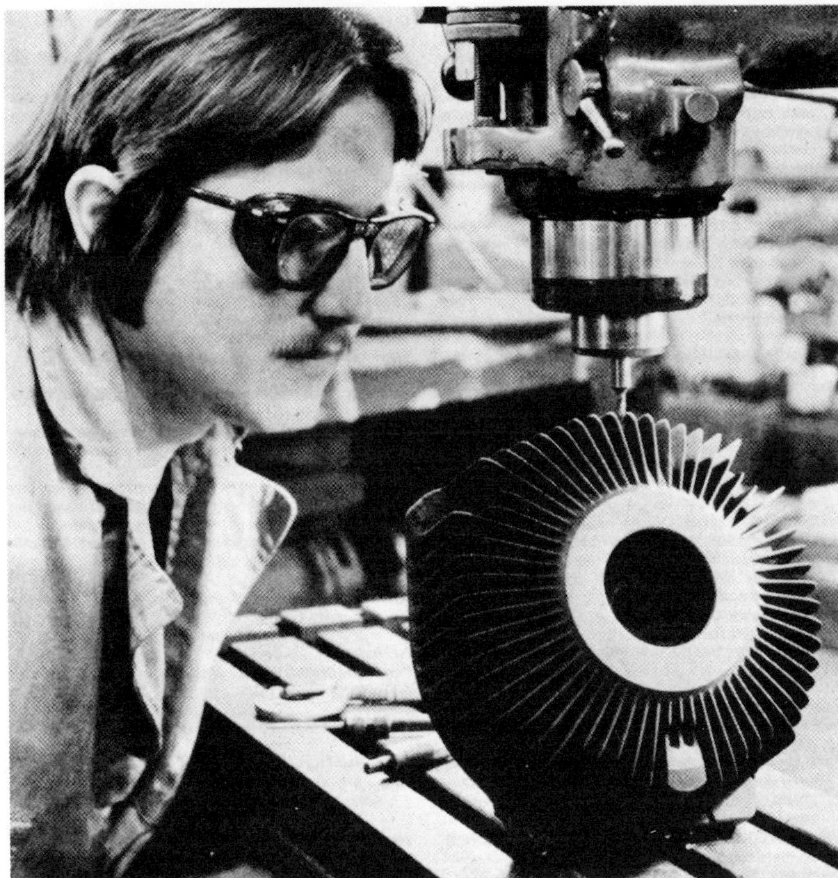

33. Most electrodes are machined on conventional mills, drills, and grinders. This multifinned cylinder-head electrode is made of graphite, easy-to-machine but abrasive

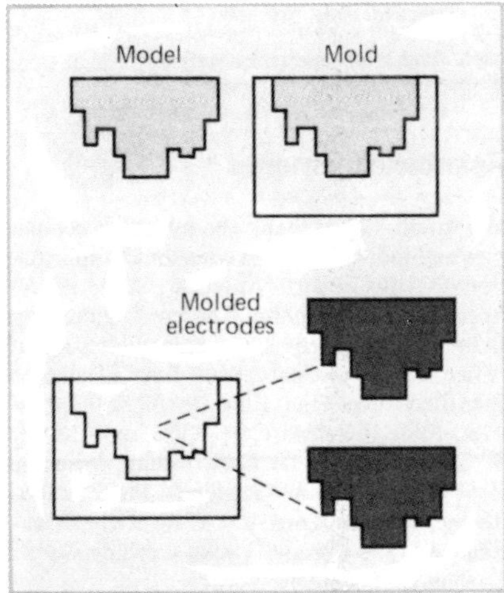

34. 3M's electrode-molding technique casts copper-tungsten electrodes

bring the problem down to size. The search for an easier way to make electrodes, though, continues.

Two methods for mass-producing electrodes without a steel mold have been developed in answer to the problem.

One is Easco-Sparcatron's Total Form Machining (TFM) system. This is an abrasive process that grinds graphite with a molded tool that contains the inverse of the desired electrode shape.

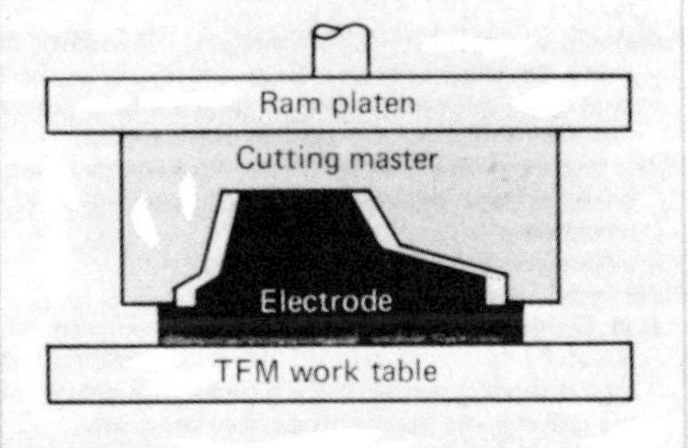

35. Easco-Sparcatron's TFM method machines graphite with a molded master

The tool, called the cutting master, can be cast from a blend of epoxy and silicon-carbide grit. Casting is done over a model, which can be practically any material, that is an oversize form of the final electrode shape. Any number of cutting masters can be made from a single mold.

The cutting masters are fixtured upside down on the platen of a special machine that looks like a small press. Graphite workpieces are fixtured under the cutting masters on a table. Several dozen of each can be accommodated, depending on size, on the 40 x 44-in. work area of the largest TFM model.

Cuting is done in a combination of orbiting and vertical-stroking movements. The cutting masters grind their way down into the workpieces, producing identical electrodes.

Surprisingly thin, fragile projections can be produced with the process. Savings over conventional machining are dramatic, sometimes increasing production by a factor of ten or more.

The other method starts with a wood, plastic, or metal model also, but it produces copper-tungsten electrodes. 3M is the developer of the process and is keeping it in-house. Customers send models to 3M and get electrodes back.

From the model, a mold for casting electrodes is made. Shrinkage of the mold is critical and must be carefully controlled if it's to produce accurate electrodes. This is a secret of 3M's process, but, whatever it is, it's effective. Electrode finish is held to 15 μin., dimensional accuracy is $\pm$0.001 in., and corner radii of 0.001 in. can be reproduced.

Glossary

Arc. A continuous flow of electricity between an electrode and workpiece. Should an arc occur in EDM, the workpiece and/or the electrode will usually be damaged. An arc is normally visually recognizable as a 'yellow flash.'

Arc guard. Same as arc suppressor or arc dampener.

Arc suppressor. A circuit in the EDM power supply that reduces the possibility of arcing.

Average current. The average value of all the minimum and maximum peaks of amperage in the spark gap, as read on the ammeter. For any given available current setting, the higher the average current, the greater the efficiency of the particular cut being made. See Machining rate.

Carbon. The raw material used to make graphite. Often used in place of the word 'graphite' to describe graphite.

Center flushing. A method of flushing dielectric through a center hole in an electrode.

Coolant. Same as dielectric fluid.

Copper graphite. An EDM electrode material made from a blend of graphite and copper.

Copper tungsten. An EDM electrode material made from a blend of copper and tungsten. It has a relatively low wear ratio and is considered to be one of the best electrode materials for close-tolerance, fine-finish work.

Core. The slug that remains after EDMing with an electrode that has a flushing hole in it.

Corner wear. In EDM, the corners of the electrode wear the most. Corner wear is the distance up the electrode that its corners show signs of breakdown.

Crater (pit). The small cavities left on the EDMed surface of the workpiece by the EDM sparks.

DC arcing. Same as arc.

De-ionization. Return of the dielectric to a nonconductive state. Failure to accomplish deionization during off-time of the spark is responsible for dc arcing.

Dielectric fluid. In EDM, usually a light oil or water in which the work zone is immersed. It fills the gap between the electrode and the workpiece and acts as an insulator until a specific gap and voltage are achieved. It then ionizes and becomes an electrical conductor, allowing a current (spark) to flow through the ionized path to the workpiece. It also serves to cool the work and to flush away the particles generated by the spark.

Dither. See vibrator.

Discharge. The EDM spark.

Dual power supply. Two EDM power supplies in a single cabinet that can operate two machines.

[continued]

Duty cycle. The percentage of one on-off cycle during which the power supply is on. Formula: On-time/(Off-time x 100) = duty cycle

Electrical-discharge grinding (EDG). A surface-machining process done with an EDM machine resembling a surface grinder but using a wheel made from electrode material. Actually misnamed, because wheel and workpiece never touch. Can also be done with a horizontal spindle attachment (mounted on the quill of a conventional EDM machine) that has a built-in motor drive for the electrode 'wheel.'

Electrode. The tool in the EDM process. It must be made from an electrically conductive material. Its form, or shape, is usually a 'mirror image' of the finished form or shape desired in the workpiece, with its dimensions adjusted to take into account the amount of overcut being used. In wirecut EDM, the electrode is a copper or brass wire.

Electrical discharge machining (EDM). A metal-removal process using an electric spark to erode material from a workpiece under carefully controlled conditions.

Electrode 'growth'. An electroplating action, occurring at certain low-wear settings, that causes workpiece material to plate on to the electrode, causing the electrode to grow in size without control. Some controversy exists about the physical process involved in growth.

Electroforming. An electroplating process used to make EDM electrodes.

Eroding. Also referred to as 'spark-eroding,' 'spark-erosion,' 'burning,' etc. Removing metal by the EDM process.

Finish cut. The final cut made with EDM on the workpiece. The finer the finish desired, the longer the finish cut will take.

Flashpoint. Temperature at which any flammable material will burst into flame. A factor in selecting dielectric fluid for EDM.

Flushing. Flowing dielectric through the gap to remove the debris caused by machining with EDM.

Flushing hole. A hole through the workpiece or electrode used to introduce dielectric to the gap for flushing purposes.

Gap (spark gap). The distance between the electrode and workpiece when the spark occurs.

Gap voltage. This can be measured as two different values during one complete 'spark cycle': The voltage that can be read across the electrode/workpiece gap before the spark current begins to flow is called the open-gap voltage (for instance, 100 V OGV). The voltage that can be read across the gap during the spark current flow is the working-gap voltage.

Graphite. An electrode material that has high heat-resistance and transfers electric current very efficiently. It is the most popular electrode material and probably the easiest to machine.

HAZ (heat affected zone). A layer in the finished work that lies under the recast layer. It refers to the depth to which the action of the EDM cut has altered the parent metal's metallurgical structure and characteristics. It ranges in depth from 0.0002 to 0.0008 in., depending on the material and the power being used to cut it.

Hunting. An erratic 'bouncing' movement of the quill of an EDM machine during a cut. It can occur from many causes, including poor flushing conditions in the gap, servo response has been set for too much sensitivity, and buildup of carbon deposits on the bottom of the cavity being EDMed.

Injection flushing. Forcing dielectric under pressure through the spark gap.

Insulator. Any material highly resistant to flow.

Insulated dieset. A dieset with upper and lower plates insulated from each other by various methods. It is used in some EDM machines to maintain accurate alignment between the electrode and workpiece.

Lateral flushing. Same as surface flushing.

Lateral wear. The length of electrode that shows signs of breakdown

Low wear. The result of certain settings for EDM machining producing a very low degree of wear on the electrode. In some cases, less than 1%.

Machining rate. The rate at which material is removed from a workpiece by EDM. It is usually measured in cubic inches per hour. For wirecut, it usually refers to lineal inches of electrode travel through the workpiece.

Metal-removal rate. Same as machining rate, except for wirecut.

Microinch. 0.000,001 in. Used in describing surface finish or roughness.

Multiple-lead (multilead) power supply. A power supply with more than one wire lead that can be connected to the electrode or workpiece and through which the total power available can be channeled. It permits the total available power to be divided into small units, each capable of being channeled through separate wire leads connected to multiple electrodes or workpieces for production work. Or all power can be put through a single wire lead by connecting all the individual leads into one.

Normal polarity. On older power supplies, 'negative' electrode polarity. The term is misleading, since either the electrode or the workpiece may be negative or positive depending on workpiece/electrode material combination being used and proper polarity required for desired results.

No wear. An operating mode in which the electrode does not decrease in size during the cut. Wear and 'growth' (see Electrode 'growth') cancel each other out.

On-time. The duration time of the EDM spark, measured in microseconds.

Off-time. The time, in microseconds, between EDM sparks, during which deionization is supposed to take place. Too low an off-time setting can result in a dc-arcing condition.

Overcut. An EDM cavity is always larger than the electrode used to machine it. The difference between the size of the electrode and the size of the cavity (or hole) is called the overcut. Diametral, or total, overcut is twice the overcut per side.

Overcut per side. One-half of diametral overcut value.

Overlap. The difference between the rough-machined hole or cavity size and the size of the electrode to be used for the next cut.

Pause duration (or time). Same as off-time.

Peak current. The maximum current available from each pulse of the power supply (generator).

Percentage of electrode wear. The volume of electrode worn away as compared to the volume of workpiece EDM'd away. Formula: % electrode wear = (volume electrode wear/volume workpiece EDMed) x 100.

Polarity. In EDM, electrical potential (positive or negative) on the electrode.

Power pack. Same as power supply.

Power supply. The part of the EDM system that supplies the current that causes the sparks or discharges between the electrode and workpiece. It is usually housed in a cabinet separate from the machine tool and connected to it by a cable. Modern power supplies produce pulsed direct current using transistor circuitry.

Pulsator. A unit added to, or built- into, an EDM machine to cause the electrode to retract periodically a short time to aid in flushing a deep or blind cavity. It causes the electrode to act like a piston in a chamber.

Pulse. Discharge of a quantity of electrical energy having preset voltage and amperage and expended over a preset time.

Pulse duration or time. Same as on-time.

Pulse length. Same as on-time.

Pulsed flushing. Flushing that is synchronized with the pulsator of the EDM machine. When a machine is set for this mode, pressure flushing takes place only when the quill retracts the electrode from the cavity.

Pulse timer. Used to set the length of on- and off-time of the spark. On some machines, on- and off-time can be set individually in microseconds.

Ram. The moving member of an EDM machine on which the electrode or electrode holder is mounted. A dovetail-guided arrangement.

RC circuit. An EDM power-supply circuit that uses capacitors to store the current that produces the spark at the gap. The capacitor is charged through a resistor and discharged across the gap when conditions are correct (gap distance, voltage, etc.). This is the original EDM circuit and is used only for finishing today.

Ram cycler. Same as pulsator.

Relaxation circuit. Same as RC circuit.

Recast layer. A layer of melted and resolidified material on the surface of the workpiece.

Resolidified layer. Same as recast layer.

Reverse 'burning'. The technique of mounting the electrode on the machine table or flush tank and the workpiece on the quill. Used in EDMing a blanking punch with a female electrode.

Reverse flush. Flushing through the flush tank.

Reverse polarity. See Normal polarity.

Rotating spindle. Either an accessory mounted on the quill or ram or a built-in machine spindle used to rotate the electrode to achieve more-uniform wear and to improve flushing conditions. Its use is limited to round electrodes. Another use for the rotating spindle is in tramming the workpiece with an indicator.

Servomechanism. The device that drives and controls the movement of the quill or ram.

Servo reaction time. The time between a signal to the servo and its physical response to the signal.

Short-circuit or short. Extremely high-current-flow condition that occurs when electrode and workpiece come in direct contact. A short stops all machining.

Silver tungsten. An EDM electrode material that is a blend of silver and tungsten. It has a relatively low wear ratio and is considered to be one of the best electrode materials for close-tolerance, fine-finish work.

Solid state power supply. A power supply that has no vacuum tubes. Current is switched on and off by transistors.

Spark. See discharge.

Spark gap. The distance between the electrode and the workpiece when discharges are occurring.

Spark erosion. Another name for EDM.

Spark generator. Same as power supply.

Spark intensity. The amount of energy in the spark.

Split-lead power supply. Same as multiple-lead power supply.

Suction flushing. Using a vacuum to draw the contaminated dielectric away from the gap instead of forcing it out with pressure.

Surface flushing. The use of hoses or nozzles to direct jets of dielectric at the cutting area to flush away the debris. Generally used with a pulsating electrode.

Swarf. The eroded particles or residue produced by electrical-discharge machining. Also called chips.

Through-hole flushing. Use of a predrilled hole in the workpiece to inject dielectric up toward the gap by injection flushing or down from the gap by suction flushing.

Time unit (timer). The electrical device used to set on- and off-time of the spark.

Vacuum flushing. See Suction flushing.

Wear ratio. The volume of electrode worn away compared with the volume of workpiece material removed by EDM. See Volumetric wear, Corner wear.

Vacuum-tube power supply. An EDM power supply that uses vacuum tubes (radio tubes) to switch the electrical machining pulses (sparks) on and off.

Vibrator. An accessory used on an EDM machine to vibrate the workpiece or electrode in order to produce flushing action by 'pumping' (as a piston in a cylinder). Used primarily for improving flushing in blind cavities.

Volumetric wear. The total or 'all-over' wear on the electrode expressed in cubic inches. See Percentage of electrode wear, Wear ratio.

Wash. Another term for surface flushing.

Woodpecker device. Same as pulsator. ■

Acknowledgements

We would like to thank the following companies and individuals who contributed information for this report: Agietron Corp, Airco-Speer Carbon-Graphite, Andrew Engineering, Belmont Equipment Co, Charmilles Corp of America, Easco-Sparcatron Inc, Electrodes Inc, Electrotools Inc, Elox Div of Colt Industries, Eltee Pulsitron Corp, GE Large Motor & Generator Div, Dr C.H. Kahng, Machine Tool Research Assn (England), Poco Graphite Inc, Raycon Corp, Dr M.R. Reda, Rust-Lick Inc, 3M Co, and Union Carbide Corp's Carbon Products Div.

Lasers in metalworking

Here is an insight into how lasers work, a rundown on their applications, and a guide to 44 laser materials-processing systems

For a long time, the tools that man used to fashion his world — to measure and manufacture—were based on mechanical methods. No matter whether the tool be an obsidian, flaked to provide a cutting edge, or a modern-day carbide insert, the concepts are similar. The art of mechanical measurement and material removal has been refined and perfected, but the fundamental principles have remained pretty much the same.

Then, within a relatively short span of time, radical new tools, based on other than mechanical principles, have appeared. Electricity and chemistry, for instance, were applied.

By George Schaffer

directly to the transformation of material into functional components, using techniques like electrical-discharge machining (EDM), electrochemical machining (ECM), and electroforming. And, more recently, electronic and electro-optical methods have been used extensively to provide digital displacement readouts (DROs).

Now perhaps the most radical new tool is beginning to make an impact on metalworking. Lasers, devices that harness electromagnetic energy—a fundamental physical phenomenon—in the form of a very special light beam, are beginning to find a host of industrial applications.

As a metalworking tool, the laser beam is multifaceted: It can be used to measure and align; to weld, drill, or cut; to heat treat and alloy. It is a tool that never gets dull, is easily manipulated for automated processes, and could eventually be piped throughout the shop from one central laser source.

What are lasers?

Just exactly what is this very special light beam generated by a laser? How is it generated? To begin with, we are talking about light in the sense of electromagnetic radiation. Specifically, we are concerned with radiation in the optical part of the electromagnetic spectrum, the part that can be treated in terms of optical theory. The optical spectrum extends from the ultraviolet to the far-infrared region and includes the visible range. Lasers can be constructed to operate at wavelengths ranging from 0.3 to 300 μm but those most commonly used in metalworking applications typically operate at wavelengths ranging from 0.6 to 10.6 μm.

Although laser beams are not necessarily visible, they can be handled essentially like visible light, with lenses and mirrors made from materials with characteristics suitable for reflection or transmission of the particular wavelength radiation involved. But laser beams are unique because they posess high intensity (not limited by the laws of black-body radiation), monochromaticity, a high degree of spatial and temporal coherence, and low divergence. Not all of these characteristics are necessarily obtained from any one laser, but each makes a particular laser most suitable for certain applications (see Table). More about these characteristics later.

First, however, let's see how laser beams are generated. LASER is an acronym for a phenomenon called Light Amplification by Stimulated Emission of Radiation. Stimulated emission, first discovered by Einstein, occurs when an atom in an upper energy state is "stimulated" by incoming radiation to give up its energy as it falls to a lower energy state.

Laser characteristics needed for various applications

Application	Monochromaticity	Low divergence	Low-order mode	Coherence	High output	Efficiency	Typical lasers
Measurement	●		●	●			He-Ne
Alignment	●	●					He-Ne
Holography	●		●	●			Ruby, Nd:yag argon, He-Ne
Drilling	●	●			●	●	Ruby, Nd:yag
Cutting	●	●			●	●	CO_2
Welding	●	●			●	●	CO_2, Nd:yag ruby

Consider it this way. In an ordinary fluorescent light tube, a stream of electrons, sent through the gas inside the tube, slams into many atoms of gas, and this "bombardment" excites the atoms to a higher energy state. As the atoms calm down, they give off radiation energy, which, in turn, is converted to visible light by the coating material of the tube. But this radiation process is random, and the resultant energy—in the form of photons—is disorganized, a disorderly conglomeration of spontaneous emission, absorption, and some stimulated emission.

It all has to do with energy levels. Light energy is emitted when an electron moves from a higher energy state to a lower energy state. This energy transition occurs when an electron moves from its orbit to an orbit closer to the nucleus of the atom, and, to move away from the nucleus, an electron must absorb energy. Stimulated emission results when, during an induced downward transition, a photon "liberates" a new photon rather than being absorbed.

For laser action to occur, the number of electrons at two energy levels has to be inverted from the normal condition so that a majority of the electrons are at the upper energy level. This condition is referred to as a population inversion and is produced through external excitation of the lasing medium, which is commonly called "pumping" and may be achieved optically or electronically.

When a population inversion is achieved, photons of spontaneous emission, corresponding to the proper energy level, hit molecules in the upper energy level, and each collision stimulates the emission of another photon. This, in turn, stimulates the emission of more photons from the series of collisions with other excited molecules at the upper energy level. And, in the laser, every successive photon is stimulated to emit in precisely the same direction and at the same frequency as the last.

This cascade of identical photons becomes the output, the coherent beam of light energy amplified by stimulated emission. In order to build up this cascade of photons, the stimulated-emission process is generated inside a resonant cavity, which acts like an amplifier by ensuring that the radiation will make many passes through the active laser medium, thereby affording maximum propagation of stimulated emissions.

So, essentially, a laser consists of a reservoir of active atoms, which can be excited to an upper energy level, a pumping source to excite the available active atoms, and a resonant cavity to provide feedback for the laser oscillations. It is an electro-optical device that converts electrical energy into electromagnetic energy.

Since its inception, many materials have been shown to exhibit lasing properties, but only a limited number of types have found application in metalworking. Lasers can be classified according to the lasing medium as solid-state, liquid, or gas types. Only solid-state and gas lasers are presently used in metalworking.

Solid-state lasers consist of a crystalline or glass host material and a doping additive to provide the reservoir of active ions (charged atoms) needed for the lasing action. A typical lasing medium is a single crystal rod of yttrium-aluminum-garnet (called yag) containing ions of neodymium. Neodymium-doped yag (Nd:yag) has replaced most other materials (including the original ruby laser) for high-energy cystal-laser applications because it allows high pulse rates, has a relatively good efficiency, and can be operated with simple cooling systems.

Solid-state lasers are pumped optically by a flashlamp mounted in a reflecting cavity, which also contains the laser rod. One efficient arrangement, employed by Raytheon, is to mount the laser rod and a linear flashlamp parallel to each other inside an elliptical gold-plated enclosure, with each at

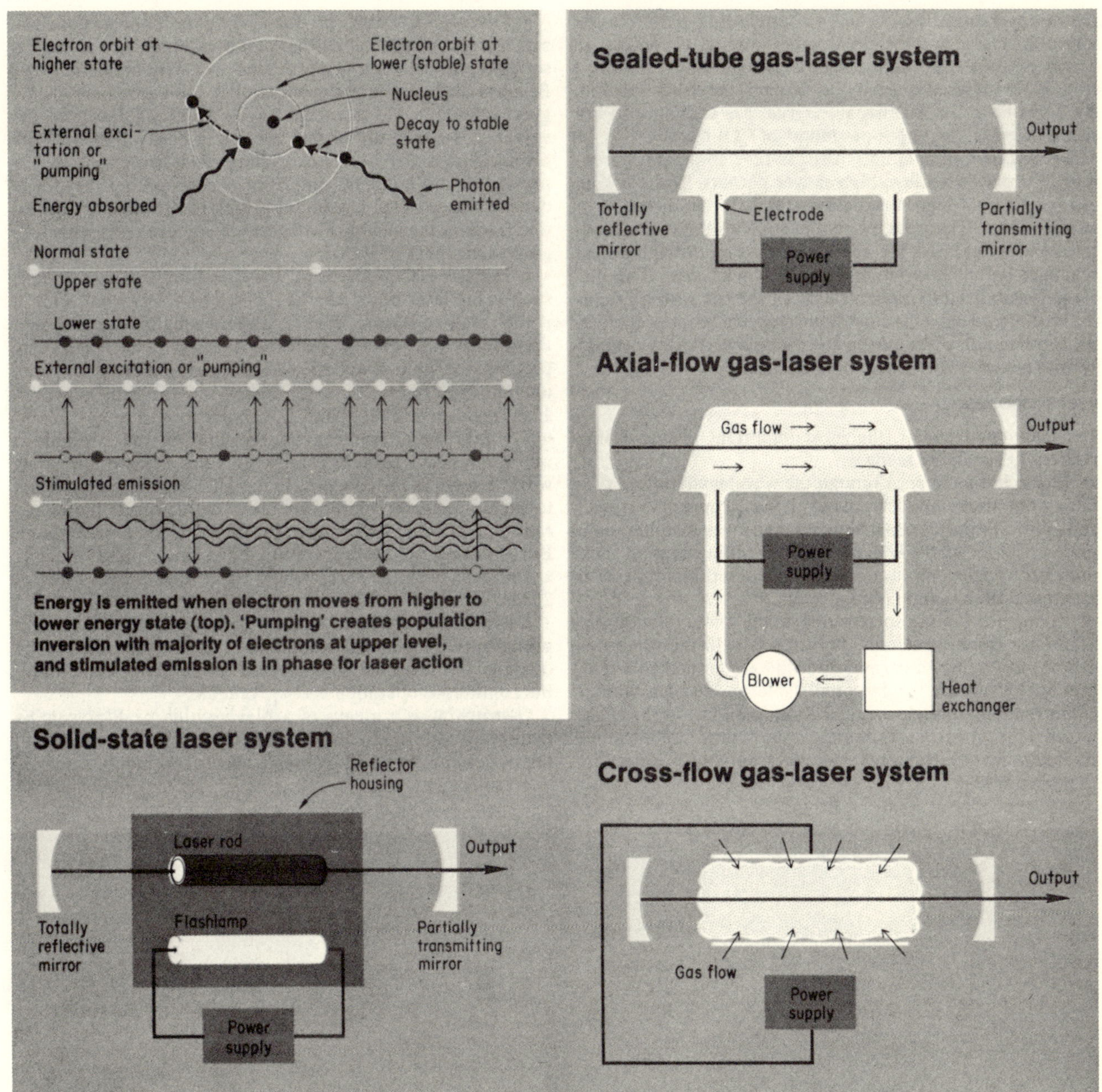

Energy is emitted when electron moves from higher to lower energy state (top). 'Pumping' creates population inversion with majority of electrons at upper level, and stimulated emission is in phase for laser action

one of the foci of the elipse. Flashlamps are usually xenon or krypton filled.

Feedback is generated by a totally reflective mirror at one end of the rod and by a partially reflective mirror at the output end. The mirrors are aligned perpendicular to the optical axis of the laser rod so that only light along the laser axis is reflected and transmitted, resulting in the preferential buildup of light. The area between the mirrors is the resonating cavity. In a typical solid-state laser, the laser rod, mirrors, flashlamp, and other optical paraphernalia are usually housed in a common laser head.

Gas lasers typically consist of an optically transparent tube filled with either a single gas or a gas mixture as the lasing medium. The principal gas lasers used commercially are He-Ne, argon, and CO_2 lasers. The pumping source is some form of electrical discharge applied by electrodes. Avco Everett Research Laboratory uses a special electron-beam arrangement for its high-power CO_2 laser.

One of the difficulties that had to be overcome with high-power gas lasers is that the available output power for a given tube length is limited by two factors: the ability to cool the gas and the proper stabilization of the gas discharge. Thermal energy upsets the lasing equilibrium of the gas, and discharge stabilization is needed to prevent electrical arcing within the lasing medium.

For example, a conventional flowing-gas CO_2 laser can develop about 50 W of output power for every meter (3.3 ft) of resonator-cavity length. On that basis, a 500-W laser would require a cavity nearly 33-ft long—rather unwieldy for an industrial arrangement. Ferranti Electric Inc gets around this by folding the active-discharge tube into 24 short tubes arranged in zig-zag fashion with reflecting mirrors at each fold. Thus, a 400-W laser (the MF 400) is achieved with an overall length of about 4 ft.

A typical CO_2 laser actually uses a mixture of three gases: carbon dioxide, helium, and nitrogen for high-power operation. The CO_2 supplies the molecular action required for photon generation; the N_2 acts to sustain and reinforce the molecular action; and the He provides intracavity cooling. In flowing-gas systems, this mixture is constantly pumped through the resonator cavity to sustain the lasing action. Rapid gas flow and additional external gas cooling also are used to increase the output-per-length of CO_2 lasers.

The Avco design eliminates the gas-tube configuration entirely in order to achieve high power outputs. The working gas is pumped at high speed at cross-flow to the direction of the laser beam. The system operates somewhat like a closed-cycle wind tunnel with the heated gas passing through a heat exchanger before it is returned to the cavity area. The discharge is stabilized by preionization of the gas with a broad-area electron beam. This cross-flow/electron-beam preionization combination allows Avco to reach power levels over 15 kW with a laser-cavity length of about 4 ft.

Laser properties

As mentioned earlier, the most important distinguishing features of laser light are these:

■ Monochromaticity. Although the wavelength output of a laser is not truly monochromatic, it has a much narrower bandwidth at considerably higher intensity than other light sources. This property is important for applications in metrology or in gaging applications in which discrimination from background illumination is important.

■ Coherence. This property refers to the phase relationship of the laser-beam waveform. Waveforms with the same frequency, phase, amplitude, and direction are termed coherent. Laser light waves are regular, predictable, and in-phase rather than in random jumbled array like radiation from other light sources. This property is extremely important for applications in holography, which stores the wavefront of an object by recording the phase relationship of two interfering beams.

■ Divergence. Lasers produce very parallel beams of light, and it is this directionality that makes it possible to collect laser light and deliver it to a localized area with high efficiency. Because the beams are almost parallel, the laser energy is not greatly dissipated as the beam travels over long distances. Divergence is a measure of the increase in beam diameter with distance from the laser's exit aperture. It is commonly expressed in milliradians (mrad), at a point where the power density is $1/e^2$ of the maximum value. (In the table of fabricating lasers at the end of this report, divergence represents full-angle divergence at $1/e^2$).

■ Intensity. The output of a well-collimated light source, such as the laser beam, can be focused to a very small spot. It is the result of low divergence, and the smaller the spot size, the higher the energy concentration. In fact, lasers are presently capable of producing light that is more than seven magnitudes brighter than the light from the sun. However, such levels are available only in very short pulses.

For fabricating applications, the total energy available is more significant because it represents the capacity for doing work. Energy is expressed in joules (J), which are equivalent to watts x seconds. A laser capable of delivering 25 joules at a rate of 1 pulse per second is considered to be a 25-W laser. But, if those 25 joules were emitted in a single pulse of only 1×10^{-3} sec, then the laser would achieve a peak power of 25,000 W.

Lasers are operated in either a continuous-wave (cw) or a pulsed mode. Solid-state lasers, such as Nd:yag, are usually operated in the pulsed mode because of flashlamp limitations, but continuous operation is also available.

Q-switching is a means of achieving high peak powers by temporarily storing some of the energy in the laser cavity and then releasing it in a short burst. This is commonly achieved

Hewlett-Packard laser interferometer set up to check table motion of machine tool. Target is in spindle

by preventing reflection from one of the end mirrors to build up the population inversion and then suddenly changing this condition to permit reflection. The sudden feedback condition then produces a high-power pulse with a rapid time rise.

Gas lasers are usually operated in a continuous-wave mode and are capable of developing higher continuous average power than are solid-state lasers.

One other property of laser beams worth mentioning is mode structure, which is the configuration of the electromagnetic field generated by the laser and which affects the intensity distribution of the beam. TEM (transverse excitation mode) is the ability of a laser to oscillate in the laser cavity along one or more paths parallel to the axis of the cavity, and subscript numbers are used to describe the distribution pattern in a plane perpendicular to the beam.

TEM_{00}, indicates the fundamental mode, in which lasing occurs only along the cavity axis. Most lasers generate multimode beams, and, although the total power of such a beam may equal that of a TEM_{00} beam, the power density at its focal point may be as much as two orders of magnitude less. And a TEM_{00} laser beam can be focused to the smallest theoretical spot. In fabricating processes, a low-order mode beam can achieve very narrow kerf widths and minimum heat-affected zones when cutting. A stable mode structure is also very important for interferometric distance measurement and holography.

Lasers can measure

One of the first practical applications of low-power lasers was the interferometer. Laser interferometers measure changes in distance by the same technique developed by Michelson. The laser beam is split into two beams, which are directed at a fixed and a movable mirror. The reflected beams are recombined at the splitter, where they interfere and form a dark or light spot, called a "fringe," depending on the phase relationship of the two light beams. The generation of fringes depends on the use of monochromatic light; hence, the use of the laser. With both mirrors stationary, the fringe patterns at the beam splitter (interferometer) will not change. However, as the movable reflector moves one half wavelength, the fringe disappears and reappears, and one full cycle of brightness can be detected at the interferometer.

Counting the number of times the light alternates on and off at the detector provides a measurement of displacement of the moving reflector. The measurement is actually a function of the speed of light; therefore, for very accurate measurement, the ambient conditions must be taken into consideration. The Hewlett-Packard 5526A laser measurement system provides automatic compensation for ambient conditions and can also automatically correct for the temperature of the part under measurement. The unit has a remote interferometer feature—a small interferometer cube—that facilitates mounting and eliminates deadpath error. H-P also offers a modular miniaturized laser transducer system, the 5501A, which can be incorporated into a machine as a dedicated measuring system compatible with digital control.

Lasers can gage, inspect, align

Because it can be focused to a small accurate spot and is readily distinguishable from other ambient light conditions, laser light can be useful for a variety of optical gaging methods, including alignment procedures. Typical laser-based gaging methods include profile imaging, beam scanning, and triangulation.

■ Profile imaging. Perhaps the simplest, this technique works like a profile projector but uses high-speed, digital techniques for recording image edge position. A laser beam, (or other light, for that matter) illuminates the edge of an ob-

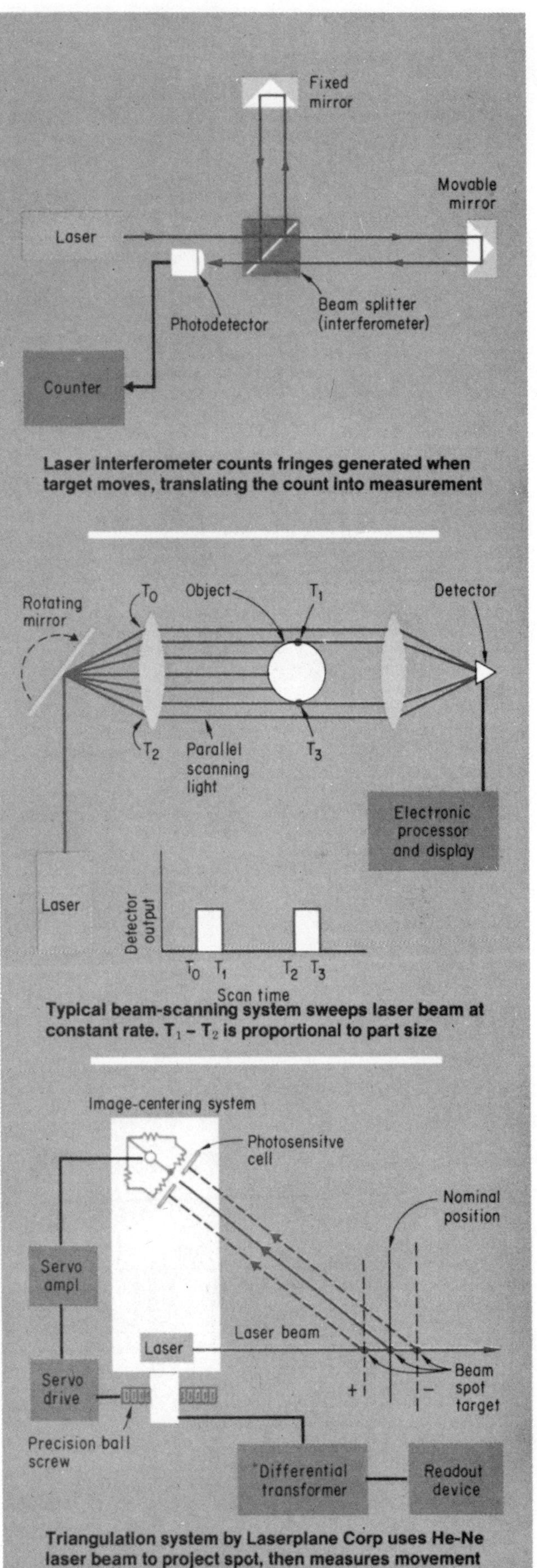

Laser interferometer counts fringes generated when target moves, translating the count into measurement

Typical beam-scanning system sweeps laser beam at constant rate. $T_1 - T_2$ is proportional to part size

Triangulation system by Laserplane Corp uses He-Ne laser beam to project spot, then measures movement

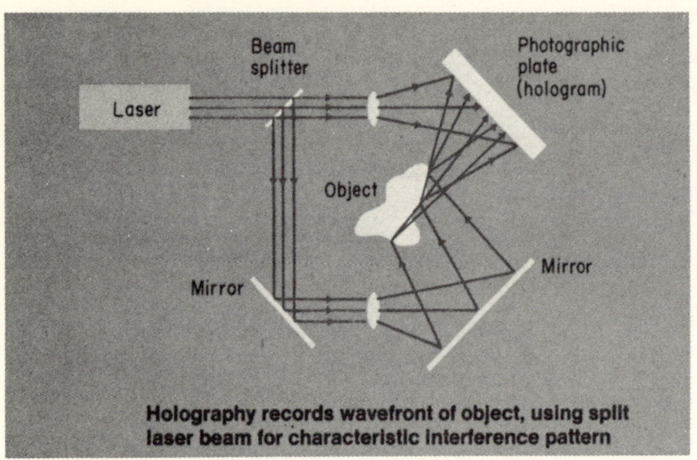

Holography records wavefront of object, using split laser beam for characteristic interference pattern

ject, and a lens forms an image of its edge on a photodiode array that is constantly scanned. An associated electrical circuit determines which element is detecting the edge, and a digital readout displays the appropriate information. This may be merely the identification of the particular diode or an actual part dimension, such as a diameter.

Diode arrays with as many as 1024 separate photodetector elements are available, and typical detector spacings are 0.001 in. Appropriate magnification can extend the resolution of such a system to 0.0001 in. Measurement speeds can reach 100 kHz with 64-element arrays. Matrix arrays are also available for two-dimensional work, and TV-type cameras have been used for low-accuracy applications.

■ Beam scanning. A further development of edge sensing involves the application of a scanning laser beam coupled with a photodetector. The Lasermike system (Autometrix Div, SRL, Dayton, Ohio), originally developed for measuring

Gas-assist nozzle on this Ferranti laser cutting machine also has height sensor to maintain cutting gap

wire diameter as it is drawn, is such a system. Here laser light scans through a lens system to produce essentially parallel sweeps of light across the part. Another lens at the detector side focuses the light at the detector. The detector, therefore, sees no light during the time that the part is blocking the laser-beam sweep. Comparing this time with an accurate oscillator, the unit translates the time-dependent function into a measure of size. Such a system is relatively unaffected by whipping or fluttering of the material under observation.

Even larger part excursions are allowable with a scanning system recently introduced by Zygo Corp, Middlefield, Conn. The system is modular and can measure multiple dimensions or part location within a relatively large measuring range.

■ Triangulation. Here the movement of a spot of light projected onto a surface is observed as a measure of surface location. The Lasermeasure (Laserplane Corp, Dayton, Ohio) is such a system. The spot is generated by a He-Ne laser beam normal to the surface under observation. The backscattered light from this spot is then observed at a specific angle to that surface by means of an electro-optical image-centering system. Because this viewing axis is at an angle to the surface, the spot moves laterally across the observation plane as the distance between light source and surface varies. In the Lasermeasure, a servo-drive system continuously adjusts this distance so that the reflected spot remains at the center of the detector. And this movement is, of course, a direct measure of the surface position.

The Autotech gaging system (Autotech Corp, Columbus, Ohio) uses a time-dependent measure in a triangulation setup. A laser spot is projected onto the surface to be gaged, but it is viewed by mirrors from two directions. These two spots are then focused on an image converter, and the spacing of the dots changes as the workpiece surface moves. An optical scanning system times the distance it takes to sweep from spot to spot, and this time is converted into an appropriate measurement. To make a thickness measurement, for example, the time separation for a reference surface is stored in a memory circuit, and, when a part is positioned on that reference surface, the change in time is proportional to the part thickness.

The low divergence and high intensity of the laser beam make it suitable for use as a reference beam in optical-alignment systems. Used in conjunction with a photodetector target, such a system can provide a graphic or numerical display of the offset between the center of the laser beam and the center of the target. Some laser alignment systems, like the Model 711 made by Hamar Laser Instruments Inc, Wilton, Conn, have provisions for accurately sweeping the beam to establish a measuring plane and are equipped with means for aligning the laser sweep to a desired plane. This system can be used over distances up to 200 ft with an accuracy of 0.001 in. at 10 ft.

Inspection for flaws, assembly faults, part alignment, and sorting are some of the other tasks performed by lasers. For the most part, such systems are custom-made for a particular application. Some use signature-analysis techniques to detect pattern differences in the backscattered light from a part and master. Others merely detect the presence or absence of a part feature, such as the proper number of holes.

Holography

The narrow line width (single frequency) and inherent coherence of laser light make optical holography a possibility. Holography is a means of recording and reproducing the wavefront of an object. The wavefront recording, which contains all the information necessary to reconstruct an accurate three-dimensional image of the object, is called a hologram.

Holograms are made by splitting a laser beam into two separate beams, one of which lights the object under observation while the other serves as a reference beam. Light from the reference beam is made to interfere with that reflected by the object, and this produces a characteristic interference pattern, which is captured on a photograhic plate to form the hologram. It's a little like the interferometer described earlier.

When the hologram is viewed under appropriate laser illumination, a three-dimensional reconstruction of the image is generated. If such an image is superimposed on the actual object while that object is subjected to varying conditions of stress, minute dimensional changes become apparent as interference patterns.

Another holographic testing technique is double-exposure holography—two holograms of the same object under varying conditions of stress are superimposed to create characteristic interference patterns. The resulting fringe patterns lend themselves to quantitative analysis similar to interferometric mea-

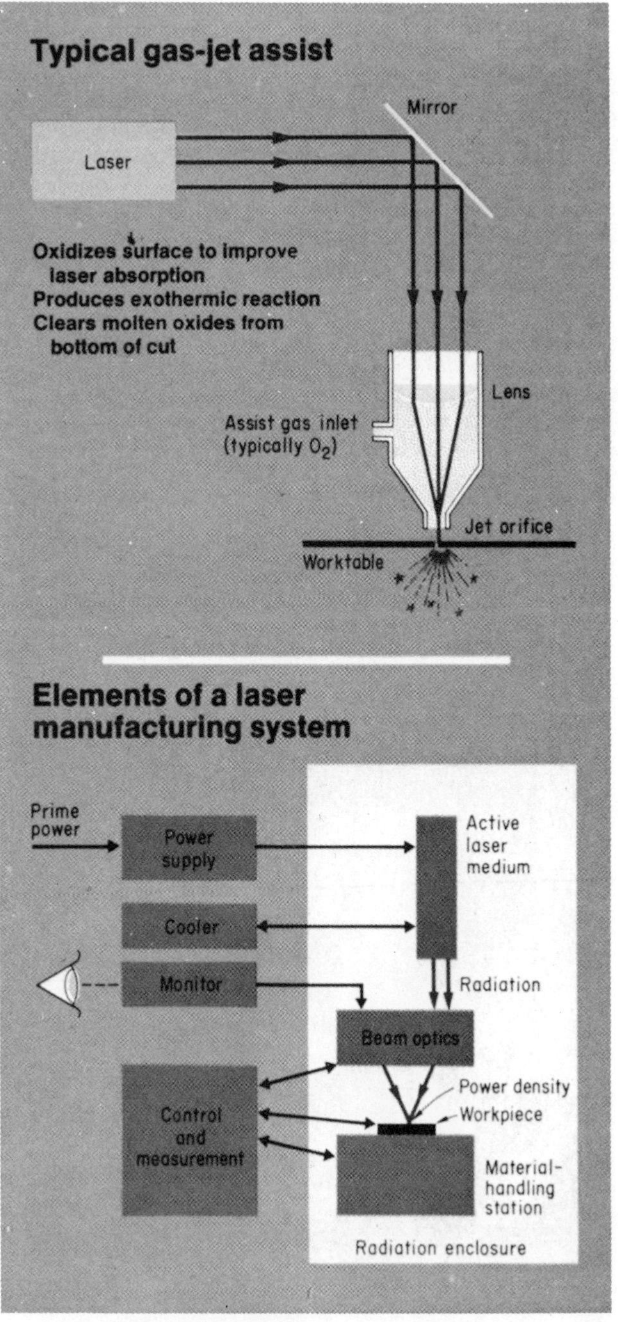

surement. The technique is being used for nondestructive testing for stress, vibration, structural defects of complex components, and surface-defect analysis.

Lasers can process materials

For materials processing, the laser acts strictly as a source of heat. Because of the high energy that can be attained in the focused output of the laser beam, it can be used for welding, drilling, cutting, and surface treatment, such as hardening and alloying. Such processes are generally governed by the following material properties:

- Properties that directly affect the way in which light is absorbed by the material. These include the surface condition, reflectivity at the wavelength of the laser beam, and the absorption coefficient of the bulk material.
- Properties that govern the flow of heat in the material, specifically thermal conductivity and diffusivity. In general, materials with high thermal diffusivity accept and conduct thermal energy very quickly, which favors the weldability of the material.
- Properties that relate to the amount of energy required to cause a desired phase change, such as melting. These include density, specific heat (heat capacity), and the latent-heat effect (heat of fusion).

Some of the problems associated with the reflection of laser energy due to the surface condition of the material can be overcome with the application of an absorptive coating, such as graphite. Once an oxide is formed, the laser's energy is absorbed more rapidly. By the same token, surface treatment with a reflective coating can be used to mask areas when a sweeping laser beam is used for surface treatment, such as skin hardening.

The effectiveness of a laser beam as a cutting tool can be improved by the use of a gas jet, commonly applied through a nozzle coaxial with the beam. A typical assist gas, such as oxygen, helps with oxidation to improve the absorption, produces an exothermic heat reaction, and also helps to clear the molten metal from the bottom of the cut. In general, laser power for cutting must be sufficient to vaporize the material partially. The heated vapor itself, as well as any assist gas used, blows the remaining molten material out of the cut.

For welding, the power of the laser system must be controlled accurately so that melting is localized. Some of the variables controlled by various systems for this purpose are pulse width, pulse-repetition rate, spot size, and power level. Short, fast pulses are needed for scribing ceramics or semiconductors. The pulse energy must be sufficient for microcracking of the material without causing heat damage to surrounding circuitry. For trimming material, such as in the production of film-type resistors, the energy level must be high enough to completely vaporize the material in question.

A typical laser system for materials processing consists of a laser head, a cooling system for the laser cavity, an optical delivery system, a work station with appropriate positioning system, and a means for targeting the laser beam at the desired location. Targeting is frequently accomplished with a binocular viewing system with appropriate interlocks that prevent radiation from reaching the observer. Closed-circuit TV (CCTV) systems or a visible He-Ne laser beam using the same optical path as the working laser beam are some other targeting methods.

Until recently, most of the fabrication applications of lasers have been in the area of micromachining, including the drilling of small holes in hard materials, such as diamond, and in the electronics industry, for such operations as automatic resistor trimming and substrate scribing.

The high power densities available with lasers make the beam an ideal tool where concern with heat migration is a factor. Because the heat input is fast and can be concentrated at a very small area, the process is self-quenching, and the

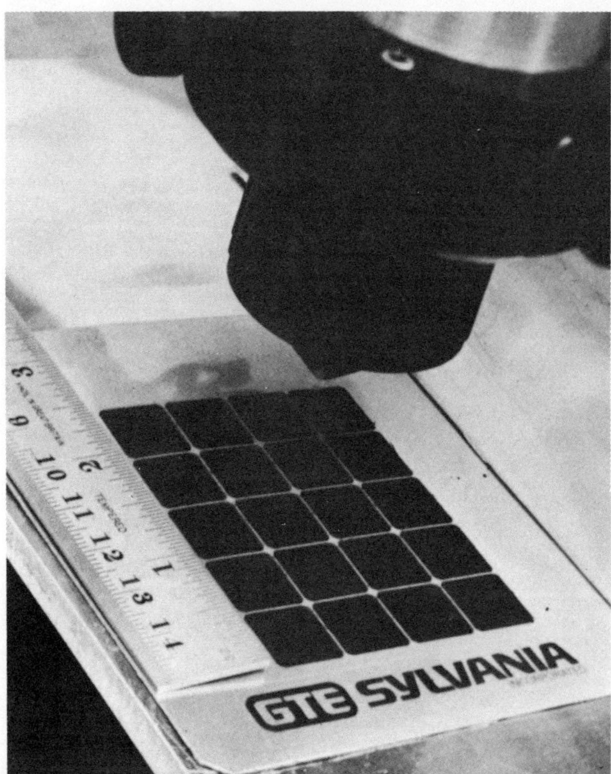

Good mode structure makes for narrow kerf width, and small heat-affected zone allows close nesting of cut parts

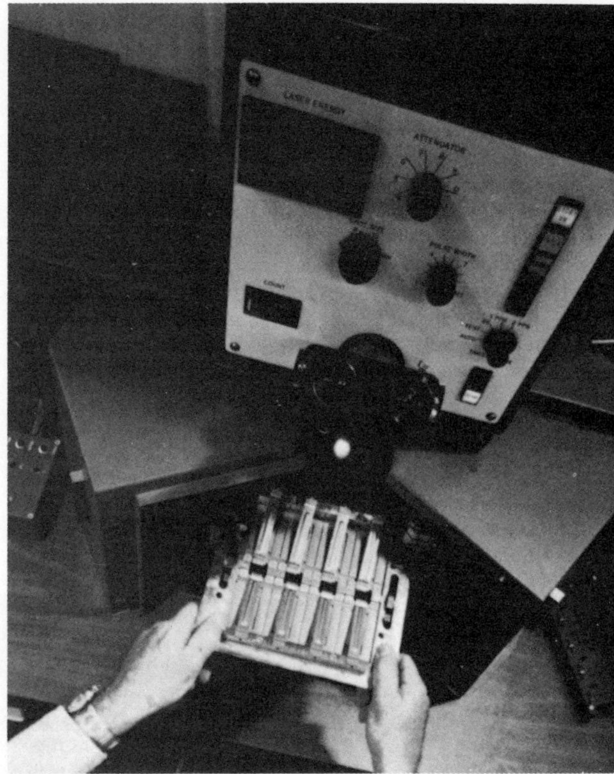

Ni-Span C and 430 stainless-steel components are joined on this Raytheon production laser spotwelding machine

heat-affected area is kept to a minimum. This is an important factor in welding dissimilar metals, a job for which lasers are finding increasing use.

Historically, some of the first applications of laser welding involved spotwelding of very small wires. Now, however, pulsed solid-state lasers producing 50 joules and more are common, and seam welds can be made by fast-repetition-rate lasers at weld rates up to 12 in. per minute. High-power cw solid-state lasers have recently been reported to produce full weld penetration to 0.125 in. at speeds of 10 ipm. Shallow welds are made at faster rates.

But the big news recently has centered around the development of higher-power CO_2-laser systems capable of developing power ranges up to 15 kW. Because the cw CO_2 laser can be operated at much higher average powers than other laser systems, it has attracted much attention as a potential production tool. Opinions regarding the best applications vary, and it will still take time to develop a cogent picture, but there are some emerging trends.

One area of great interest is heat treatment and alloying. The big factor here is the selective nature of the laser beam, which can be easily manipulated to cover only the desired portion of a surface. And, with appropriate optical delivery systems, the beam can be directed at hard-to-reach places. Alloying holds much fascination because the laser would make it possible to confer properties on a material at the time of manufacture. In a time of material shortages, this means that you could buy readily available material at the best price and then treat it at the time of manufacture to achieve such properties as localized hardness, surface properties, or corrosion resistance. The idea is to provide metallurgical changes *in situ* and take advantage of what is available most economically at the mill.

The automobile industry has been looking into the application of lasers for some time, and at present there are some production applications of surface hardening to improve the wear characteristics of engine components. Typically, such an operation is the last in the fabrication sequence, and, because the process is literally distortion-free, no subsequent reworking is required.

Ford Motor Co has done investigative work on welding underbodies using a Hamilton Standard 6-kW CO_2 laser head and an associated beam-transport system that provides four axes of motion: two for translation in X and Y and two for rotating the beam. The entire system, including the functions of the transfer table for handling the stamped panels that form the underbody are computer-controlled.

Most of the welds performed with the equipment are burn-through-type lap welds joining two sheets of 0.035-in.-thick material. Such welds have been performed at 400-450 ipm and the resulting continuous weld has been good. The main problem has been one of part fitup. Because the depth of field of a laser beam is limited and the welding process is autogenous, parts must fit closely, something not readily achieved when panels must be stacked before processing.

General Motors, in concert with Avco Everett Research Laboratory, has investigated high-powered laser heat treating. One interesting application, involving the continuous heat treatment of a shaft, makes use of the fact that, before optical treatment, the laser beam produced by the Avco HPL system is hollow, shaped like a donut. If the spinning shaft is surrounded with a toric focusing mirror, shaft-like parts can be heat treated very uniformly—the beam literally surrounds the part, and the entire periphery of the shaft is treated at one time. With translation of the shaft as it spins, the shaft surface can be treated progressively on a continuous basis.

The aircraft industry is looking into the possibility of using lasers for automatic profile cutting of sheetmetal. Because a laser beam can be easily manipulated, it readily lends itself to computer control, and it is thought that nesting programs can

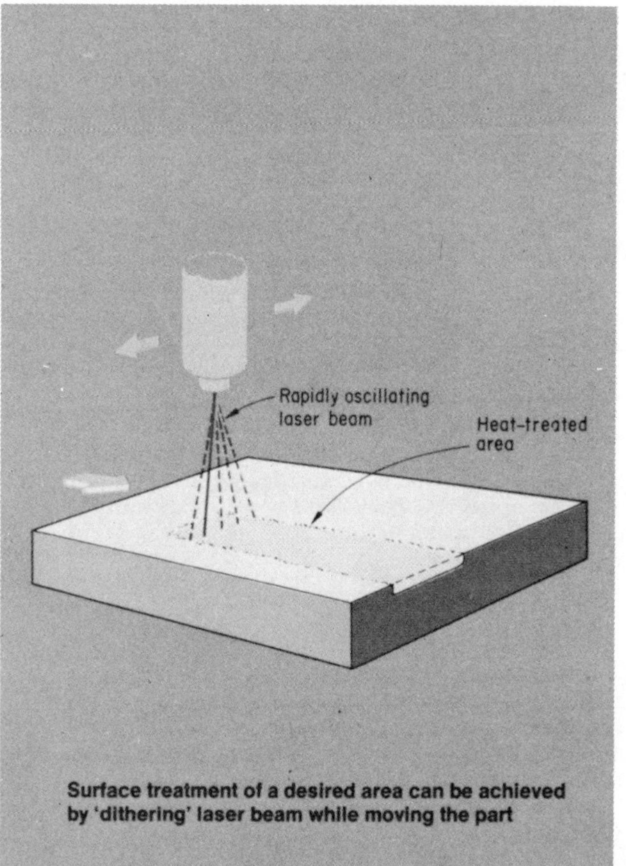

Surface treatment of a desired area can be achieved by 'dithering' laser beam while moving the part

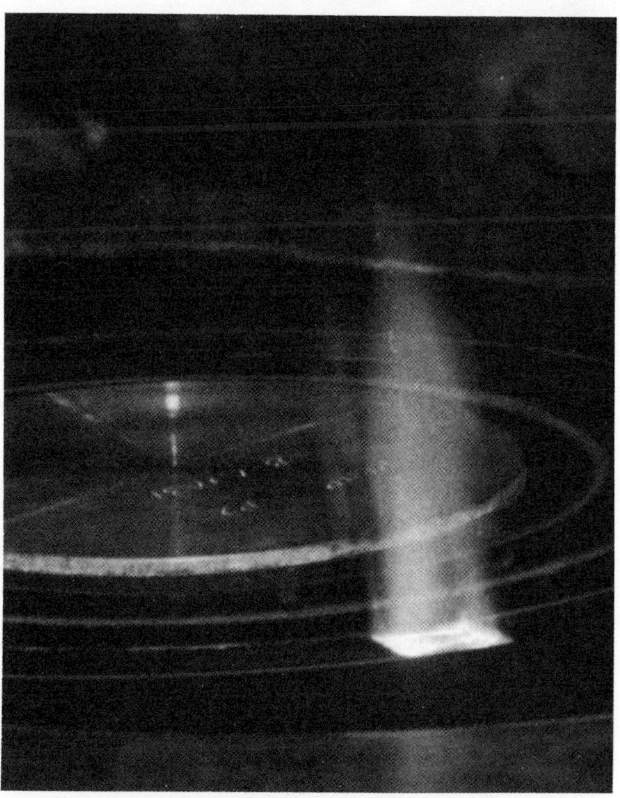
Avco CO_2 laser beam is oscillated, and work is rotated to produce annular heat-treat area on automotive part

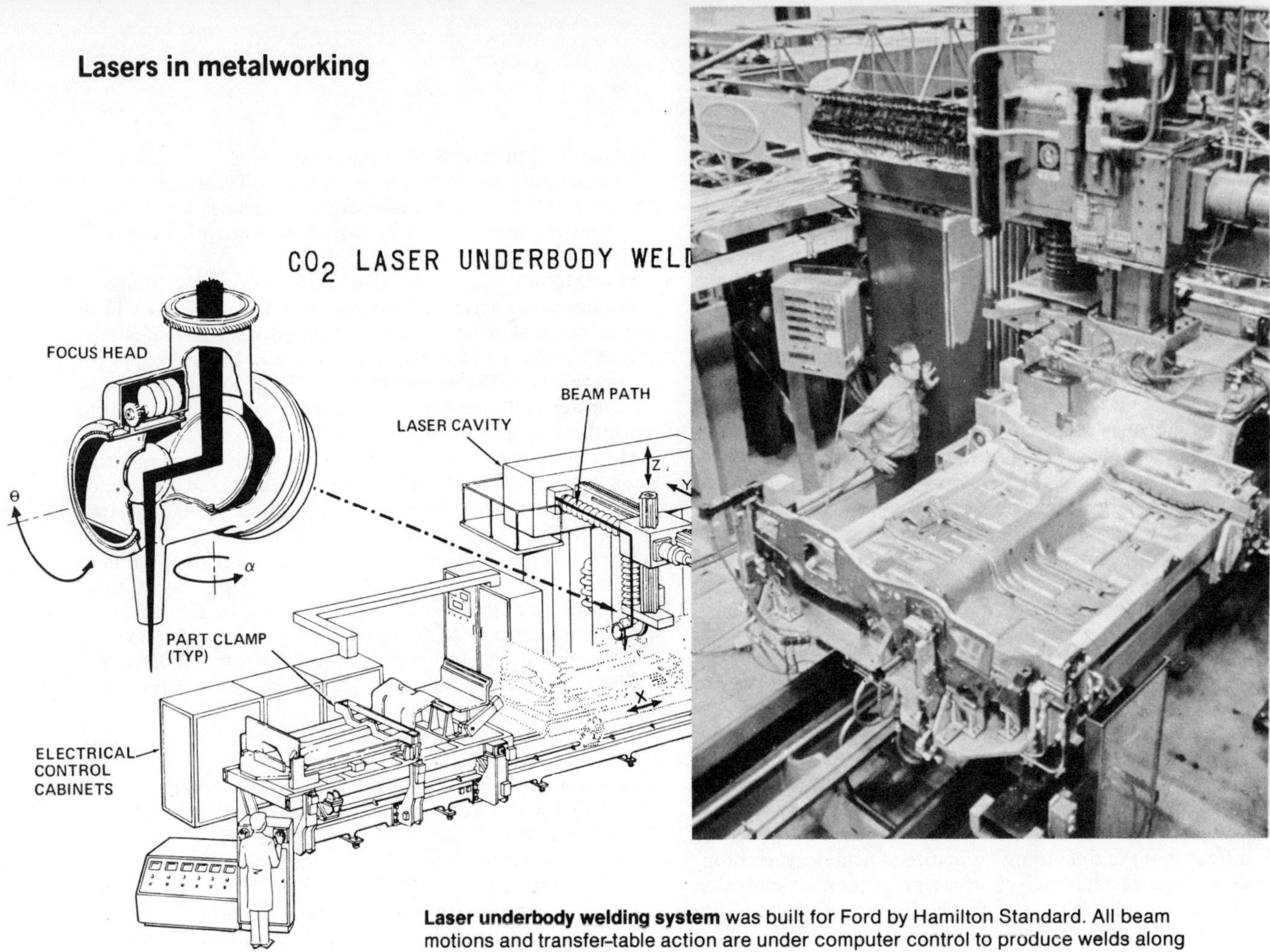

Laser underbody welding system was built for Ford by Hamilton Standard. All beam motions and transfer-table action are under computer control to produce welds along preprogrammed paths at selected speeds

effectively arrange the parts to be cut, thereby minimizing the amount of material wasted between parts. Such a system would also eliminate the need for cutting dies, provide faster production turnaround, and permit the rapid implementation of design changes. The Industrial Products Div of Hughes Aircraft Co, which has supplied similar systems for cutting cloth in the apparel industry, is now looking into applying its know-how to such sheetmetal profiling. In fact, several companies are presently cooperating with the Aerospace Industries Assn to study whether laser-cut aluminum aircraft panels are acceptable as cut, without need for any further trimming to remove any undesirable edge conditions.

Guide to lasers for materials processing

In order to provide an overview of laser systems for processing materials, the rest of this report is devoted to a survey of systems available for metalworking applications.

This guide lists the major characteristics and specifications of 42 different laser systems currently available. Information from 14 different suppliers is listed, but that does not necessarily include all manufacturers with equipment applicable to metalworking. Questionnaires were sent to companies claiming to offer systems with industrial processing capability. Some companies may have been overlooked; some did not respond to the questionnaire. But the guide is a fair representation of what is available.

Although most of the terms used in the listings have been covered elsewhere in this report, here are some explanations and remarks for some of the characteristics listed:

Pulse width. Time duration of energy burst of pulsed or Q-switched laser. Usually measured at half-power points.

Rep rate. Repetition rate of pulsed output, usually expressed in pulses per second (pps).

Exit beam diameter. Laser-beam diameter at exit aperture of cavity, usually before optical delivery system.

Divergence. The increase in the diameter of the laser beam with distance from the laser's exit aperture. The values given are the full angle at a point where the laser energy density is $1/e^2$ of the maximum available.

Operating services. The utility services, including water, as well as special gases needed to operate the system.

Multiple work stations. A laser beam can be split or diverted (shared) for use at more than one work station, using appropriate optical systems. Manufacturers were asked to indicate whether multiple work stations were possible and whether splitting or sharing is used.

Targeting method. The means provided to establish the point of focus of the laser beam.

Positioning system. Provisions for positioning of work.

Consumables. Flashlamp cost is a factor in the operation of solid-state laser systems, with typical prices from $90 to $300. In gas lasers, gas consumption may be a factor. In CO_2 lasers, nearly 75% of the gas is usually He, the most significant expense in replenishing the gas mixture.

Base price. The price of a minimum system, including a work station.

Company	Apollo Lasers Inc.	Apollo Lasers Inc.	Apollo Lasers Inc.	Apollo Lasers Inc
Address	6357 Arizona Circle Los Angeles, Calif. 90045	6375 Arizona Circle Los Angeles, Calif. 90045	6357 Arizona Circle Los Angeles, Calif. 90045	6357 Arizona Circle Los Angeles, Calif. 90045
Telephone	213/776-3343	213/776-3343	213/776-3343	213/776-3343
Model designation	**Lasertrim, sealed-off system**	**Lasertrim, flowing-gas system**	**Welder-driller**	**System 300**
Principal application	Thick-film resister trimming	Thick-film resister trimming	Weld/drill dissimilar metals	Scribing ceramic substrates
Type; wavelength	CO_2/pulsed; 10.6 μm	CO_2/pulsed; 10.6 μm	Ruby/Nd:yag, Nd:glass;*	CO_2/pulsed; 10.6 μm
Output (av); stability	1 W; 2%	2 W; 2%	10-20 kW peak	50 W; 2%
Pulse width; rep rate	250 μs; 0-100 pps	250 μs; 0-100 pps	1-10 ms; 1 pps max	250 μs; 0-100 pps
Exit beam dia; divergence	3 mm; diffraction limited	3 mm; diffraction limited	15 mm; 0.5 mrad	7 mm; 2 mrad
Spot dia; focus depth	0.005 in.*; 0.025 in.	0.005 in.*; 0.025 in.	Variable	0.003-0.030 in./customer
Operating services	115 V, 3 A	115 V, 3 A, gas mixture	110 V, 30 A	115 V, 20 A; water: 1 gpm gas mixture
Cavity cooling	Internal heat sink	Internal heat sink	Recirc refrigerated water	Water
Work station	2 x 2 x 1 in., integral	2 x 2 x 1 in., integral	Per customer, integral	4 x 4 x 2 in., integral
Multiple work stations	No	No	No	No
Targeting method	Binocular microscope	Binocular microscope	Microscope, CCTV optional	Optical magnifying CCTV
Controlled variables	Rep rate, spot size, power level	Rep rate, spot size, power level,	Pulse width, rep rate, spot size, power level	Rep rate, spot size, power level
Positioning system			Manual X-Y table	Summit Engineering stages for X-Y motion
Typical working rates	0.2-10 ipm trim speed, 350 trims/hr, manual feed	0.2-10 ipm trim speed, 350 trims/hr, manual feed	Depends on material	1 ips typical depending on material
Power use standby/run	100/300 VA	100/300 VA	1500/30,000 VA	550/2300 VA
Consumables		Gas mixture: 1 cfh	Flashlamp: 50,000 shots	Gas mixture: 4 cfm
Servicing requirements	Sealed-off plasma tube refill, periodic lubrication	Periodic table lubrication		
Safety provisions			Interlock	Plexiglass shield, interlock
Principal options	Flowing-gas conversion, feeder, hi-res optics	High resolution optics, substrate feeder,	Interchangeable laser head, system design	Computer control, CCTV for backside scribing
Technical services			Installation engineer avail	Installation engineer avail
Special features	Cutting capability	Cutting capability	Ultra-stable mirror mounts**	Modular for system upgrading
Warranty	6 mo laser, 1 yr controls	2 yr laser, 1 yr controls	1 yr system; 50,000 shots	2 yr laser, 1 yr controls
Base price	$7000	$8500	$30,000	$35,000
Remarks	*0.002 with high resolution optics	*0.002 with high resolution optics.	*0.6943 μm, 1.06 μm **double-ignitron triggering.	Cw operating feature for other materials including paper cutting

Company	Avco Everett Research Laboratory Inc.	Coherent Radiation Inc.	Coherent Radiation Inc.	Ferranti Electric Inc.
Address	2385 Revere Beach Pkwy. Everett, Mass. 02149	3210 Porter Dr. Palo Alto, Calif. 94304	3210 Porter Dr. Palo Alto, Calif. 94304	P. O. Box 245 Sturbridge, Mass. 01566
Telephone	617/389-3000	415/493-2111	415/493-2111	617/347-7316
Model designation	**HPL-10, HPL-20**	**Model 41**	**Model 43**	**MF-400**
Principal application	Wld, cut, surf & heat treat*	Wld, cut, drill thin gage	Wld, cut, drill thin gage	Cutting
Type; wavelength	CO_2/cw; 10.6 μm	CO_2/cw/pulsed; 10.6 μm	CO_2/cw/pulsed; 10.6 μm	CO_2/cw; 10.6 μm
Output (av); stability	10 kW, 20 kW; ±1%	250 W; 5%	500 W; 5%	500 W; ±3%
Pulse width; rep rate		1 ms-10 s; 1-100 pps	0.1 ms-10 s; 1-1000 pps	
Exit beam dia; divergence	4 mm focused; NA	8 mm; 1.7 mrad	12 mm; 1.4 mrad	7 mm; 2.5 mrad
Spot dia; focus depth	0.015-0.5 in.; 0.05-2 in.	0.012 in.; 0.50 in.	0.012 in.; 0.50 in.	Depends on lens focal length
Operating services	370 kVA, water, He, CO_2, N_2, CO	208 V, 30 A, 2 gpm water, gas mixture	12 kVA, 2 gpm water, gas mixture	440 V, 1.5 gpm water, gas mixture
Cavity cooling	Water, refrigeration	Water	Water to air	Water
Work station	Per customer requirements	Custom, integral	Custom, integral	Can be integral
Multiple work stations	Beam splitting, beam sharing	Beam splitting, beam sharing	Beam splitting, beam sharing	Beam splitting, beam sharing
Targeting method	Coaxial He-Ne laser	Off-axis microscope	Off-axis microscope	Alignment beam
Controlled variables	Pulse width, rep rate, spot size, power level, power slope	Pulse width, rep rate, spot size, power level, power slope	Pulse width, rep rate, spot size, power level, power slope	Power level, spot size by changing lens
Positioning system	Sciaky positioning systems, NC, CNC	Aerotech, Icon, custom, NC available	Aerotech, Icon, custom, NC available	Adaptable to commercial products, can be NC
Typical working rates	Butt weld 0.5-in. ss at 40 ipm, heat treat steel to 0.025 in. at 50 in.^{2}/min	Cut 0.25-in. titanium at 2 ips	Weld 0.030-in. ss at 2.5 ipm with full penetration	Cut 0.125-in. ss at 40 ipm
Power use standby/run	5/100 kW	6240 kW/hr max, running	12 kVA max, running	11 kVA, running
Consumables	Makeup gas	He: 3.8 cfh, N_2: 0.4 cfh, CO_2: 0.2 cfh	He: 8.4 cfh, N_2: 1.5 cfh, CO_2: 0.5 cfh	Gas mixture: 19 gpm
Servicing requirements	Preventive maintenance at 2-mo intervals	Clean optics 24-48 hr	Clean optics 24-48 hrs	
Safety provisions	Fully enclosed, interlocked	Fully interlocked	Fully interlocked	Interlocked and shielded
Principal options	Optical and tooling packages	Pulser, beam delivery systems, gas assist	Pulser, beam delivery systems	Optical delivery systems
Technical services	R&D, application studies	Custom development	Custom development	Operator training program
Special features	Programmable power control			Light weight, compact
Warranty	90 days, service contract opt.	1 yr	1 yr	1 yr
Base price	$500,000	$65,000	$95,000	On quotation
Remarks	*Also alloying. Marketed in U. S. and Canada by Sciaky Bros. (see entry elsewhere)	Suited for work on steels, nickel alloys, titanium	Suited for work on steels, nickel alloys, titanium	

Company	GTE Sylvania Electro-Optics Organization	GTE Sylvania Electro-Optics Organization	Hamilton Standard Div. United Aircraft Corp.*	Hamilton Standard Div. United Aircraft Corp.*
Address	P. O. Box 188 Mountain View, Calif. 94040	P. O. Box 188 Mountain View, Calif. 94040	Bradley Field Rd. Windsor Locks, Conn.	Bradley Field Rd. Windsor Locks, Conn.
Telephone	415/966-3678	415/966-3678	203/623-1621, 565-4321	203/623-1621, 565-4321
Model designation	**Model 1610 YAG**	**Model 971 GTL**	**3.5 kW**	**6 kW**
Principal application	Drill, cut, sldr, wld, ht trt	Wld, cut, heat treat, hd face	Wld, cut, heat treat, alloy	Wld, cut, heat treat, alloy
Type; wavelength	Nd:yag/pulsed; 1.06 μm	CO_2/cw; 10.6 μm	CO_2/cw; 10.6 μm	CO_2/cw; 10.6 μm
Output (av); stability	150 W; 5%	1kW rated, 1.5 kW max; 5%	3.5 kW; 2%	6 kW; 5%
Pulse width; rep rate	0.6-8 ms; 150 pps	80 ms (opt); 10 pps (opt)		
Exit beam dia; divergence	6 mm; 8 mrad	12 mm; less than 1.3 mrad	75 mm; 0.8 mrad	75 mm; 0.65 mrad
Spot dia; focus depth	0.010 in. w/4-in. f.l. lens	0.003 in. w/2.5-in. f.l. lens	0.035 in. typical	0.023 in.; 0.2 in. typical
Operating services	10 kW, 4.5 gpm water, shieldgas	40 kW, water, gas mixture	460 V, 100 A, 30 gpm water at 70 F, gas mixture	460 V, 270 A, 65 gpm water at 70 F, gas mixture
Cavity cooling	Water	Water	Water	Water
Work station	56 x 25 x 56 in., integral		Per customer requirements, separate unit	Per customer requirements, separate unit
Multiple work stations	Beam splitting, beam sharing	Beam splitting, beam sharing	Beam splitting, beam sharing	Beam splitting, beam sharing
Targeting method	Microscope, CCTV (opt)	He-Ne laser*	He-Ne laser beam	He-Ne laser beam
Controlled variables	Pulse width, rep rate, spot size, power level power slope	Pulse width (opt), rep rate, spot size, power level, power slope (opt)	Spot size (variable focus), power level, power slope	Spot size (variable focus), power level, power slope
Positioning system			Per customer requirements	Per customer requirements
Typical working rates	Weld up to 0.040 in. deep to 90 ipm, drill 0.200 in. deep in ss	Cut 0.12-in. low-carbon steel at 100 + ipm, other data available	Weld 0.20-in. ss at 50 ipm, 0.20 lead at 200 ipm	Weld 0.060-in. rimmed steel at 500 ipm, low carbon steel at 35 ipm
Power use standby/run	500/8000 W	1/30 kW		
Consumables	Flashlamp: 10^6 -10^7 shots	He: 0.9 cfm, N_2: 0.2 cfm, CO_2: 0.04 cfm	Gas mixture: 24 cfh	Gas mixture: 48 cfh
Servicing requirements	Filters, deionizers/6 mo	Lube and oil blower and pump every 500 hrs	Varies with configuration of system, and use	Varies with configuration of system and use
Safety provisions	Interlocks, enclosed path	Interlocks, enclosed path	Interlocks, enclosed path	Interlocks, enclosed path
Principal options	Lenses, nozzles, customized work stations	Power meter, shutter, rotating optics, focusing assy	Focusing optics, power level/ beam quality	Focusing optics, power level/ beam quality
Technical services	Applications lab, consulting	Applications lab, consulting		
Special features	Adjustable pulse shape,*	TEM_{00} mode, compact	Solid output window	Solid output window
Warranty	1 yr parts and labor	1 yr parts and labor	1 yr	1 yr
Base price	$42,500	$64,500	Per quotation	Per quotation
Remarks	*Closed-loop output energy control	*Microscope and CCTV optional, transverse gas transport system. Marketed worldwide by Union Carbide (see entry)	*In cooperation with United Aircraft Research Lab	*In cooperation with United Aircraft Research Lab

Company	Holobeam Laser Inc.	Holobeam Laser Inc.	Holobeam Laser Inc.	Holobeam Laser Inc.
Address	560 Winters Ave. Paramus, N. J. 07652	560 Winters Ave. Paramus, N. J. 07652	560 Winters Ave. Paramus, N. J. 07652	560 Winters Ave. Paramus, N. J. 07652
Telephone	201/265-5335	210/265-5335	201/265-5335	201/265-5335
Model designation	**Model 940**	**Model 980**	**Model 930**	**Model 990**
Principal application	Micro-precision welding	Drilling	Welding, drilling	Purging sealed containers
Type; wavelength	Nd:yag/pulsed; 1.06 μm	Nd:yag/pulsed; 1.06 μm	Nd:yag/pulsed; 1.06 μm	Nd:glass pulsed; 1.06 μm
Output (av) stability	0.15 J; 5%	1 J; 5%	100 J max; 5%	100 J; 5%
Pulse width; rep rate	300 μs; 50 pps	0.1-1 ms; 50 pps	0.5-12 ms; 1 pps	1-10 ms; 20 ppm
Exit beam dia; divergence	6 mm; 3 mrad	0.125 in.; 3 mrad	0.375 in.; 2 mrad	0.375 in.; 8 mrad
Spot dia; focus depth	0.0005 in. min	Depends on optics	Depends on optics	0.01-0.06 in.
Operating services	2 gpm water	220 V; 2 gpm water	220 V; 2 gpm	220 V, 3 gpm water
Cavity cooling	Water-to-water or refrig	Water-to-water	Water-to-water	Water-to-water
Work station	12 x 8 x 3 in., integral	Per application, integral	24 x 24 x 12 in., integral	Per application, integral
Multiple work stations				
Targeting method	Binocular microscope & CCTV	Binocular microscope,	Binocular microscope*	Monocular microscope or CCTV
Controlled variables	Pulse width, rep rate, spot size	pulse width, rep rate, power level	Pulse width, rep rate, spot size, power level	Pulse width, spot size, power level
Positioning system	Manual, NC optional			
Typical working rates		Drill 0.005-in. dia in ss at 50 holes/sec		Drill 0.01-in. dia in 0.08-in. Zircalloy and reweld at 2 per min
Power use standby/run		3 kW, intermittent	5 kW, intermittent	3 kW, intermittent
Consumables	Flashlamp: 10^6 shots	Flashlamp: 10^6 shots	Flashlamp: 10^6 shots	Flashlamp: 10^6 shots
Servicing requirements	Flashlamp change	Flashlamp change	Flashlamp change	Flashlamp change
Safety provisions	Interlocked	Interlocked	Interlocked	Interlocked
Principal options	Frequency doubler, cw or Q-switched operation	CCTV viewing, rotating optics for large-dia hole drilling	Ruby or Ni:glass laser, optics, work handling	
Technical services	Applications lab	Applications lab	Applications lab	Applications lab
Special features	Micro-precision machine tool		One of modular family	Single fixture
Warranty	1 yr	1 yr	1 yr	1 yr
Base price		$30,000	$20,000	$28,000
Remarks			*CCTV & He-Ne setup laser also available	

Company	Holobeam Laser Inc.	Holobeam Laser Inc.	International Laser Systems	Korad Div. Hadron Inc.
Address	560 Winters Ave. Paramus, N. J. 07652	560 Winters Ave. Paramus, N. J. 07652	30404 N. Orange Blossom Trail Orlando, Fla. 32804	2520 Colorado Ave. Santa Monica, Calif. 90404
Telephone	201/265-5335	201/265-5335	305/843-4731	213/829-3377
Model designation	**Model 950**	**Model 900**	**ML-1**	**KRT**
Principal application	Metal engraving, serializing	Continuous welding	Hole drilling	Thin film trimming
Type; wavelength	Nd:yag/pulsed; 1.06 μm	Nd:yag/cw; 1.06 μm	Yag; 1.06 μm	Yag/pulsed; 1.06 μm
Output (av); stability	80 W; 5%	400 W; 5%	0.5 W; ±5%	10 W; ±2%
Pulse width; rep rate	0.2 μs; 100-75,000 pps		0.04 ms; 55-60 pps	0.0003 ms, 0-10,000 pps
Exit beam dia; divergence	8 mm; 6 mrad	10 mm; 10 mrad	10 mm; 10 mrad	3 mm; 2-5 mrad
Spot dia; focus depth	Depends on optics	Depends on optics	0.066 mm; 2.5 mm	0.0003-0.1 in.; ±0.08 in. max
Operating services	240 V; 3 gpm water	20 kW, 3 gpm	115 V	
Cavity cooling	Water-to-water or refrig	Water-to-water	Air	Water
Work station	Per application, integral	Integral	20 x 15 x 3 in., integral	24 x 20 x 13-40 in., integral
Multiple work stations		Beam splitting, beam sharing	No	Beam splitting
Targeting method	CCTV	He-Ne laser beam*	Coaxial microscope	Microscope or CCTV
Controlled variables	Rep rate, spot size, power level	Spot size, power level, power slope	Rep rate, spot size, power level	Pulse width, rep rate, spot size, power level
Positioning system	Icon 380M or equivalent	Autonumerics NC, 3-axis	Icon 350T & 1520 XYD, drill-on-fly per encoder	Manual or servo X-Y beam positioning
Typical working rates	Generates 10 characters/sec	Butt weld 0.025 ss at 120 ipm	3 + ips for 0.04-in. spacing in 0.003-in. nickel steel	Trim 0.0001-in. Nichrome, 6000 trims/hr
Power use standby/run	3/5 kW	10/20 kW	2000 VA running	2.5 kW running
Consumables	Flashlamp: 200 hr	Flashlamp: 200 hr	Coolant filter at 4 mo, flashlamp: 15 x 10⁶ shots	Flashlamps: 200 hrs
Servicing requirements	Flashlamp change		Flashlamp change	Preventive maintenance, monthly
Safety provisions	Interlocked	Interlocked, meets OSHA	Interlocks, indicator/alarm	Meets ANSI Z-136.1
Principal options	NC, CNC, cw unit for heavy-duty applications	800 W power add-on module		CCTV, probes, bridges, analog programmer, lenses
Technical services	Applications lab	Applications lab	Custom design available	Applications lab, training
Special features		Modular system	Accurate drill-on-fly	Analog programmer, expandable
Warranty	1 yr	1 yr	90 days parts and labor	1 yr
Base price	$50,000-100,000	$78,000	$75,000	$24,000
Remarks		*Viewing optics available	5-min flashlamp change	Suitable for metals and non-metals. Sample processing available

Company	Korad Div. Hadron Inc.	Korad Div. Hadron Inc.	Korad Div. Hadron Inc.	Laser Inc.
Address	2520 Colorado Ave. Santa Monica, Calif. 90404	2520 Colorado Ave. Santa Monica, Calif. 90404	2520 Colorado Ave. Santa Monica, Calif. 90404	P. O. Box 537 Sturbridge, Mass. 01566
Telephone	213/829-3377	213/829-3377	213/829-3377	617/347-7314
Model designation	**KWD**	**KMT**	**KSS3/KSS4**	**Model 7Y**
Principal application	Welding, drilling	Micromachining	Scribing	Hole drilling
Type; wavelength	Yag, ruby/plsd; 1.06, 0.69 μm	Yag/cw; 1.06 μm	Yag/cw; 1.06 μm	Yag/pulsed; 1.06 μm
Output (av); stability	40 W; ±3%	200 W; ±2%	200 W; ±2%	5 J; ±4%
Pulse width; rep rate	1-7 ms; 0-2 pps	Cw-0.0003 ms; 0-50,000 pps	Cw-0.0003 ms; 0-50,000 pps	0.5 ms; 0-5 pps
Exit beam dia; divergence	2-5 mrad	5 mm; 2-15 mrad	5 mm; 2-15 mrad	6 mm; 5 mrad
Spot dia; focus depth	0.0005-0.1 in.; ±0.08 max	0.0003-0.1 in.; ±0.08 in. max	0.0003-0.1 in.; ±0.08 in. max	
Operating services		240 V, 15 gpm water	240 V, 30 A	208/240 V
Cavity cooling	Water or refrigeration	Water	Water	Closed-cycle water
Work station	24 x 17 x 13-30 in., integral	24 x 20 x 10-40 in., integral	5 x 5 x 1 in., integral	Separate
Multiple work stations	Beam splitting	Beam splitting	No	Beam splitting, beam sharing
Targeting method	Microscope and/or CCTV	Microscope or CCTV	Microscope and/or CCTV	Aiming spot built in
Controlled variables	Pulse width, rep rate, spot size, power level	Pulse width, rep rate, spot size, power level	Pulse width, rep rate, spot size, power level	Rep rate, spot size*, power level
Positioning system	Slo-Syn, Summit, or other NC	Slo-Syn, Summit, or other NC	Korad hydraulic/pneumatic/ stepping motor w/leadscrew	Slo-Syn or Cambion
Typical working rates	Weld 0.05-in. 300 series ss at 4 ipm	Cut 0.025 in. 6Al-4V Titanium alloy at 60 ipm	Scribe 0.016 in. deep in silicon	
Power use standby/run	5000 W running	7 kW running	6 kW running	300/1500 kW
Consumables	Flashlamp: 10⁶-10⁷ shots	Flashlamp: 400-1000 hr	Flashlamp: 400-1000 hr	Flashlamp: 5 x 10⁵ shots
Servicing requirements	Preventive maintenance, monthly	Preventive maintenance, monthly	Preventive maintenance, monthly	Flashlamp change, deionizing filters 6 mo
Safety provisions	Meets ANSI Z-136.1	Meets ANSI Z-136.1, Class IV	Meets ANSI Z-136.1, Class IV	Meets ANSI Z-136.1
Principal options	CCTV, power monitor, lenses	CCTV	Back scribing	
Technical services	Applications lab, training	Applications lab, training	Applications lab, training	
Special features		Power meter, gas-jet lens	Pre-alignment station	Optimized specs for drilling
Warranty	1 yr	1 yr	1 yr	6-mo, parts and labor
Base price	$25,000-30,000	$33,000	$67,500	$20,250
Remarks	Sample processing available			*Spot size continuously variable

Company	Laser Inc.	Laser Inc.	Optimation Inc.	Optimation Inc.
Address	P. O. Box 537 Sturbridge, Mass. 01566	P. O. Box 537 Sturbridge, Mass. 01566	358 Baker Ave Concord, Mass. 01742	358 Baker Ave Concord, Mass. 01742
Telephone	617/347-7314	617/347-7314	617/369-8428	617/369-8428
Model designation	**Model 11**	**Model 14**	**OPTI-60**	**OPTI-100**
Principal application	Drill or weld	Welding	Micromachining & trimming	Cut & trim thin metal
Type; wavelength	Nd:glass/pulsed; 1.06 μm	Nd:yag/pulsed; 1.06 μm	Yag/cw; 1.06 μm	Yag/cw; 1.06 μm
Output (av); stability	30 J; ±3%	30 J; ±3%	30 W; 2%	100 W; 2%
Pulse width; rep rate	0.3-8 ms; 0-1 pps	1, 3, or 5 ms; 0-2 pps		
Exit beam dia; divergence	6 mm; 6 mrad	3 x 9 mm; 10 mrad	3 mm; 5 mrad	6 mm; 5 mrad
Spot dia; focus depth			0.0005 in.	0.0015 in.
Operating services	208/240 V	208/240 V	208 V, deionized water	208 V
Cavity cooling	Closed-cycle water	Closed-cycle water	Water	Water
Work station	Separate	Separate	Integral	Integral
Multiple work stations	Beam splitting, beam sharing	Beam splitting, beam sharing	Beam splitting, beam sharing	Beam splitting, beam sharing
Targeting method	Aiming spot built in	Aiming spot built in	Optional viewer	Optional viewer
Controlled variables	Pulse width, rep rate, spot size, power level	Pulse width as above, rep rate, spot size, power level	Pulse width, rep rate, spot size, power level	Pulse width, rep rate
Positioning system	Slo-Syn or Cambion	Slo-Syn or Cambion	Aerotec or equivalent	Customer supplies
Typical working rates				Cut 0.040-in. metals w/oxy assist at 1 ips
Power use standby/run	300/1200 W	300/1500 kW		
Consumables	Flashlamp: 3.5 x 10^6 shots	Flashlamp: 2.5 x 10^6 shots	Flashlamp: 200 hr	Flashlamp: 200 hr
Servicing requirements	Flashlamp change, deionizing filters 6 mo	Flashlamp change, deionizing filters 6 mo	User maintenance	User maintenance
Safety provisions	Meets ANSI Z-136.1	Meets ANSI Z-136.1	Interlocked	
Principal options	Limited-access door, drill or weld or both	Limited-access door, drill or weld or both	Q-switch, mode lock, viewer, aperture wheel	Q-switch, mode lock, viewer, apertures
Technical services			Contract work	Contract work
Special features	High energy/pulse	Rectangular beam format*	For industrial use	Low-cost optics
Warranty	6-mo, parts and labor	6-mo, parts and labor	1 yr*	1 yr
Base price	$16,695	$17,970	$8,400	$14,000
Remarks		*Rectangular beam for best seam weld results	*No warranty on optics, lamps, or crystal	New product as of Feb. '75

Company	Photon Sources Inc.	Photon Sources Inc.	Photon Sources Inc.	Photon Sources Inc.
Address	37100 Plymouth Rd. Livonia, Mich. 48150	37100 Plymouth Rd. Livonia, Mich. 48150	37100 Plymouth Rd. Livonia, Mich. 48150	37100 Plymouth Rd. Livonia, Mich. 48150
Telephone	313/261-5400	313/261-5400	313/261-5400	313/261-5400
Model designation	**Model 104**	**Model 130**	**Model 108**	**Model 300**
Principal application	Perforate, cut, mark Fe, Ni	Weld ss or nickel alloys	Perforate, cut, weld	Perforate, cut, weld
Type; wavelength	CO_2/pulsed; 10.6 μm	CO_2/pulsed; 10.6 μm	CO_2/cw/pulsed; 10.6 μm	CO_2/cw.pulsed; 10.6 μm
Output (av); stability	50 W; 5%	75 W; 10%	100 W; 5%	375 W, 2%
Pulse width; rep rate	0.1 ms min; 2500 pps max	0.05 ms min; 300 pps max	0.1 ms min; 2500 pps max	0.1 ms min; 1000 pps max
Exit beam dia; divergence	6.5 mm; 2.1 mrad	15 mm; 4 mrad	7.4 mm; 1.85 mrad	10.1 mm; 1.35 mrad
Spot dia; focus depth	0.002 in. min; ±0.2 in. max	0.002 in. min; ±0.2 in. max	0.002 in. min; ±0.2 in. max	0.002 in. min; ±0.2 in. max
Operating services	2 kVA, 0.5 gpm water, gas mixture	2.4 kVA, water, gas mixture	3 kVA; 0.5 gpm water, gas mixture	7.6 kVA; 2.5 gpm water, gas mixture
Cavity cooling	Water	Water/oil	Water	Water/oil
Work station	2 x 2 x 4 ft to 8 x 8 x 8 ft, separate or integral	2 x 2 x 4 ft to 8 x 8 x 8 ft, separate or integral	2 x 2 x 4 ft to 8 x 8 x 8 ft, separate or integral	2 x 2 x 4 ft to 8 x 8 x 8 ft, separate or integral
Multiple work stations	Beam sharing	Beam sharing	Beam sharing	Beam sharing
Targeting method	He-Ne laser beam*	He-Ne laser beam*	He-Ne laser beam*	He-Ne laser beam*
Controlled variables	Pulse width, rep rate, spot size, power level, power slope	Pulse width, rep rate, spot size, power level, power slope	Pulse width, rep rate, spot size, power level, power slope	Pulse width, rep rate, spot size, power level, power slope
Positioning system	Icon 350 or 380 series, Vega IIIG series, PSI CNC	Icon 350 or 380 series, Vega IIIG series, PSI CNC	Icon 350 or 380 series, Vega IIIG series, PSI CNC	Icon 350 or 380 series, Vega IIIG series, PSI CNC
Typical working rates	Perforate 0.003 in. dia in 0.002-in. Ni, 20/s. Mark steel 0.004 in. at 60 ipm	Weld 0.01-in. ss at 0.2 ips. Drill 0.012-in. dia in 0.025-in. Al_2O_3, 5/sec	Cut 0.01-in. 1010 steel at 60 ipm. Drill 0.01-in. dia in 0.02-in. steel in 50 ms**	Cut 0.042-in. Hastelloy X at 90 ipm. Weld 0.01-in. 302 ss at 240 ipm**
Power use standby/run	2/2 kVA		3/3 kVA	5/7.6 kVA
Consumables	Gas mixture: 1.77 cfh	Gas mixture: 14 cfh	Gas mixture: 1.77 cfh	Gas mixture: 8.29 cfh
Servicing requirements	Pump lube 200 hr, mirror inspect 100 hr	Pump lube 200 hr, mirror inspect 1000 hr	Pump lube 200 hr, mirror inspect 1000 hr	Pump lube 200 hr, mirror inspect 1000 hr
Safety provisions	Meets ANSI Z 136.1, Class V	Meets ANSI Z-136.1, Class V	Meets ANSI Z-136.1, Class V	Meets ANSI Z-136.1, Class V
Principal options	NC, CNC, pulse & slope contr., coincident microscope	NC, CNC, pulse & slope contr., coincident microscope	NC, CNC, pulse & slope contr., coincident microscope	NC, CNC, pulse & slope contr., coincident microscope
Technical services	Process dev., eval., samples	Process dev., eval., samples	Process dev., eval., samples	Process dev., eval., samples
Special features	Meets JIC	Meets JIC	Meets JIC	Meets JIC
Warranty	1 yr	1 yr	1 yr	1 yr
Base price	$35,000	$80,000	$40,000	$60,000
Remarks	*Thermal screen or coincident microscope available	*Thermal screen or coincident microscope avail	*Thermal screen or coincident microscope avail. **Weld 0.01-in. 303 ss at 30 ipm. Scribe Al_2O_3 at 480 ipm	*Thermal screen or coincident microscope avail. **Drill 0.06-in. dia in 0.1-in. 1080 steel, 100 ms

Company	Photon Sources Inc.	Photon Sources Inc.	Quantronix Corp.	Quantronix Corp.
Address	37100 Plymouth Rd. Livonia, Mich. 48150	37100 Plymouth Rd. Livonia, Mich. 48150	225 Engineers Rd. Smithtown, N. Y. 11787	225 Engineers Rd. Smithtown, N. Y. 11787
Telephone	313/261-5400	313/261-5400	516/273-6900	516/273-6900
Model designation	**Model 500**	**Model 1000**	**Model 603**	**Model 114**
Principal application	Surface harden, cut, drill	Surface harden, cut, drill	Scribing	Scribing, drilling
Type; wavelength	CO_2/cw/pulsed; 10.6 μm	CO_2/cw/pulsed; 10.6 μm	Yag/pulsed; 1.06 μm	Yag/pulsed; 1.06 μm
Output (av); stability	525 W; 2%	1 kW; 2%	5-50 W; 5%	5-100 W, 5%
Pulse width; rep rate	0.1 ms min; 1000 pps max	0.1 min; 1000 pps max	0.1-0.5 μs; to 50,000 pps	0.1-0.5 μs; to 50,000 pps
Exit beam dia; divergence	9.6 mm; 1.45 mrad	8.8 mm; 1.55 mrad	1-3 mm; 1 mrad	1-3 mm; 1 mrad
Spot dia; focus depth	0.002 in. min; ±0.2 in. max	0.002 in. min; ±0.2 in. max	0.0002-0.002 in.; 0.05 in. max	0.0002-0.002 in.; 0.05 in. max
Operating services	7.6 kVA, 2.5 gpm water, gas mixture	11 kVA, 3.75 gpm water, gas mixture	208 V, 2 gpm water	208 V, 2 gpm water
Cavity cooling	Water/oil or water/Freon/oil	Water/oil or water/Freon/oil	Water	Water
Work station	2 x 2 x 4 ft to 8 x 8 x 8 ft, separate or integral	2 x 2 x 4 ft to 8 x 8 x 8 ft, separate or integral	4 x 4 x ⅛ in., integral	Separate
Multiple work stations	Beam sharing	Beam sharing	No	No
Targeting method	He-Ne laser beam*	He-Ne laser beam*	Visual viewing screen	
Controlled variables	Pulse width, rep rate, spot size, power level, power slope	Pulse width, rep rate, spot size, power level, power slope	Rep rate, spot size, power level	Rep rate, spot size power level
Positioning system	Icon 350 or 380 series, Vega IIIG series, PSI CNC	Icon 350 or 380 series, Vega IIIG series, PSI CNC	Integral	
Typical working rates	Surface harden nodular iron to 0.015 in. x 0.125 wide at 60 ipm.**	Cut 0.04-in. 1010 steel at 180 ipm, 0.25-in. at 18 ipm. Harden 0.015 nodular iron**	0.002-in. scribe in semicond. at 8 ips	Micromachine 0.001 in. in Al at 1-5 ips
Power use standby/run	5/7.6 kVA	8.5/11 kVA	2/4-7 kW	2/4-7 kW
Consumables	Gas mixture: 8.78 cfh	Gas mixture: 14.85 cfh	Flashlamp: 200 hr	Flashlamp: 200-400 hr
Servicing requirements	Pump lube 200 hr, mirror inspect 1000 hr	Pump lube 200 hr, mirror inspect 1000 hr	Flashlamp replacement	Flashlamp replacement
Safety provisions	Meets ANSI Z-136.1, Class V	Meets ANSI Z-136.1, Class V	Interlocks, safety filters	Totally enclosed, interlocks
Principal options	NC, CNC, pulse & slope contr., coincident microscope	NC, CNC, pulse & slope contr., coincident microscope	1 or 2 lamp system	1 or 2 lamps
Technical services	Process dev., eval., samples	Process dev., eval., samples	Service package	Service package
Special features	Meets JIC	Meets JIC	Computer controlled	
Warranty	1 yr	1 yr	1 yr or 3000 hr	1 yr
Base price	$67,000	$85,000		
Remarks	*Thermal screen or coincident microscope avail. **Weld Mg 0.012 in. at 120 ipm, Sn-Ti 0.02 in. at 240 ipm	*Thermal screen or coincident microscope avail. **120 ipm, 0.125 in. wide. Weld 0.030-in. Al at 270 ipm		

Company	Raytheon Co. Industrial Laser Systems	Raytheon Co. Industrial Laser Systems	Raytheon Co. Industrial Laser Systems	Raytheon Co. Industrial Laser Systems
Address	130 Second Ave. Waltham, Mass. 02154	130 Second Ave. Waltham, Mass. 02154	130 Second Ave. Waltham, Mass. 02154	130 Second Ave. Waltham, Mass. 02154
Telephone	617/890-8080	617/890-8080	617/890-8080	617/890-8080
Model designation	**SS-323**	**SS-117**	**SS-347**	**SS-380**
Principal application	Microdrilling	Microdrill, scribe, cut	Drilling, cutting	Welding
Type; wavelength	Yag/pulsed; 1.06 μm	Yag/pulsed; 1.06 μm	Yag/pulsed; 1.06 μm	Yag/pulsed; 1.06 μm
Output (av); stability	2.5 W; 1%	2.5 W; 1%	10 W; 2%	40 W; 2.5%
Pulse width; rep rate	20-120 ms*; 1, 5, 10, 20 pps**	50-150 ms; 1, 5, 10, 20 pps*	1-500 μs; 1, 5, 10, 20 pps*	2-10 ms; 1, 2, 4, 6 pps, manual
Exit beam dia; divergence	5 mm; 0.0015-0.003 mrad	5 mm; 0.0035 mrad max	6.35 mm; 5 mrad	6.35 mm; 10 mrad
Spot dia; focus depth	0.001-0.01 in.	0.001-0.01 in.	0.005-0.015 in.	0.02-0.06 in.; variable
Operating services	208-240 V	208-240 V	208-220 V	208-240 V
Cavity cooling	Water/air, self-contained	Water/air, self-contained	Water/air, self-contained	Water/air, self-contained
Work station	10 x 5 x 7 in., integral, expandable	10 x 5 x 7 in., integral, expandable	10 x 5 x 7 in., integral expandable	16 x 16 x 10 in., integral, expandable
Multiple work stations	Beam split or share optional	Beam split or share optional	Beam split or share optional	Beam split or share optional
Targeting method	CCTV	Binocular microscope	CCTV	Binocular microscope
Controlled variables	Pulse width, rep rate, spot size, power level	Pulse width, rep rate, spot size, power level,	Pulse width, rep rate, spot size, power level	Pulse width, rep rate, spot size, power level
Positioning system	Chessman positioner, NC adaptable	Gaertner X-Y translation stage	X-Y-Z micropositioner, die-centering fixt, NC or CNC	NC or CNC, various models
Typical working rates	Drill 0.0008-0.005-in. dia in 0.01-in. silicon wafer	Drill 0.001-0.005-in. dia to 0.06 in. deep in metal, 0.1-in. in diamond crystal	Drills diamond wire-drawing dies to 0.06 dia. w/coring. 0.15-in. dia in 0.15-in. metl	To 6 spot welds/sec, or 3.6 ipm seam weld in metal up to 0.03-in. thick
Power use standby/run	1.4/2.5-6.4 kVA	1.4/2.5-6.4 kVA	2/6.5 kW	1.5/3.7 kV
Consumables	Flashlamp: 10^6 shots, deionizer cartridges	Flashlamp: 10^6 shots, deionizer cartridges	Flashlamp: 10^6 shots, deionizer cartridges	Flashlamp: 10^6 shots, deionizer cartridges
Servicing requirements	Periodic maintenance	Periodic maintenance	Periodic maintenance	Periodic maintenance
Safety provisions	Interlocks, special enclosure	Interlocks	Interlocks, safety enclosure	Interlocks, safety enclosure
Principal options	Fine-hole drilling kit, CCTV, remote control	Fine-hole drilling kit, CCTV, remote control	Various die fixtures, X-Y-Z micropositioner	Water-to-water heat exchanger, pulse-train selector*
Technical services	Training, applications lab	Training	Training, applications lab	Training, applications lab
Special features	Front-panel control, modular	Front-panel control, modular	Controlled divergence avail	No pulse-forming network
Warranty	1 yr	1 yr	1 yr	1 yr
Base price	On quotation	$19,000	$35,950	$21,500
Remarks	*In 20-ms intervals. **Manual pulsing included. Price includes installation and training	*Manual pulsing included. Price includes installation and training	*Manual pulsing included. Price includes installation and training	*Output energy monitor, spot-size control, others. Price includes instal & training. Tooling development avail

Company	**Sciaky Bros., Inc.**
Address	4915 W. 67th St. Chicago, Ill. 60638
Telephone	312/594-3800
Model designation	**HPL-10, HPL-20**
Principal application	Wld, cut, surf & heat treat*
Type; wavelength	CO_2/cw; 10.6 μm
Output (av); stability	10 kW, 20 kW; ±1%
Pulse width; rep rate	
Exit beam dia; divergence	4 mm focused; NA
Spot dia; focus depth	0.015-0.5 in.; 0.05-2 in.
Operating services	370 kVA, water, He, CO_2, N_2, CO
Cavity cooling	Water, refrigeration
Work station	Per customer requirements
Multiple work stations	Beam splitting, beam sharing
Targeting method	Coaxial He-Ne laser
Controlled variables	Pulse width, rep rate, spot size, power level, power slope
Positioning system	Sciaky positioning systems, NC, CNC
Typical working rates	Butt weld 0.5-in. ss at 40 ipm ht trt 1040 steel to 0.025 in. at 50 in.^{2}/min
Power use standby/run	5/100 kW
Consumables	Makeup gas
Servicing requirements	Prevented maintenance at 2-mo intervals
Safety provisions	Fully enclosed, interlocked
Principal options	Optical and tooling packages
Technical services	R&D, application studies
Special features	Programmable power control
Warranty	90 days, service contract opt
Base price	$500,000
Remarks	*Also alloying. Manufactured by Avco Everett Research Lab (see entry elsewhere)

Company	**Union Carbide Advanced Systems**
Address	5700 W. Raymond St. Indianapolis, Ind. 46241
Telephone	317/247-1586
Model designation	**Model 971 GTL**
Principal application	Wld, cut, heat treat, hd face
Type; wavelength	CO_2/cw; 10.6 μm
Output (av); stability	1 kW rated, 1.5 kW max; 5%
Pulse width; rep rate	80 ms (opt); 10 pps (opt)
Exit beam dia; divergence	12 mm; less than 1.3 mrad
Spot dia; focus depth	0.003 in. w/2.5-in. f.1. lens
Operating services	40 kW, water, gas mixture
Cavity cooling	Water
Work station	
Multiple work stations	Beam splitting, beam sharing
Targeting method	He-Ne laser*
Controlled variables	Pulse width (opt), rep rate, spot size, power level, power slope (opt)
Positioning system	
Typical working rates	Cut 0.12-in. low-carbon steel at 100 + ipm, other data available
Power use standby/run	1/30 kW
Consumables	He: 0.9 cfm, N_2: 0.2 cfm, CO_2: 0.04 cfm
Servicing requirements	Lube and oil blower and pump every 500 hr
Safety provisions	Interlocks, enclosed path
Principal options	Power meter, shutter, rotating optics, focusing assy
Technical services	Applications lab, consulting
Special features	TEM_{00} mode, compact
Warranty	1 yr parts and labor
Base price	$64,500
Remarks	*Microscope and CCTV optional, transverse gas transport system. Manufactured by GTE (see entry elsewhere).

Typical metalcutting rates for CO_2 lasers

Metal	Gage (in.)	Power (kW)	Rate (ipm)	Metal	Gage (in.)	Power (kW)	Rate (ipm)
Aluminum, alloy	0.24	3.8	1.2	Steel, high-speed	0.280	0.5	3
Aluminum	0.50	10	40	Steel, low-alloy	0.024	0.85	23
Brass	0.005	0.28	156	Steel, maraging	0.189	0.5	3
Inconel, 718	0.50	11	50	Steel, mild	0.039	0.4	177
Mn-Ni alloy	0.003	0.5	67	Steel, mild	0.051	0.5	142
Molybdenum	0.002	0.5	16	Steel, mild	0.063	0.5	98
Monel mesh	2.00	0.5	4	Steel, mild	0.118	0.35	60
Nickel	0.005	0.28	156	Steel, mild	0.126	0.5	40
Niobium	0.126	0.5	8	Steel, mild	0.252	0.5	20
Nimonic 90	0.39	0.5	96	Steel, stainless	0.004	0.5	197
Nimonic 90	0.059	0.25	23	Steel, stainless	0.012	0.5	146
Nimonic 90	0.059	0.85	90	Steel, stainless	0.039	0.5	65
Nimonic 75	0.002	0.5	98	Steel, stainless	0.059	0.4	18
Nimonic 75	0.047	0.5	51	Steel, stainless	0.063	0.5	75
Nimonic 75	0.079	0.5	31	Steel, stainless	0.110	0.4	47
Steel, galvanized	0.039	0.4	177	Steel, stainless	0.118	0.4	45
Steel, galvanized	0.051	0.4	142	Steel, stainless	0.126	0.5	35
Steel, galvanized	0.118	0.4	60	Steel, stainless	0.252	0.5	20
Steel, high-speed	0.093	0.5	35	Steel, tool (H25)	0.063	0.5	31
Steel, high-speed	0.134	0.5	24	Titanium	0.252	0.5	142
Steel, high-speed	0.165	0.5	16	Titanium	0.80	11	100
Steel, high-speed	0.205	0.5	27	Tungsten carbide	0.071	0.5	1.8
Steel, high-speed	0.220	0.5	23	Tungsten carbide	0.189	0.5	1.8

Note: The rates listed are not necessarily maximum for any given thickness, because kerf width, edge condition, or heat effect, for example, may have been the governing factors.

Source: Avco Everett, Ferranti

Typical welding rates for CO_2 lasers

Metal	Depth (in.)	Power (kW)	Rate (ipm)	Metal	Depth (in.)	Power (kW)	Rate (ipm)
Aluminum	0.08	5	100	Steel, rimmed	0.2	2.5	24
Aluminum	0.44	10	24	Steel, rimmed	0.2	6	190
Lead	0.08	1	400	Steel, stainless	0.1	2.5	100
Lead	0.139	3	400	Steel, stainless	0.8	20	50
Lead	0.2	1.5	200	Titanium-Al	0.12	5	100
Steel, low-carbon	0.04	2.5	200	Ti-Al-V	0.12	10	270
Steel, low-carbon	0.55	10	24	Ti-Al-V	0.5	10	24

Source: SME Technical Paper MR74-954

The following sources were used in the preparation of this report:

Industrial Lasers and their Applications, John E. Harry, McGraw-Hill, London, 1974.

Lasers in Industry, edited by S S. Charschan, Van Nostrand Rheinhold, New York, 1972.

Various Technical Papers published by the Society of Manufacturing Engineers (SME).

INDEX